Skin Tissue Engineering and Regenerative Medicine

Skin Tissue Engineering and Regenerative Medicine

Mohammad Z. Albanna
James H. Holmes IV
Medical Center Boulevard Winston-Salem, NC, USA

ELSEVIER

AMSTERDAM • BOSTON • HEIDELBERG • LONDON
NEW YORK • OXFORD • PARIS • SAN DIEGO
SAN FRANCISCO • SINGAPORE • SYDNEY • TOKYO

Academic Press is an imprint of Elsevier

Academic Press is an imprint of Elsevier
125 London Wall, London EC2Y 5AS, UK
525 B Street, Suite 1800, San Diego, CA 92101-4495, USA
50 Hampshire Street, 5th Floor, Cambridge, MA 02139, USA
The Boulevard, Langford Lane, Kidlington, Oxford OX5 1GB, UK

ISBN: 978-0-12-801654-1

British Library Cataloguing-in-Publication Data
A catalogue record for this book is available from the British Library

Library of Congress Cataloging-in-Publication Data
A catalog record for this book is available from the Library of Congress

For information on all Academic Press publications visit our website at http://store.elsevier.com/

Acquisition Editor: Mica Haley
Editorial Project Manager: Lisa Eppich
Production Project Manager: Karen East and Kirsty Halterman
Designer: Maria Ines Cruz

Typeset by TNQ Books and Journals
www.tnq.co.in

Printed and bound in the United States of America

Dedication

To those, without whom this book would not have been achieved…

To my lovely wife Ruba, my princess Layann, my handsome boys Qusai, Adam, and Jad, Mom and Dad who kept me in their prayers every day and taught me to learn and teach, my brother Ahmad and his family and kids, the dedicated contributors of this book, and all passionate scientists.

Mohammad Z. Albanna

To my wife, Susan, my daughters, Lane and Dickinson, my mother, Judy, and my late father, James, for their love and support. You make the journey meaningful.

James H. Holmes IV

Contents

7. Advances in Acellular Extracellular Matrices (ECM) for Wound Healing

Rhiannon Harries, Jared Torkington and Keith Harding

8. Natural Biomaterials for Skin Tissue Engineering

Mihail Climov, Tripp Leavitt, Joseph Molnar and Dennis Orgill

9. Synthetic Biomaterials for Skin Tissue Engineering

Mohammed Hussein Kailani, Hanan Jafar and Abdalla Awidi

List of Contributors

Mohammad Z. Albanna Wake Forest Baptist Medical Center (WFBMC), Department of General Surgery, Winston-Salem, NC, USA; Pinnacle Transplant Technologies, Research & Development, Phoenix, AZ, USA

B. Lynn Allen-Hoffmann Stratatech Corporation, Madison, WI, USA; Department of Pathology and Laboratory Medicine, School of Medicine and Public Health, University of Wisconsin, Madison, WI, USA; Department of Surgery, School of Medicine and Public Health, University of Wisconsin, Madison, WI, USA

Abdalla Awidi Cell Therapy Center, University of Jordan, Amman, Jordan

Kyle Binder Department of Neurology, Wake Forest University, Winston-Salem, NC, USA

Steven Boyce Department of Surgery, University of Cincinnati, Cincinnati, OH, USA

Katie Bush Clinical Sciences & Research, TEI Medical, Boston, MA, USA

Anders H. Carlsson Quality Skin Collaborative for Advanced Reconstruction and Regeneration (Q-SCARRTM), United States Army Institute of Surgical Research, San Antonio, TX, USA; Dental and Craniofacial Trauma Research and Tissue Regeneration, San Antonio, TX, USA

Jeffrey E. Carter Department of Surgery, Wake Forest University School of Medicine, Winston-Salem, NC, USA

Rodney Chan Quality Skin Collaborative for Advanced Reconstruction and Regeneration (Q-SCARRTM), United States Army Institute of Surgical Research, San Antonio, TX, USA; Plastic and Reconstructive Surgery, Clinical Division and Burn Center, United States Army Institute of Surgical Research, San Antonio, TX, USA

Richard A.F. Clark Departments of Dermatology and Biomedical Engineering, Stony Brook University, Stony Brook, NY, USA

Mihail Climov Tissue Engineering and Wound Healing Laboratory, Division of Plastic Surgery, Department of Surgery, Brigham and Women's Hospital, Harvard Medical School, Boston, MA, USA

Christopher R. Davis Hagey Laboratory for Regenerative Medicine, Stanford University, Stanford, CA, USA; Division of Plastic Surgery, Stanford University School of Medicine, Stanford, CA, USA

Idris El-Amin Max Planck Florida Institute for Neuroscience, Jupiter, FL, USA

William J. Ennis Section of Wound Healing and Tissue Repair, University of Illinois Hospital and Health Sciences System, Chicago, IL, USA

Justine Fenner Departments of Dermatology and Biomedical Engineering, Stony Brook University, Stony Brook, NY, USA

Mark E. Furth Wake Forest Innovations, Wake Forest Baptist Medical Center, Winston-Salem, NC, USA

Arthur A. Gertzman Musculoskeletal Transplant Foundation, Edison, NJ, USA

Ursula Graf-Hausner Zurich University of Applied Sciences, Waedenswil, Switzerland

John E. Greenwood Royal Adelaide Hospital, Adelaide, SA, Australia

Edward M. Gronet Division of Plastic Surgery, Baylor Scott and White, Temple, TX, USA

Geoffrey C. Gurtner Department of Surgery, Stanford University School of Medicine, Stanford University, Stanford, CA, USA

Keith Harding Cardiff University, Medical Director of Welsh Wound Innovation Initiative, Cardiff, UK

Rhiannon Harries School of Medicine, Cardiff University, Cardiff, UK

David A. Hart Department of Surgery, University of Calgary, Calgary, AB, Canada

Danielle Hill Section of Wound Healing and Tissue Repair, University of Illinois Hospital and Health Sciences System, Chicago, IL, USA

James H. Holmes IV Department of Surgery, Wake Forest University School of Medicine, Winston-Salem, NC, USA

Glicerio Ignacio David H. Murdock Research Institute, SOS Division, Working Buildings, LLC, Kannapolis, NC, USA

Hanan Jafar Cell Therapy Center, University of Jordan, Amman, Jordan

Mohammed Hussein Kailani Cell Therapy Center, University of Jordan, Amman, Jordan; Department of Chemistry, Faculty of Science, University of Jordan, Amman, Jordan

Ferdinand V. Lali Blond McIndoe Research Foundation, Queen Victoria Hospital, West Sussex, UK; The Brighton Centre for Regenerative Medicine, The University of Brighton, Brighton, UK

Tripp Leavitt Tissue Engineering and Wound Healing Laboratory, Division of Plastic Surgery, Department of Surgery, Brigham and Women's Hospital, Harvard Medical School, Boston, MA, USA

Yella H. Martin Blond McIndoe Research Foundation, Queen Victoria Hospital, West Sussex, UK; The Brighton Centre for Regenerative Medicine, The University of Brighton, Brighton, UK

Stephanie Mathes Zurich University of Applied Sciences, Waedenswil, Switzerland

Vince Mendenhall Wake Forest University School of Medicine, Winston-Salem, NC, USA

Anthony D. Metcalfe Blond McIndoe Research Foundation, Queen Victoria Hospital, West Sussex, UK; The Brighton Centre for Regenerative Medicine, The University of Brighton, Brighton, UK

Joseph Molnar Department of Plastic Surgery, Wake Forest University School of Medicine, Winston-Salem, NC, USA

Sean V. Murphy Wake Forest Institute for Regenerative Medicine (WFIRM), Winston-Salem, NC, USA

Dennis Orgill Tissue Engineering and Wound Healing Laboratory, Division of Plastic Surgery, Department of Surgery, Brigham and Women's Hospital, Harvard Medical School, Boston, MA, USA

Shadi A. Qasem Department of Pathology, Wake Forest University School of Medicine, Winston-Salem, NC, USA

Peggy J. Rooney Stratatech Corporation, Madison, WI, USA

Lloyd F. Rose Quality Skin Collaborative for Advanced Reconstruction and Regeneration (Q-SCARRTM), United States Army Institute of Surgical Research, San Antonio, TX, USA; Dental and Craniofacial Trauma Research and Tissue Regeneration, San Antonio, TX, USA

Heinz Ruffner Novartis Institutes for BioMedical Research, Basel, Switzerland

Saahil Sanon Wound Healing Research Group, Department of Surgery, Faculty of Medicine and Dentistry, University of Alberta, Edmonton, AB, Canada

Aleksander Skardal Wake Forest Institute for Regenerative Medicine, Wake Forest School of Medicine, Winston-Salem, NC, USA; Virginia Tech-Wake Forest School of Biomedical Engineering and Sciences, Wake Forest University Health Sciences, Winston-Salem, NC, USA

Dorothy Supp Department of Surgery, University of Cincinnati, Cincinnati, OH, USA; Research Department, Shriners Hospitals for Children - Cincinnati, Cincinnati, OH, USA

Peter A. Than Hagey Laboratory for Regenerative Medicine, Stanford University, Stanford, CA, USA; Department of Surgery, Stanford University School of Medicine, Stanford University, Stanford, CA, USA

Jared Torkington Cardiff and Vale Health Board, Cardiff, UK

Edward E. Tredget Wound Healing Research Group, Department of Surgery, Faculty of Medicine and Dentistry, University of Alberta, Edmonton, AB, Canada; Critical Care, University of Alberta Hospital, Edmonton, AB, Canada

Fiona Wood University of Western Australia, Crawley, WA, Australia

Mustafa Q. Yousif Department of Pathology, Wake Forest University School of Medicine, Winston-Salem, NC, USA

Foreword

A.J. Russell
Disruptive Health Technology Institute, Carnegie Mellon University, Pittsburgh, PA, United States

References to the concepts that we now call tissue engineering and regenerative medicine have surrounded us for millennia. Early Indian writings, Greek mythology, centuries of fiction, and decades of films have made the miracle of using cells and materials to restore the form and function of tissues and organs an almost expected outcome from the march of scientific discovery. The repair of skin damage, by orchestrating regenerative responses to injury, was demonstrated even before the terms tissue engineering and regenerative medicine were defined. Indeed, if one includes wound closure in the remit of tissue engineering, then the long history of sutures and biomaterials can be traced back to the Neolithic period. Before cells, proteins, and biomaterials were discovered and described, our yearning for the miracle of regeneration was ever present. Today, skin tissue engineering and regenerative medicine are compelling targets of discovery because of the scale and complexity of the need for scarless healing after injury. No clinical or scientific expertise is needed to recognize the tragic clinical sequelae of a serious burn. Reversing the dramatic loss of form and function during healing of the skin has turned out to be a truly vexing challenge for scientists committed to a better future. There are many definitions of tissue engineering and regenerative medicine, but all agree that these fields are highly interdisciplinary and are mechanistically agnostic. Tissue engineering of the skin is an example of regenerative medicine, and it refers to the clustering of tools and techniques that would cause the form and function of lost or damaged skin to be restored. Indeed, the skin is a vascularized organ that is as complex in its own way as the internal organs. The three-dimensional reengineering of skin is the ideal playground for tissue engineers of today.

Evidence for human use of replacement skin can be traced back 3000 years. The availability of autologous and allogeneic epidermal, dermal, and dermo-epidermal living or nonliving skin equivalents builds on this foundation of early science. This book on *Skin Tissue Engineering and Regenerative Medicine* describes the start of the art in the heart of the field as well at the vital periphery. Skin tissue engineering, as noted above, is no different than tissue engineering of an internal organ. One must balance the need to create a diffusionally functional graft that can act as a barrier to infection, while it reengineers the three-dimensional, multilayered environment that will restore form and function. The perfect skin replacement is clearly *not* going to be a perfect replica of human skin. If this were the case, then human skin grafts would heal without loss of form and function. Instead, the real challenge is in knowing what *not* to try to deliver at the start of the healing process and what the body must build for itself while it restores skin. Definitions become murky when one considers the boundaries between tissue engineering and regenerative medicine. For example, many would agree that tissue-engineered skin would imply that living cells were used to create a material that when applied to the skin would lead to restoration of form and function. The earliest tissue engineers knew at the outset that any cells that they chose to supply with that graft material may not be the cells that finally reside in the restored tissue. Indeed, we now know that the biologic factors that those cells release are the key drivers of the regenerative response. Therefore, for example, would a barrier cream consisting of just those factors engineered to release at the right rate and location be a tissue-engineered skin or a biopharmaceutical? Debate could ensue on the topic, but all would agree that this would be regenerative medicine.

Another fundamental challenge for the skin tissue engineer is how to incorporate their engineered grafts into the healing milieu. Most approaches today rely on the recipient to send blood vessels and nerves to the skin equivalents, but almost all engineered tissues are subject to inflammatory processes that drive scarring at anastomoses. One cannot describe skin tissue engineering without first describing the anatomy and physiology of what one seeks to replace. Skin has been called two dimensional and simple, but successful tissue engineering of the skin must begin with the recognition that skin is richly complex and far from two dimensional. It is no consequence that the first chapter in this book describes skin in all of its complexity and diversity. What becomes immediately clear is that skin tissue engineers must build with biology versus seek to replicate it.

Using biology as the building blocks for functional repair of damaged skin and looking deeply into how scarring and wound healing are inexorably linked, the book begins to move through the tools and techniques that clinically driven scientists have in their quiver today. Regenerative medicine should be tool agnostic, since it is focused on a clinical deliverable (restoration of form and function). The book elegantly walks through acellular (Chapters 6–9) and “living” skin matrices (Chapters 10 and 11) before focusing on the clinical management of skin wounds (Chapters 12–16). As described above, we have learned enough to know that the biologic signals that cells generate are more important than anything else during healing. It is therefore particularly apt that, before turning to the future in the final chapters of this book, the authors focus on the delivery of biologic signals in biopharmaceutical regenerative medicine.

Replacing and repairing the skin is like resurfacing the window to our souls. The need is great and desire has been with us for millennia. That said, the real science of how to drive scarless wound healing of the skin is still in its early days, and this book provides a marvelous stake in the ground that defines the path ahead from the context of where we are today. This book proves to be a timeless roadmap to a bright future for skin regeneration research and clinical practice.

Chapter 1

Anatomy, Physiology, Histology, and Immunohistochemistry of Human Skin

Justine Fenner, Richard A.F. Clark

Departments of Dermatology and Biomedical Engineering, Stony Brook University, Stony Brook, NY, USA

Chapter Outline

INTRODUCTION

The largest organ in the human body is the skin. It composes 16% of a person's body weight with an average weight of 4 kg, while encompassing a surface area of 1.8 m^2. The skin is a metabolically active organ with a variety of vital functions essential to maintenance of homeostasis and protection of the body. It acts as a barrier to chemical and physical agents, prevents the loss of body fluids, and helps to regulate body temperature [1]. The skin also serves as a sensory organ and provides a surface for one to grip. It plays a vital role in vitamin D production and makes antimicrobial peptides [2]. Skin is continuous with the

Skin Tissue Engineering and Regenerative Medicine. http://dx.doi.org/10.1016/B978-0-12-801654-1.00001-2

mucous membranes of our respiratory system, digestive tract, and urogenital tract and gives rise to nails, hair, and sweat glands.

SKIN ANATOMY, HISTOLOGY, AND PHYSIOLOGY

Skin is composed of three layers: the epidermis, dermis, and hypodermis (Figure 1). Skin is also often classified as being either thick or thin depending on the width of the epidermis. Thick skin has an epidermis thickness of 0.8–1.5 mm, whereas thin skin has an epidermis thickness of 0.07–0.15 mm [3]. Thick, non-hair-bearing (glabrous) skin is found on palmar and plantar surfaces and has no hair, arrector pili muscles, or sebaceous glands [4]. Thinner skin is found over the rest of the body, but is especially thin over the eyelids, and is composed of less cellular layers [3].

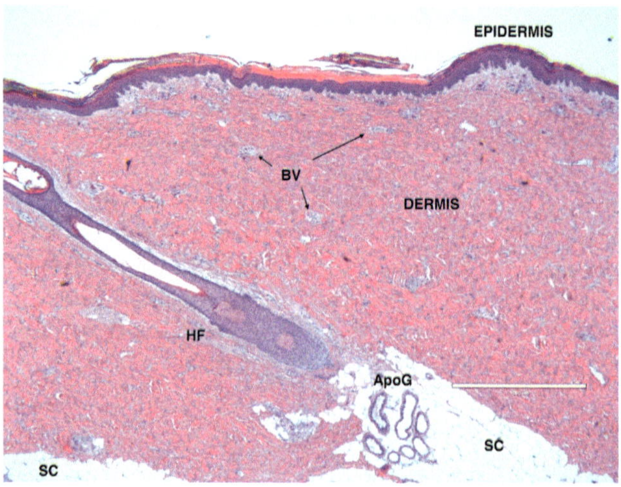

FIGURE 1 **Normal porcine back skin fixed and stained with H&E.** Pig skin, of all animals, is most like human in terms of architecture, thickness, lack of a panniculus carnosis (muscle layer under the subcutaneous adipose tissue), and sparseness of hair; and therefore, a superb model for evaluating tissue-engineered constructs. This specimen of porcine skin comes from the back and is stained with hematoxylin and eosin that renders nuclei blue and keratin and collagen pink. The keratinized EPIDERMIS (pink) is composed of corneocytes that along with the underlying granular cell layer (not seen at this low magnification) provide the permeability barrier to the skin. The underlying stratum spinosum and basal cell layer (blue) are the differentiating and proliferating layers of the skin. Normal skin in humans regenerates every 28 days. The DERMIS is mostly composed of collagen (pink), which provides tensile strength, but contains many blood vessels (BV), nerves (not well-visualized with this stain), and appendageal structures like hair follicles (HF), apocrine glands (ApoG), sebaceous glands that secrete oil, and eccrine glands that generate sweat. The latter are on most skin surfaces in humans but only reside in specialized areas of pig skin, for example, the snout. The hypodermis or subcutaneous adipose tissue (SC) provides a cushioning effect from blunt trauma, as well as insulation.

EPIDERMIS

The epidermis is the outermost layer of the skin and ranges in thickness from 0.05 mm on the eyelids to 1.55 mm on palms and soles [5]. It is composed mostly of stratified squamous epithelium, with the innermost layer consisting of a single row of columnar cells called basal cells that are attached to the basement membrane. The epidermis is constantly regenerating, making itself a durable keratinized boundary [3].

The epidermis is further divided into four layers: the stratum basale, the stratum spinosum, the stratum granulosum, and the stratum corneum. A fifth layer, the stratum lucidum, is found between the stratum corneum and stratum granulosum, only in the thick skin of the palms and soles [6].

The stratum basale is the innermost layer of the epidermis and the location of cell division. Some basal cells are stem cells that slowly generate other basal cells and suprabasal cells that rapidly divide (transient-amplifying cells) to generate more keratinocytes. Human skin regenerates about every 28 days [7]. Basal cells also produce antimicrobial proteins that are imperative in the skin's defensive role [8].

Melanocytes comprise about 5–10% of the basal cell population [3]. Their principle role is in the production of melanosomes that are transferred to keratinocytes. The type and abundance of melanosomes determine the pigment intensity of the skin [4]. Merkel cells are found, albeit infrequently, in the basal cell layer as well. They are closely associated with terminal filaments of cutaneous nerves and are believed to play a role in sensation [3]. They are found in especially high concentration in areas associated with cutaneous nerves and touch sensation, such as the fingers and lips [9].

Superficial to the stratum basale is the stratum spinosum. The stratum spinosum contains a high concentration of keratin filaments and desmosomes that tightly adhere adjacent cells to one another [5]. During H&E staining the filaments between desmosomes shrink resulting in a "spiny" appearance, thus the name stratum spinosum. Keratinization begins in the basal cells, but the keratin type switches as keratinocytes differentiate and transit into the stratum spinosum [10]. Keratinization, or cornification, is the process of cell differentiation in which keratinocytes transition from their postgerminative state in the stratum basale and suprabasal cell layer to terminally differentiated, hardened cells filled with protein in the stratum corneum [11]. In the stratum basale, keratinocytes also begin to produce lamellar bodies in the golgi [12]. Lamellar bodies are tubulovesicular secretory organelles related to lysosomes. They secrete their contents including lipids, protease inhibitors, hydrolases, and antimicrobial peptides into the upper layers of the epidermis [10]. Consequently, they are important in forming a boundary that prevents the loss of fluids while providing antimicrobial protection [3]. Langerhans cells reside for the most part in the stratum spinosum. These cells are antigen-presenting cells that serve an immunologic role in the skin [4].

Above the stratum spinosum is the stratum granulosum. At the interface between the stratum granulosum and the stratum corneum, keratinocytes become flattened and lose their nuclei. It is here that lamellar bodies secrete their contents forming a lipid barrier. Keratohyalin granules are also formed in the stratum granulosum where they bind to keratin filaments [3]. This binding creates large aggregations that form the electron dense masses within the cytoplasm of keratinocytes resulting in a "granular" appearance [4].

The most superficial layer, the stratum corneum, contains cells completely devoid of nuclei and organelles [13]. The keratinocytes become elongated and flattened to form a lamellar array of corneocytes [4]. Dense bodies, remnants of desmosomes, along with a lipid glue partially derived from lamellar granules, hold the corneocytes together [10]. Corneocytes are enveloped in a layer of protein and filled with keratin proteins. This layer is most important in creating a protective barrier against the environment.

Four different cell types reside in these layers of the epidermis: keratinocytes, melanocytes, Langerhans cells, and Merkel cells.

Keratinocytes

Keratinocytes are the principal cells of the epidermis. These ectoderm-derived cells are squamous and originate in the bottommost stem cell pools of the stratum basale. During the process of keratinzation they migrate up from the basement membrane toward the stratum corneum [14]. Under basal conditions, they require about 2 weeks to exit the nucleated compartment and an additional 2 weeks to move through the stratum corneum. Keratinocytes mature as they move through the cell layers and are dead by the time they reach the stratum corneum [15]. Once they reach the stratum corneum, they are either sloughed off or rubbed off by friction in a process called desquamization [16].

Keratinocytes are derived from undifferentiated cells in the stratum basale of the epidermis. The process of keratinization occurs in two stages, a synthetic stage and a degradative stage. The synthetic stage begins after the basal stem cells divide. Half of the daughter cells remain in the basal cell layer, while the other half of the cells progress toward the surface and differentiate [4]. It is believed that histones control cellular differentiation in the epidermis. Keratinocytes have the capacity to increase their rate of replication during periods of inflammation, disease, or injury [15].

In the stratum spinosum the cells begin to change from columnar to polygonal. At this stage, keratinocytes begin to synthesize keratin, insoluble proteins that act as intermediate filaments and tether half desmosomes from one side of the cell to another [5]. Desmosomes are specialized cadherin molecules, called desmogleins, and desmocollins, and function to bind epidermal cells together [16]. Thus, the combination of desmosomes acting as spot-welds between keratinocytes and keratin intermediate filaments act as intracellular cables that tie desmosomes together intracellularly. If formed, this creates an incredibly effective tension-resistant system that prevents shear force from tearing apart the epidermis. Keratin filaments also secure basal cells to the basement membrane by connecting to hemidesmosomes that contain proteins that link to the basement membrane [4].

Keratin filaments are retained by keratinocytes to eventually become a major component of the stratum corneum. Keratin is also the structural protein of hair and nails. Keratin is always the product of two subfamilies of keratin proteins, one acidic and one basic keratin, which combine to form the multiple keratins found in many epithelial tissues [4]. The presence of various keratin types can be used to detect the type and degree of differentiation of epithelial cells in general [17].

Keratinocytes gradually travel to and through the stratum granulosum, where enzymes induce degradation of their nuclei and organelles but not their keratin. These terminally differentiating keratinocytes contain keratohyalin granules composed of profilaggrin, a precursor to filaggrin that causes keratin filament aggregation [8]. Conversion to filaggrin occurs in the granular layer, resulting in the formation of an electron dense interfilamentous protein matrix containing keratin and several other structural proteins including involucrin. Involucrin encases a group of keratin macrofibers that have been aggregated by filaggrin [8]. Keratohyalin is important in the formation of so-called soft flexible keratin. In the absence of keratohyalin, the keratin formed is hard and rigid, as seen in hair and nails. In the stratum granulosum, membrane-coating granules attach to the cell membrane and release a viscous lipid substance that contributes to cell adhesion, thus creating the permeability barrier [4].

The keratinocytes continue to travel up to their last stop, the stratum corneum. At this stage the cells have become flattened and dead, with thick cornified envelopes containing keratin, filaggrin, and involucrin [18]. Additionally, the disulfide bonds of keratin provide strength to the stratum corneum. As desmosomal intercellular adhesion and lamellated lipid are lost, the cells shed from the skin. This programmed maturation that ultimately results in cell death is called terminal differentiation [16].

Melanocytes

Melanocytes are neural crest-derived cells that reside in the stratum basale at the frequency of 1 in every 10 basal keratinocytes. Certain areas of the body including the face, shins, and genitalia have an even greater density of melanocytes. In these areas the frequency approaches 1:1 [4]. The principal role of the melanocyte is in making melanosomes. Melanocytes transfer melanosomes to keratinocytes by way of cytocrine secretion. Melanosomes are elongated, membrane-bound, pigment-producing granules within the keratinocytes that determine differences in skin color [12].

Melanocytes are dendritic cells with cellular processes, or dendrites that allow them to contact many keratinocytes, over long distances. While their cell bodies may be between the stratum basale and basement membrane, their dendritic processes extend to reach many keratinocytes throughout the stratum basale and stratum spinosum [19]. Together, keratinocytes and melanocytes form a melanin unit. Melanocytes produce melanosomes in the golgi zone of the cell. The melanosomes are then moved to the tips of their cellular processes. Keratinocytes then ingest the tips of the melanocytic dendrites by phagocytosis, allowing them to take in the melanosomes. This process is called apocopation [13]. Melanin granules then form a protective cap over the nucleus of the keratinocyte protecting the nucleus from the photo damage of UV light [4]. Tyrosinase acts on a variety of melanin precursors, including tyrosine, to make melanin [10]. The melanocortin 1 receptor plays a prominent role in melanin production [4].

The number of melanocytes does not determine skin color. Instead, the number, size, and distribution of the melanosomes determine skin color. People with pale skin have fewer melanosomes that are packaged in membrane-bound complexes. People with darker skin have a greater number of melanosomes that are larger and more widely distributed. Chronic sun exposure stimulates melanocytes to make melanosomes in a pattern similar to what is found in people with darker skin. In addition, melanocytes in people with red hair tend to be more round and produce more phaeomelanin. Leukoderma, or diseases that cause whitening of the skin, can be caused by a decrease in the number of melanocytes [4]. Vitiligo is a result of autoimmune destruction of melanocytes. Albinism, on the other hand, is caused by a deficiency of fully pigmented melanosomes. Increases in pigment can be caused by a variety of different factors [12]. Freckles result when a normal number of melanocytes increase their production of pigment. Black "sunburn" or "ink spots" result from basilar hyperpigmentation and increased melanin content in the stratum corneum. Lastly, nevi are the result of benign proliferations in melanocytes, while melanomas are the result of malignancies [4].

Langerhans Cells

Langerhans cells constitute the first line of immunologic defense in the skin [10]. These cells are derived from the bone marrow and can normally be found scattered among the keratinocytes of the stratum spinosum. Langerhans cells makes up 3–5% of the cells in the stratum spinosum [4]. Langerhans cells and melanocytes are connected to adjacent cells by desmosomes the same way keratinocytes are connected to one another. Langerhans cells are derived from the monocyte lineage and function in the afferent limb of the immune response. These antigen-presenting cells take up foreign invaders and process them to present to T cells. Once they present antigens, they migrate to lymph nodes to activate T cells [20]. These cells are essential for the induction of delayed-type hypersensitivity reactions. Hyaluronan is important in the maturation and migration of Langerhans cells. Prolonged exposure to UV radiation causes the skin to lose its ability to be sensitized until the entire population of Langerhans cells has been replenished. In Langerhans cell-depleted skin, it is the macrophages that present antigen, which can ultimately lead to immune tolerance [4].

Merkel Cells

Merkel cells can be found in the stratum basale of the palms, soles, oral and genital mucosa, and nail bed. Found directly above the basement membrane, they contain intracytoplasmic dense core neurosecretory-like granules, as well as neurofilaments and keratin [4]. These cells are often associated with neuritis, as they act as slow adapting touch receptors [8]. Merkel cells are found closely associated with the terminal filaments of cutaneous nerves and play a role in sensation. Desmosomes connect Merkel cells to adjacent keratinocytes [4].

DERMOEPIDERMAL JUNCTION

The dermoepidermal junction is found at the boundary between the epidermal and dermal layers that on electron microscopy has been divided into the lamina lucida and lamina densa. Epidermal basal cells are attached to the basement membrane, which lies below, by hemidesmosomes, collagen XVII, integrin $\alpha6\beta4$, and laminin 332 (laminin 5 by old nomenclature); the latter three proteins creating the so-called microfilaments that run from the hemidesmosomes through the lamina lucida and connect into the lamina densa. Keratinocytes make type IV collagen and laminin 111, the major structural components of the lamina densa. In fact, most components of the dermoepidermal junction come from keratinocytes, with a minor contribution from dermal fibroblasts. The basement membrane is attached to underlying types I and III collagen bundles of the papillary dermis by anchoring fibrils made of type VII collagen that attach to the underside of the lamina densa. The basement membrane is a porous and semipermeable filter, which allows for the exchange of nutrients and fluids between the epidermal and dermal layers. The basement membrane regulates adhesion, movement, and growth of keratinocytes and fibroblasts. It also provides structural support for the epidermis, while holding together the epidermis and dermis [4].

DERMIS

The layer of skin found between the epidermis and hypodermis (subcutaneous tissue) is the dermis (Figure 1). The dermis varies in thickness from 0.3 mm on the eyelid to 3.0 mm on the back, making it much thicker than the epidermis [5]. Blood vessels and nerves course through the dermis providing both nutrition and sensation. Various appendages including sweat glands, hair follicles, and sebaceous glands can also be found in this layer [10]. The dermis provides both nutritional and structural support to the epidermis.

The dermis is derived from mesenchyme and is composed of two main layers: the papillary dermis and the reticular dermis [21]. The thin papillary layers lie below the epidermis and interdigitate with epidermal rete ridges via dermal papillae [8]. This contour resembles an egg carton, appears more complex in thick skin than thin skin, and helps the dermal–epidermal junction zone resist shear stress [3]. The thicker reticular layer extends from the base of the papillary layer to the hypodermis [5].

The main components of the dermal matrix are collagen fibers, elastic fibers, and extrafibrillary matrix, which are all made by dermal fibroblasts. Collagen fibers make up about 70% of the dermis [8]. The papillary layer is composed of thin loosely woven collagen fibers, while the reticular layer is composed of thicker more densely packed and coarse collagen fibers arranged parallel to the skin surface. This is a reflection of the general trend that, as you move down the dermis toward the hypodermis, the collagen fibers become thicker and coarser. The papillary layer contains loose connective tissue composed of thin bundles of type I and III collagen, elastic fibers, connective tissue cells, and type VII collagen-anchoring fibrils, while the reticular dermis is composed mostly of large, densely packed type I collagen bundles that approach 100 μm in diameter, are organized in a basket weave fashion, and arranged in a net parallel to the skin surface, giving the skin its ability to resist tensile forces coming from any tangential direction. Elastic fibers are mainly located in the reticular layer and give the skin its elasticity [3]. Extrafibrillary matrix fills the space between fibers and is mostly a mucopolysaccharide gel composed of proteoglycans and hyaluronan. Thus, the extrafibrillary matrix imparts a hydrogel property upon the dermis that facilitates the movement of fluids, molecules, and inflammatory cells and resists compression [3]. Dermal dendritic cells and wandering macrophages accumulate hemosidernin, melanin, and cellular debris during episodes of inflammation [5]. Mast cells located along the microvasculature induce marked vasopermeability by release of histamine during inflammation [8].

Vasculature

The arterial side of the dermal vasculature acts as a conduit for oxygen, nutrients, and inflammatory cells to the skin, and the venous side returns blood depleted of oxygen to the central cardiovascular system. The epidermis has no intrinsic blood supply and relies on diffusion from the microvasculature of the papillary dermis for oxygen and nutrients [12]. The vasculature of the dermis is divided into the superficial and deep dermal vascular plexi. The superficial plexus is found at the junction between the papillary and reticular dermis and the deep plexus at the interface between the reticular dermis and subcutaneous tissue (hypodermis). The deep plexus contains larger blood vessels than the superficial plexus and is supplied by branches of cutaneous and musculocutaneous arteries coursing through the hypodermis [4]. Branches of this plexus supply the superficial hypodermis, hair follicles, and the secretory portions and ducts of sweat and apocrine glands. Branches of the deep dermal plexus also supply the superficial plexus. The superficial plexus contains a rich supply of capillaries, terminal arterioles, and postcapillary venules to supply enough oxygen and nutrients to the epidermis through diffusion.

Muscles

Smooth muscle is distributed in the skin in special areas of the body, including the areolas of the nipple, the tunica dartos of the scrotum, and around hair follicles. Smooth muscle found in the arrector pili (erectors of hair) attach to the hair follicle below the sebaceous glands. When they contract, they pull the hair follicle upward resulting in gooseflesh [10]. Smooth muscle cells are also found in the blood vessels of dermis and hypodermis. Arteries have smooth muscle cells that form concentric rings within the arterial wall that are critical for control of perfusion. Veins have small bundles of smooth muscle cells that criss-cross at right angles. Skeletal muscle can be found in the skin of the face and anterolateral neck (platysma) where they facilitate facial expression.

Nerves

Nerves course through the dermis in nerve bundles, along arterioles and venules. These neurovascular bundles travel parallel to the surface of the skin. Touch and pressure sensation in the skin are mediated by Meissiner's and Vater-Pacini corpuscles found in the dermal papillae. Sensation of temperature, itch, and pain are received by unmyelinated nerve endings in the papillary dermis. High-intensity stimulation by inflammation results in pain, whereas low-intensity stimulation results in itching. The autonomic system supplies motor innervation to the smooth muscle in the skin. Eccrine sweat glands receive cholinergic innervation. Apocrine glands, hair erector muscles, and blood vessels receive adrenergic innervation. Sebaceous glands are regulated by the endocrine system, not autonomic fibers [5].

Skin Appendages

The major appendages found in the skin are hairs, erector pili, adenexal glands, and nails. The four major types of adenexal glands in the skin are eccrine, apocrine, apoeccrine, and sebaceous [10].

Eccrine sweat glands play an important role in regulating body temperature. They can be found all over the body, but are especially concentrated in the palms and soles. They are not found in the clitoris, labia minora, external ear canal, and lips [10]. Eccrine sweat glands are composed of a coiled secretory portion and a long duct that courses through the dermis to open into the epidermis. They can be activated by either thermal or emotional stimuli and excrete sweat onto the surface of the skin. Eccrine sweat is secreted by exocytosis and is usually clear and odorless [18].

Apocrine sweat glands are located mainly in the axillary and anogenital regions and are associated with hair follicles. They were named "apocrine" because it was originally believed that their method of secretion involved shedding of the apical cytoplasm from the cell. It is now known that they also secrete by exocytosis [18]. Secretion occurs via "decapitation secretion" in which the free luminal end of the cell is lost with the secretory products [22]. Secretory ducts empty into the upper part of the hair follicle just above the sebaceous glands. Apocrine glands depend on androgens for development and become functional at puberty. Apocrine sweat is acted on by skin bacteria causing the characteristic smell of body odor [18]. The apocrine sweat may function in producing pheromones for mate selection. Apoeccrine glands are found in the axillae of adults. They are so named because they have features of both apocrine and eccrine glands. They open directly onto the skin rather than onto the hair follicle [10].

Sebaceous glands are intimately associated with the hair follicles of the dermis [18]. Sebaceous glands secrete an oily substance that allows hair to grow with less resistance. It also makes hair less brittle and skin suppler. Sebaceous glands are under androgen control and secrete via a holocrine mechanism [23]. In holocrine secretion, whole acinar cells disintegrate and slough into the duct to form an oily sebum. Sebum is composed of various lipid types. Sebaceous glands play a role in acne, and when they work in excess, it results in oily skin.

Human hair has a variety of different biological functions. Scalp hair protects against skin cancer, while eyelashes, eyebrows, and nose hairs protect the body from airborne particles. Hair follicles are found all over the body except for on the palms and soles. The activity of melanocytes in the matrix of the hair bulb determines hair color [24]. The hair shaft is composed of a medulla, cortex, and cuticle. The cortex is made of cornified hair matrix cells, analogous to the stratum corneum. The cuticle forms an outer layer around the hair shaft [5]. The cortex interdigitates with the cuticle, anchoring the hair in the follicle. There are two different types of hair follicles—vellus and terminal. Vellus hair is short, fine, and light-colored. Terminal hair or sexual hair is thicker, longer, and darker. Terminal hair is hormonally regulated and typically does not appear until puberty. Beard, pubic, and axillary hair are terminal hair. Sebaceous glands are found at the base of hair follicles [10].

Human nails are important tools and even sometimes weapons. Nails are made of a nail matrix, nail plate, nail bed, and periungual skin [10]. The nail matrix is the area of nail growth. The nail matrix is located beneath the proximal nail fold. The nail plate is made of an inner thick and elastic cellular layer and an outer hard layer of flat, densely arranged corneocytes. The cuticle seals the proximal part of the nail plate, while the lateral part is sealed by the nail folds. The onychodermal band marks where the nail plate is no longer attached to the nail bed [25]. Beyond this point the nail appears white because of the underlying air [10].

HYPODERMIS

The hypodermis (subcuticular adipose tissue), found beneath the dermis and above the muscle and composed mostly of adipocytes (fat cells), serves a variety of functions in the body. It provides insulation from the cold, cushions deep tissue from violent trauma, provides buoyancy, is a repository for energy, and even acts as an endocrine organ [10]. Adipocytes contain fat lobules that are separated by fibrous septa composed of collagen and large blood vessels. A network of septae keeps the lobules of fat in place while providing support to the structure. This collagen is continuous with the collagen found in the dermis. A rich microvascular network runs through the septae providing oxygenation and nutrient exchange [4].

WOUND HEALING AND IMMUNOHISTOCHEMISTRY

It has been estimated that each year 35.2 million cases of significant skin loss in the US require major therapeutic intervention [26]. About five million of these wounds become chronic. Causes of skin loss include trauma and chronic ulcerations secondary to diabetes mellitus, pressure, and venous stasis. Every year in the US there are about 2 million cases of chronic diabetic ulcers, many of which eventually necessitate amputation. Pressure ulcers and leg ulcers, including venous ulcers, affect another 3 million people in the US. Taken together, treatment costs are as high as $8 billion annually [27]. Cost of chronic wound care is expected to increase significantly as the population ages [28]. Of acute wounds, burns are one of the most devastating injuries known to mankind. Each year in the US about 450,000 patients with burns present to emergency departments [29]. The human, social, and economic burden of burns is considerable with an estimated annual expenditure of >$2 billion direct costs and much greater indirect costs. In 2011, a survey estimated the US market for advanced wound care products, including biological and synthetic dressings, at about $3 billion [30,31]. Despite huge advances in our understanding of wound-healing processes at the cell and molecular levels and how they can go awry, *no cost-effective, transformative therapy has reached the market for either chronic wounds or catastrophic acute wounds such as burns*. Tissue engineering is a potential solution to this pressing need [32]. A brief synopsis of acute cutaneous wound injury and healing or conversion to a nonhealing ulcer is portrayed in Figure 2.

Wound Morphometrics

In order to validate the efficacy of a tissue-engineered construct or any wound treatment on healing, it is fundamentally important to set standards for healing in the animal models under study. For excisional wounds that heal by secondary intention, the most critical parameter is time to complete reepithelization, a function of species and wound size [33]. In addition, healing rate is also a useful parameter. Wound size decrease can be easily calculated from wound

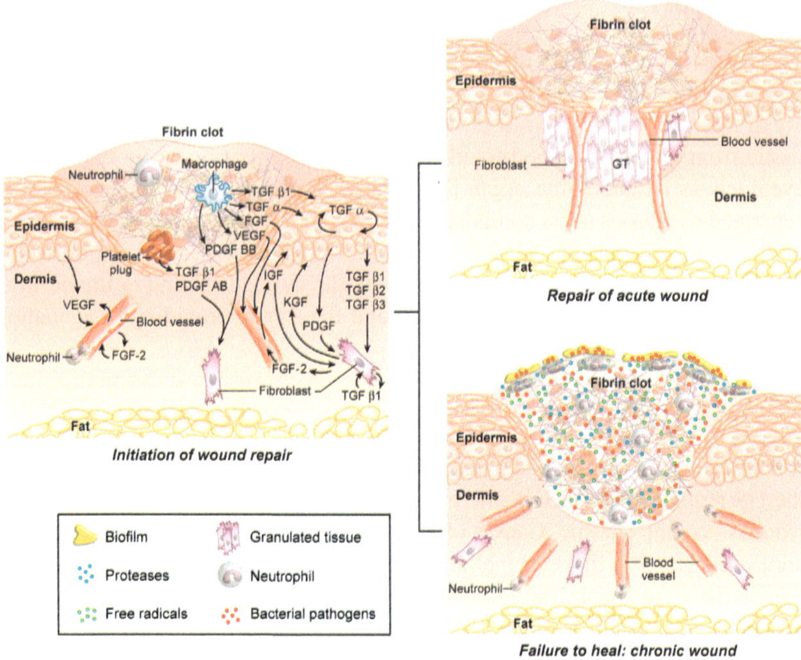

FIGURE 2 Schematics of cutaneous injury, healing, or conversion to a chronic ulcer. Cutaneous injury (left panel) results in an acute wound-healing response under normal physiologic conditions (right upper panel) but may fail when underlying pathobiology or microbial invasion interferes with the healing process, thereby creating a chronic wound (right lower panel). Tissue injury precipitates blood clotting, platelet aggregation, and migration of leukocytes, including neutrophils and macrophages, to the site of injury. Initially, the blood clot is composed of fibrin and fibronectin, which provide a scaffold for cell migration and aggregated platelets, which release growth factors into the surrounding tissue. By 3 days the clot has synergized (contracted) and accumulated numerous neutrophils, which phagocytose and kill microorganisms, and macrophages, which produce and secrete growth factors into the wound environment. Although epidermal migration along the interface between the clot and surrounding normal tissue begins within 24 h after injury, no tissue cells have invaded the clot even by 3 days. Nevertheless by 3 days, fibroblasts and endothelial cells in the periwound stroma are activated to express the appropriate integrins (extra cellular matrix (ECM) receptors) for migrating on fibrin and fibronectin, and to secrete growth factors. Epidermal cells also participate in this cytokine network of growth factor release at 3 days (upper right panel). Under normal physiologic conditions, the wound continues to heal at 5 days with an ingrowth of granulation tissue composed of fibroblasts, additional macrophages, and neovasculature. The migrating epidermis now changes its course and migrates over the newly forming tissue. Proteases restricted to the leading edge of migrating tissue cells are critical for cellular invasion of the clot and ingrowth into the wound. These proteases include urokinase-type plasminogen activator (uPA), tissue-type plasminogen activator (tPA), and matrix metalloprotineases 1, 2, 3, 9, and 13 (MMP 1, 2, 3, 9, and 13). The latter include collagenases (MMP 1 and 13), gelatinases (MMP 2 and 9), and stromolysin (MMP 3). Many growth factors released by platelets and secreted by macrophages during the first phase of healing have been sequestered in the provisional matrix and stimulate tissue cells as they move into the wound. To these are added growth factors now secreted by tissue cells themselves, including epidermal cells, fibroblasts, and endothelial cells, as well as macrophages (lower right panel). When underlying pathobiology or microorganism invasion interrupts the wound-healing

diameter grossly or histologically (Figure 3(A)), assuming that the wound is circular. Alternatively, wound size can be directly measured from planimetry of the wound surface. At least two time points prior to closure should be measured, since the healing rate slows near completion of healing. The rate of healing can be expressed as the rate in the reduction of wound diameter or area, with the initial wound size included. In addition, a calculation of the percentage of closure due to wound contraction versus epithelization should also be included based on histological analysis at the time of wound closure. The latter is critically important when using animals with a panniculus carnosis that contracts the wound within a few days.

Another useful measurement of wound healing is the accumulation of granulation tissue ingrowth into the wound space. If wounds are circular, harvested with adjacent normal tissue, bisected, and fixed and stained with H&E or trichrome, this can be easily quantified by dividing the area of granulation tissue by the area of total wound space on the bisected histologic specimen (Figure 3(B)).

Immunohistochemistry to Define Elements within the Skin

Angiogenesis within the wound granulation tissue can be quantified if stained for an antigen that specifically reveals blood vessels from the surrounding cells and extracellular matrix elements. Classic stains for blood vessels include the common determinant antigen CD31 (PECAM) that stains endothelial cells and laminin 111 that stains basement membranes (Figure 4) [34]. The total area of blood vessels by morphometry divided by the total area of granulation tissue (Figure 3(B)) times 100 gives the percent blood vessels in the granulation tissue.

Reestablishment of epidermal maturation after tissue injury can be investigated by staining for keratins 5 and 14, which should be localized to basal cells, and keratins 1 and 10, which should be localized to epidermal cells throughout the stratum spinosum. Involucrin is also a marker for epidermal maturation in the suprabasal layers of the epidermis. Markers of later epidermal differentiation include loricrin and fillagrin that appear in the stratum granulosum [35].

process, a failure to heal occurs, often leading to a chronic wound (ulcer). Underlying pathobiology known to interfere with acute wound healing includes venous insufficiency that results in fluid transudation and fibrin cuffing of venules secondary to high hydrostatic pressure in the venous system; diabetes mellitus that results in high glucose and both cell and ECM dysfunction from nonenzymatic glycation; arterial occlusion or high external pressure that results in tissue hypoxia and cell dysfunction or death. Bacteria colonizing the wound often produce a biofilm composed of a wide variety of polysaccharides. The biofilm protects these colonies of mixed microorganisms, as it is relatively impervious to phagocytic cells and impermeable to antibiotics. Frustrated phagocytes release a plethora of proteases and toxic oxygen radials into the wound milieu making a bad situation worse, as these agents destroy tissue cells, extracellular matrix, and growth factors in the wound. Not surprisingly, such chronic wounds lack epidermal migration and ingrowth of granulation tissue. *Reprinted from Ref. [32] with permission.*

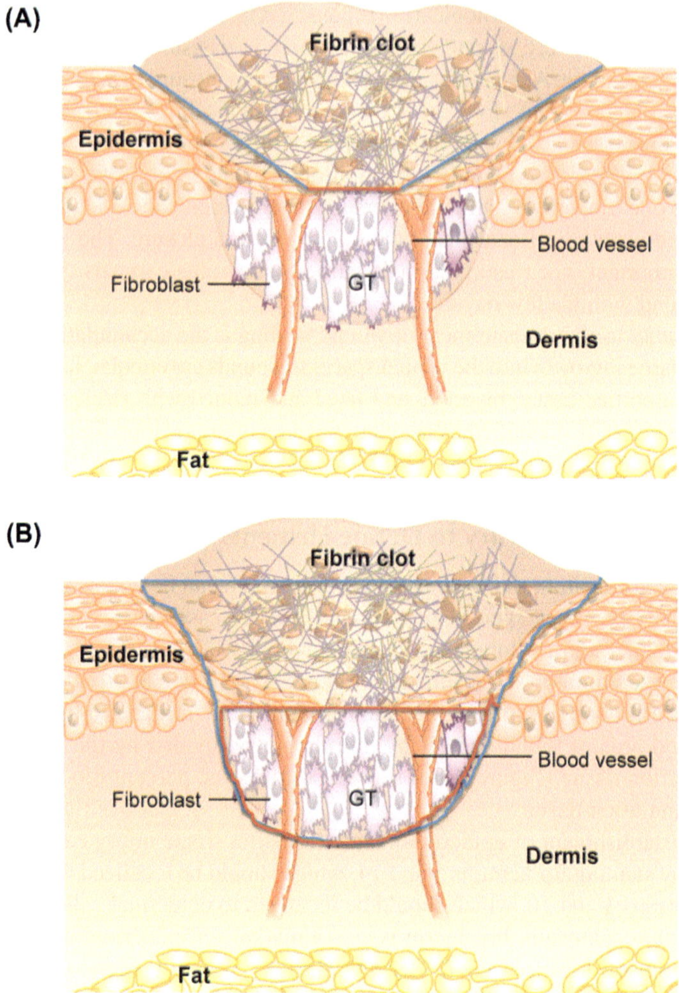

FIGURE 3 **Morphometrics of cutaneous wound healing.** (A) If wounds are circular, re-epithelialization can be quantified morphometrically by carefully bisecting a wound specimen, fixing and staining with either H&E or Masson trichrome, and measuring the reepithelialized portion of the bisected wound surface (blue line) and the nonepithelialized surface (red line). Percent reepithelialization can then be calculated as ((blue line)/blue line + red line)) × 100. NB: Wound diameter is closely related, but not identical, to the blue line + red line measurement. (B) Granulation tissue can be quantified from the same bisected specimen by measuring the total wound area (blue line) and the total granulation tissue area. Percent granulation tissue is calculated as ((area within red line)/(area within blue line)) × 100. *Modified from Ref. [32] with permission.*

Basement membrane integrity in the healed skin or tissue-engineered skin is an important attribute secondary to its critical function of binding the overlying epidermis to the underlying dermis. Biomarkers of such integrity include α6β4 and type XVII collagen that link the basal cell hemidesmosomes to the basement membrane, laminin 332 (laminin 5) that interacts with α6β4 and the

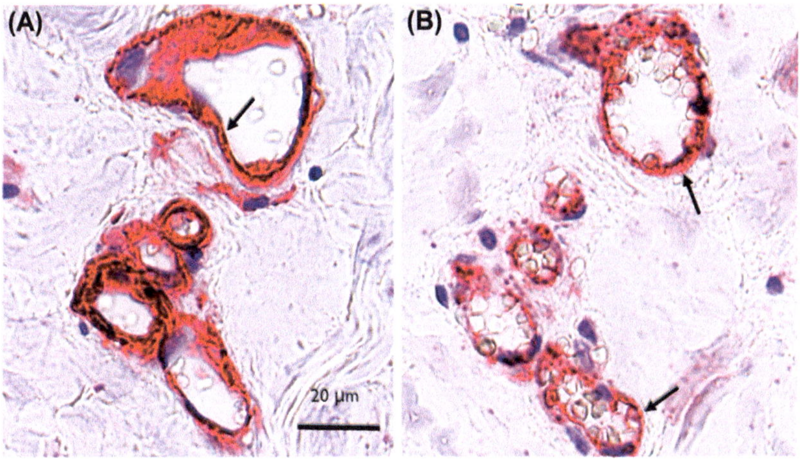

FIGURE 4 **Elucidation of blood vessels within the wound.** (A) Blood vessels of porcine skin stained for laminin 111 (laminin 1 by old nomenclature) that stains basement membrane. (B) Blood vessels of porcine skin stained for the common determinant antigen CD31 that stains endothelial cells. *Modified from Ref. [34] with permission.*

underlying lamina densa, and type IV collagen and laminin 111 (laminin 1) that are the major structural proteins of the lamina densa. Ultimately, however, transmission electron microscopy is required to determine when the fine ultrastructure of the basement membrane has been established [36].

The establishment of nonepidermal cells in the epidermis, such as melanocytes and Langerhans cells, is important for overall epidermal function. Melanocytes can be identified by antibodies to S-100, a biomarker for neural crest-derived cells, coupled with Fontana-Masson stain for melanin. Langerhans cells are also positive for S-100 but can be differentiated from melanocytes in the basal layer and they reside in the stratum spinosum. In addition, specific markers, such as CD1a, can be used to identify Langerhans cells. In mouse skin, however, Langerhans cells lack CD1a, and thus, biomarkers for MHC class II antigens are commonly used [15].

During an inflammatory response to injury or to an adverse reaction to tissue-engineered implant material, specific inflammatory cell biomarkers can be used to identify specific cell types. The biomarkers to be used depend on the animal species and the type of inflammatory cell of interest.

REFERENCES

[1] Telser AG, Young JK, Baldwin KM. Elsevier's integrated histology. Philadelphia (PA): Mosby, Inc.; 2007.

[2] Charruyer A, Ghadially R. Stem cells and tissue-engineered skin. Skin Pharmacol Physiol 2009;22(2):55.

[3] Ovalle WK, Nahirney PC. Netter's essential histology. Elsevier Health Sciences; 2013.

[4] James WD, Berger TG, Elston DM. Andrews' diseases of the skin clinical dermatology. 11th ed. Elsevier Inc.; 2011.

[5] Habif TP. Clinical dermatology: a color guide to diagnosis and therapy. 5th ed. Hanover (NH): Elsevier Inc.; 2010.
[6] Kumar V, Abbas AK, Fausto N, Mitchell RN. Robbins basic pathology. Elsevier Health Sciences; 2012.
[7] Weston WL. Color textbook of pediatric dermatology. Elsevier Inc.; 2007.
[8] Gawkrodger DJ, Ardern-Jones MR. Dermatology: an illustrated colour text. 5th ed. Elsevier Ltd; 2012.
[9] Seung-Hyun EAL, Patapoutian A. Merkel cells and neurons keep in touch. 2014.
[10] Marks JG. Lookingbill and Marks' principles of dermatology. Elsevier Inc.; 2013.
[11] Shetty S, Gokul S. Keratinization and its disorders. Oman Med J 2012;27(5):348–57.
[12] Gartner LP, Hiatt JL. Concise histology. Philadelphia (PA): Saunders; 2011.
[13] Fitzpatrick JE, Morelli JG. Dermatology secrets plus. 4th ed. Philadelphia (PA): Mosby, Inc.; 2011.
[14] The skin barrier as an innate immune element. In: Elias PM, editor. Seminars in immunopathology. Springer; 2007.
[15] Bolognia JL, Jorizzo JL, Schaffer JV. Dermatology. 3rd ed. Elsevier Limited; 2012.
[16] Kierszenbaum AL. Histology and cell biology: an introduction to pathology. 3rd ed. Philadelphia (PA): Saunders; 2012.
[17] Moll R, Divo M, Langbein L. The human keratins: biology and pathology. Histochem Cell Biol 2008;129(6):705–33.
[18] Mescher AL. In: Mescher AL, editor. Junqueira's basic histology: text & atlas. 13th ed. New York (NY): McGraw-Hill; 2013.
[19] Wasmeier C, Hume AN, Bolasco G, Seabra MC. Melanosomes at a glance. J Cell Sci 2008;121(24):3995–9.
[20] Lipscomb MF, Masten BJ. Dendritic cells: immune regulators in health and disease. Physiol Rev 2002;82(1):97–130.
[21] Adkinson NF, Bochner BS, Burks AW, Busse WW, Holgate ST, Lemanske RF, et al. Middleton's allergy: principles and practice. Philadelphia (PA): Saunders; 2014.
[22] Scrivener Y, Cribier B. Morphology of sweat glands. Morphologie 2002;86(272):5–17.
[23] Thody AJ, Shuster S. Control and function of sebaceous glands. Physiol Rev 1989;69(2):383–416.
[24] Elder DE. Lever's histopathology of the skin. Lippincott Williams & Wilkins; 2014.
[25] de Berker D. Nail anatomy. Clin Dermatol 2013;31(5):509–15.
[26] Bickers DR, Lim HW, Margolis D, Weinstock MA, Goodman C, Faulkner E, et al. The burden of skin diseases: 2004 a joint project of the American Academy of Dermatology Association and the Society for Investigative Dermatology. J Am Acad Dermatol 2006;55(3):490–500.
[27] Supp DM, Boyce ST. Engineered skin substitutes: practices and potentials. Clin Dermatol 2005;23(4):403–12.
[28] Ashcroft GS, Mills SJ, Ashworth JJ. Ageing and wound healing. Biogerontology 2002;3(6):337–45.
[29] American-Burn-Association. Burn incidence and treatment in the US: 2013 fact sheet. National Burn Repository; 2013.
[30] Grinnell F, Lamke CR. Reorganization of hydrated collagen lattices by human skin fibroblasts. J Cell Sci 1984;66:51–63.
[31] Yanai H, Budovsky A, Tacutu R, Fraifeld VE. Is rate of skin wound healing associated with aging or longevity phenotype? Biogerontology 2011;12(6):591–7.
[32] Clark RA, Ghosh K, Tonnesen MG. Tissue engineering for cutaneous wounds. J Invest Dermatol 2007;127(5):1018–29.

[33] Kim DJ, Mustoe T, Clark RAF. Cutaneous wound healing in aging small mammals: a systematic review. Wound Repari Regen 2015;33(3):318–39.

[34] Lanier ST, McClain SA, Lin F, Singer AJ, Clark RA. Spatiotemporal progression of cell death in the zone of ischemia surrounding burns. Wound Repair Regen 2011;19(5):622–32.

[35] Rangarajan A, Talora C, Okuyama R, Nicolas M, Mammucari C, Oh H, et al. Notch signaling is a direct determinant of keratinocyte growth arrest and entry into differentiation. EMBO J 2001;20(13):3427–36.

[36] Harriger MD, Warden GD, Greenhalgh DG, Kagan RJ, Boyce ST. Pigmentation and microanatomy of skin regenerated from composite grafts of cultured cells and biopolymers applied to full-thickness burn wounds. Transplantation 1995;59(5):702–7.

Chapter 2

Molecular and Cellular Biology of Wound Healing and Skin Regeneration

Saahil Sanon[1], David A. Hart[2], Edward E. Tredget[1,3]
[1]*Wound Healing Research Group, Department of Surgery, Faculty of Medicine and Dentistry, University of Alberta, Edmonton, AB, Canada;* [2]*Department of Surgery, University of Calgary, Calgary, AB, Canada;* [3]*Critical Care, University of Alberta Hospital, Edmonton, AB, Canada*

Chapter Outline

INTRODUCTION

As the outer covering of the human body, skin is the body's largest organ system. It plays an integral role in protecting the body from its surrounding environment and maintaining homeostasis through a variety of specialized cell types contained within its structure. Anatomically, human skin is composed of three distinct regions: epidermis, dermis, and hypodermis. A disruption of this organization, whether through injury or disease, initiates a process of wound healing and regeneration to restore proper skin function. A multitude of cellular and molecular components are involved in ensuring that this mechanism is properly

Skin Tissue Engineering and Regenerative Medicine. http://dx.doi.org/10.1016/B978-0-12-801654-1.00002-4

carried out, abnormalities of which often lead to chronic wounding and/or fibroproliferative disorders of the skin. An understanding of both ideal and irregular wound healing provides additional insight into the underlying biology of the skin and can be used to guide the future of skin tissue engineering and regenerative medicine.

Normal Adult Wound Healing

Upon injury to the skin, wound healing is immediately initiated in order to begin the process of repair and reinstate the skin's protective barrier. While in some organisms this mechanism regenerates an identical copy of the injured tissue, the wound healing process in adult humans most often leads to fibrosis and scar formation [1]. This process involves a complex interplay of cells and bioactive molecules that can be grouped into four overlapping phases: hemostasis, inflammation, proliferation, and remodeling [1] (Figure 1).

Hemostasis, the first phase of wound healing, takes place in the first few hours following an injury. Successful hemostasis will produce a fibrin clot to stop any bleeding and will initiate the subsequent inflammatory mechanism [1]. Formation of the fibrin clot begins when von Willebrand factor binds to the exposed tissue, resulting in platelet aggregation and the formation of a platelet plug [2]. This plug, in addition to the clotting factors released from injured tissue, activates the various clotting pathways responsible for producing thrombin that subsequently converts fibrinogen into a fibrin matrix [1,2]. Platelets in the fibrin clot release chemotactic factors that are responsible for recruiting leukocytes to the injury site, thus beginning the inflammatory immune response [1].

Inflammation begins with the arrival of neutrophils into the wound site, which are present for about 2–5 days unless an infection occurs [1]. Neutrophils release chemical mediators like tumor necrosis factor (TNF)-α, interleukin (IL)-1β, and IL-6, in order to further amplify the immune response [1]. Monocytes are then

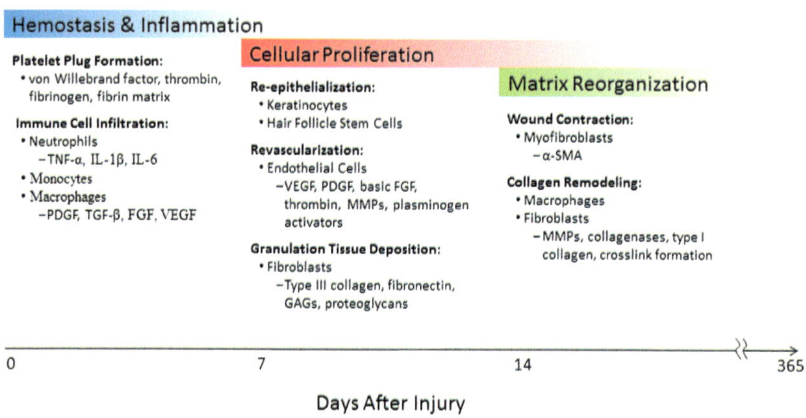

FIGURE 1 Cellular and molecular components involved in the normal wound healing process.

recruited to the wound site after about 3 days, where they differentiate into macrophages and engulf pathogens and cellular debris [1,2]. Macrophages also secrete a variety of growth factors and cytokines, such as IL-1, platelet-derived growth factor (PDGF), transforming growth factor (TGF)-β, TGF-α, fibroblast growth factor (FGF), insulin-like growth factor-1, and vascular endothelial growth factor (VEGF), which indicates that these cells play a crucial role in transitioning from the inflammation to the proliferation phase [1,2].

Proliferation is characterized by reepithelialization, neovascularization, and formation of granulation tissue. Reepithelialization occurs within hours after injury and aims to cover the wound surface [3]. Cells at the wound edge release cytokines and growth factors, such as epidermal growth factor, keratinocyte growth factor (KGF), and TGF-β, in order to stimulate fibroblasts and keratinocytes to migrate into the wound bed from the wound periphery [3]. In addition, stem cells from the hair follicle bulge can differentiate into epidermal progenitor cells, further contributing to the restoration of epidermis, though the fact that they are eventually eliminated from the resulting tissue suggests that they only play a role in initiating the reepithelialization process [4,5]. Furthermore, hair follicles and other skin appendages are not usually regenerated after injury, suggesting that the restorative potential of these stem cells is muted by the lack of an appropriate niche [3]. Neovascularization, also known as angiogenesis, is critical to maintaining the newly formed tissue by providing nutrients and gas exchange [1]. VEGF, PDGF, basic FGF, and thrombin that are present at the site of injury bind to receptors on the endothelial cells of existing blood vessels, thus initiating sprout formation [1]. These vascular sprouts migrate toward the wound bed with the help of matrix metalloproteinases (MMPs) secreted by vascular endothelial cells at the leading edge, as well as urokinase-type plasminogen activators associated with these endothelial cells, and eventually differentiate into new vasculature [1,6]. Finally, granulation tissue begins to form in the wound bed about 4 days after injury [3]. Fibroblasts—stimulated by the presence of PDGF, TGF-β, and FGF—begin to replace the provisional fibrin matrix in the wound bed with type III collagen and extracellular matrix (ECM) molecules like fibronectin, glycosaminoglycans (GAGs), and proteoglycans [1,2]. Because angiogenesis is still occurring when granulation tissue forms, this highly vascularized tissue presents clinically with a characteristic red color [1].

The final phase of wound healing involves remodeling and reorganization of the granulation tissue into a mature scar. This stage typically begins about 2–3 weeks after injury and may last for several months [2]. Fibroblasts begin to differentiate into myofibroblasts—contractile cells which are characterized by a high expression of alpha-smooth muscle actin (α-SMA) [2]. MMPs and other collagenases secreted by fibroblasts, macrophages, and the other remaining cell types remodel the matrix, signifying the development of a mature scar [2]. It should be noted that collagen found in this mature scar tissue is organized into bundles oriented parallel to the skin surface instead of the normal basket-weave orientation of uninjured skin [2]. As the collagen reorganization continues, cross-links are formed between adjacent collagen bundles resulting in an increase in the tensile strength of the scar [2]. However, at maximum, the scar only regains about 70% of its preinjured mechanical strength [2].

Scarless Fetal Healing

Remarkably, human fetuses early in gestation have the unique ability to fully regenerate cutaneous wounds without scar formation. Regenerative behavior after injury appears to be only present during the first 24 weeks of gestation and varies with injury size—with larger fetal injuries healing by a reparative, adult-like mechanism instead [7]. Regeneration was initially thought to be a consequence of the sterile environment produced by the amniotic fluid, but this was determined not to be the case after scarless fetal healing was also witnessed in marsupials, which undergo fetal development in the mother's nonsterile pouch [8]. Additionally, human fetal skin transplanted onto adult athymic mice and subsequently wounded showed a scarless healing behavior, further suggesting that scarless healing results from an internal mechanism unique to the fetus as opposed to the environmental factors present in the womb [9]. Although the underlying mechanism of this process is not yet fully understood, key differences between the adult and fetal wound healing processes have been elucidated, particularly with respect to the inflammatory progression and the presence of various growth factors and ECM molecules.

Fetal scarless wound healing displays a decreased inflammatory response which is manifested both at cellular and at molecular level. Early embryonic wounds have lower levels of immune cells, such as neutrophils, monocytes, and macrophages, which are less differentiated and shorter lived than their adult counterparts [10,11]. Decreased recruitment of these immune cells to the site of fetal injury is thought to be due to the abnormal platelet degranulation and aggregation behavior seen in fetal wound healing [12,13]. This may also explain the rapidly diminishing levels of PDGF found in fetal wounds [12]. In addition, adult wounds have a high expression of IL-6 and IL-8, while fetal wounds have significantly lower levels of these pro-inflammatory mediators [7]. Interestingly, fetal wounds exhibit a significantly increased expression of IL-10, which prevents inflammatory cell migration and inhibits IL-6 and IL-8 expression, and lentiviral-mediated delivery of IL-10 to large fetal wounds has enabled regenerative instead of reparative wound healing [7]. Expression of the three main isoforms of TGF-β is also altered in fetal skin: TGF-β1 and TGF-β2, which are pro-fibrotic by nature, are expressed in lower levels in fetal skin, while the anti-fibrotic TGF-β3 is significantly increased [14]. Additionally, TGF-β receptor (TGFBR) I and TGFBRII have been found in reduced numbers early in gestation compared to their expression levels postnatally [14].

Fetal fibroblasts exhibit slightly different cellular functions than adult fibroblasts, leading to the development of an anti-fibrotic ECM environment. These fibroblasts tend to synthesize more collagen III than adult fibroblasts, evident by the increased collagen III:collagen I ratio found in uninjured fetal tissue [15,16]. Fetal fibroblasts also express higher prolyl hydrolase activity—the rate-limiting enzyme in synthesizing collagen—which may explain their ability to proliferate and deposit collagen at the same time after cutaneous injury [17]. This postregenerative collagen is also organized in a manner similar to the normal orientation of uninjured gestational skin, which is thought to occur because of the increased fetal MMP:TIMP ratio (TIMP, tissue inhibitor of metalloproteinase) [18]. Uninjured fetal skin shows increased expression of MMP-1, MMP-2, MMP-14, and TIMP-2 compared to adult skin [18]. During fetal wound healing, MMP-1, MMP-9, and MMP-14 display increased expression, with MMP-1 and MMP-9 peaking more rapidly and at higher levels than in adult wounds [18]. Furthermore, scarless wound healing did not exhibit increased expression of TIMP-1 and TIMP-3, as did their adult counterparts [18]. Additional studies suggest that MMP-13 also plays a role in fetal regeneration via bimodal up-regulation with MMP-9 during early and late wound healing [19]. A higher MMP:TIMP ratio would yield more matrix degradation and turnover, thus promoting the scarless healing behavior seen in fetal fibroblasts. Most importantly, fetal ECM is characterized by an increased level of hyaluronic acid (HA), one of the main GAGs found in the ECM, which promotes fibroblast metabolic activity and increases the synthesis of both collagen and noncollagen proteins [20]. Studies show that exogenous HA reduces collagen matrix contraction in vitro by adult fibroblasts and promotes the reduction of scar formation in skin incision wounds [21,22]. These downstream effects might be further enhanced in fetal regeneration due to the increase in HA receptors on the cell surface of fetal fibroblasts [20].

Scarless Wound Healing in the Adult

The above discussions have focused on adult and fetal skin; however, healing of full-thickness defects in mucosal surfaces, such as those in the oral cavity, appear to heal in the adult to a near regeneration state in spite of the presence of a large concentration of microorganisms (discussed in detail elsewhere) [23–26]. Comparison of skin and oral mucosal wounds have revealed differences between the two sites at the molecular level that may provide insights into the consequences of altered regulation of inflammation and cellular contributions.

FIBROPROLIFERATIVE DISORDERS OF THE SKIN

As demonstrated, the normal wound healing process involves quite a delicate balance of cellular and molecular components. Though it appears that redundancies exist to prevent minor aberrations from completely disturbing this mechanism, larger disruptions typically lead to excessive scarring upon repair. Fibroproliferative scars are both cosmetically and functionally deforming, affecting a patient's quality of life both at physiological and at psychological level. As such, an understanding of abnormal wound progression, particularly the molecular and cellular biology of hypertrophic scars (HTS), may provide new insights into the wound healing process and lead to novel therapeutic solutions.

HTS is a fibroproliferative disorder unique to humans that often occurs after surgery, trauma, burn injury, or excessive inflammation [27]. As a whole, it has an incidence varying from 32% to 77% and is generally more prevalent in lesions caused by burn injury, as compared to surgical or traumatic injury, with a frequency as high as 91% postburn [28–30]. While fibrotic disorders occur in all races, more than 75% of nonwhite individuals develop HTS postburn, indicating skin pigmentation as a significant risk factor [31]. HTS presents clinically as raised, red, itchy, and firm scars that form within the confines of the injury site [27] (Figure 2). They usually appear about 4–8 weeks after injury,

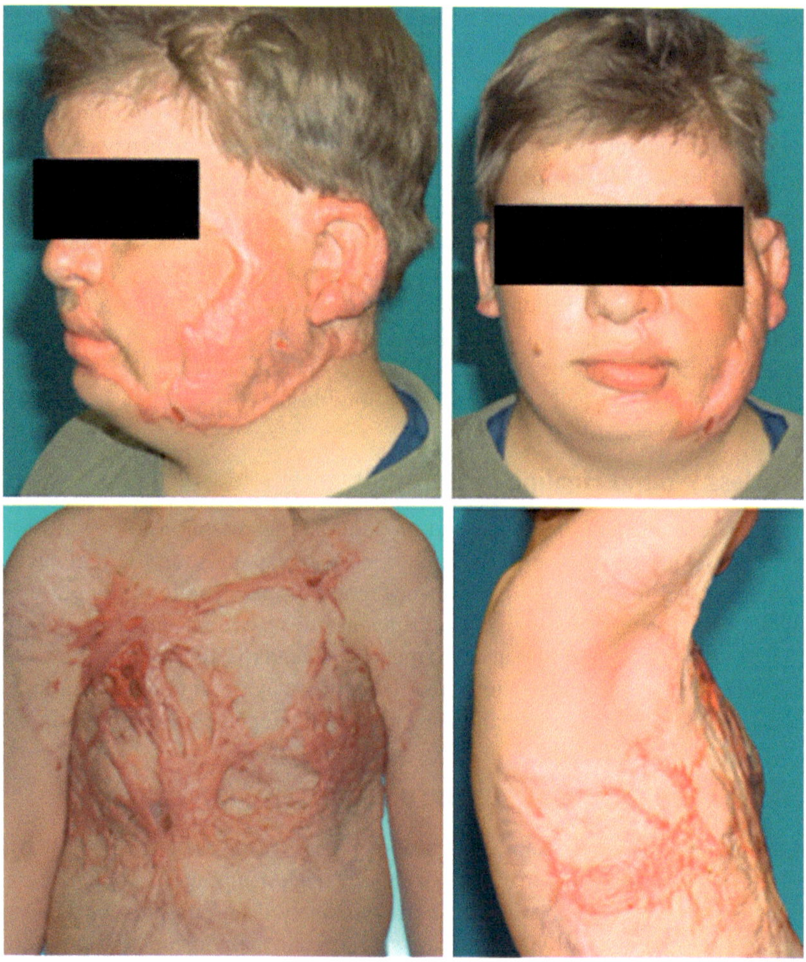

FIGURE 2 HTS on a 12-year-old male 29 months after burn injury to the face and trunk. Raised, erythematous lesions displaying contracture are evident.

reach peak growth after 6 months, and then gradually remodel over time into normal scar [32]. HTS are often highly contractile in nature, limiting patient mobility and range of motion when they develop over joints and mobile body regions like the neck, wrists, elbows, knees, and digits [33]. Histologically, HTS displays whorled nodules of fine collagen in the dermis that gradually disappear during scar maturation as the collagen chains reorient themselves parallel to the skin surface [33,34]. The ratios of different collagen types are notably altered in HTS. While normal skin and mature scar have about 80% type I collagen, 10–15% type III collagen, and minimal amounts of type V collagen, HTS tissue contains about 33% type III collagen and up to 10% type V collagen [35]. Collagens III and V have both been shown to modify type I collagen fibril diameter, thus providing an explanation for the thinner collagen fibrils seen in HTS [35,36]. Also, an increased density of blood vessels exists within HTS lesions, with blood vessels organized perpendicular to the skin surface [37]. Similar to normal wound healing, skin appendages like sebaceous glands, hair follicles, and rete ridges are typically absent from HTS tissue.

Cellular Basis of HTS

HTS have a high density of cells within their collagen-abundant structures, particularly fibroblasts [38]. The total cell number peaks during the immature phase of HTS and gradually declines as the HTS remodels into a mature scar [39]. Furthermore, the HTS fibroblasts in immature scar exhibit an abnormal behavior, when compared to fibroblasts present in normal skin. HTS fibroblasts show a significantly decreased mRNA expression of MMP-1, resulting in decreased MMP-1 protein levels, as compared to site-matched normal fibroblasts from the same patient [40]. These HTS fibroblasts also express higher levels of profibrotic TGF-β1, which concurrently stimulates excessive production of collagen in fibroblasts [40]. Moreover, HTS fibroblasts display a decreased synthesis of nitric oxide, a molecule produced spontaneously by normal fibroblasts that exerts its anti-proliferative effect through collagenase regulation [41]. These altered fibroblastic phenotypes provide a fundamental explanation for the disproportionate amounts of collagen seen in HTS.

In addition, increased numbers of myofibroblasts exist in HTS tissue [33]. These differentiated fibroblasts promote additional synthesis of ECM and contain vast amounts of α-SMA and microfilament bundles, thus resulting in contractile behavior [33]. Furthermore, myofibroblasts in HTS have lower basal apoptotic rates and express higher levels of BCL-2, an important anti-apoptotic marker, than those in normal skin [42]. This finding helps to clarify the reasons for increased levels of myofibroblasts in HTS and contraction of HTS tissue beyond the point of wound closure.

Notably, fibroblasts in normal skin display a heterogeneous profile; that is, fibroblasts isolated from different depths of the skin show a different phenotype and expression profile when compared with one another [43]. In particular, reticular fibroblasts from the deep dermis exhibit characteristics that are very similar to those seen in HTS fibroblasts: higher production of type I and III procollagen paired with a decreased expression of collagenase, resistance to apoptosis and migration, and amplified expression of α-SMA leading to increased collagen matrix contraction [44–46]. Furthermore, deep dermal fibroblasts produce more TGF-β1 and its downstream effector, connective tissue growth factor (CTGF), than papillary fibroblasts located in the superficial layers of the dermis, as well as higher levels of heat shock protein 47—a chaperone protein responsible for the proper formation of type I collagen [47]. Like HTS fibroblasts, deep dermal fibroblasts proliferate at a slower rate and are morphologically larger than normal and superficial fibroblasts [48]. As such, deep dermal fibroblasts have been implicated as a significant driver of HTS development [47]. This idea is further supported by research reporting that HTS incidence is highly related to the depth of injury. A study by Dunkin and colleagues found a critical depth for HTS by using an experimental jig, which created a linear dermal scratch wound with varying depth—from no injury to full-thickness injury—within the same wound lesion [49]. The threshold depth beyond which the resulting scar became hypertrophic in the patients was determined as 0.56 ± 0.03 mm or 33% of the lateral hip thickness, a result which was confirmed by another similar study [49,50].

In addition to depth of injury, prolonged duration and increased severity of inflammation in burn patients have been shown to be reliable predictors of HTS formation [48]. Recent studies suggest that the type of cells involved in the inflammatory response dictates the fibrotic progression. For example, immune cells known as fibrocytes contribute to HTS development. These cells, which only make up about 0.1–0.5% of the peripheral blood mononuclear cells in normal circulation, account for about 10% of the cells infiltrating acute wounds and express collagens I and III, CD13, CD34, and the bone marrow-derived surface marker CD45 [51,52]. Fibrocytes have been implicated in the synthesis of ECM and collagenase, as antigen-presenting cells responsible for priming naive T cells, and in the stimulation of dermal fibroblasts toward pro-fibrotic behavior through paracrine signaling and secretion of cytokines like TGF-β1, TNF-α, IL-6, and IL-8 [48,53]. They can also differentiate into myofibroblasts through the SMAD2/3 and JNK MARK pathway upon activation by pro-fibrotic factors like TGF-β1, further supporting their significant role in HTS development [52]. Double immunostaining with antibodies to leukocyte-staining protein 1 and type I collagen allowed for identification of fibrocytes from fibroblasts and other leukocytes in burn patient tissue samples and showed that fibrocytes exist in higher levels in HTS tissue than in mature scar and normal skin [54].

$CD4^+$ T helper (Th) cells also show an increased presence in HTS dermal tissue, as compared to normal skin [55]. $CD4^+$ T lymphocytes can transition to either a Th1 or a Th2 subtype, each having a unique cytokine expression profile [56]. Th1 cells express anti-fibrotic cytokines (e.g., IL-2, IL-12, and interferon

(IFN)-γ), while Th2 cells express pro-fibrotic cytokines (e.g., IL-4, IL-5, IL-10, and IL-13) [52,57]. In addition, Th1 cells induce fibroblast apoptosis and activate nitric oxide synthase in order to promote collagenase activity, further supporting their anti-fibrotic role, while Th2 cells induce ECM production and up-regulation of procollagens I, III, and V and TIMP-1, adding to their pro-fibrotic role [57,58]. A study of Th1 and Th2 cytokines in the serum of burn patients with HTS has shown elevated levels of IL-4 and IL-10 with decreased levels of IL-12 and an absence of IFN-γ [59]. Another study found an increased frequency of $CD4^+$/TGF-β^+ cells in the peripheral blood of burn patients with HTS, suggesting that Th2 cells may influence dermal fibroblasts toward pro-fibrotic differentiation and behavior [55]. The recent discovery that Th1 presence inhibits fibrocyte differentiation into myofibroblasts, while Th2 presence promotes this behavior has further confirmed the influence that these cells have on HTS [60].

Mast cells have also been implicated in HTS development, due to their increased presence in HTS tissue [61]. Since mast cells contribute to the immune response by releasing a number of mediators including histamine, which causes vasodilation and the itchiness associated with allergies, their presence in HTS tissue explains the pruritic behavior of HTS. Mast cells have also been implicated in promoting pro-fibrotic behavior in skin fibroblasts, in part via gap junction-mediated communication [62]. Moreover, some pro-inflammatory mast cell functions in HTS are the result of prostanoid and leukotriene release, hence blocking the mast cell synthesis of these mediators may provide a therapeutic effect on HTS formation [63].

Finally, keratinocytes may help to control abnormal wound healing through paracrine signaling. When fibroblasts were cultured in keratinocyte-conditioned media, they had an increased mRNA and protein expression of MMP-1 that was attributed to a keratinocyte-derived anti-fibrogenic factor known as 14-3-3 sigma, a signal transduction progression factor in the cell cycle [64]. In addition, fibroblasts that were either co-cultured with keratinocytes or grown in keratinocyte-conditioned medium demonstrated a reduction in mRNA and protein expression of type I collagen [65]. Furthermore, keratinocytes have displayed a regulatory role on the pro-fibrotic molecular expression profile of dermal fibroblasts, by down-regulating CTGF, collagens I and III, fibronectin, plasminogen activator inhibitor-1, TIMP-2, TIMP-3, and α-SMA [66]. Recently, it has been shown that keratinocytes reduced TGF-β1 production of deep dermal fibroblasts cultured in a tissue-engineered skin substitute and inhibited their differentiation into myofibroblasts [67]. Moreover, release of TGF-α, IL-1, and KGF from keratinocytes improved epidermal formation in these co-cultured constructs [67]. However, keratinocytes may not completely attenuate the fibrotic behavior of fibroblasts, as high levels of PDGF and IL-6 were also noted in matrices containing deep dermal fibroblasts and keratinocytes [67]. Other evidence suggests that keratinocytes isolated from HTS tissue may influence fibroblasts toward a fibrotic phenotype during co-culture, perhaps due to an increased secretion of TIMP-1 [68,69].

Molecular Basis of HTS

The three mammalian isoforms of TGF-β—TGF-β1, TGF-β2, and TGF-β3—appear to play a central role in HTS development. The pro-fibrotic isoforms, TGF-β1 and TGF-β2, are known to increase fibronectin, collagen, and TIMP expression in fibroblasts while also inhibiting collagenase induction [70–72]. On the other hand, TGF-β3 is anti-fibrotic and acts as an antagonist toward the other two isoforms [72]. It has been suggested that the ratio of two TGF-β cell surface receptors, TGFBRI and TGFBRII, dictates fibrotic development, with higher TGFBRI:TGFBRII ratios promoting fibrosis [73]. A recent study has shown that involving siRNA inhibition of TGFBRI expression in HTS fibroblasts significantly suppressed cellular proliferation; decreased mRNA levels of collagen I, collagen III, fibronectin, and CTGF; and reduced collagen I and fibronectin protein expression [74]. The TGF-β receptors are involved in a particularly important TGF-β1 signaling pathway known as the SMAD pathway. SMADs are intracellular regulatory proteins that can be subclassified as either receptor-regulated SMADs (R-SMAD), common-mediator SMADs (Co-SMADs), or inhibitory SMADs (I-SMADs) [52]. When TGF-β1 binds to TGFBRII, it dimerizes with TGFBRI and then phosphorylates the R-SMADs [52]. Phosphorylated R-SMADs combine with Co-SMAD 4 to form a SMAD complex, which enters the nucleus and functions as a controller of genetic expression [52]. Of particular importance is R-SMAD 3, the predominantly phosphorylated SMAD in HTS fibroblasts [75]. In R-SMAD 3 knockout mice, TGF-β1-mediated fibrosis was absent for up to 28 days [76]. Moreover, I-SMAD 7 inhibits R-SMAD phosphorylation by binding with TGFBRI, overexpression of which resulted in the inhibition of α1 procollagen and α-SMA expression along with decreased collagen gel contraction in both HTS and normal fibroblasts, suggesting a potential therapeutic use [75,77].

Pro-fibrotic TGF-β stimulation of TIMP synthesis by fibroblasts manifests as a reduction of MMP levels in HTS tissue [71]. TIMP-1 serum levels in burn patients were found to be significantly higher than that of normal patients and correlated with total body surface area (TBSA) burned [78]. In addition, MMP-2 levels were significantly increased in HTS, while MMP-9 levels were low or undetectable [79]. However, MMP-2 up-regulation in both normal and scarless healing suggests that MMP-2 increase after injury is not indicative of fibrosis [18,78]. Furthermore, MMP-1 mRNA and protein expression were found to be significantly decreased in HTS-derived fibroblasts [80]. Since MMP-1 and MMP-9 were increased in the scarless fetal healing process (Table 1), increasing these levels in HTS may prove to be therapeutic. Recent studies subjecting HTS fibroblasts to FGF-2 treatments have significantly up-regulated MMP-1 levels and led to a reduction in hydroxyproline content [80]. Hydroxyproline is a posttranslationally modified form of the amino acid proline in the collagen alpha helix and induces turns or kinks in the collagen molecule, critical for normal fibrillogenesis [81]. Recall that extracellular secretions from normal

TABLE 1 Comparison of Fetal Scarless Healing and Hypertrophic Scar Features

	Fetal Scarless Healing	Hypertrophic Scar
Immune cells	Low recruitment with less-differentiated, shorter-lived cells	Fibrocyte infiltration; presence of Th2 cells, mast cells
PDGF	Rapidly diminishing levels; abnormal platelet degranulation	Increased expression found in scar tissue
Interleukins	High expression of IL-10; low expression of IL-6, IL-8	Elevated IL-4, IL-10, decreased IL-12 in burn patient serum
TGF-β1:TGF-β3 ratio	Low ratio; significantly increased TGF-β3 production	High ratio; raised TGF-β1 levels are characteristic
TGF-β receptors	Reduced numbers of TGFBRI and TGFBRII are present	Increased TGFBRI:TGFBRII ratio promotes fibrosis
Collagen I: Collagen III ratio	Low ratio; higher synthesis of type III collagen	High ratio; more type III collagen than normal skin, mature scar
MMPs	MMP-1, MMP-9 peak rapidly; MMP-2, MMP-13, MMP-14 are also increased	Decreased expression of MMP-1, MMP-9; MMP-2 up-regulated
TIMPs	TIMP-1, TIMP-3 not elevated during healing; TIMP-2 higher than normal adult skin	TIMP-1, TIMP-2, and TIMP-3 secreted by dermal fibroblasts
GAG profile	High levels of HA present; fetal fibroblasts have more HA cell-surface receptors	Twofold elevation in total GAGs; increased sulfated GAGs; HA content similar to normal skin
Collagen orientation	Normal, basket-weave orientation upon regeneration	Whorled nodules; collagen reorients parallel to skin surface during maturation

keratinocytes were also shown to increase MMP-1 and reduce type I collagen expression in fibroblasts. Some studies suggest that keratinocytes may increase fibroblast MMP-9 levels as well, further supporting the need for reepithelialization upon wound healing [82].

As a consequence of changes in the molecular expression of HTS fibroblasts, the ECM environment is also altered. Structural changes are present, including a significant reduction of elastin and fibrillin-1 in HTS tissue, two major microfibrils involved in providing elasticity to the skin [83]. The reduction of fibrillin-1 may play a role in enhancing the abnormal behavior of TGF-β in HTS, as fibrillin-1 is reported to regulate TGF-β activation and signaling [83]. In addition, a twofold increase in the total GAG content has been found in the scar tissue, which facilitates ECM swelling since GAGs are highly polar and attract water [84]. In particular, this GAG increase is attributed to an elevated amount of the sulfated GAGs—particularly dermatan sulfate, chondroitin sulfate, and heparin sulfate—which are found bound to a protein core in the ECM [84]. On the other hand, HA (an unbound, nonsulfated GAG) is present in HTS at similar levels to those found in normal skin [84].

The higher levels of sulfated GAGs suggest that modified amounts of proteoglycans exist in HTS tissue. Of importance to HTS are three of the small leucine-rich proteoglycans (SLRPs): decorin, biglycan, and fibromodulin. SLRPs are responsible for binding to type I collagen in order to regulate fibrillogenesis, modulating the activity of various growth factors, and influencing cellular behavior and growth through autocrine and paracrine signaling [85,86]. Decorin, biglycan, and fibromodulin are able to interact with all three isoforms of TGF-β, indicating that they play a role in HTS development [87]. In HTS tissue, decorin and fibromodulin are found in significantly lower quantities than in normal skin, while biglycan is significantly up-regulated [86]. Adding recombinant decorin to HTS fibroblasts results in an inhibition of cellular proliferation, down-regulated TGF-β1 production, and impaired collagen synthesis, while fibromodulin delivery displays decreased TGF-β1 and TGF-β2 levels and increased TGF-β3 and TGFBRII expression [88,89]. Moreover, decorin has been shown to induce apoptosis and to decrease collagen gel contraction, thus deterring hypercellularity and supporting its anti-fibrotic role [45,90]. An additional proteoglycan of importance to HTS is versican, a large chondroitin sulfate proteoglycan with a cartilage-like phenotype, which is an order of magnitude larger in size than the SLRPs [91]. While almost absent in normal skin tissue, it was found to be present in elevated levels throughout the connective tissue of HTS [84,91]. It is thought that versican disrupts collagen bundle formation by positioning itself between the collagen fibrils in order to expand the ECM network, leading to the increased scar volume seen clinically [35,91]. Fibrocytes have recently been shown to modulate the proteoglycan environment of the ECM as well, as they have higher mRNA levels of versican and perlecan and very low mRNA levels for decorin [92]. Co-culture of deep dermal fibroblasts with normal keratinocytes also resulted in increased decorin and fibromodulin expression, further supporting the anti-fibrotic role of reepithelialization [67].

Chemokines, small 8–10kDa proteins that induce local chemotaxis after injury, are also present in HTS, particularly those from the CXC and CC subfamilies [52]. CXCR4 is a unique CXC chemokine receptor because it only binds to one CXC chemokine, stromal cell-derived factor (SDF)-1 [93]. This chemokine pathway is important for recruiting bone marrow-derived cells and stimulating angiogenesis [93]. As such, increased SDF-1 was found in HTS tissue, burn patient serum, and deep dermal fibroblasts and is thought to contribute to HTS pathogenesis by recruiting bone marrow-derived cells to the injury site, where they then differentiate into fibrocytes and myofibroblasts [93]. An important CC chemokine involved in HTS is monocyte chemotactic protein (MCP)-1. MCP-1 is secreted by the cells involved in normal wound healing, in order to recruit monocytes and T cells to the injury site [94]. MCP-1 is found to be up-regulated in HTS fibroblasts, where it stimulates collagen overproduction and increases endogenous TGF-β expression via Th2 polarization [94,95].

Toll-like receptors (TLRs) have only recently been implicated in fibrosis [94]. TLRs are known to play a key role in the innate immune system by recognizing highly conserved molecules on microbial surfaces and are now thought to play a modified role in fibrotic conditions, such as liver fibrosis [48]. Moreover, TLR-2 and TLR-4 have been shown to be aberrantly modulated by ECM proteoglycans, such as decorin, biglycan, versican, and HA [96]. HTS tissue and fibroblasts were shown to have an increased expression of TLR-4 surface receptors, suggesting that fibroblasts contain a mechanism to up-regulate inflammation through TLR activation and immune cell recruitment [94].

Lessons Learned from Preclinical Porcine Models of Abnormal Skin Healing

Porcine models of skin wound healing offer a number of important features that allow for the assessment of normal and abnormal wound healing, while providing a target species to assess the efficacy of tissue-engineered constructs in wound healing. This animal model shares a number of metabolic features with human wound healing, offers an understanding of tight skin wound healing, and provides the genetic diversity required to better understand genetic factors in healing due to the availability of different breeds of pigs [97]. Moreover, the porcine genome has been sequenced, which allows for comparisons to features of human conditions [98].

The red Duroc model of skin wound healing was first described in the 1970s but has been "rediscovered" more recently [97,99,100]. Full-thickness, but not partial-thickness, excisional wounds on the dorsum of skeletally immature red Duroc pigs (females and castrated males) heal with the formation of hypercontracted, hyperpigmented scars, in contrast to hypopigmented human scars. Features of HTS (e.g., collagen whorls) are present, but to be accurate, the healing is likely more fibrogenic than HTS like [97]. In this model, the fibrogenic response to dorsal injury exhibits an abnormal molecular expression pattern with a second wave of expression of inflammatory molecules [97]. Not unlike abnormal human wound healing, in the porcine model the abnormal wound healing phenotype is controlled by a complex set of genetic factors [99]. However, the genetic contributions have not been extensively studied yet, so the extent of the complexity is currently undefined.

While the mechanisms responsible for the fibrogenic response in the red Duroc pig remain somewhat undetermined, some insights have been elucidated via studies with drug interventions. Oral treatment of red Duroc pigs with the asthma drug ketotifen fumerate immediately after injury led to an abrogation of the hypercontracted phenotype [101]. Ketotifen fumerate is a mast cell stabilizer that inhibits degranulation of mast cells and has been used in both adult and pediatric populations for over 20 years. Initiating drug treatment 28 days post injury was without effect, suggesting that the events critical for the abnormal wound phenotypes occur early after injury. Stopping drug treatment at 28 days post injury did not lead to a reactivation of the abnormal features of the healing wounds, confirming that there is a critical "window" for drug treatment. Analysis of wound sites with and without drug treatment revealed decreased numbers of nerves, mast cells, and myofibroblasts in the drug-treated animals. These findings have generated the concept that a bidirectional abnormal "nerve-mast cell–myofibroblast axis" contributes to the abnormal healing phenotype [102,103].

Additionally, treatment of similar skin wounds in Yorkshire pigs, which heal with a normal phenotype, with ketotifen fumerate did not detectably influence wound healing [101]. Thus, while nerves, mast cells, and myofibroblasts are present in skin wounds in Yorkshire pigs, the drug did not compromise normal healing. Moreover, in vitro studies with dermal fibroblasts of red Duroc pigs have revealed that cells derived from the dorsum exhibit different features than those from the ventral surface. In contrast, cells from the dorsum and ventral surface of Yorkshire animals are indistinguishable [104]. As cells from the dorsal and ventral surface arise from different cellular lineages during development, it appears that in the red Duroc pig the abnormality arises from the lineage that contributes to formation of the dorsal surface. At present, it is not clear whether skin wounds on the ventral surface of red Duroc pigs heal with a different phenotype from those on the dorsal surface, as there are no reports in the literature focused on this issue. However, the finding of potential differences between ventral and dorsal healing may be relevant to healing in humans, where risk for abnormal healing can vary.

As mast cells and their activation products have been detected in burn patients, the findings with ketotifen fumerate in these preclinical models may have the potential, either alone or in combination with other interventions, to prevent HTS formation and treat subsets of HTS patients with chronic pruritus. Furthermore, such drug interventions may have efficacy after surgical removal of HTS to prevent reoccurrence of an HTS-like scar.

MEDICAL THERAPIES FOR SKIN REGENERATION

HTS after burn injury continues to be a significant cause of patient morbidity and presents a major clinical challenge for medical providers. Typically, noninvasive treatment modalities, like pressure garments and silicone gel sheeting, are applied upon wound closure in order to prevent HTS development. A supplementary therapy regimen of injection-based treatment with the corticosteroid triamcinolone acetonide is recommended for patients with continuing hypertrophy after 6 months of treatment, and surgical scar revision is recommended for patients with HTS after 12 months of treatment; although, those with functional impairments or severe injury may undergo earlier surgical treatment [105]. In burn treatment, for example, an autologous skin graft is often used as a preventative measure in order to close the wound and reestablish the skin's barrier function. Z-plasty revisional surgery may be carried out in order to relieve tension and contracture in the scar and promote collagen remodeling [48]. Although widely used, the classic clinical treatments for HTS have been ineffective at alleviating and preventing the fibroproliferative condition, providing a need for alternative strategies. New information about the inflammatory mechanisms involved in HTS has led to the development of injection-based therapies involving IFNs and chemokines [48]. Furthermore, pulsed dye and fractional CO_2 laser surgical therapies, as well as novel gene and drug-delivery methodologies, have resulted in improved HTS scar characteristics post treatment [106,107]. Recent advances in skin tissue engineering and regenerative medicine also appear to provide promising new solutions to the HTS problem.

Commercially Available Tissue-Engineered Skin Solutions for HTS

Tissue-engineered skin substitutes came into being following the advent of serial subculture of keratinocytes in 1975 [108]. Not long after, keratinocytes from burn patients started being cultured into epithelial sheets, known as cultured epithelial autografts (CEAs), in order to assist the treatment of extensive third-degree burns [109]. However, CEAs have suffered from fragility (making them difficult for the physician to handle and graft), high cost, lengthy production times, and a varied "take"—the adherence and integration of a graft into the recipient's skin [110]. Nevertheless, many commercial CEA products have come to market—Epicel™ (Genzyme Corporation; Cambridge, MA), MySkin™ (Regenerys Ltd; Sheffield, United Kingdom), and EpiSkin™ (SkinEthic Laboratories; Nice, France), to name a few—some of which attempt to improve wound engraftment using dermal substitutes like silicone, hydrogel, fibrin, or decellularized collagen scaffolds [111–113]. Recently, sprayed epithelial products like CellSpray™ (Avita Medical; Woburn, MA) and ReCell™ (Avita Medical; Woburn, MA) have been developed [113]. These keratinocyte suspensions reduce the culture time needed before clinical use and increase cell proliferation, minimize enzymatic degradation of cell surface proteins, and produce a more robust epithelial layer [113].

In addition to epidermal replacements, numerous dermal substitutes have been created in order to replicate the dermis in full-thickness wounds and promote reepithelialization, revascularization, and restoration of an appropriate growth factor and cytokine profile [114]. Many of these products are simply a decellularized collagen matrix from either human (AlloDerm™ (LifeCell Corporation; Bridgewater, NJ), SureDerm™ (HansBiomed Corporation; Seoul, South Korea), KaroDerm™ (Karocell Tissue Engineering AB; Stockholm, Sweden)), porcine (Permacol™ (Medtronic; Minneapolis, MN), EZDerm™ (Molnlycke Health Care US, LLC; Norcross, GA), OASIS Wound Matrix™ (Cook Biotech, Inc.; West Lafayette, IN)), or bovine (Integra DRT™ (Integra LifeSciences Corporation; Plainsboro, NJ), MatriDerm™ (MedSkin Solutions Dr. Suwelack AG; Billerbeck, Germany), PriMatrix™ (TEI Medical Inc.; Waltham, MA)) sources [111,113]. Since these matrices are acellular, they must recruit their cells and growth factors from the underlying wound bed [112]. Other dermal replacements, like Dermagraft™ (Organogenesis, Inc.; Canton, MA) and TransCyte™ (Organogenesis, Inc.; Canton, MA), are embedded with allogeneic cells that express growth factors, cytokines, and inflammatory mediators into the wound bed upon engraftment to stimulate a natural wound healing response [112,114]. Although immune rejection is an inherent risk to the efficacy of these products, it seems to be a negligible issue [114].

Advances in tissue engineering have led to the development of commercially available bilayered skin substitutes that are composed of both keratinocytes and fibroblasts, such as Apligraf™ (Organogenesis, Inc.; Canton, MA) and Orcel™ (Forticell Bioscience, Inc.; New York, NY) [111]. The co-culture of keratinocytes with fibroblasts in these skin substitutes promotes the formation of a basement membrane between the epidermal and dermal layers of the product, encouraging increased epidermal attachment [112]. The basement membrane is a rich source of additional growth factors, cytokines, and cell signals, so its development replicates the behavior seen in normal skin [115]. While they may seem ideal as a replacement for split-thickness skin grafts, these products are generally used on a temporary basis as biological wound dressings due to their limited lifespan and high cost [111,112]. Other bilayered skin substitutes, like PermaDerm™ (Amarantus BioScience Holdings, Inc.; San Francisco, CA), PolyActive™ (HC Implants BV; Leiden, The Netherlands), and the TissueTech Autograft System™ (Anika Therapeutics, Inc.; Bedford, MA), have improved upon the above products by incorporating autologous keratinocytes and fibroblasts from the patient into their scaffolds, although their efficacy, cost, and ease of use are still not ideal [111]. Notably, PolyActive™ and the TissueTech Autograft System™ use noncollagen matrix materials, with PolyActive™ using a synthetic polyethylene oxide terephthalate/polybutylene terephthalate block co-polymer and the TissueTech Autograft System™ incorporating its cells into an HA ester matrix [111]. The latter product seems to show improved cellular proliferation and enhanced graft take, which is not surprising given the benefit of increased HA levels in fetal scarless healing [110].

Recent Developments in Dermal Engineering

The preceding tissue-engineered skin products have certainly improved patient outcomes in recent years, with increased survival rates seen in patients with greater than 60% TBSA burned who were treated with skin substitutes and excellent/good outcomes reported for neck contracture in 27 of 28 burn patients treated with a dermal substitute/skin graft combination [116,117]. However, a recent multicenter study of skin substitute effectiveness in burn patients only has shown an average take of 49%, while another study reported almost negligible take at 2 weeks, suggesting that the harsh burn wound environment may present a significant challenge to engraftment of skin substitutes [118,119]. Decreased take may also be due to the inadequate revascularization rates plaguing current skin substitutes [111]. Other problems include varied consistency, loss of sensation because of insufficient innervation, and the formation of scar at the graft margins [111,112]. Furthermore, hypopigmentation due to the absence of melanocytes and the lack of various skin appendages, like sweat glands and hair follicles, in these products are not only aesthetically disfiguring but also functionally limiting [111,112] (Figure 3).

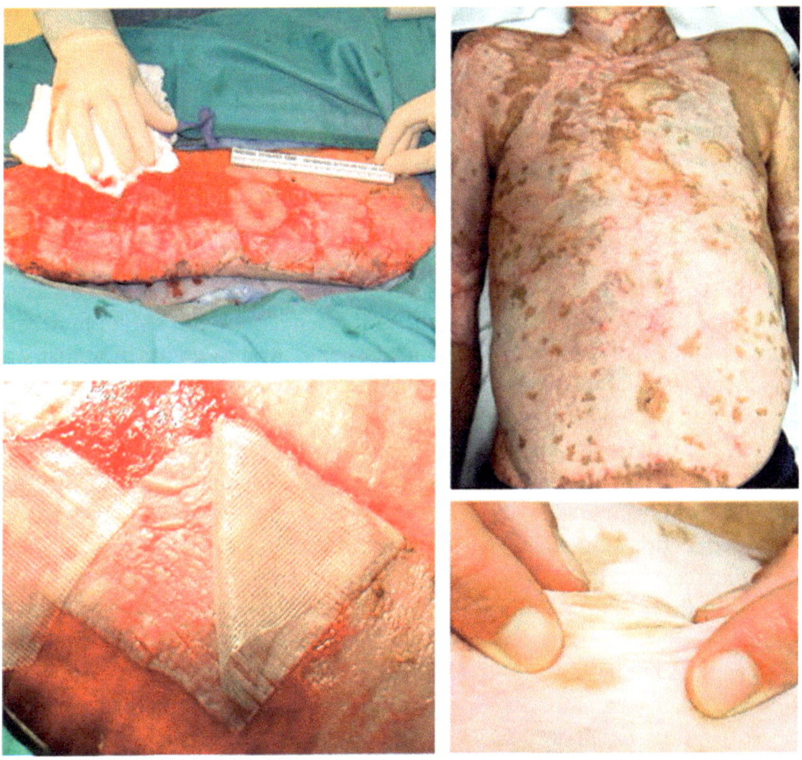

FIGURE 3 A 28-year-old male suffered injuries to 93% TBSA, with 85% TBSA full-thickness burns, after a closed space explosion. Autografts were administered to the patient's back within the first week and the remaining 60% TBSA burned was covered with Integra™. Due to the extent of injury, 700 cm^2 of cultured skin substitutes containing autologous keratinocytes and fibroblasts (Steven T. Boyce, University of Cincinnati, Shriners Hospitals for Children) were grafted over 11 procedures and allowed for successful resurfacing of 50% TBSA. High-quality coverage was noted upon follow up 7 years post injury, though hypopigmentation is clearly present. *From Medina A, Tredget EE. Strategies to increase flap survival in nasal reconstruction in patients with deep panfacial burns. J Burn Care Res January–February 2013;34(1):42–7; with permission.*

Recent advances have been attempting to overcome some of these limitations and improve the efficacy of tissue-engineered skin.

The collagen–GAG matrix has been the "gold-standard" scaffold material used for skin tissue engineering since it was developed in 1980 [120]. Recent developments in scarless fetal healing research have led to the incorporation of HA into these collagen matrices, which have shown good proliferation and migration of seeded fibroblasts in vitro, as well as restoration of normal skin structure in an in vivo rat scar model [121,122]. Another material that has recently been incorporated into collagen scaffolds is chitosan, a linear polysaccharide with thrombogenic and antibacterial properties derived from the shells of crustaceans, which produces scaffolds with good biocompatibility and fibroblast proliferation behavior [123]. A recent study has functionalized these chitosan-containing matrices with siRNA ligands and reported down-regulation of scar-related TGF-β1, α-SMA, and collagen expression by seeded fibroblasts, suggesting they may be useful for dermal engineering as a gene therapy delivery vehicle [124]. Additionally, synthetic polymers are on the rise for dermal tissue engineering, as they show an increased capability for tuning matrix properties, as compared to natural materials like collagen, chitosan, and HA [125]. Common synthetic materials being explored are poly-L-lactic acid, polycaprolactone, and polyglycolic acid due to their established biocompatibility—they are often found in resorbable sutures—and approval for biomaterials use by the US Food and Drug Administration [125]. A study using poly(lactic-*co*-glycolic acid) copolymers for delivery of cell sheets to dermal wounds in rats showed an excellent take rate, as well as contraction that was statistically similar to autografts, supporting the use of these materials for skin tissue engineering [126].

The rise of synthetic materials for dermal engineering has also led to alternative methods for making matrix, notably electrospinning [125]. Electrospinning uses a voltage gradient in order to deposit a suspension of matrix solution onto a collection vehicle in a controlled manner, allowing for a highly tunable fiber diameter and orientation (random or parallel) [125,127]. Coaxial electrospinning allows for the production of synthetic fibers with a continuous shell coating of bioactive polymer, thus improving cellular adhesion and proliferation, while retaining the ability to tune scaffold mechanics [127]. Recently, a gelatin-coated, coaxially electrospun polycaprolactone scaffold has allowed for scalable tissue-engineered skin mechanics with no adverse effect on the epidermal–dermal organization in vitro, thus creating a mechanism to allow for biomechanical matching of different anatomical skin sites [127]. Furthermore, electrospinning also offers an ideal way to incorporate elastin fibers into tissue-engineered skin substitutes, which may allow for improved pliability of regenerated skin and provides a tool for combating the increased skin contractures seen in recovering burn patients [128]. Recently developed electrospun scaffolds containing 80% human tropoelastin and 20% type I collagen supported fibroblast infiltration, collagen deposition, and angiogenesis and behaved similarly to Integra™ in a subcutaneous implantation model [128]. Moreover, electrospun scaffolds provide another means for delivering drugs and growth factors into healing wound sites. A recent study using acid-responsive, ibuprofen-loaded polylactic acid scaffolds in an in vivo rat wound model has attenuated inflammation and reportedly accelerated wound healing [129]. The ratios of collagen I:collagen III and TGF-β1:TGF-β3 were reduced, while the MMP:TIMP ratio was increased, in addition to a regulated collagen, α-SMA, and basic FGF expression profile [129]. Another electrospun scaffold containing collagen, HA, and gelatin with pro-angiogenic growth factors loaded to allow for staged release—with basic FGF and endothelial growth factor released first, followed later by VEGF and PDGF—showed accelerated wound closure and enhanced maturation of newly formed wound vasculature, providing a potential solution to the reduced angiogenesis seen in currently available tissue-engineered skin substitutes [130].

Recent steps to improve innervation in tissue-engineered skin have been successful. The incorporation of Schwann cells into tissue-engineered skin resulted in a 1.7–1.8-fold increase in the nerve fiber numbers post engraftment [131]. Additionally, these grafts had a perception threshold similar to normal skin and promoted myelin sheath formation [131]. Nerve regeneration also appears to have the capacity to improve revascularization of tissue-engineered skin constructs. Transplanted skin analogs in immunocompromised rats showed the formation of new nerve fibers after 10 weeks, which were found to be in close proximity to $CD31^+$ blood vessels [132]. Notably, lymphatic vessels were also found in the regenerated site, although not alongside the aforementioned blood vessels [132]. Moreover, a 40–80% increase in capillary-like tube formation was reported in an in vitro tissue-engineered angiogenesis model upon addition of various neurotrophic growth factors, further supporting the capacity of nervous tissue to restore blood vasculature [133].

Stem cells and endothelial cells have also been incorporated into tissue-engineered skin in order to improve angiogenesis [134]. Recently, a novel endothelialized skin substitute incorporating dermal fibroblasts, adipose-derived stem cells, and human umbilical vein endothelial cells into a fibrin matrix showed a promotion of capillary-like structures, enhanced tissue repair, and the formation of a keratinized epidermis [134]. Furthermore, transplantation of bone marrow-derived stem cells and epidermal stem cells into an in vivo rat wound model via a porous gelatin-based scaffold-accelerated reepithelialization, collagen synthesis, and blood vessel formation. Interestingly, these skin substitutes also resulted in regeneration of hair follicles in the wound site, an important step for further replicating normal skin structure via tissue engineering [135]. Stem cells have shown a capacity to differentiate in multiple skin types in wounds, including keratinocytes, endothelial cells, pericytes, and monocytes, and seem to provide a mechanism for the restoration of skin appendages as well [135,136]. However, future studies involving stem cells should be mindful of their tumorigenic capacity in vivo.

Melanocytes have also been incorporated into newer tissue-engineered skin constructs in order to overcome issues of hypopigmentation post engraftment. Interestingly, a lower number of blood vessels were formed in tissue-engineered skin with melanocytes as compared to matrices without melanocytes, suggesting that melanocytes may be hindering revascularization [137]. No noticeable difference was reported between lymphatic vessel formation in pigmented and nonpigmented tissue-engineered skin [137]. As melanocytes are also the source of cancerous melanomas, their tumorigenicity has been recently evaluated in athymic mice transplanted with engineered skin substitutes [138]. Fortunately, no detectable tumors were formed when melanocytes were used from normal skin, although melanocytes isolated from melanomas induced tumor formation in these mice [138]. Therefore, normal melanocytes appear to be safe for use in skin tissue engineering [138].

FUTURE DIRECTIONS

In the past four decades, much progress has been made with regard to building an understanding of the biological underpinnings of HTS and developing new treatment modalities to combat and prevent the resulting morbidity, particularly after burn injury. Despite these advancements, the complete pathobiological picture of HTS has not yet been fully realized, resulting in a need for continued research into the biological mechanisms of HTS to help improve emerging clinical therapies. Interestingly, review of literature suggests that a majority of the future research on HTS will fall into one of two categories: a cell-based approach—in which the goal would be to modify the cellular recruitment at the wound site in order to modulate the fibrotic response—or a matrix-based approach—where fibrosis would be affected by changes to the extracellular environment. The first of these avenues is exemplified by studies exploring the immune cells involved in HTS propagation, leading to the recent discovery of the fibrocyte, as well as studies incorporating additional cell types into tissue-engineered skin, like mesenchymal stem cells and melanocytes. The second avenue has manifested itself through studies involving the ECM of scarless fetal healing and research on the differing proteoglycan expression of HTS fibroblasts.

The ultimate tissue engineering goal of creating a highly organotypic skin substitute aligns well with both of these approaches and would reduce the current dependence on allografts and autografts in HTS burn therapy. However, there are many limitations to this therapeutic strategy that hinder its ability to be realized in the short term. Current co-culture techniques seem restricted to sustaining just a few cell types at a time and have so far been unable to result in regeneration of the various skin appendages present in normal skin, which is both cosmetically and functionally limiting. Moreover, the speed at which these devices can currently be created in the laboratory is quite slow and therefore limits their therapeutic potential. Recent advances in 3D printing of tissue-engineered scaffolds and cells, along with improvements to bioreactor technologies, may allow for generation of complex tissue-engineered skin architectures within a shorter amount of time. However, even if highly organotypic skin substitutes could be created quickly and grafted to burn wounds, the enhanced inflammatory environment of the burn site may still result in graft rejection or induce fibrosis in the skin substitute. Therefore, it may be useful to simultaneously explore methods of modulating the pro-fibrotic environment of the wound area in future tissue-engineered skin constructs. Advanced tissue-engineered structures that make use of effective anti-inflammatory drug delivery and gene therapy strategies may provide one solution. Furthermore, exploration of these treatment modalities may result in new therapies for those burn patients who are dealing with morbidity from existing HTS.

LIST OF ABBREVIATIONS

α-SMA Alpha-smooth muscle actin
CEA Cultured epithelial autograft
Co-SMAD Common-mediator SMAD
CTGF Connective tissue growth factor
ECM Extracellular matrix
FGF Fibroblast growth factor
GAG Glycosaminoglycan
HA Hyaluronic acid
HTS Hypertrophic scar
IFN Interferon
I-SMAD Inhibitory SMAD
IL Interleukin
KGF Keratinocyte growth factor
MMP Matrix metalloproteinase
PDGF Platelet-derived growth factor
R-SMAD Receptor-regulated SMAD
SDF Stromal cell-derived factor
SLRP Small leucine-rich proteoglycan
TBSA Total body surface area
TGF Transforming growth factor
TGFBR Transforming growth factor-β receptor
TIMP Tissue inhibitor of metalloproteinase
TNF Tumor necrosis factor
VEGF Vascular endothelial growth factor

ACKNOWLEDGMENTS

This work was supported by grants from the Canadian Institutes of Health Research, the Calgary Foundation-Grace Glaum Professorship, funds from the McCaig Professorship, the Firefighters' Burn Trust Fund of the University of Alberta Hospital, and the Alberta Heritage Trust Fund for Medical Research.

REFERENCES

[1] Reinke JM, Sorg H. Wound repair and regeneration. Eur Surg Res 2012;49(1):35–43.

[2] Clark RAF, Singer AJ. Wound repair: basic biology to tissue engineering. In: Lanza RP, Langer R, Vacanti J, editors. Principles of tissue engineering. 2nd ed. San Diego (CA): Academic Press; 2000. p. 857–78.

[3] Wu Y, Tredget EE. Pathology of tissue regeneration repair: skin regeneration. In: McManus LM, Mitchell RN, editors. Pathobiology of human disease. 1st ed. San Diego (CA): Elsevier; 2014. p. 558–66.

[4] Levy V, Lindon C, Zheng Y, Harfe BD, Morgan BA. Epidermal stem cells arise from the hair follicle after wounding. FASEB J May 2007;21(7):1358–66.

[5] Ito M, Liu Y, Yang Z, Nguyen J, Liang F, Morris RJ, et al. Stem cells in the hair follicle bulge contribute to wound repair but not to homeostasis of the epidermis. Nat Med December 2005;11(12):1351–4.

[6] Brodsky S, Chen J, Lee A, Akassoglou K, Norman J, Goligorsky MS. Plasmin-dependent and -independent effects of plasminogen activators and inhibitor-1 on ex vivo angiogenesis. Am J Physiol Heart Circ Physiol October 2001;281(4):H1784–92.

[7] Morris Jr MW, Allukian III M, Herdrich BJ, Caskey RC, Zgheib C, Xu J, et al. Modulation of the inflammatory response by increasing fetal wound size or interleukin-10 overexpression determines wound phenotype and scar formation. Wound Repair Regen May–June 2014;22(3):406–14.

[8] Armstrong JR, Ferguson MWJ. Ontogeny of the skin and the transition from scar-free to scarring phenotype during wound-healing in the pouch young of a marsupial, *Monodelphis domestica*. Dev Biol May 1995;169(1):242–60.

[9] Lorenz HP, Lin RY, Longaker MT, Whitby DJ, Adzick NS. The fetal fibroblast – the effector cell of scarless fetal skin repair. Plast Reconstr Surg November 1995;96(6):1251–9.

[10] Naik-Mathuria B, Gay AN, Zhu X, Yu L, Cass DL, Olutoye OO. Age-dependent recruitment of neutrophils by fetal endothelial cells: implications in scarless wound healing. J Pediatr Surg January 2007;42(1):166–71.

[11] Cowin AJ, Brosnan MP, Holmes TM, Ferguson MWJ. Endogenous inflammatory response to dermal wound healing in the fetal and adult mouse. Dev Dyn July 1998;212(3):385–93.

[12] Ferguson MWJ, O'Kane S. Scar-free healing: from embryonic mechanisms to adult therapeutic intervention. Philos Trans R Soc Lond B Biol Sci May 29, 2004;359(1445):839–50.

[13] Olutoye OO, Alaish SM, Carr ME, Paik MY, Yager DR, Cohen IK, et al. Aggregatory characteristics and expression of the collagen adhesion receptor in fetal porcine platelets. J Pediatr Surg December 1995;30(12):1649–53.

[14] Chen W, Fu X, Ge S, Sun T, Zhou G, Jiang D, et al. Ontogeny of expression of transforming growth factor-β and its receptors and their possible relationship with scarless healing in human fetal skin. Wound Repair Regen January–February 2005;13(1):68–75.

[15] Merkel JR, Dipaolo BR, Hallock GG, Rice DC. Type-I and type-III collagen content of healing wounds in fetal and adult-rats. Proc Soc Exp Biol Med April 1988;187(4):493–7.

[16] Knight KR, Lepore DA, Horne RSC, Ritz M, Hurley JV, Kumta S, et al. Collagen content of uninjured skin and scar tissue in fetal and adult sheep. Int J Exp Pathol December 1993;74(6):583–91.

[17] Zgheib C, Xu J, Liechty KW. Targeting inflammatory cytokines and extracellular matrix composition to promote wound regeneration. Adv Wound Care (New Rochelle) April 1, 2014;3(4):344–55.

[18] Dang CM, Beanes SR, Lee H, Zhang XL, Soo C, Ting K. Scarless fetal wounds are associated with an increased matrix metalloproteinase-to-tissue-derived inhibitor of metalloproteinase ratio. Plast Reconstr Surg June 2003;111(7):2273–85.
[19] Gawronska-Kozak B. Scarless skin wound healing in FOXN1 deficient (nude) mice is associated with distinctive matrix metalloproteinase expression. Matrix Biol May 2011;30(4): 290–300.
[20] Mast BA, Diegelmann RF, Krummel TM, Cohen IK. Hyaluronic-acid modulates proliferation, collagen and protein-synthesis of cultured fetal fibroblasts. Matrix November 1993;13(6):441–6.
[21] Huanglee LLH, Wu JH, Nimni ME. Effects of hyaluronan on collagen fibrillar matrix contraction by fibroblasts. J Biomed Mater Res January 1994;28(1):123–32.
[22] Hu M, Sabelman EE, Cao Y, Chang J, Hentz VR. Three-dimensional hyaluronic acid grafts promote healing and reduce scar formation in skin incision wounds. J Biomed Mater Res B Appl Biomater October 15, 2003;67B(1):586–92.
[23] Wong JW, Gallant-Behm C, Wiebe C, Mak K, Hart DA, Larjava H, et al. Wound healing in oral mucosa results in reduced scar formation as compared with skin: evidence from the red Duroc pig model and humans. Wound Repair Regen September–October 2009;17(5):717–29.
[24] Eslami A, Gallant-Behm CL, Hart DA, Wiebe C, Honardoust D, Gardner H, et al. Expression of integrin αvβ6 and TGF-β in scarless vs scar-forming wound healing. J Histochem Cytochem June 2009;57(6):543–57.
[25] Larjava H, Wiebe C, Gallant-Behm C, Hart DA, Heino J, Hakkinen L. Exploring scarless healing of oral soft tissues. J Can Dent Assoc 2011;77:b18.
[26] Mak K, Manji A, Gallant-Behm C, Wiebe C, Hart DA, Larjava H, et al. Scarless healing of oral mucosa is characterized by faster resolution of inflammation and control of myofibroblast action compared to skin wounds in the red Duroc pig model. J Dermatol Sci December 2009;56(3):168–80.
[27] Tredget EE. Pathophysiology and treatment of fibroproliferative disorders following thermal injury. In: Occupational electrical injury: an international symposium. vol. 888. 1999. p. 165–82.
[28] Lawrence JW, Mason ST, Schomer K, Klein MB. Epidemiology and impact of scarring after burn injury: a systematic review of the literature. J Burn Care Res January–February 2012;33(1):136–46.
[29] Gangemi EN, Gregori D, Berchialla P, Zingarelli E, Cairo M, Bollero D, et al. Epidemiology and risk factors for pathologic scarring after burn wounds. Arch Facial Plast Surg March–April 2008;10(2):93–102.
[30] Lewis WHP, Sun KKY. Hypertrophic scar: a genetic hypothesis. Burns June 1990;16(3):176–8.
[31] Bombaro KM, Engrav LH, Carrougher GJ, Wiechman SA, Faucher L, Costa BA, et al. What is the prevalence of hypertrophic scarring following burns? Burns June 2003;29(4):299–302.
[32] Gauglitz GG, Korting HC, Pavicic T, Ruzicka T, Jeschke MG. Hypertrophic scarring and keloids: pathomechanisms and current and emerging treatment strategies. Mol Med January–February 2011;17(1–2):113–25.
[33] Ehrlich HP, Desmouliere A, Diegelmann RF, Cohen IK, Compton CC, Garner WL, et al. Morphological and immunochemical differences between keloid and hypertrophic scar. Am J Pathol July 1994;145(1):105–13.
[34] Linares HA, Kischer CW, Dobrkovsky M, Larson DL. The histiotypic organization of the hypertrophic scar in humans. J Invest Dermatol October 1972;59(4):323–31.
[35] Kwan P, Hori K, Ding J, Tredget EE. Scar and contracture: biological principles. Hand Clin November 2009;25(4):511–28.

[36] Kischer CW. Collagen and dermal patterns in the hypertrophic scar. Anat Rec May 1974;179(1):137–45.

[37] Lee JYY, Yang CC, Chao SC, Wong TW. Histopathological differential diagnosis of keloid and hypertrophic scar. Am J Dermatopathol October 2004;26(5):379–84.

[38] Kose O, Waseem A. Keloids and hypertrophic scars: are they two different sides of the same coin? Dermatol Surg March 2008;34(3):336–46.

[39] Nedelec B, Shankowsky H, Scott PG, Ghahary A, Tredget EE. Myofibroblasts and apoptosis in human hypertrophic scars: the effect of interferon-α2b. Surgery November 2001; 130(5):798–808.

[40] Ghahary A, Shen YJ, Nedelec B, Wang R, Scott PG, Tredget EE. Collagenase production is lower in post-burn hypertrophic scar fibroblasts than in normal fibroblasts and is reduced by insulin-like growth factor-1. J Invest Dermatol March 1996;106(3):476–81.

[41] Wang R, Ghahary A, Shen YJ, Scott PG, Tredget EE. Human dermal fibroblasts produce nitric oxide and express both constitutive and inducible nitric oxide synthase isoforms. J Invest Dermatol March 1996;106(3):419–27.

[42] Moulin V, Larochelle S, Langlois C, Thibault I, Lopez-Valle CA, Roy M. Normal skin wound and hypertrophic scar myofibroblasts have differential responses to apoptotic inductors. J Cell Physiol March 2004;198(3):350–8.

[43] Sorrell JM, Caplan AI. Fibroblast heterogeneity: more than skin deep. J Cell Sci February 15, 2004;117(5):667–75.

[44] Ali-Bahar M, Bauer B, Tredget EE, Ghahary A. Dermal fibroblasts from different layers of human skin are heterogeneous in expression of collagenase and types I and III procollagen mRNA. Wound Repair Regen March–April 2004;12(2):175–82.

[45] Honardoust D, Ding J, Varkey M, Shankowsky HA, Tredget EE. Deep dermal fibroblasts refractory to migration and decorin-induced apoptosis contribute to hypertrophic scarring. J Burn Care Res September–October 2012;33(5):668–77.

[46] Varkey M, Ding J, Tredget EE. Differential collagen-glycosaminoglycan matrix remodeling by superficial and deep dermal fibroblasts: potential therapeutic targets for hypertrophic scar. Biomaterials October 2011;32(30):7581–91.

[47] Wang J, Dodd C, Shankowsky HA, Scott PG, Tredget EE, Wound Healing Research Grp. Deep dermal fibroblasts contribute to hypertrophic scarring. Lab Invest December 2008;88(12):1278–90.

[48] Tredget EE, Levi B, Donelan MB. Biology and principles of scar management and burn reconstruction. Surg Clin North Am August 2014;94(4):793–815.

[49] Dunkin CSJ, Pleat JM, Gillespie PH, Tyler MPH, Roberts AHN, McGrouther DA. Scarring occurs at a critical depth of skin injury: precise measurement in a graduated dermal scratch in human volunteers. Plast Reconstr Surg May 2007;119(6):1722–32.

[50] Honardoust D, Varkey M, Marcoux Y, Shankowsky HA, Tredget EE. Reduced decorin, fibromodulin, and transforming growth factor-β3 in deep dermis leads to hypertrophic scarring. J Burn Care Res March–April 2012;33(2):218–27.

[51] Bucala R, Spiegel LA, Chesney J, Hogan M, Cerami A. Circulating fibrocytes define a new leukocyte subpopulation that mediates tissue repair. Mol Med November 1994;1(1): 71–81.

[52] Zhu Z, Ding J, Shankowsky HA, Tredget EE. The molecular mechanism of hypertrophic scar. J Cell Commun Signal December 2013;7(4):239–52.

[53] Wang JF, Jiao H, Stewart TL, Shankowsky HA, Scott PG, Tredget EE. Fibrocytes from burn patients regulate the activities of fibroblasts. Wound Repair Regen January–February 2007;15(1):113–21.

[54] Yang L, Scott PG, Dodd C, Medina A, Jiao H, Shankowsky HA, et al. Identification of fibrocytes in postburn hypertrophic scar. Wound Repair Regen July–August 2005;13(4):398–404.

[55] Wang J, Jiao H, Stewart TL, Shankowsky HA, Scott PG, Tredget EE. Increased TGF-β-producing $CD4^+$ T lymphocytes in postburn patients and their potential interaction with dermal fibroblasts in hypertrophic scarring. Wound Repair Regen July–August 2007;15(4):530–9.

[56] Mosmann TR, Coffman RL. TH1 and TH2 cells: different patterns of lymphokine secretion lead to different functional properties. Annu Rev Immunol 1989;7:145–73.

[57] Wynn TA. Fibrotic disease and the T(H)1/T(H)2 paradigm. Nat Rev Immunol August 2004;4(8):583–94.

[58] Armour A, Scott PG, Tredget EE. Cellular and molecular pathology of HTS: basis for treatment. Wound Repair Regen September–October 2007;15(Suppl. 1):S6–17.

[59] Tredget EE, Yang L, Delehanty M, Shankowsky H, Scott PG. Polarized Th2 cytokine production in patients with hypertrophic scar following thermal injury. J Interferon Cytokine Res March 2006;26(3):179–89.

[60] Shao DD, Suresh R, Vakil V, Gomer RH, Pilling D. Pivotal advance: Th-1 cytokines inhibit, and Th-2 cytokines promote fibrocyte differentiation. J Leukoc Biol June 2008;83(6): 1323–33.

[61] Tredget EE, Shankowsky HA, Pannu R, Nedelec B, Iwashina T, Ghahary A, et al. Transforming growth factor-β in thermally injured patients with hypertrophic scars: effects of interferon α-2b. Plast Reconstr Surg October 1998;102(5):1317–28. Discussion 1329–30.

[62] Foley TT, Ehrlich HP. Through gap junction communications, co-cultured mast cells and fibroblasts generate fibroblast activities allied with hypertrophic scarring. Plast Reconstr Surg May 2013;131(5):1036–44.

[63] Wilgus TA, Wulff BC. The importance of mast cells in dermal scarring. Adv Wound Care (New Rochelle) April 1, 2014;3(4):356–65.

[64] Ghahary A, Marcoux Y, Karimi-Busheri F, Li Y, Tredget EE, Kilani RT, et al. Differentiated keratinocyte-releasable stratifin (14-3-3 sigma) stimulates MMP-1 expression in dermal fibroblasts. J Invest Dermatol January 2005;124(1):170–7.

[65] Ghaffari A, Kilani RT, Ghahary A. Keratinocyte-conditioned media regulate collagen expression in dermal fibroblasts. J Invest Dermatol February 2009;129(2):340–7.

[66] Koskela A, Engstrom K, Hakelius M, Nowinski D, Ivarsson M. Regulation of fibroblast gene expression by keratinocytes in organotypic skin culture provides possible mechanisms for the antifibrotic effect of reepithelialization. Wound Repair Regen September–October 2010;18(5):452–9.

[67] Varkey M, Ding J, Tredget EE. Fibrotic remodeling of tissue-engineered skin with deep dermal fibroblasts is reduced by keratinocytes. Tissue Eng Part A February 2014;20(3–4):716–27.

[68] Bellemare J, Roberge CJ, Bergeron D, Lopez-Valle CA, Roy M, Moulin VJ. Epidermis promotes dermal fibrosis: role in the pathogenesis of hypertrophic scars. J Pathol May 2005;206(1):1–8.

[69] Simon F, Bergeron D, Larochelle S, Lopez-Valle CA, Genest H, Armour A, et al. Enhanced secretion of TIMP-1 by human hypertrophic scar keratinocytes could contribute to fibrosis. Burns May 2012;38(3):421–7.

[70] Ignotz RA, Massague J. Transforming growth factor-β stimulates the expression of fibronectin and collagen and their incorporation into the extracellular matrix. J Biol Chem March 25, 1986;261(9):4337–45.

[71] Edwards DR, Murphy G, Reynolds JJ, Whitham SE, Docherty AJ, Angel P, et al. Transforming growth factor beta modulates the expression of collagenase and metalloproteinase inhibitor. EMBO J July 1987;6(7):1899–904.

[72] Shah M, Foreman DM, Ferguson MW. Neutralisation of TGF-β 1 and TGF-β 2 or exogenous addition of TGF-β 3 to cutaneous rat wounds reduces scarring. J Cell Sci March 1995;108(Pt 3): 985–1002.

[73] Bock O, Yu H, Zitron S, Bayat A, Ferguson MW, Mrowietz U. Studies of transforming growth factors beta 1-3 and their receptors I and II in fibroblast of keloids and hypertrophic scars. Acta Derm Venereol 2005;85(3):216–20.

[74] Wang YW, Liou NH, Cherng JH, Chang SJ, Ma KH, Fu E, et al. siRNA-targeting transforming growth factor-β type I receptor reduces wound scarring and extracellular matrix deposition of scar tissue. J Invest Dermatol July 2014;134(7):2016–25.

[75] Kopp J, Preis E, Said H, Hafemann B, Wickert L, Gressner AM, et al. Abrogation of transforming growth factor-β signaling by SMAD7 inhibits collagen gel contraction of human dermal fibroblasts. J Biol Chem June 3, 2005;280(22):21570–6.

[76] Bonniaud P, Kolb M, Galt T, Robertson J, Robbins C, Stampfli M, et al. Smad3 null mice develop airspace enlargement and are resistant to TGF-β-mediated pulmonary fibrosis. J Immunol August 1, 2004;173(3):2099–108.

[77] Nakao A, Afrakhte M, Moren A, Nakayama T, Christian JL, Heuchel R, et al. Identification of Smad7, a TGFβ-inducible antagonist of TGF-β signalling. Nature October 9, 1997;389(6651):631–5.

[78] Ulrich D, Noah EM, von Heimburg D, Pallua N. TIMP-1, MMP-2, MMP-9, and PIIINP as serum markers for skin fibrosis in patients following severe burn trauma. Plast Reconstr Surg April 1, 2003;111(4):1423–31.

[79] Neely AN, Clendening CE, Gardner J, Greenhalgh DG, Warden GD. Gelatinase activity in keloids and hypertrophic scars. Wound Repair Regen May–June 1999;7(3):166–71.

[80] Eto H, Suga H, Aoi N, Kato H, Doi K, Kuno S, et al. Therapeutic potential of fibroblast growth factor-2 for hypertrophic scars: upregulation of MMP-1 and HGF expression. Lab Invest February 2012;92(2):214–23.

[81] Alberts B, Bray D, Lewis J, Raff M, Roberts K, Watson JD. Molecular biology of the cell. 1st ed. New York (NY): Garland Publishing, Inc.; 1983.

[82] Sawicki G, Marcoux Y, Sarkhosh K, Tredget EE, Ghahary A. Interaction of keratinocytes and fibroblasts modulates the expression of matrix metalloproteinases-2 and -9 and their inhibitors. Mol Cell Biochem January 2005;269(1–2):209–16.

[83] Amadeu TP, Braune AS, Porto LC, Desmouliere A, Costa AMA. Fibrillin-1 and elastin are differentially expressed in hypertrophic scars and keloids. Wound Repair Regen March–April 2004;12(2):169–74.

[84] Kozma EM, Olczyk K, Glowacki A, Bobinski R. An accumulation of proteoglycans in scarred fascia. Mol Cell Biochem January 2000;203(1–2):103–12.

[85] Kresse H, Schonherr E. Proteoglycans of the extracellular matrix and growth control. J Cell Physiol December 2001;189(3):266–74.

[86] Honardoust D, Varkey M, Hori K, Ding J, Shankowsky HA, Tredget EE. Small leucine-rich proteoglycans, decorin and fibromodulin, are reduced in postburn hypertrophic scar. Wound Repair Regen May–June 2011;19(3):368–78.

[87] Hildebrand A, Romaris M, Rasmussen LM, Heinegard D, Twardzik DR, Border WA, et al. Interaction of the small interstitial proteoglycans biglycan, decorin and fibromodulin with transforming growth factor beta. Biochem J September 1, 1994;302(Pt 2):527–34.

[88] Zhang Z, Li XJ, Liu Y, Zhang X, Li YY, Xu WS. Recombinant human decorin inhibits cell proliferation and downregulates TGF-β1 production in hypertrophic scar fibroblasts. Burns August 2007;33(5):634–41.

[89] Stoff A, Rivera AA, Mathis JM, Moore ST, Banerjee NS, Everts M, et al. Effect of adenoviral mediated overexpression of fibromodulin on human dermal fibroblasts and scar formation in full-thickness incisional wounds. J Mol Med (Berl) May 2007;85(5):481–96.

[90] Zhang Z, Garron TM, Li XJ, Liu Y, Zhang X, Li YY, et al. Recombinant human decorin inhibits TGF-β1-induced contraction of collagen lattice by hypertrophic scar fibroblasts. Burns June 2009;35(4):527–37.

[91] Scott PG, Dodd CM, Tredget EE, Ghahary A, Rahemtulla F. Immunohistochemical localization of the proteoglycans decorin, biglycan and versican and transforming growth factor-β in human post-burn hypertrophic and mature scars. Histopathology May 1995;26(5):423–31.

[92] Bianchetti L, Barczyk M, Cardoso J, Schmidt M, Bellini A, Mattoli S. Extracellular matrix remodelling properties of human fibrocytes. J Cell Mol Med March 2012;16(3):483–95.

[93] Ding J, Hori K, Zhang R, Marcoux Y, Honardoust D, Shankowsky HA, et al. Stromal cell-derived factor 1 (SDF-1) and its receptor CXCR4 in the formation of postburn hypertrophic scar (HTS). Wound Repair Regen September–October 2011;19(5):568–78.

[94] Wang J, Hori K, Ding J, Huang Y, Kwan P, Ladak A, et al. Toll-like receptors expressed by dermal fibroblasts contribute to hypertrophic scarring. J Cell Physiol May 2011;226(5):1265–73.

[95] Gu L, Tseng S, Horner RM, Tam C, Loda M, Rollins BJ. Control of TH2 polarization by the chemokine monocyte chemoattractant protein-1. Nature March 23, 2000;404(6776):407–11.

[96] Frey H, Schroeder N, Manon-Jensen T, Iozzo RV, Schaefer L. Biological interplay between proteoglycans and their innate immune receptors in inflammation. FEBS J May 2013;280(10):2165–79.

[97] Gallant CL, Olson ME, Hart DA. Molecular, histologic, and gross phenotype of skin wound healing in red Duroc pigs reveals an abnormal healing phenotype of hypercontracted, hyperpigmented scarring. Wound Repair Regen May–June 2004;12(3):305–19.

[98] Groenen MA, Archibald AL, Uenishi H, Tuggle CK, Takeuchi Y, Rothschild MF, et al. Analyses of pig genomes provide insight into porcine demography and evolution. Nature November 15, 2012;491(7424):393–8.

[99] Gallant-Behm CL, Hart DA. Genetic analysis of skin wound healing and scarring in a porcine model. Wound Repair Regen January–February 2006;14(1):46–54.

[100] Zhu KQ, Carrougher GJ, Gibran NS, Isik FF, Engrav LH. Review of the female Duroc/Yorkshire pig model of human fibroproliferative scarring. Wound Repair Regen September–October 2007;15(Suppl. 1):S32–9.

[101] Gallant-Behm CL, Hildebrand KA, Hart DA. The mast cell stabilizer ketotifen prevents development of excessive skin wound contraction and fibrosis in red Duroc pigs. Wound Repair Regen March–April 2008;16(2):226–33.

[102] Monument MJ, Hart DA, Salo PT, Befus AD, Hildebrand KA. Posttraumatic elbow contractures: targeting neuroinflammatory fibrogenic mechanisms. J Orthop Sci November 2013;18(6):869–77.

[103] Monument MJ, Hart DA, Salo PT, Befus AD, Hildebrand KA. Neuroinflammatory mechanisms of connective tissue fibrosis: targeting neurogenic and mast cell contributions. Adv Wound Care (New Rochelle) March 2015;4(3):137–51.

[104] de Hemptinne I, Gallant-Behm CL, Noack CL, Parreno J, Hart DA. Dermal fibroblasts from red Duroc and Yorkshire pigs exhibit intrinsic differences in the contraction of collagen gels. Wound Repair Regen January–February 2008;16(1):132–42.

[105] Meaume S, Le Pillouer-Prost A, Richert B, Roseeuw D, Vadoud J. Management of scars: updated practical guidelines and use of silicones. Eur J Dermatol August 1, 2014;24(4):435–43.

[106] Anderson RR, Donelan MB, Hivnor C, Greeson E, Ross EV, Shumaker PR, et al. Laser treatment of traumatic scars with an emphasis on ablative fractional laser resurfacing: consensus report. JAMA Dermatol February 2014;150(2):187–93.
[107] Chaudhary K, Moore H, Tandon A, Gupta S, Khanna R, Mohan RR. Nanotechnology and adeno-associated virus-based decorin gene therapy ameliorates peritoneal fibrosis. Am J Physiol Renal Physiol October 1, 2014;307(7):F777–82.
[108] Rheinwald JG, Green H. Serial cultivation of strains of human epidermal keratinocytes: the formation of keratinizing colonies from single cells. Cell November 1975;6(3):331–43.
[109] Grafting of burns with cultured epithelium prepared from autologous epidermal cells. Lancet January 10, 1981;1(8211):75–8.
[110] Atiyeh BS, Hayek SN, Gunn SW. New technologies for burn wound closure and healing – review of the literature. Burns December 2005;31(8):944–56.
[111] Kamel RA, Ong JF, Eriksson E, Junker JP, Caterson EJ. Tissue engineering of skin. J Am Coll Surg September 2013;217(3):533–55.
[112] Tissue Engineering. American Burn Association Postgraduate Course A – Scars: Biology and Management; 2007.
[113] Fang T, Lineaweaver WC, Sailes FC, Kisner C, Zhang F. Clinical application of cultured epithelial autografts on acellular dermal matrices in the treatment of extended burn injuries. Ann Plast Surg November 2014;73(5):509–15.
[114] Catalano E, Cochis A, Varoni E, Rimondini L, Azzimonti B. Tissue-engineered skin substitutes: an overview. J Artif Organs December 2013;16(4):397–403.
[115] Breitkreutz D, Mirancea N, Nischt R. Basement membranes in skin: unique matrix structures with diverse functions? Histochem Cell Biol July 2009;132(1):1–10.
[116] Lamy J, Yassine AH, Gourari A, Forme N, Zakine G. The role of skin substitutes in the surgical treatment of extensive burns covering more than 60 % of total body surface area. A review of patients over a 10-year period at the Tours University Hospital. Ann Chir Plast Esthet October 30, 2013;60(2):131–9.
[117] Seo DK, Kym D, Hur J. Management of neck contractures by single-stage dermal substitutes and skin grafting in extensive burn patients. Ann Surg Treat Res November 2014;87(5): 253–9.
[118] Gomez C, Galan JM, Torrero V, Ferreiro I, Perez D, Palao R, et al. Use of an autologous bioengineered composite skin in extensive burns: clinical and functional outcomes. A multicentric study. Burns June 2011;37(4):580–9.
[119] Nanchahal J, Dover R, Otto WR. Allogeneic skin substitutes applied to burns patients. Burns May 2002;28(3):254–7.
[120] Yannas IV, Burke JF. Design of an artificial skin. I. Basic design principles. J Biomed Mater Res January 1980;14(1):65–81.
[121] Wang W, Zhang M, Lu W, Zhang X, Ma D, Rong X, et al. Cross-linked collagen-chondroitin sulfate-hyaluronic acid imitating extracellular matrix as scaffold for dermal tissue engineering. Tissue Eng Part C Methods April 2010;16(2):269–79.
[122] Park SN, Lee HJ, Lee KH, Suh H. Biological characterization of EDC-crosslinked collagen-hyaluronic acid matrix in dermal tissue restoration. Biomaterials April 2003;24(9):1631–41.
[123] Lin YC, Tan FJ, Marra KG, Jan SS, Liu DC. Synthesis and characterization of collagen/hyaluronan/chitosan composite sponges for potential biomedical applications. Acta Biomater September 2009;5(7):2591–600.
[124] Liu X, Ma L, Liang J, Zhang B, Teng J, Gao C. RNAi functionalized collagen-chitosan/silicone membrane bilayer dermal equivalent for full-thickness skin regeneration with inhibited scarring. Biomaterials March 2013;34(8):2038–48.

[125] MacNeil S. Biomaterials for tissue engineering of skin. Materials Today May 2008;11(5): 26–35.

[126] Ng KW, Hutmacher DW. Reduced contraction of skin equivalent engineered using cell sheets cultured in 3D matrices. Biomaterials September 2006;27(26):4591–8.

[127] Blackstone BN, Drexler JW, Powell HM. Tunable engineered skin mechanics via coaxial electrospun fiber core diameter. Tissue Eng Part A October 2014;20(19–20):2746–55.

[128] Rnjak-Kovacina J, Wise SG, Li Z, Maitz PK, Young CJ, Wang Y, et al. Electrospun synthetic human elastin:collagen composite scaffolds for dermal tissue engineering. Acta Biomater October 2012;8(10):3714–22.

[129] Yuan Z, Zhao J, Chen Y, Yang Z, Cui W, Zheng Q. Regulating inflammation using acid-responsive electrospun fibrous scaffolds for skin scarless healing. Mediators Inflamm 2014;2014:858045.

[130] Lai HJ, Kuan CH, Wu HC, Tsai JC, Chen TM, Hsieh DJ, et al. Tailored design of electrospun composite nanofibers with staged release of multiple angiogenic growth factors for chronic wound healing. Acta Biomater October 2014;10(10):4156–66.

[131] Blais M, Grenier M, Berthod F. Improvement of nerve regeneration in tissue-engineered skin enriched with schwann cells. J Invest Dermatol December 2009;129(12):2895–900.

[132] Biedermann T, Klar AS, Bottcher-Haberzeth S, Schiestl C, Reichmann E, Meuli M. Tissue-engineered dermo-epidermal skin analogs exhibit de novo formation of a near natural neuro-vascular link 10 weeks after transplantation. Pediatr Surg Int February 2014;30(2):165–72.

[133] Blais M, Levesque P, Bellenfant S, Berthod F. Nerve growth factor, brain-derived neuro-trophic factor, neurotrophin-3 and glial-derived neurotrophic factor enhance angiogenesis in a tissue-engineered in vitro model. Tissue Eng Part A August 2013;19(15–16):1655–64.

[134] Sanchez-Munoz I, Granados R, Holguin Holgado P, Garcia-Vela JA, Casares C, Casares M. The use of adipose mesenchymal stem cells and human umbilical vascular endothelial cells on a fibrin matrix for endothelialized skin substitute. Tissue Eng Part A August 19, 2014;21(1–2):214–23.

[135] Peng LH, Mao ZY, Qi XT, Chen X, Li N, Tabata Y, et al. Transplantation of bone-marrow-derived mesenchymal and epidermal stem cells contribute to wound healing with different regenerative features. Cell Tissue Res June 2013;352(3):573–83.

[136] Sasaki M, Abe R, Fujita Y, Ando S, Inokuma D, Shimizu H. Mesenchymal stem cells are recruited into wounded skin and contribute to wound repair by transdifferentiation into multiple skin cell type. J Immunol February 15, 2008;180(4):2581–7.

[137] Klar AS, Bottcher-Haberzeth S, Biedermann T, Schiestl C, Reichmann E, Meuli M. Analysis of blood and lymph vascularization patterns in tissue-engineered human dermo-epidermal skin analogs of different pigmentation. Pediatr Surg Int February 2014;30(2):223–31.

[138] Boyce ST, Zimmerman RL, Supp DM. Tumorigenicity testing in athymic mice of cultured human melanocytes for transplantation in engineered skin substitutes. Cell Transplant July 23, 2014;24(8):1423–9.

Chapter 3

Tissue Processing and Staining for Histological Analyses

Mustafa Q. Yousif, Shadi A. Qasem
Department of Pathology, Wake Forest University School of Medicine, Winston-Salem, NC, USA

Chapter Outline

INTRODUCTION

Proper tissue processing and staining is paramount in understanding the structure of tissue and identifying pathological processes. In this chapter, we provide a basic introduction to tissue processing and preservation into paraffin. In addition, we discuss staining techniques and their utility in histopathology.

TISSUE FIXATION

This is the first and most important step in histological preparation. It serves to preserve tissue, inactivate infectious agents, stabilize tissue components, and enhance avidity for dyes. Other effects include alteration of protein structure, changing solubility of tissue components, shrinkage of tissue, and DNA/RNA degradation [1]. Careful consideration should be taken before fixing tissue,

Skin Tissue Engineering and Regenerative Medicine. http://dx.doi.org/10.1016/B978-0-12-801654-1.00003-6

especially if frozen sections, cytogenetic studies, immunofluorescence (IF), and flow cytometry are to be considered.

Appropriate fixation depends on multiple factors, including the use of an adequate amount of fixative (typically 20 times the volume of the tissue), accessibility of the fixative to the tissue (large pieces of tissue should be sliced or pinned out on a paraffin block), duration of fixation (depending on size, tissue may need 6–8 h in formaldehyde for adequate fixation), and temperature (most fixatives can be used at room temperature) [2,3].

Although there is not a single fixative that is good for all the specimens, 10% phosphate buffered formalin (4% formaldehyde) is the standard fixative used in most laboratories. Other fixatives include Bouin's solution (especially for small biopsies), B-plus (used for lymph node, spleen, and other lymphoid tissue), Zenker's acetic fixative (used for bone marrow biopsies and soft tissue tumors), glutaraldehyde (used to preserve tissue for electron microscopy), and alcohol (used for synovial specimens) [4].

TISSUE PROCESSING AND EMBEDDING

The processing of tissue into paraffin is an important step in histologic preparation. It stiffens the tissue and enables very thin cuts to be made. Initially the tissue is put in cassettes. The standard cassette measurements are 3 × 2.5 × 0.4 cm. Tissue thickness should be no more than 0.3 cm, in order to fit in the cassette properly. Subsequently, the tissue undergoes multiple steps for several hours in order to be processed [2]. The nonaqueous paraffin cannot penetrate the water-containing tissue; therefore, the tissue should be dehydrated first by replacing the water with alcohol, then alcohol is replaced with a clearing agent (xylene) to make the tissue receptive to paraffin. Xylene is miscible with both alcohol and paraffin and has a high refractive index; the tissue becomes transparent/clear. Next xylene is replaced with paraffin [4,5].

The next step is embedding. The tissue is removed from the cassette and put into a metallic mold. After proper orientation of the tissue, liquid paraffin is poured into the mold followed by cooling to solidify the paraffin. The tissue is now ready for cutting (Box 1) (Figure 1).

Box 1 Tissue-Processing Steps

1. Sectioning of tissue into 2–3-mm-thick sections
2. The tissue is placed into cassettes
3. Processing through dehydration, clearing, and infiltration of paraffin (often done overnight using a tissue processor)
4. Embedding (making a paraffin block)

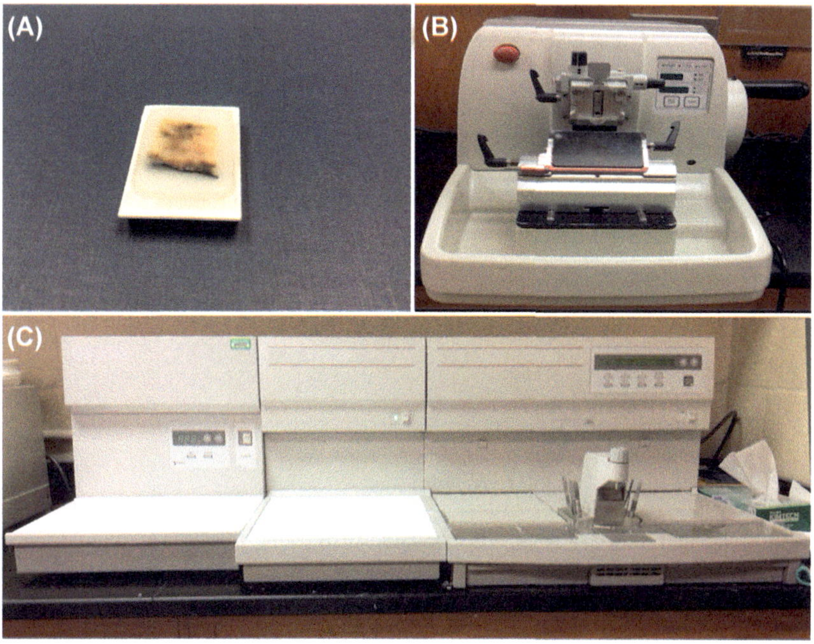

FIGURE 1 (A) Tissue block. Tissue is embedded in paraffin wax. (B) Microtome. This device is used to cut paraffin blocks. (C) Embedding station used to transfer processed tissue to a paraffin block as in (A).

HEMATOXYLIN AND EOSIN STAIN

A microtome is used to make thin slices of tissue (usually 4 μm but can be 2–10 μm). The cut tissue is floated over a water bath, in order to eliminate wrinkles and distortion in the tissue, and picked up on a slide. The temperature of the water bath depends on the type of wax and is typically 5–9 °C below the melting point for the wax. The slides can be either charged or uncharged depending on the type of tissue being cut and the staining protocol for the tissue. A tissue sample 0.3 cm thick can yield up to 750 glass slides. Next, the slides are baked and will be ready for staining [5].

Hematoxylin and eosin (H&E) staining technique is used to stain the nuclei blue to purple (hematoxylin), and counterstain the cytoplasm pink to red (eosin). Hematoxylin is extracted from the bark of wood, oxidized to hematin (a purple die) and positively charged with mordant (a metallic salt). Eosin is an anionic xanthene dye that combines to cytoplasmic components. This technique is essential and remains unchanged for a half century. It is relatively quick, inexpensive, suitable for most situations, and easy to master. Most histologic processes including neoplasms, infections, tissue injury, and repair can be identified using this stain. It can be used to evaluate frozen

Box 2 Steps for H&E Staining

- Absolute alcohol, 60 s
- Rinse in tap water, dip until clear
- Hematoxylin (Harris), 60 s
- Rinse in tap water
- Dip until clear blue in Scott's water
- Rinse in tap water, dip until clear
- Eosin–phloxine, 30 s
- 50% alcohol, five dips
- 70% alcohol, five dips
- 95% alcohol, five dips
- Absolute alcohol, five dips
- Xylene, dip until clear
- Mount and coverslip

sections, touch preparations, smears, and crush preparations, and can be ready in less than 3 min [5] (Box 2).

BRIGHT FIELD MICROSCOPY

Light microscopes date at least to 1595, when Zacharias Jansen (1580–1638) of Holland invented a compound light microscope, one that used two lenses, with the second lens further magnifying the image produced by the first one [6]. Microscopes use a visible light and magnifying lenses to detect small objects that are not visible by the naked eye. This simple and common form of microscopy is called bright field microscopy. The benefit of the microscope is that it produces a better resolution than the naked eye. The higher the resolution, the better is our ability to detect two distinct objects without any blurring of the image. As the technology advances, the resolution becomes better [7].

To achieve Kohler illumination, place the slide on the stage, adjust the eyepieces, and use the 10× objective to focus on the slide. Microscopes with one adjustable eyepiece can be focused by closing the adjustable eye and using the adjustable knob to focus on the fixed eye. Once the fixed eye is focused, use the eyepiece ring adjustment to focus the adjustable eyepiece. Diaphragm condenser aperture should be wide open and in the center of the field [7]. Tissue can be viewed with objectives ranging from 2× to 100× (Box 3) (Figure 2).

IMMUNOFLUORESCENCE

It was first described in 1941 by Coons, Creech, and Jones to identify pneumococcal antigens in tissue [8–10]. Florescence has luminescence properties when it absorbs light and it emits it at a different wavelength. The wavelength of the emitted light is longer than the excitation light, because the emitted florescence

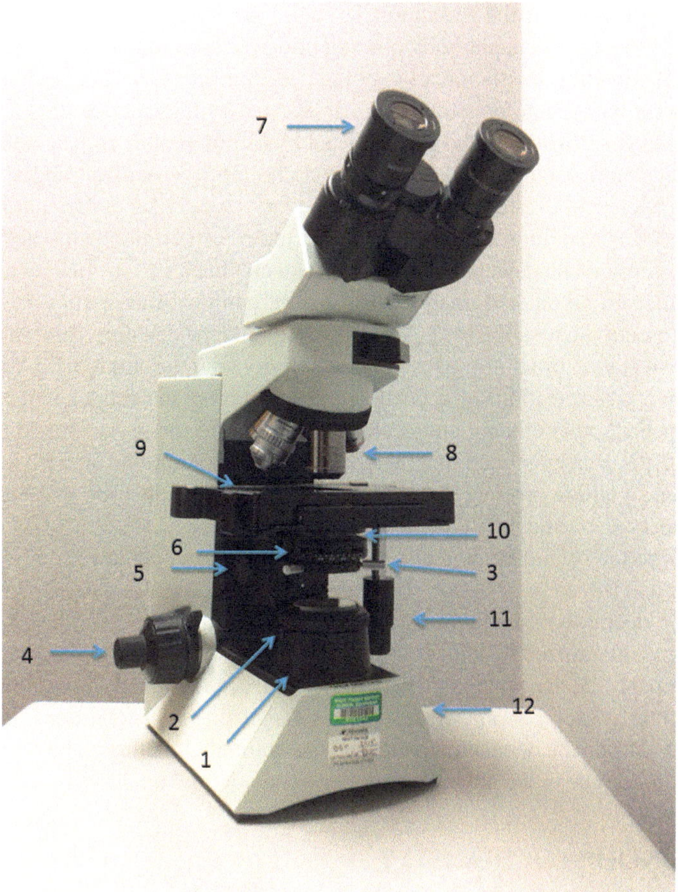

FIGURE 2 Bright field microscope.

Box 3 Microscope Parts

1. Light source (a hidden bulb)
2. Field diaphragm (adjusting the width of the diaphragm by the knurled ring)
3. Screws to center the condenser on each side
4. Coarse and fine focus knobs
5. Knob to adjust the condenser
6. Flip knob to move the condenser and to achieve a low-power light
7. Eyepieces with adjustment ring on one or both eyepieces
8. Objectives
9. Slide stage
10. Substage condenser with an aperture diaphragm
11. Knob to move the stage left, right, or forward and backward
12. Sliding bar to adjust the light intensity

energy is lower than the excitation light. IF simply detects antigen in tissue using the same antigen–antibody binding principle as in immunohistochemistry (IHC). Antibodies are chemically conjugated to a florescent dye such as fluorescein isothiocyanate or tetramethyl rhodamine isothiocyanate. IF is divided into two types, **direct IF** (the method used by Coons and his colleagues) where a primary antibody conjugated with fluorochrome is visualized under a fluorescent microscope, and **indirect IF**. In the latter type, a secondary antibody is conjugated with a fluorochrome and is used to detect a primary antibody [8,11].

Antigens can be detected using different techniques such as flow cytometry, array scanner, automated imaging instrument, confocal microscopy, or immunofluorescent microscopy [10]. Antigens should not be denatured and should be soluble in situ, in order not to lose their reactivity with the antibodies. Air-dried, unfixed, cryostat sections of skin allow the detection of the antigen in the tissue. Another method is ethanol and acetone fixation, and the tissue should be kept in a 7.0 pH transport medium (Michel's transport solution) and frozen after washing. Avoiding prolonged exposure to the transport medium will minimize autofluorescence and background staining [12].

Skin tissue should be mounted in optimal cutting temperature compound on a chuck and frozen to −25 °C. Using positively charged slides will help in adherence and decrease tissue fragmentation during washing stages. With skin tissue, IF is typically utilized to detect the deposition of IgM, IgG, IgA, kappa, lambda, complement C3, and complement C1q. Skin IF, in conjunction with light and electron microscopy, is a powerful tool for diagnosing skin conditions. IF of the skin depends on the deposition pattern of the immunoglobulins in the tissue. The deposition can be in the upper dermis, dermo–epidermal junction, or dermis [8].

IMMUNOHISTOCHEMISTRY

IHC is based on an antigen–antibody reaction process in which an antigen is visualized by an enzyme-labeled antibody. The results are often interpreted as being either positive or negative, indicating the presence or absence of the desired antigen, respectively. It is important to compare the positive cells of IHC with a known control. IHC can be used for the classification of tumors, identification of in situ lesions versus invasive carcinoma, prognostic factors, predictive factors, biomarkers, identification of extracellular material, and identification of infectious agents [8].

Immunoreactivity varies with the type of tissue, the antibody used, and the assay. IHC depends on multiple factors including type of fixative, length of fixation time, decalcifying agents, antigen retrieval procedure, type of antibody, incubation time, incubation temperature, dilution of antibody, and antibody detection procedure [1,3,4,13]. Adjusting these factors is necessary in order to achieve optimal specific staining with less interference and background staining. Special charged slides are also used in order to prevent the tissue from sliding due to high temperature [8].

Antibodies Commonly Used for Skin Disorders

Cytokeratin is used to differentiate epithelial and mesothelial cells from mesenchymal cells in normal and tumor tissue. CD1a is used to differentiate Langerhans cells from interdigitating cells. CD20, CD3, CD4, and CD8 are various lymphocytic markers. CD31 and CD34 are used to highlight endothelial cells and vasculature. HMB 45, MART-1, and tyrosinase are often used to stain melanocytes in melanoma and other benign skin conditions [14] (Figure 3).

HISTOCHEMICAL STAINS

Cell structures and extracellular matrix become transparent after tissue processing, and different cell parts have different biochemical properties and take up stains with different avidity. Although the color of the stain is not the real color of a particular tissue, tissue will show different colors using different stains, enabling the separation of the tissue components and/or the identification of certain structures.

Trichrome Stain

This stain is used to highlight collagen and differentiate collagen from smooth muscle in tumors or diseases, where collagen deposition is increased. The stain has three dye components: the nuclei stain black; cytoplasm, muscle, and erythrocytes stain red; and collagen stains blue. Reagents used in the procedure include Bouin's fixative, Biebrich scarlet, Weigert's iron hematoxylin, phosphotungstic/phosphomolybdic acid solution, aniline blue, and 1% acetic acid [15].

Procedure: Treat the tissue with Bouin's solution, microwave for 1 min, allow to stand for 15 min, wash in running tap water to remove the picric acid for 5 min, Weigert's working hematoxylin for 10 min, blue in running tap water for 10 min, rinse in distilled water, Biebrich scarlet for 5 min, phosphotungstic/phosphomolybdic acid for 10 min, transfer directly into aniline blue for 5 min, rinse in distilled water, 1% acetic acid for 1 min, rinse in distilled water, dehydrate, clear, and coverslip [16].

GMS Stain (Methenamine Silver—Grocott's)

This stain is used for fungal identification. Fungal organisms contain mucopolysaccharide in their cell walls that will oxidize to release an aldehyde group. Silver nitrite reacts with the aldehyde group reducing it to metallic silver, making them visible. The fungi cell walls are outlined black/brown. Background stain is green. Reagents used in the procedure include 2% chromic acid, 1% metabisulfite, 5% borax, methenamine silver, 0.5% gold chloride, and 0.2% light green [15].

Procedure: Deparaffinize and hydrate in distilled water, 2% chromic acid, microwave for 45 s, allow to stand for 5 min, wash in tap water, rinse in distilled water, 1% sodium metabisulfite for 1 min, wash in tap water, rinse in distilled

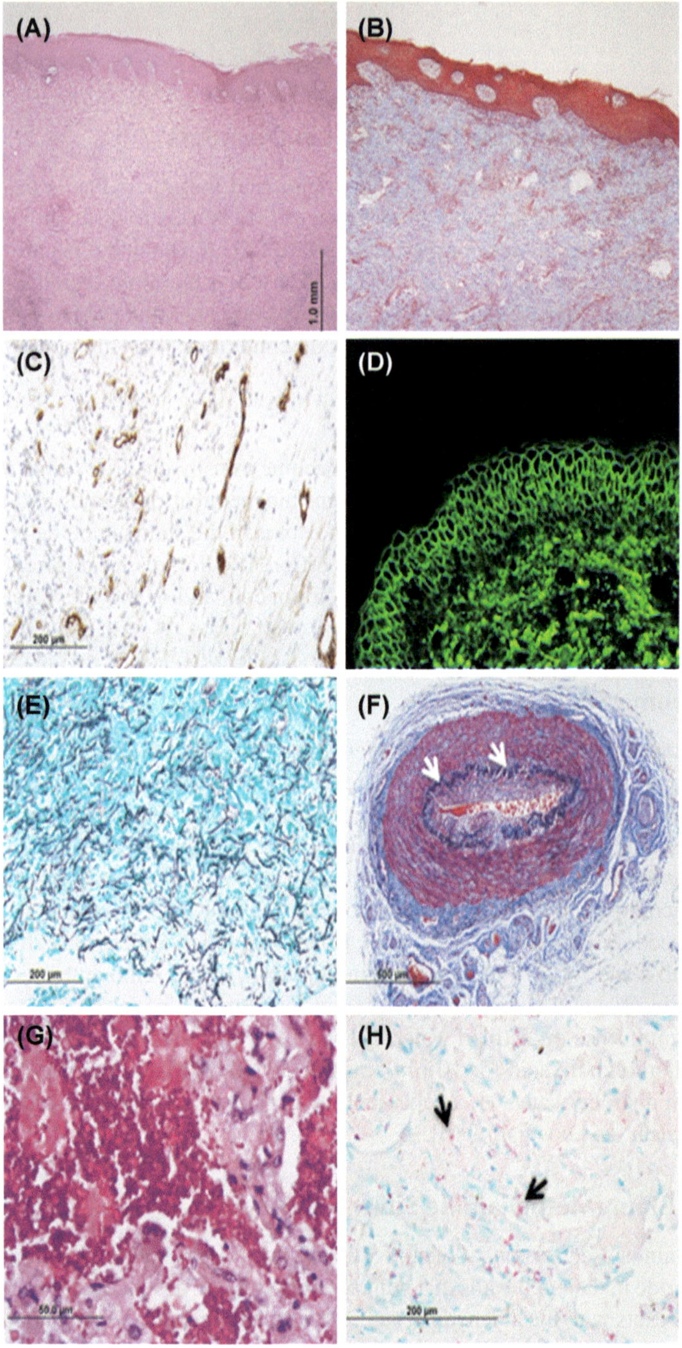

FIGURE 3 (A) Histologic sections of skin construct using bioprinter. Notice the epidermis on the top and dermis underneath (H&E, 4×). (B) Trichrome stain of the same tissue in (A), highlighting dermal collagen (blue) and blood vessels (red) (4×). (C) CD34 immunostain highlighting dermal capillary vessels (40×). (D) Immunofluorescence stain for IgA staining epidermal keratinocytes (40×). (E) GMS stain showing fungal hyphae (20×). (F) Elastic trichrome stain of an artery highlighting internal elastic lamina (arrow heads) (10×). (G) PAS stain showing fungal yeast forms (60×). (H) AFB stain showing tiny pink organisms consistent with *Mycobacterium tuberculosis* (arrow heads) (40×). *Image D is courtesy of Dr Omar Sangueza, Wake Forest University School of Medicine, Winston-Salem, NC, USA-Salem, NC, USA.*

water, methenamine silver solution, microwave for 70s (the tissue color should be like a brown paper bag), agitate the slide in hot solution, rinse in distilled water, put the slide in 0.5% gold chloride 1 min or until gray, wash in distilled water, put the slide in 5% hypo for 3min, wash in tap water, rinse in distilled water, light green for 1 min, rinse in distilled water, dehydrate, clear, and coverslip [16].

Phosphotungstic Acid–Hematoxylin, Mallory's

This stain is used to highlight muscle cross-striations and identify fibrin. It can also be used to highlight contraction bands in necrosis and show gliosis in the central nervous system. This is a polychrome stain (i.e., one stain gives more than one color). Cross-striations, fibrin, glial fibers, and nuclei stain blue; collagen stains red-brown; and elastic fibers stain purple. Reagents used in the procedure include Lugol's, 5% hypo, phosphotungstic acid–hematoxylin (PTAH) stain, 0.25% potassium permanganate, 5% oxalic acid solution, Zenker's fixative, and glacial acetic acid [15].

Procedure: Deparaffinize and hydrate in distilled water, Zenker's fixative, microwave for 45s, let stand for 5min, wash with tap water, Lugol's iodine for 5min, wash in tap water, 5% hypo for 1–2min, wash in tap water for 10min, oxidize in 0.25% potassium permanganate for 5min, wash in tap water, bleach in 5% oxalic acid until white, wash in tap water, rinse in distilled water, PTAH stain overnight at room temperature, dehydrate, and coverslip [17].

PAS Stain (McManus' Periodic Acid–Schiff's)

Structures like connective tissues, mucus, glycocalyx, and basal laminae are rich with carbohydrate macromolecules (glycogen, glycoprotein, proteoglycans) and stain with periodic acid–Schiff's (PAS). The magenta color of the stain is formed by the reaction of the periodic acid with the tissue and oxidizing the carbon–carbon bonds, subsequently forming aldehydes. The later react with the fuchsin–sulfurous acid to form the color. Glycogen and fungus will stain magenta. Nuclei will stain blue. Reagents used include 0.5% periodic acid, hematoxylin, GILL-3, and Lillie's cold Schiff's reagent [15].

Procedure: Deparaffinize and hydrate in distilled water, place slides in 0.5% periodic acid for 5min, rinse in distilled water, Schiff's reagent, microwave for 45–60s (until deep magenta), wash in running tap water for 5min, counterstain in hematoxylin for 3min, wash in tap water, blue, rinse in distilled water, dehydrate in alcohol, clear, and coverslip [16].

Elastic Tissue Fibers—Verhoeff's Van Gieson

Elastin stain is useful to stain elastic fibers in arteries and veins and demonstrate damage to blood vessels. It can also be used to highlight pathological changes in elastic fibers like atrophy of elastic tissue in emphysema, thinning or loss in arteriosclerosis, and age-related changes like reduplication, breaks, and splitting that lead to skin wrinkles. The basic principle behind the stain is that elastic tissue has strong affinity for iron–hematoxylin complex so that it will retain the dye longer than other tissue elements when other elements are decolorized. The most commonly used counterstain is Van Gieson's solution. Elastic fibers and nuclei will stain black; collagen will stain red; and other tissue elements will stain yellow. Reagents include alcoholic hematoxylin, 10% ferric chloride, Lugol's iodine, Verhoeff's hematoxylin, differentiating solution (2% ferric chloride), and Van Gieson's solution [15].

Procedure: Deparaffinize and hydrate in distilled water, Verhoeff's hematoxylin for 30 min, wash in tap water, differentiate in 2% ferric chloride solution, rinse in water, hypo for 1 min to remove iodine, wash in water, counterstain in Van Gieson's for 5 min, dehydrate, clear in xylene, and coverslip [16].

Acid-Fast Bacteria—Ziehl–Neelsen Stain

This stain is used to identify *Mycobacterium tuberculosis*, the causative agent of tuberculosis.

Acid-fast organisms have a lipoid capsule that has a high molecular weight and is waxy at room temperature. This makes the organism impenetrable by aqueous-based staining solutions. The lipoid capsule of an acid-fast organism stains with carbol-fuchsin and resists decolorization with dilute acid rinse. The acid-fast bacilli will stain bright red, and the background will stain blue. Reagents used in the procedure include Ziehl–Neelsen carbol-fuchsin solution, 1% acid alcohol, and methylene blue solution [15].

Procedure: Deparaffinize and hydrate in distilled water, stain with carbol-fuchsin solution, microwave for 45 s, wash in running tap water, put the slide in 1% acid alcohol until light pink, and color stops running, wash in running tap water for 5 min, rinse in distilled water, stain with methylene blue for 30 s, rinse in water, dehydrate, clear, and coverslip [16].

CONCLUSION

Proper tissue processing and staining is important for any clinical or research work that involves tissue analysis. Understanding changes in tissue structure requires experience in performing these techniques and knowledge in normal histology, as well as abnormal/pathologic processes.

REFERENCES

[1] Chung JY, Braunschweig T, Williams R, Guerrero N, Hoffmann KM, Kwon M, et al. Factors in tissue handling and processing that impact RNA obtained from formalin-fixed, paraffin-embedded tissue. J Histochem Cytochem 2008;56(11):1033–42.

[2] Hewitt SM, Lewis FA, Cao Y, Conrad RC, Cronin M, Daenberg KD, et al. Tissue handling and specimen preparation in surgical pathology: issues concerning the recovery of nucleic acids from formalin-fixed, paraffin-embedded tissue. Arch Pathol Lab Med December 2008;132(12):1929–35.

[3] van Maldegem F, de Wit M, Morsink F, Musler A, Weegenaar J, van Noesel CJ. Effects of processing delay, formalin fixation, and immunohistochemistry on RNA recovery from formalin-fixed paraffin-embedded tissue sections. Diagn Mol Pathol March 2008;17(1):51–8.

[4] Westra WH, Hruban RH, Phelps TH, Isacson C. Surgical pathology dissection: an illustrated guide. 2nd ed. Springer; 2009. p. 2–21.

[5] Rosai J. Rosai and Ackerman's surgical pathology. 10th ed. Mosby; June 2011. p. 37–73.

[6] Albert VH, Dupre S, van Gent R, Zuidervaart H. The origins of the telescope. Amsterdam: Knaw Press; 2010. p. 28–43.

[7] Molavi DW. The practice of surgical pathology: a beginner's guide to the diagnostic process. Springer; 2008. p. 1–5.

[8] Bancroft JD, Layton C, Suvarna KS. Bancroft's theory and practice of histological techniques. 7th ed. Churchill Livingstone; 2012. p. 382–434.

[9] Coons AH, Creech HJ, Norman Jones R. Immunological properties of an antibody containing a fluorescent group. Proc Soc Exp Biol Med 1941;47:200–2.

[10] Coons AH, Kaplan MH. Localization of antigen in tissue cells. II. Improvements in a method for the detection of antigen by means of fluorescent antibody. Journal of Exp Med 1950;91(1):1–13.

[11] Allan VJ. Protein localization by fluorescence microscopy. Oxford: Oxford University Press; 2000.

[12] Carson F, Hladik C. Histotechnology: a self-instructional text. 1st ed. ASCP, Ill; 1990. p. 142–3. pp. 156–191.

[13] Adegboyega PA, Gokhale S. Effect of decalcification on the immunohistochemical expression of ABH blood group isoantigens. Appl Immunohistochem Mol Morphol June 2003;11(2):194–7.

[14] Orchard GE. Melan A (MART-1): a new monoclonal antibody for malignant melanoma diagnosis. Br J Biomed Sci 1998;55:9.

[15] Sheehan D, Hrapchak BB. Theory and practice of histotechnology. 2nd ed. (Ohio): Battelle Press; 1980. p. 188–246.

[16] Luna L. Manual of histologic staining methods of the AFIP. 3rd ed. (NY): McGraw-Hill; 1968. p. 94–5. pp. 230–231.

[17] Vargas SO, Sampson BA, Schoen FJ. Pathologic detection of early myocardial infarction: a critical review of the evolution and usefulness of modern techniques. Mod Pathol 1999;12(6):635–45.

Chapter 4

Clinical Management of Wound Healing and Hypertrophic Scarring

Peter A. Than[1,2], Christopher R. Davis[1,3], Geoffrey C. Gurtner[2]

[1]Hagey Laboratory for Regenerative Medicine, Stanford University, Stanford, CA, USA; [2]Department of Surgery, Stanford University School of Medicine, Stanford University, Stanford, CA, USA; [3]Division of Plastic Surgery, Stanford University School of Medicine, Stanford, CA, USA

Chapter Outline

WOUND HEALING AND THE BIOMEDICAL BURDEN OF ITS DYSFUNCTION

The integumentary system serves several key functions including protecting deeper structures, regulating temperature, providing a barrier against external pathogens, excreting wastes, and acting as a sensory interface with the outside world. The importance of this system is underscored by the complex repair mechanisms that have evolved to restore tissue integrity after injury. Wound repair requires the immediate activation of numerous overlapping pathways and

Skin Tissue Engineering and Regenerative Medicine. http://dx.doi.org/10.1016/B978-0-12-801654-1.00004-8

cell types to synchronously clear debris, produce extracellular matrix (ECM), and revascularize the injured area. Just as importantly, these processes must be effectively shut down once the wound is healed.

The enormous biomedical and financial burden of wounds emphasizes the importance of understanding the mechanisms governing wound healing and designing effective therapies to improve outcomes. Surgical incisions are performed during an estimated 80 million and 250 million operations in the United States and worldwide, respectively, with an additional 12 million lacerations treated in US emergency departments each year [1–3]. Each injury, whether iatrogenic or traumatic in etiology, activates a common pathway of wound healing, where deposition of cells and ECM restores skin integrity. However, the deposited tissue is predominantly composed of fibroblasts and collagen, resulting in a patch of nonfunctional fibrotic tissue, known as scar. This tissue protects deeper structures from desiccation and foreign pathogens but lacks the original architecture and function (glands, nerves, hair follicles, etc.) present in unwounded skin.

In contrast, injured fetal skin up to 6 months of gestation and some eukaryotic organisms retain the ability to regenerate skin with structure and function nearly identical to the original tissue [4,5]. These regenerative processes are not present in adult humans, with evolution favoring accelerated wound closure and subsequent scar formation over restoration of function, architecture, and appearance. This is likely due to the survival benefits associated with rapid restoration of tissue integrity, minimization of blood loss, and prevention of infection. The net effect in adult humans is that even wounds that heal "perfectly" are inferior to uninjured skin. The resulting scar possesses 80% of the tensile strength of innate skin and, if anatomically located in proximity to joints or orbits, may result in functional impairment as the scar contracts [6].

The mechanism of cutaneous wound healing with the resultant deposition of fibrotic tissue is similar to the response to injury found in tissue throughout the body. While ECM deposition in the skin can cause problematic "scar," fibrosis in organs such as the heart, lung, or liver can result in end-stage organ dysfunction. The molecular mechanisms discovered through cutaneous wound healing research therefore have potential impact in virtually every organ system. This chapter provides an overview of normal and impaired wound healing, highlights some of the clinical and research challenges, and describes current and developing therapies to improve the response to injury.

STAGES OF WOUND HEALING

Wound healing occurs through three distinct but overlapping stages: inflammation, proliferation, and remodeling [5]. The inflammatory stage, which begins immediately after injury, is critical in mitigating blood loss, clearing debris, and modifying the microenvironment to promote subsequent tissue repair. Platelet degranulation and activation of the complement and clotting cascades are the first responses in this process, forming a platelet plug that helps to achieve hemostasis. Secreted ECM proteins form a fibrin matrix that functions as a scaffold to support the migration and proliferation of inflammatory cells. These infiltrating cells—including neutrophils, monocytes/macrophages, and lymphocytes—are recruited through complement-mediated pathways and function to remove damaged tissue, reduce the risk of infection, and modify the microenvironment to affect downstream cellular activity.

The proliferative phase occurs 2–10 days after injury and results in tissue deposition and closure of the defect. Progenitor cells migrate into the wound with subsequent proliferation, differentiation, and ECM production yielding new tissue. Early inflammatory infiltrate provides chemotactic signals that recruit keratinocyte progenitors from the wound edges and their niche within hair follicles, and stimulates their migration across the provisional ECM. The chemotactic stimulus mediated by the inflammatory milieu drives the advancing keratinocyte border to proliferate and mature, forming the basis for restoration of the epithelial barrier.

The role of the inflammatory infiltrate is complex and changes over time. As the inflammatory phase concludes, the cellular constituents develop a phenotype supportive of tissue deposition. Macrophages display an anti-inflammatory role, producing cytokines such as TGF-β, which recruit fibroblasts from surrounding tissue to deposit collagen and other components of ECM. Fibroblasts are therefore an essential component of the proliferative phase. Fibroblasts are activated by signals produced during the inflammatory phase, such as TGF-β1, and ECM components such as fibronectin. Activated fibroblasts are called myofibroblasts because of their expression of α-smooth muscle actin (α-SMA). α-SMA-mediated contraction of myofibroblasts brings the edges of the wound together to assist in closure. In this way, fibroblasts and myofibroblasts affect the mechanical forces, cytokine signaling, and structural composition that contribute to scar formation [7].

Angiogenesis, the formation of a vascular network, is vital to new tissue formation. Granulation tissue, the bed of sprouting capillaries at the wound base, provides a pathway for circulating progenitor cells to infiltrate the wound and delivers the necessary metabolic substrates to sustain new tissue formation. Signaling molecules such as vascular endothelial growth factor-A (VEGF-A) and fibroblast growth factor-2 (FGF-2) are potent stimuli for angiogenesis [8]. Replacement of these deficient factors, in some instances, is sufficient to promote angiogenesis and facilitate healing of chronic wounds [5].

Remodeling, the third stage of wound repair, begins 10 days after injury and can continue for over a year. The myriad of cells recruited during the earlier phases migrate out of the wound or undergo apoptosis as proliferative processes are inactivated. The proliferative stage, characterized by an influx of cells and heightened metabolic activity, requires a more dense vascular supply to deliver the necessary circulating cells and provide the substrate for metabolically active processes. As these processes cease and the cellular

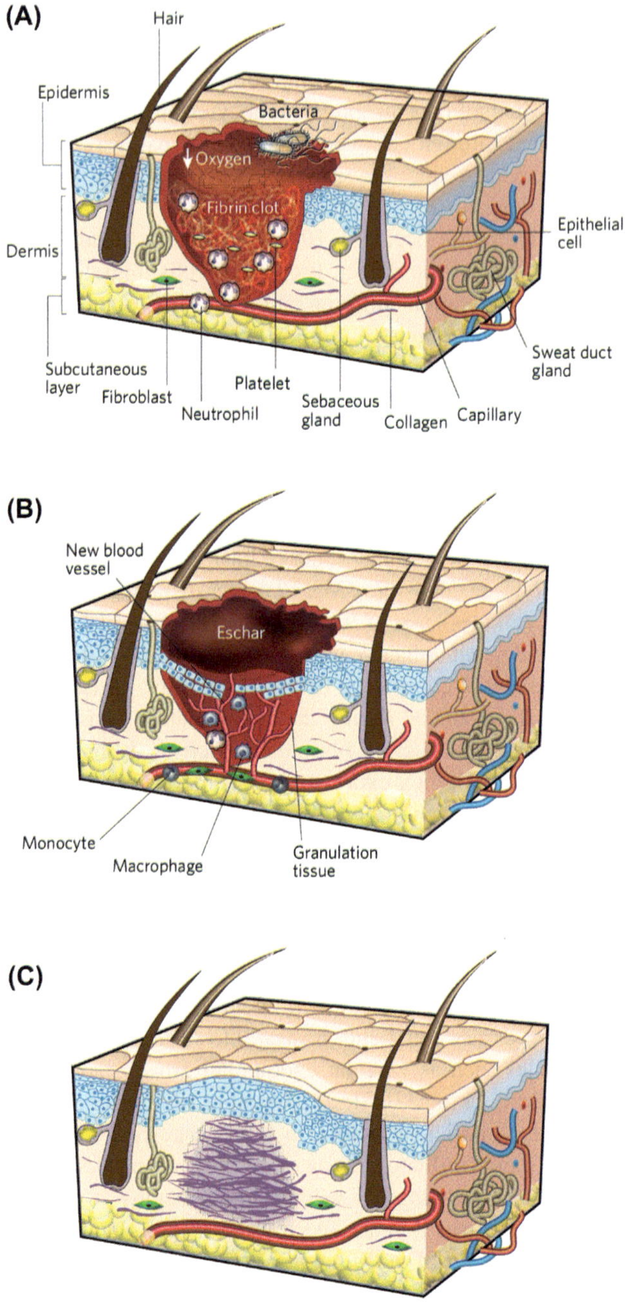

FIGURE 1 **Stages of wound repair.** There are three classic stages of wound repair. (A) Inflammation. This stage lasts until about 48 h after injury. Depicted is a skin wound at about 24–48 h after injury. The wound is characterized by a hypoxic (ischemic) environment in which

constituents of the wound exit, the vasculature regresses. The predominant tissue in a healed wound is acellular collagen with an epithelial covering. Collagen III is initially deposited by fibroblasts during the second stage of wound healing. In the remodeling phase, remaining cells (fibroblasts, macrophages, and endothelial cells) produce matrix metalloproteinases (MMP) that degrade disorganized collagen type III and slowly replace it with stronger and more durable collagen type I, the predominant collagen type in a healed adult wound [9] (Figure 1).

PATHOLOGIC WOUND HEALING

Mature scar that forms through the process of normal wound healing, while fundamentally different from unwounded skin, accomplishes the goal of restoring the barrier function of the integument. Normal wound healing represents a "middle ground" in the spectrum of healing responses. Underhealing occupies one end of the spectrum and is represented clinically by patients with chronic nonhealing ulcers. At the other end of the spectrum, overhealing is seen in fibroproliferative diseases, including hypertrophic scars and keloids. These pathologies result from failure of key regulatory components governing wound healing and cause the balance of cell and ECM deposition to shift to one extreme or the other.

CHRONIC WOUNDS

Millions of patients in the US alone suffer significant morbidity from chronic nonhealing wounds, with an annual healthcare cost of $15 billion [10]. The most ubiquitous clinical example is the diabetic ulcer. Diabetes mellitus is a massive global health burden, with an increasing incidence worldwide. Persistent hyperglycemia results in systemic nonenzymatic glycosylation, leading to cardiovascular disease, cerebrovascular disease, retinopathy, peripheral neuropathy, renal dysfunction, and many other complications. Peripheral neuropathy increases the likelihood of developing traumatic foot injuries, because of insensate plantar surfaces. Peripheral vascular disease impairs wound healing

a fibrin clot has formed. Bacteria, neutrophils, and platelets are abundant in the wound. Normal skin appendages (such as hair follicles and sweat duct glands) are still present in the skin outside the wound. (B) Proliferation. This stage occurs about 2–10 days after injury. Depicted is a skin wound at about 5–10 days after injury. An eschar (scab) has formed on the surface of the wound. Most cells from the previous stage of repair have migrated from the wound, and new blood vessels now populate the area. The migration of epithelial cells can be observed under the eschar. (C) Remodeling. This stage lasts for a year or longer. Depicted is a skin wound about 1–12 months after repair. Disorganized collagen has been laid down by fibroblasts that have migrated into the wound. The wound has contracted near its surface, and the widest portion is now the deepest. The re-epithelialized wound is slightly higher than the surrounding surface, and the healed region does not contain normal skin appendages. *Adapted with permission from Gurtner et al. [5].*

and increases the risk of infection, exacerbating the situation. This cascade of events frequently leads to the need for limb amputations, patient morbidity, and mortality [11].

A number of co-morbidities are associated with poor wound healing including advanced age, diabetes, vascular disease, malnutrition, smoking, renal dysfunction, radiation exposure, infection, and immunosuppression. These factors compromise wound healing through a number of pathways, but are all characterized by an impaired progenitor cell response and insufficient angiogenesis. The subsequently hypoxic wound environment is prone to infection and a protracted inflammatory phase. The persistent inflammatory cells produce toxin-free radicals that breakdown ECM, damage fibroblasts and keratinocytes, and do not allow the normal switch to a fibroproliferative environment. Consequently, the overall balance favors MMP and other enzymes that degrade ECM proteins and inhibit fibroblast and keratinocyte proliferation, with a resulting wound that fails to heal [12].

The majority of underlying causes of chronic wounds can be simulated in the laboratory setting and studied using animal models. Stented excisional wound models can be applied to diabetic mice, ischemia, and irradiated skin to describe just a few examples [13]. Consequently, research in wound healing has been largely aimed at understanding and developing strategies to ameliorate underhealing. To this end, investigators have developed cell therapies, gene therapies, protein delivery methods, biomimetic dermal scaffolds, specialized bandages, and other medical devices.

WOUND HEALING THERAPIES

Negative Pressure Therapy

Topical negative pressure (TNP) therapy utilizes a permeable foam dressing and mechanical suction to exert negative pressure on a wound. First described in 1995 by Fleischmann, and trialed in 1997 by Argenta and Morykwas, TNP is a ubiquitous wound-healing adjunct [14,15]. Clinical indications are broad, spanning acute or dehiscent wounds, chronic ulcers, scalp defects, and skin grafts [16–18]. The mechanism of action involves stimulation of granulation tissue and wound vascularity, removal of interstitial fluid, and reduction of bacterial colonization [19]. Further refinements are expected, as biocompatible foam materials are developed to deliver antimicrobial molecules through the dressing.

Cytokines and Growth Factors

Cell signaling has a fundamental role in coordinating wound healing. Chronic wounds are thought to develop as a result of processes that alter the cellular constituents and normal signaling pathways. Delivery of growth factors and cytokines to normalize wound healing has been in the preclinical armamentarium for decades. However, only conditioned media and other nonspecific substances were available. Understanding the molecular signals involved in wound healing at a granular level has allowed investigators to apply more targeted molecular therapies via advanced delivery methods.

Deferoxamine (DFO) is one example that effectively exploits known molecular targets to improve wound healing. Hypoxia-inducible factor-1 (HIF-1) is a transcription factor that promotes a number of downstream targets, though its effects on wound healing are primarily related to VEGF-mediated angiogenesis and SDF-1-mediated recruitment of circulating progenitor cells [20,21]. HIF-1 is degraded by prolyl hydroxylases in the presence of oxygen and iron. DFO, an iron chelator, stabilizes HIF-1 at the wound site through iron depletion and prolyl hydroxylase inhibition, which improves wound closure in diabetic models [22]. Advances in biomaterials fabrication have allowed DFO delivery within slow-release patches.

Other important cytokines including VEGF-A, FGF-2, and PDGF have demonstrated potential in accelerating wound healing through their effects on progenitor cell recruitment and improved angiogenesis. The frontiers of growth factor and cytokine delivery involve identifying potent stimuli of wound healing and developing effective delivery methods. For instance, gene therapy, biologic dressings that slowly release cytokines, and others are avenues that require further investigation.

Cell-Based Therapy

The number of cell types and the complexity of signaling pathways that evolve over time to produce normal wound healing are difficult to modulate through the delivery of single-agents. One way to address this complexity is to deliver cells, which can secrete multiple factors and produce different secretory profiles in response to an evolving wound environment. Skin grafting, the most clinically established form of cell therapy, has been widely used in surgery for decades. Here, a partial (part of the dermis) or full-thickness (includes the complete dermis) graft of skin is harvested from a healthy donor site and applied to a wound. The skin graft remains viable initially by diffusion of oxygen and metabolites from the wound bed, known as imbibition. The donor site spontaneously re-epithelializes. This technique is used frequently for burn patients, but is limited if the area of burned skin exceeds the availability of donor sites.

Cultured epithelial autografts (CEA) allow surgeons to expand autologous keratinocytes in vitro and deliver them to wounds, thereby minimizing donor site size and allowing cell therapy to proceed when the injured skin surface area is large. The main challenge that CEA therapy highlights is the complexity of wound healing itself. Keratinocytes alone are unable to compensate for extensive deep dermal injury, as they are insufficient to provide the cytokines and growth factors that are normally present when fibroblasts, inflammatory cells, and others are present. Identification of the optimal cell types necessary to deliver the required signals and techniques in delivery (spray, sheets, hydrogel, etc.) remain to be addressed.

Application of stem cells to wounds is an extension of cell therapy that has demonstrated promise. Stem cells are self-renewing, multipotent cells available from a variety of sources. They are relatively immune privileged and possess immunomodulatory properties. Initially thought to contribute to wound healing through direct differentiation, the evolving understanding of their role favors modification of the microenvironment in favor of wound healing through secretion of anti-inflammatory molecules and their influence on other cells. The rapid pace of stem cell investigation is shedding light on the best cell types for use, identifying subpopulations with greater wound-healing potential, and developing gene therapies and other manipulations that allow improved engraftment and differentiation.

Biomimetic Scaffolds

The concept of providing the "seed and soil" suggests that progenitor cells, biomaterials, and cell signals are all required for tissue regeneration and effective wound healing [5]. Natural dermal matrices, such as AlloDerm (LifeCell Corporation, NJ, USA), are human-derived ECM, widely used in clinical practice. Large burns are a primary indication, but their use is expanding to donor sites and other cutaneous injuries. Dermal matrices provide immediate wound coverage, and function as a scaffold for resident cell infiltration and granulation tissue formation. A variety of commercially available products exist, from both human and animal sources. These naturally sourced scaffolds are processed to remove cells and antigens, often with variable results. Their composition may be different to the needs of the healing tissue into which they are placed, and issues with biocompatibility arise. Limited clinical trials to distinguish between performance differences, as well as financial cost, make choosing the "best" product a difficult task [23].

Synthetic acellular scaffolds, such as Integra (Integra LifeSciences Corp, NJ, USA), offer greater control over consistency and allow tailoring of functional composition for specific applications. These materials generally provide immediate wound coverage and serve as a scaffold for proliferating cells, ultimately facilitating restoration of integumentary barrier function. Histologically, the tissue formed contains collagen in an appropriate arrangement. However, other dermal elements such as elastic fibers and nerves are often missing [24–26]. Newer, synthetic biomimetic hydrogels offer a promising alternative that may serve as a dermal regenerative template. The porosity of these materials can be carefully controlled to modulate cell behavior and function [24–27]. Moreover, these hydrogels can serve as a delivery vehicle for cell-based wound therapies, allowing delivery of progenitor cells and functioning as a regenerative matrix for proliferating cells [28–30].

Cell-laden hydrogels are emerging in wound therapy as a new class of "biologic dressing." These materials can be used as traditional dressings to maintain moisture in the wound site, protect against contaminants, and provide some debridement function. The addition of other factors such as cells, growth factors, and antimicrobial agents allows them to modify the environment and promote wound closure through a number of biologic pathways [29,31–35]. The optimal dressing structure and composition, delivered molecules, and cell types is an area of intense investigation.

FIBROPROLIFERATIVE DISEASE

Overhealing or fibroproliferative disease, including hypertrophic scarring and keloid disease, is characterized by abnormally large scar formation. In addition to esthetic consequences, patients can suffer significant morbidity from scars that are frequently pruritic, painful, and psychologically debilitating. When large scars form or contract in vital areas, significant functional impairments can occur.

Distinction between keloid disease and hypertrophic scar is made primarily through clinical observation, namely that keloids extend beyond the boundaries of the original wound [36]. Clinical differences are summarized in Table 1. Keloid formation appears related to genetic and environmental factors, while hypertrophic scar often develops in response to deep dermal injury or burns, implicating a protracted inflammatory stage of wound healing [36–38]. The different clinical behavior and histologic findings have led to ongoing controversy as to whether keloid disease and hypertrophic scarring represent distinct diseases or different manifestations along a spectrum of dysfunctional wound healing [39,40]. Because of their symptoms and more aggressive nature, keloid disease is generally of more clinical interest and more frequently investigated than hypertrophic scar.

The pathophysiology causing excessive scar formation remains incompletely understood. Theories implicate a dysfunctional inflammatory stage, mechanical forces, hormones, and genetics [37,41–44]. As fibroproliferative disease appears unique to humans, the lack of available animal models makes investigation challenging. Much of what is known about excess scar formation comes from microarray and single-cell transcriptional analyses of human fibroblasts from patients with hypertrophic scar and keloid disease [38]. Microarray studies show that fibroblasts from fibroproliferative wounds are phenotypically different from those found in normal wounds, demonstrating protracted periods of activity, resistance to apoptosis, and increased collagen production [45–49]. They produce higher quantities of fibroproliferative cytokines such as TGF-β1, and lower levels of MMP, with a net increase in ECM production. Mechanical tension is another important stimulus that activates inflammatory pathways, chemokine signaling, and cell–matrix interactions to result in increased scar formation [50–55]. Taken together, inflammatory and mechanical forces result

TABLE 1 Clinical Manifestations of Keloid and Hypertrophic Scars

	Keloid	Hypertrophic Scar
Age distribution	10–30 years of age at higher risk	All equally affected
Ethnic groups	African, Asian, Hispanic, Mediterranean predominance	All equally affected
Location	Chest, cheek, shoulders, neck, ears	Extensor surfaces
Characteristics	Extends beyond wound border, raised, firm, telangiectatic, pigmented, does not regress	Contained within wound border, soft, often regresses
Symptoms	Pruritic, painful	Usually less bothersome
Histology	Highly proliferative fibroblasts, thickened disorganized collagen, abnormally high ratio of collagen type I to III	Fine collagen bundles parallel to the dermis, relatively senescent fibroblasts
Risk factors	Genetic predisposition, epigenetic factors	Deep dermal injuries, burns, protracted inflammatory phase of wound healing
Regression	No	Yes

in excessive and disorganized matrix production with impaired matrix-degradation pathways unable to reduce the scar formation.

SCAR REDUCTION THERAPIES

The complexity of fibroproliferative disease, with its overlapping and redundant pathways, makes intervention difficult. Previously developed therapies aimed at scar reduction are largely untargeted and limited in success. Recent insight into the pathogenesis of fibroproliferative disease has allowed the emergence of targeted therapies. However, few randomized controlled studies evaluate efficacy of scar reduction therapy. Of the studies with level I evidence in this arena, only one has shown significant reduction in scar size [56]. Here, we will review currently employed scar reduction strategies, discuss the evidence, and focus on novel targeted therapies.

General Approaches

Surgeons have traditionally relied on clinical observations to develop scar reduction strategies. General approaches include minimization of surgical wounds through minimally invasive procedures. Surgical techniques that minimize the inflammatory response (e.g., sharp instruments, atraumatic handling of tissue) are also recommended. Surgical planning may direct incision placement to areas of low mechanical tension to facilitate a reduced fibroproliferative response. Few techniques exist to *prevent* fibroproliferative scar formation, though this is recently coming into focus as a therapeutic goal. The majority of therapies, however, are aimed at reducing the size and symptoms of *existing* scar.

Scar Revision Surgery

Surgical excision of an existing scar is generally a late-stage treatment option, as it again exposes patients to potentially dysfunctional wound healing that precipitated the initial fibroproliferative scar. Excision is usually performed in conjunction with other approaches for refractory scars. The goal in scar excision is to optimize the environment to allow the fresh skin edges to heal normally, *preventing* reformation of scar. In this modality, all abnormal tissue is sharply excised. The procedure is done as atraumatically as possible to minimize the amount of postprocedure inflammation, and the wound edges are reapproximated with minimal tension. Surgeons often avoid the use of absorbable sutures in favor of permanent sutures, which incite less inflammatory response. Unfortunately, surgical excision alone is rarely effective, with recurrence rates as high as 45–100% [57]. This should be avoided if possible, because excision makes the wound longer; and with a high recurrence rate, a larger fibroproliferative scar is possible.

Radiotherapy

Impaired wound healing following oncological irradiation is a common complication affecting cancer patients. As a scar reduction modality, radiotherapy exploits this mechanism by impairing the fibroproliferative process in surgical wounds by destroying fibroblasts, reducing collagen synthesis, and shifting the balance toward the wound-healing response of nondiseased skin [58]. Apoptosis of proliferating cells in the scar tissue is therefore promoted [59]. Infrequently used as monotherapy, radiation is more efficacious as an adjuvant therapy in the weeks following scar excision in order to target the proliferating fibroblasts [60]. Benefits are modest, with a recurrence rate of 33% [61]. Despite advances to reduce the effective radiation dose, risks from overexposure include impaired wound healing, fibrosis, or malignant change in adjacent tissue. Targeted therapy using brachytherapy is a potential solution to the deleterious effects from external beam radiation, with significant improvement in esthetic outcomes [60]. Radiation should be avoided, if possible, in pediatric patients with keloid scars, due to potential metaphyseal irradiation with secondary growth impairment [62].

Laser Therapy

Laser therapy is a newer approach to scar therapy that aims to use high-energy photons to ablate tissue. Lasers may be classified as ablative or nonablative fractional resurfacing devices, with the most common types emitting carbon dioxide (CO_2) or NdYaG in differing pulsation regimens. CO_2 laser can also be used in combination with IFN [63]. CO_2 dampens the inflammatory response and reduces fibroblast-mediated collagen production, while IFN-α2b induces selective glycosaminoglycan and collagen production [63,64]. Efficacy data for laser treatment are limited, but early data suggest that it frequently results in failure. While more efficacy data are needed, this nontargeted modality does not seem to address any fundamental derangements associated with excessive wound healing.

Cryotherapy

Cryotherapy, typically in the form of liquid nitrogen, is used to ablate scar through freezing. The freezing process is thought to induce vascular damage that may lead to anoxia, tissue necrosis, and subsequent flattening of keloid scar [65]. It is also hypothesized to modify collagen synthesis and facilitate fibroblast differentiation [66]. Treatment tends to be directed toward smaller keloid areas, and regimens typically include two 15–20 s thaw cycles on multiple occasions for maximal efficacy. This seems to work better for small lesions and requires multiple treatments for maximal efficacy [62]. However, cryotherapy can be painful and poorly tolerated and may cause hypopigmentation and blistering, thus reducing compliance. More recent advances include intra-lesional cryosurgery using a needle probe [67].

Moisture-Retaining Dressings

Silicone sheet placement is another nontargeted scar reduction therapy. Silicone sheets are often used as dressings following scar excision. While the mechanism of its action is poorly understood, it is likely that these dressings prevent desiccation and soften scars, resulting in decreased scar size, pruritus, and pain. These features do not seem unique to silicone, as nonsilicone dressings with properties that allow them to maintain moisture within the wound are equally as effective as silicone. Further, silicone in forms such as sprays or gels that do not form an occlusive sheet over the wound have been shown ineffective in clinical trials [68,69]. These dressings are applied after epithelialization is complete and require frequent dressing changes for over a month.

Antiproliferative Agents

Attenuating the fibroblast response is a target that has emerged from improved understanding of the fibroproliferative response. Ginsenoside Rg3 (G-Rg3) when delivered in a poly(L-lactide) (PLLA) scaffold has been used to induce apoptosis of hypertrophic scar fibroblasts in a rabbit ear model of hypertrophic scar, effectively reducing scar formation [33]. Clinical trials are needed to more rigorously assess its efficacy, but the introduction of this regimen is a promising introduction of a targeted scar therapy.

Anti-inflammatory Therapies

A heightened inflammatory stage appears important in excess scar formation, as inflammatory cells produce pro-fibrotic cytokines and perpetuate the fibroproliferative response. Therapies that target the inflammatory phase may hold promise in reducing scar formation. Corticosteroids are currently the first-line treatment agents employed. Their immunosuppressant mechanism nonspecifically decreases inflammation and fibroblast proliferation, and promotes vasoconstriction. A more targeted mechanism of action is via α2-macroglobulin (an inhibitor of fibrinolysis) inhibition, resulting in net collagen breakdown within scar [70]. Treatment regimens for hypertrophic scar typically begin with local therapy, through the administration of topical immunosuppressant creams. However, absorption through the epidermis and penetration into scar is poor. Injection of corticosteroids directly into scars, using triamcinolone acetonide, can improve tissue penetration while still limiting the effect to local tissue. The efficacy of this regimen is mixed, with 50–100% of patients responding to treatment, and recurrence rates of 9–50% [41]. Future therapeutic directions to prevent keloid recurrence may be revealed by ongoing clinical trials comparing steroid therapy with verapamil [71].

Knowledge of the molecular underpinnings of excess scar formation has led to the use of several targeted immunomodulatory agents. Imiquimod, IFN-α2b, 5-flouricil, and retinoic acid are examples of agents under investigation. Imiquimod, for example, is a topical immunomodulatory agent initially developed for treatment of genital warts. It stimulates the release of IFN-α, TNF-α, IL-1, IL-6, and IL-8, which act to decrease excessive collagen formation by fibroblasts. It further serves to increase apoptosis-related genes in keloid fibroblasts. IFN-α2b is an antiproliferative cytokine that is thought to improve dermal fibrosis through direct inhibition of fibroblast-mediated collagen production, TGF-β and histamine inhibition, and reduction in cellular messenger ribonucleic acid [72]. Intra-lesional injection in preliminary studies suggests that it can reduce scar size by 50% after 9 days of treatment [73]. Used as an adjuvant therapy after excision, recurrence rates decrease from 51% to 19% [74]. 5-fluorouracil (5-FU) is a pyrimidine analogue used in chemotherapy and anti-inflammatory regimens. In vitro evidence suggests that 5-FU inhibits TGF-β signaling and decreases type I collagen expression by fibroblasts. Clinical trial results of 5-FU alone, or in combination with triamcinolone, have been positive with significant scar and symptom reduction [75–79]. Retinoids are used widely in dermatologic medicine. They inhibit fibroblast proliferation and increase epidermal thickness. However, their application to scar therapy is relatively new, and their efficacy is currently not known.

TGF-β occupies an important role in scar formation. Three isoforms exist, designated as TGF-β1, -β2, and -β3, each with a unique corresponding receptor. TGF-β1 overactivity is associated with fibroproliferative disease. TGF-β3 has been shown to exert paracrine effects upon keratinocytes and is necessary for epithelialization [80]. Renovo (Manchester, United Kingdom) developed an injectable recombinant TGF-β3 called avotermin (marketed as Juvista) that was investigated for use in scar minimization following surgical incisions. Despite favorable preliminary results, a phase III clinical trial showed that it failed to meet the study endpoints [56,81,82].

Reducing Mechanical Tension

Mechanical tension plays a major regulatory role in scar formation. Focal adhesion kinase (FAK) has been identified as a central mediator between mechanical force and fibrosis, through its action on inflammatory pathways. Tension on wound edges activates FAK, which has multiple downstream targets. One target of particular importance in mechanotransduction is ERK (MAPK14). These mechanotransduction-mediated pathways converge on increased TGF-β1, MCP-1, and other pro-fibrotic cytokines, with the end-result being a thickened, fibroproliferative scar. Several therapies aim to take advantage of these mechanotransduction properties. Small molecule FAK inhibitors, for example, are under investigation for use in scar reduction in cutaneous applications, as well as in reduction of fibrosis in organs such as the lung [68,83–87].

Botulinum toxin is a popular injectable, temporary, muscle-paralyzing agent. Injection of botulinum toxin around wounds has been shown to facilitate wound healing with reduced scar size [88]. This is likely mediated by a combination of factors including reduction in tensile force on the wound edges and regulation of the balance between fibroblast proliferation and apoptosis [89,90]. The embrace device (Neodyne Biosciences, CA, USA) is a silicone elastomeric dressing that was developed for application to surgical incisions *prior* to scar formation. The mechanomodulatory properties of this device reduce tension on wound edges as they heal. The result of decreased mechanical force facilitated wound healing with minimal scar in several clinical trials [56,91,92].

REGENERATIVE HEALING

Understanding the differences between fetal and adult skin provides a lens through which to study tissue regeneration. The higher hyaluronic acid content found in fetal skin gives it a more permeable character, which facilitates greater influx of fibroblasts in response to injury. The porous structure provided by hyaluronic acid places greater distance between cells, reducing their susceptibility to inhibitory signals from neighboring cells. The ratio of collagen type III to type I is higher in fetal skin than in adult skin but decreases with age. The cytokines produced in response to injury are also different, with altered ratios of TGF-β isoforms thought to play an important role. Furthermore, fetal fibroblasts do not produce collagen in the same TGF-β1-induced pathway as adult fibroblasts. The ratio of ECM-degrading molecules such as MMP, relative to their inhibitors, is higher in fetal wounds, shifting the overall balance more toward remodeling than collagen deposition [93].

The increasingly granular understanding of wound healing may allow the development of regenerative strategies. Progress in stem cell biology, biomaterials, gene therapy, and many others allow increasingly precise manipulations of cell microenvironments. With improved understanding of wound healing and convergence of these fields, tissue regenerative strategies may be unlocked (Figure 2).

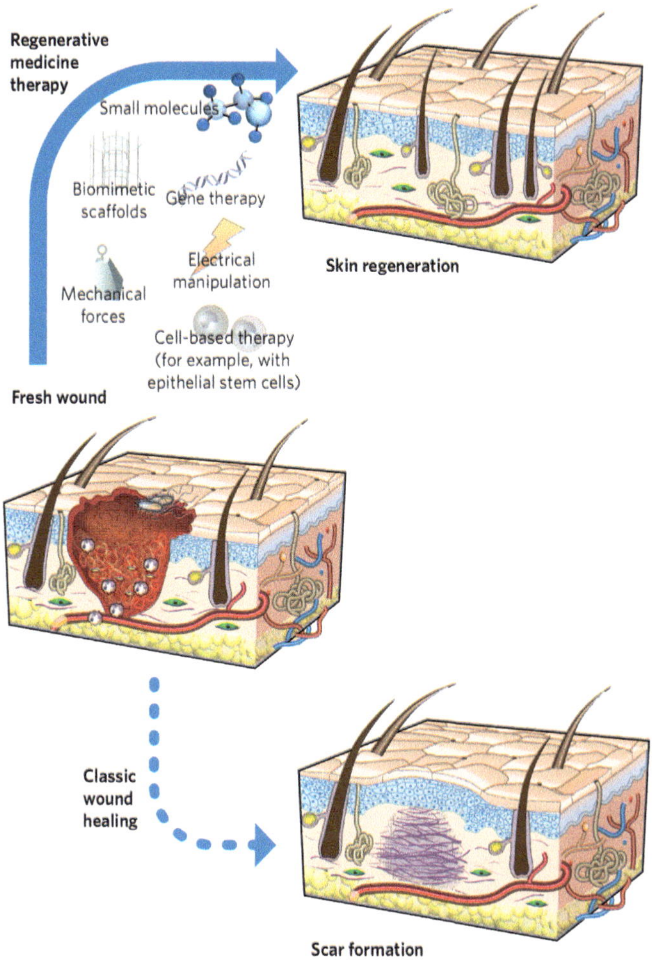

FIGURE 2 **Potential therapies for reducing scar formation during wound repair.** To manipulate wound repair to become more regenerative than scar forming, strategies include the use of biomimetic scaffolds, the manipulation of the mechanical environment (e.g., negative-pressure wound therapy to increase healing) or the electrical environment, the administration of small molecules, the use of gene-therapy approaches, and the use of cell-based strategies (including administration of epithelial stem cells). All of these elements have been demonstrated to have an effect on in vitro and in vivo models of wound healing as single-agent therapies. In theory, many of these elements could be combined to recreate a receptive environment (or "soil") to promote regeneration. Combining these with the appropriate stem cells (or "seed") will undoubtedly alter the result of the wound healing in humans. *Adapted with permission from Gurtner et al. [5].*

CONCLUSION

Only recently have investigators begun to rigorously unravel the tremendous complexity of wound healing and develop an understanding of how mechanical forces, cytokine signaling, progenitor cells, and extracellular components affect tissue repair at the molecular level. This granular understanding has provided precise therapeutic targets that are transforming how wound care is provided. With this, we are seeing the development of targeted signaling, cell-based therapies, biomimetic scaffolds, and techniques in modulating mechanical force. Given the complexity and overlapping nature of wound healing pathways, it is likely that future therapeutics will need to combine these modalities to improve efficacy. Similarly, as the ability to manipulate the wound microenvironment increases in precision, it is likely that regenerative strategies with fine control over biological behavior will become accessible.

LIST OF ABBREVIATIONS

5-FU 5-Fluorouracil
α-SMA α-Smooth muscle actin
CEA Cultured epithelial autografts
DFO Deferoxamine
ECM Extracellular matrix
FAK Focal adhesion kinase
FGF-2 Fibroblast growth factor-2
IL-1 Interleukin-1
IL-6 Interleukin-6
IL-8 Interleukin-8
IFN-α2b Interferon-α2b
MMP Matrix metalloproteinases
PDGF Platelet-derived growth factor
TGF-β Transforming growth factor-β
TNF-α Tumor necrosis factor-α
VEGF Vascular endothelial growth factor

REFERENCES

[1] Cullen KA, Hall MJ, Golosinskiy A. Ambulatory surgery in the United States, 2006. Natl Health Stat Rep 2009;(11):1–25.
[2] Weiser TG, Regenbogen SE, Thompson KD, et al. An estimation of the global volume of surgery: a modelling strategy based on available data. Lancet 2008;372(9633):139–44.
[3] Singer AJ, Hollander JE, Quinn JV. Evaluation and management of traumatic lacerations. N Engl J Med 1997;337(16):1142–8.
[4] Lorenz HP, Whitby DJ, Longaker MT, Adzick NS. Fetal wound healing. The ontogeny of scar formation in the non-human primate. Ann Surg 1993;217(4):391–6.
[5] Gurtner GC, Werner S, Barrandon Y, Longaker MT. Wound repair and regeneration. Nature 2008;453(7193):314–21.
[6] Levenson SM, Geever EF, Crowley LV, Oates 3rd JF, Berard CW, Rosen H. The healing of rat skin wounds. Ann Surg 1965;161:293–308.

[7] Werner S, Krieg T, Smola H. Keratinocyte-fibroblast interactions in wound healing. J Invest Dermatol 2007;127(5):998–1008.
[8] Werner S, Grose R. Regulation of wound healing by growth factors and cytokines. Physiol Rev 2003;83(3):835–70.
[9] Lovvorn 3rd HN, Cheung DT, Nimni ME, Perelman N, Estes JM, Adzick NS. Relative distribution and crosslinking of collagen distinguish fetal from adult sheep wound repair. J Pediatr Surg 1999;34(1):218–23.
[10] Markova A, Mostow EN. US skin disease assessment: ulcer and wound care. Dermatol Clin 2012;30(1):107–11. ix.
[11] Hambleton IR, Jonnalagadda R, Davis CR, Fraser HS, Chaturvedi N, Hennis AJ. All-cause mortality after diabetes-related amputation in Barbados: a prospective case-control study. Diabetes Care 2009;32(2):306–7.
[12] Eming SA, Krieg T, Davidson JM. Inflammation in wound repair: molecular and cellular mechanisms. J Invest Dermatol 2007;127(3):514–25.
[13] Galiano RD, Michaels J, Dobryansky M, Levine JP, Gurtner GC. Quantitative and reproducible murine model of excisional wound healing. Wound Repair Regen 2004;12(4):485–92.
[14] Fleischmann W, Becker U, Bischoff M, Hoekstra H. Vacuum sealing: indication, technique, and results. Eur J Orthop Surg Traumatol 1995;5(1):37–40.
[15] Argenta LC, Morykwas MJ. Vacuum-assisted closure: a new method for wound control and treatment: clinical experience. Ann Plast Surg 1997;38(6):563–76; discussion 77.
[16] Webster J, Scuffham P, Sherriff KL, Stankiewicz M, Chaboyer WP. Negative pressure wound therapy for skin grafts and surgical wounds healing by primary intention. Cochrane Database Syst Rev 2012;4:Cd009261.
[17] Dumville JC, Hinchliffe RJ, Cullum N, et al. Negative pressure wound therapy for treating foot wounds in people with diabetes mellitus. Cochrane Database Syst Rev 2013;10:Cd010318.
[18] Davis CR, Mitsala G, Malcolm G, Orlando A. Safe application of topical negative pressure dressings to exposed brain before definitive reconstruction. J Plast Reconstr Aesthet Surg 2014;67(8):e204–5.
[19] Morykwas MJ, Argenta LC, Shelton-Brown EI, McGuirt W. Vacuum-assisted closure: a new method for wound control and treatment: animal studies and basic foundation. Ann Plast Surg 1997;38(6):553–62.
[20] Chang EI, Loh SA, Ceradini DJ, et al. Age decreases endothelial progenitor cell recruitment through decreases in hypoxia-inducible factor 1alpha stabilization during ischemia. Circulation 2007;116(24):2818–29.
[21] Thangarajah H, Vial IN, Grogan RH, et al. HIF-1alpha dysfunction in diabetes. Cell Cycle 2010;9(1):75–9.
[22] Thangarajah H, Yao D, Chang EI, et al. The molecular basis for impaired hypoxia-induced VEGF expression in diabetic tissues. Proc Natl Acad Sci USA 2009;106(32):13505–10.
[23] Rennert RC, Rodrigues M, Wong VW, et al. Biological therapies for the treatment of cutaneous wounds: phase III and launched therapies. Expert Opin Biol Ther 2013;13(11):1523–41.
[24] Wong VW, Rustad KC, Galvez MG, et al. Engineered pullulan-collagen composite dermal hydrogels improve early cutaneous wound healing. Tissue Eng Part A 2011;17(5–6):631–44.
[25] Rustad KC, Wong VW, Sorkin M, et al. Enhancement of mesenchymal stem cell angiogenic capacity and stemness by a biomimetic hydrogel scaffold. Biomaterials 2012;33(1):80–90.
[26] Wong VW, Rustad KC, Glotzbach JP, et al. Pullulan hydrogels improve mesenchymal stem cell delivery into high-oxidative-stress wounds. Macromol Biosci 2011;11(11):1458–66.
[27] Wong VW, Levi B, Rajadas J, Longaker MT, Gurtner GC. Stem cell niches for skin regeneration. Int J Biomater 2012;2012:926059.

[28] Nauta A, Seidel C, Deveza L, et al. Adipose-derived stromal cells overexpressing vascular endothelial growth factor accelerate mouse excisional wound healing. Mol Ther 2013;21(2):445–55.

[29] Garg RK, Rennert RC, Duscher D, et al. Capillary force seeding of hydrogels for adipose-derived stem cell delivery in wounds. Stem Cells Transl Med 2014;3(9):1079–89.

[30] Lam MT, Nauta A, Meyer NP, Wu JC, Longaker MT. Effective delivery of stem cells using an extracellular matrix patch results in increased cell survival and proliferation and reduced scarring in skin wound healing. Tissue Eng Part A 2013;19(5–6):738–47.

[31] Huang G, Ji S, Luo P, et al. Accelerated expansion of epidermal keratinocyte and improved dermal reconstruction achieved by engineered amniotic membrane. Cell Transplant 2013;22(10):1831–44.

[32] Maan ZN, Rennert RC, Koob TJ, Januszyk M, Li WW, Gurtner GC. Cell recruitment by amnion chorion grafts promotes neovascularization. J Surg Res 2014;193(2):953–62.

[33] Cui W, Cheng L, Hu C, Li H, Zhang Y, Chang J. Electrospun poly(L-lactide) fiber with ginsenoside rg3 for inhibiting scar hyperplasia of skin. PLoS One 2013;8(7):e68771.

[34] Wu J, Zheng Y, Song W, et al. In situ synthesis of silver-nanoparticles/bacterial cellulose composites for slow-released antimicrobial wound dressing. Carbohydr Polym 2014;102:762–71.

[35] Meaume S, Truchetet F, Cambazard F, et al. A randomized, controlled, double-blind prospective trial with a Lipido-Colloid Technology-Nano-OligoSaccharide Factor wound dressing in the local management of venous leg ulcers. Wound Repair Regen 2012;20(4):500–11.

[36] Kose O, Waseem A. Keloids and hypertrophic scars: are they two different sides of the same coin? Dermatol Surg 2008;34(3):336–46.

[37] Bella H, Heise M, Yagi KI, Black G, McGrouther DA, Bayat A. A clinical characterization of familial keloid disease in unique African tribes reveals distinct keloid phenotypes. Plast Reconstr Surg 2011;127(2):689–702.

[38] Wong VW, You F, Januszyk M, Gurtner GC, Kuang AA. Transcriptional profiling of rapamycin-treated fibroblasts from hypertrophic and keloid scars. Ann Plastic Surg 2014;72(6):711–9.

[39] Gauglitz GG, Korting HC, Pavicic T, Ruzicka T, Jeschke MG. Hypertrophic scarring and keloids: pathomechanisms and current and emerging treatment strategies. Mol Med 2011;17(1–2):113–25.

[40] Gauglitz GG. Management of keloids and hypertrophic scars: current and emerging options. Clin Cosmet Investig Dermatol 2013;6:103–14.

[41] Robles DT, Berg D. Abnormal wound healing: keloids. Clin Dermatol 2007;25(1):26–32.

[42] Song C. Hypertrophic scars and keloids in surgery: current concepts. Ann Plast Surg 2014;73(Suppl. 1):S108–18.

[43] Chike-Obi CJ, Cole PD, Brissett AE. Keloids: pathogenesis, clinical features, and management. Semin Plast Surg 2009;23(3):178–84.

[44] Robles DT, Moore E, Draznin M, Berg D. Keloids: pathophysiology and management. Dermatol Online J 2007;13(3):9.

[45] Chen W, Fu X, Sun X, Sun T, Zhao Z, Sheng Z. Analysis of differentially expressed genes in keloids and normal skin with cDNA microarray. J Surg Res 2003;113(2):208–16.

[46] Tsou R, Cole JK, Nathens AB, et al. Analysis of hypertrophic and normal scar gene expression with cDNA microarrays. J Burn Care Rehabil 2000;21(6):541–50.

[47] Wu J, Ma B, Yi S, et al. Gene expression of early hypertrophic scar tissue screened by means of cDNA microarrays. J Trauma 2004;57(6):1276–86.

[48] Satish L, Lyons-Weiler J, Hebda PA, Wells A. Gene expression patterns in isolated keloid fibroblasts. Wound Repair Regen 2006;14(4):463–70.

[49] Smith JC, Boone BE, Opalenik SR, Williams SM, Russell SB. Gene profiling of keloid fibroblasts shows altered expression in multiple fibrosis-associated pathways. J Invest Dermatol 2008;128(5):1298–310.
[50] Wong VW, Rustad KC, Akaishi S, et al. Focal adhesion kinase links mechanical force to skin fibrosis via inflammatory signaling. Nat Med 2012;18(1):148–52.
[51] Wong VW, Paterno J, Sorkin M, et al. Mechanical force prolongs acute inflammation via T-cell-dependent pathways during scar formation. FASEB J 2011;25(12):4498–510.
[52] Januszyk M, Wong VW, Bhatt KA, et al. Mechanical offloading of incisional wounds is associated with transcriptional downregulation of inflammatory pathways in a large animal model. Organogenesis 2014;10(2):186–93.
[53] Duscher D, Maan ZN, Wong VW, et al. Mechanotransduction and fibrosis. J Biomech 2014;47(9):1997–2005.
[54] Rustad KC, Wong VW, Gurtner GC. The role of focal adhesion complexes in fibroblast mechanotransduction during scar formation. Differentiation 2013;86(3):87–91.
[55] Wong VW, Longaker MT, Gurtner GC. Soft tissue mechanotransduction in wound healing and fibrosis. Semin Cell Dev Biol 2012;23(9):981–6.
[56] Longaker MT, Rohrich RJ, Greenberg L, et al. A randomized controlled trial of the embrace advanced scar therapy device to reduce incisional scar formation. Plast Reconstr Surg 2014;134(3):536–46.
[57] Mustoe TA, Cooter RD, Gold MH, et al. International clinical recommendations on scar management. Plast Reconstr Surg 2002;110(2):560–71.
[58] Edriss AS, Mestak J. Management of keloid and hypertrophic scars. Ann Burns Fire Disasters 2005;18(4):202–10.
[59] Watters D. Molecular mechanisms of ionizing radiation-induced apoptosis. Immunol Cell Biol 1999;77(3):263–71.
[60] Guix B, Henriquez I, Andres A, Finestres F, Tello JI, Martinez A. Treatment of keloids by high-dose-rate brachytherapy: a seven-year study. Int J Radiat Oncol Biol Phys 2001;50(1):167–72.
[61] Ogawa R, Miyashita T, Hyakusoku H, Akaishi S, Kuribayashi S, Tateno A. Postoperative radiation protocol for keloids and hypertrophic scars: statistical analysis of 370 sites followed for over 18 months. Ann Plast Surg 2007;59(6):688–91.
[62] Kelly AP. Medical and surgical therapies for keloids. Dermatol Ther 2004;17(2):212–8.
[63] Conejo-Mir JS, Corbi R, Linares M. Carbon dioxide laser ablation associated with interferon alfa-2b injections reduces the recurrence of keloids. J Am Acad Dermatol 1998;39(6):1039–40.
[64] Sinclair RJ, Sinclair PJ. Carbon Dioxide laser in the treatment of cutaneous disorders. Australas J Dermatol 1991;32(3):165–71.
[65] Rusciani L, Rossi G, Bono R. Use of cryotherapy in the treatment of keloids. J Dermatol Surg Oncol 1993;19(6):529–34.
[66] Rusciani L, Paradisi A, Alfano C, Chiummariello S, Rusciani A. Cryotherapy in the treatment of keloids. J Drugs Dermatol 2006;5(7):591–5.
[67] Goldenberg G, Luber AJ. Use of intralesional cryosurgery as an innovative therapy for keloid scars and a review of current treatments. J Clin Aesthet Dermatol 2013;6(7):23–6.
[68] Liu W, Wu X, Gao Z, Song N. Remodelling of keloid tissue into normal-looking skin. J Plast Reconstr Aesthet Surg 2008;61(12):1553–4.
[69] Reish RG, Eriksson E. Scar treatments: preclinical and clinical studies. J Am Coll Surg 2008;206(4):719–30.
[70] McCoy BJ, Diegelmann RF, Cohen IK. In vitro inhibition of cell growth, collagen synthesis, and prolyl hydroxylase activity by triamcinolone acetonide. Proc Soc Exp Biol Med 1980;163(2):216–22.

[71] Wood FM. Verapamil vs steroid to prevent keloid recurrence. 2012.
[72] Jimenez SA, Freundlich B, Rosenbloom J. Selective inhibition of human diploid fibroblast collagen synthesis by interferons. J Clin Invest 1984;74(3):1112–6.
[73] Berman B, Duncan MR. Short-term keloid treatment in vivo with human interferon alfa-2b results in a selective and persistent normalization of keloidal fibroblast collagen, glycosaminoglycan, and collagenase production in vitro. J Am Acad Dermatol 1989;21(4 Pt 1): 694–702.
[74] Berman B, Flores F. Recurrence rates of excised keloids treated with postoperative triamcinolone acetonide injections or interferon alfa-2b injections. J Am Acad Dermatol 1997;37(5 Pt 1): 755–7.
[75] Wu XL, Liu W, Cao YL. Clinical study on keloid treatment with intralesional injection of low concentration 5-fluorouracil. Zhonghua Zheng Xing Wai Ke Za Zhi 2006;22(1):44–6.
[76] Nanda S, Reddy BS. Intralesional 5-fluorouracil as a treatment modality of keloids. Dermatol Surg 2004;30(1):54–6; discussion 6–7.
[77] Darougheh A, Asilian A, Shariati F. Intralesional triamcinolone alone or in combination with 5-fluorouracil for the treatment of keloid and hypertrophic scars. Clin Exp Dermatol 2009;34(2):219–23.
[78] Asilian A, Darougheh A, Shariati F. New combination of triamcinolone, 5-Fluorouracil, and pulsed-dye laser for treatment of keloid and hypertrophic scars. Dermatol Sur 2006; 32(7):907–15.
[79] Fitzpatrick RE. Treatment of inflamed hypertrophic scars using intralesional 5-FU. Dermatol Surg 1999;25(3):224–32.
[80] Le M, Naridze R, Morrison J, et al. Transforming growth factor Beta 3 is required for excisional wound repair in vivo. PLoS One 2012;7(10):e48040.
[81] Ferguson MW, Duncan J, Bond J, et al. Prophylactic administration of avotermin for improvement of skin scarring: three double-blind, placebo-controlled, phase I/II studies. Lancet 2009;373(9671):1264–74.
[82] Zielins ER, Atashroo DA, Maan ZN, et al. Wound healing: an update. Regen Med 2014;9(6):817–30.
[83] Kinoshita K, Aono Y, Azuma M, et al. Antifibrotic effects of focal adhesion kinase inhibitor in bleomycin-induced pulmonary fibrosis in mice. Am J Respir Cell Mol Biol 2013; 49(4):536–43.
[84] Mustoe TA. Evolution of silicone therapy and mechanism of action in scar management. Aesthetic Plast Surg 2008;32(1):82–92.
[85] Chen R, Zhang Z, Xue Z, et al. Focal adhesion kinase (FAK) siRNA inhibits human hypertrophic scar by suppressing integrin alpha, TGF-beta and alpha-SMA. Cell Biol Int 2014;38(7):803–8.
[86] Lagares D, Busnadiego O, Garcia-Fernandez RA, et al. Inhibition of focal adhesion kinase prevents experimental lung fibrosis and myofibroblast formation. Arthritis Rheum 2012;64(5):1653–64.
[87] Lagares D, Kapoor M. Targeting focal adhesion kinase in fibrotic diseases. BioDrugs 2013;27(1):15–23.
[88] Davison SP, Dayan JH, Clemens MW, Sonni S, Wang A, Crane A. Efficacy of intralesional 5-fluorouracil and triamcinolone in the treatment of keloids. Aesthet Surg J 2009;29(1):40–6.
[89] Gassner HG, Brissett AE, Otley CC, et al. Botulinum toxin to improve facial wound healing: a prospective, blinded, placebo-controlled study. Mayo Clin Proc 2006;81(8):1023–8.
[90] Viera MH, Amini S, Valins W, Berman B. Innovative therapies in the treatment of keloids and hypertrophic scars. J Clin Aesthet Dermatol 2010;3(5):20–6.

[91] Lim AF, Weintraub J, Kaplan EN, et al. The embrace device significantly decreases scarring following scar revision surgery in a randomized controlled trial. Plast Reconstr Surg 2014;133(2):398–405.

[92] Wong VW, Beasley B, Zepeda J, et al. A mechanomodulatory device to minimize incisional scar formation. Adv Wound Care 2013;2(4):185–94.

[93] Wong VW, Gurtner GC, Longaker MT. Wound healing: a paradigm for regeneration. Mayo Clin Proc 2013;88(9):1022–31.

Chapter 5

Process Development and Manufacturing of Human and Animal Acellular Dermal Matrices

Katie Bush[1], Arthur A. Gertzman[2]
[1]*Clinical Sciences & Research, TEI Medical, Boston, MA, USA;* [2]*Musculoskeletal Transplant Foundation, Edison, NJ, USA*

Chapter Outline

INTRODUCTION

Derived from dermis, acellular dermal matrices (ADMs) provide an ideal scaffold for regenerative medicine and tissue engineering strategies aimed at generating dermal tissue in full-thickness defects [1,2]. A variety of ADMs are commercially available that utilize different source of skin and processing methodologies. These variations in starting material and processing are known to impact the biochemical and structural properties of the final scaffold and

Skin Tissue Engineering and Regenerative Medicine. http://dx.doi.org/10.1016/B978-0-12-801654-1.00005-X

biological response [3–5]. This chapter reviews various decellularization, cross-linking, and sterilization methods used to process human and mammalian tissues focusing on ADMs developed for dermal replacement and wound healing applications. Additionally, the biological responses and tissue generation post-ADM implantation are discussed.

CLINICAL NEED

Skin is the largest organ in the mammalian body and has great importance to body form and survival. The skin is responsible for complex physiological functions that maintain homeostasis of the body with the surrounding environment including protection against environmental insult, sensory detection, and providing thermal and hydration regulation. These functions are carried out through integration of specialized cells and structures found in the dermal and epidermal components of the skin [6,7].

When adult mammalian skin is injured, the body undergoes a cascade of wound healing events aimed at reestablishing the form and function of skin. Wounds affecting the epidermis and superficial dermis typically heal within 2–3 weeks with restoration of skin function and a low risk of hypertrophic scarring. Unlike the epidermis, which heals by regeneration, the dermis heals by repair. In the absence of an established collagen matrix, granulation tissue is deposited, which is subsequently remodeled to form scar tissue [6–8]. In extensive deep burns and other complex skin wounds, the size, depth, exposed structures, or imbalance of the biochemical environment can preclude the wound from healing naturally, leaving the body susceptible to infection. In order to reduce morbidity and mortality, and to improve clinical outcome for these full-thickness defects, medical intervention is necessary [9].

Currently, the "gold standard" for surgical intervention to achieve permanent wound closure is the use of an autologous, split-thickness skin graft (STSG), which consists of the epidermis and a small portion of the papillary dermis [10]. Although this procedure accomplishes wound closure, it is often associated with contraction of the wound site, due to the lack of dermis [11,12]. Increasing the thickness of the dermal component of the graft has been shown to decrease the degree of scarring of the reconstructed site [13,14]. However, increasing the thickness of the autograft increases the trauma at the donor site and is associated with delayed donor site healing, scar formation, altered pigmentation, increased pain, and risk of donor site infection [11,15]. Additionally, patients with massive burns or trauma have limited skin availability, and therefore, the thickness of the STSG and the amount of dermal tissue transplanted must be kept to a minimum to allow for repeated harvesting and timely wound closure [8]. For patients with limited donor sites, cultured epithelial autografts (CEAs) can be used for wound coverage, but this technique often results in poor CEA take or poor-quality wound healing, due to the lack of an adequate layer of dermal tissue [16].

DEVELOPMENT OF ADMs

The recognition of the importance of the dermis in preventing wound contraction and scar formation during reconstructive procedures has led to the development of biological scaffolds focused at restoration of the dermal component of skin. Native scaffolds composed of extracellular matrix (ECM) components have increasingly gained attention as ideal constructs in regenerative medicine and tissue engineering strategies to replace lost dermal tissue [4,16–20].

The ECM of all tissues and organs is an aggregate of distinct collagenous and non-collagenous protein components arranged in a complex three-dimensional organization that influences cell mitogenesis, chemotaxis, and directs cell differentiation [2,21–23]. The composition and ultrastructure of the ECM varies among different tissues as a function of the tissue-specific biochemistries, mechanical properties, physiological needs, and required microenvironmental niche conditions [24–27].

In dermis, 85% of the ECM is collagen, with six different types detected [2,28]. Collagen types I and III are considered to be the major interstitial, fiber-forming collagens in the normal mammalian dermis [2]. These collagens form a basketweave structure in healthy skin, with collagen fibers intersecting at about 90° angles [29,30]. This architectural structure is unique to healthy dermis and is lost in physiological conditions including scarring, fibrosis, and connective tissue disorders [30]. The collagen fibrils provide a scaffold for anchoring other proteins, including other collagens as well as small proteoglycans, such as decorin and fibromodulin. Within this network elastin fibers are distributed along with glycoproteins, proteoglycans, and glycosaminoglycans [2,31,32].

Given current knowledge of the potential advantage of tissue specificity in promoting cell phenotype and selected cell functions, the ideal dermal substitute is one that most closely approximates the structure and function of the native ECM it is replacing, thus replacing "like with like"[1,3,5]. Human and other mammalian donor skin provide the natural source of the native ECM of dermis and can be used as a biological scaffold to provide the proper biologic and structural cues needed to induce dermal tissue generation at multiple hierarchical levels [17,18,27].

ADM REQUIREMENTS

In order to utilize the naturally occurring ECM of the dermis as a biological scaffold, the skin must undergo a decellularization process to remove all immunogenic factors from the donor tissue [33,34]. Although ECM components of the skin are highly conserved and well tolerated among different species, cellular materials from a donor are detected as foreign to the host [34–36]. Improper removal of the immunogenic materials will initiate both innate and adaptive immune responses leading to ultimate rejection, as seen in unprocessed cadaveric and mammalian skin used as a temporary dressing in excised burn wounds [8,17,37]. Antigenic materials include cells, cell debris, and chromosome fragments from cells that express foreign major histocompatibility complex (MHC) antigens. Cells present in the dermis that express these complexes include keratinocytes present in the epidermis and fibroblasts and endothelial cells present in the dermis [33,38,39].

Challenges exist in the decellularization of ADMs. Removal of all cellular and nuclear material from the ECM requires that the dermal tissue undergoes a series of processing steps that can affect the dermal collagen architecture and native biochemistry [5,16,27,40]. Additionally, structural and biochemical alterations can occur during viral inactivation, crosslinking, preservation, and terminal sterilization. Retaining the dermal collagen architecture of the ECM is important, as the natural basket weave pattern of the collagen fibers occupying the dermis may aid in directing the generation of a naturally oriented dermis by the infiltrating fibroblasts, thus inhibiting the deposition of abnormal collagen patterns typical for scars [16]. At the nanoscale, maintaining the native biochemistry and appropriate binding sites for various growth factors, cytokines, and other membrane-bound cellular components is critical for cell–cell interactions, cell–ECM interaction, and new dermal tissue formation [41,42]. Furthermore, any damage to collagen or non-collagenous proteins, growth factors, and glycosaminoglycans may promote a strong inflammatory profile, leading to rapid resorption of the implant [3].

PROCESSING METHODOLOGIES

The goal of decellularization is to remove antigenic material, while preserving the ECM biochemistry and structure. Methods of processing including decellularization, crosslinking, preservation, and sterilization of ECMs have been extensively described in the literature [4,27,43]. Depending on the ADM source, xenograft or allograft, different methods have been employed. This section is aimed at providing an overview of the most commonly used methods, including physical, chemical, and enzymatic approaches for decellularization of ADMs developed specifically for dermal replacement.

Tissue Acquisition

The recovery technique of skin acquired from either animal or human source has a significant effect on the final properties of the implantable ADM. The native proteases and/or the endogenous bacterial load (bioburden) in the tissue will begin enzymatic degradation of the collagen, compromising the dermis. Collagen molecular weight and integrity will change with such enzymatic digestion and can be controlled by close monitoring of process parameters, such as time and temperature [44].

Animal Dermis

For animal tissue, processing of the hide begins in the abattoir setting. Hides from fetal/neonatal, as well as adult, animals can be sourced as a byproduct of the meat industry. The animal is humanely killed, and the hide is recovered by manual separation. The hide can then be frozen or stored in a concentrated salt solution until further processed [45].

Sourcing tissue from animals allows for the manufacturer of the ADM to select the specific age of the animal or stage of development [37]. This is an important factor to consider, as it relates to tissue consistency as well as tissue composition. Fetal and neonatal mammals have an enhanced healing capacity over that of adult mammals. It has been suggested that this greater regenerative ability may be attributed to observable differences between the fetal/neonatal and adult dermal ECM composition [31,37,46]. During fetal development, the localization and expression of several proteins is different from that of adult skin [47–49]. In the fetal dermis, type III collagen initially accounts for over half of the total collagen, serving as a factor in the deposition and fiber growth of type I fibrils during fetal development as well as in wound healing [21,46,50]. In mammalian fetal dermis, the ratio of type I/type III collagen is 3:1, which reduces to about 5–6:1 in adult dermal tissue [2,48,51]. Additionally, differences in elastin expression exist between the fetal and adult ECM, as well as the localization of fibronectin and glycosaminoglycans including chondroitin sulfate.

Human Dermis

Human allograft dermis is recovered by not-for-profit tissue banks obligated to follow the procedures of both the Food and Drug Administration (FDA) and the American Association of Tissue Banks (AATB). Donor eligibility criteria is set forth by the FDA and AATB and includes the screening of a combination of 18 disease states or viruses. Procedures known as Current Good Tissue Practices (CGTPs) mandate tissue handling conditions, processing, and storage controls [52–54].

Full family consent to recover the human cadaveric tissue is a legal requirement. Donor ages are controlled; however, tissue banks vary in the age limits for accepting tissue donors. While no standards for donor age have been set by either the FDA or AATB, leading tissue banks in the United States accept donors ranging in age from 16 to 65 years. Recovery is typically performed in an operating room environment using aseptic processes, and the tissue is usually full thickness. Removal of the skin from the cadaver is performed with dermatomes by trained technicians. The skin is stored in a saline solution and shipped on wet ice at 0 °C. Processing is typically performed in an aseptic manor using ISO Class 4 clean rooms [52,54] (Table 1).

Decellularization Techniques

The first studies evaluating techniques to remove the antigenic components of skin used trypsin, freeze–thawing, and lengthy incubations, during which endogenous enzymes acted on the dermis [55]. These initial attempts

TABLE 1 Processing Methods of Acellular Dermal Matrices

	Process and Methods		Mode of Action	Examples
Decellularization	Physical methods	Snap freezing	Formation of intracellular ice crystals disrupt cell membrane	High-capacity freezer
		Pressure	Application of direct pressure to burst cells	Pressure/vacuum pulse equipment
		Mechanical agitation	Agitation causes cell lysis in presence of chemical agents	Low shear agitation
	Chemical methods	Detergents (ionic, nonionic, and Zwitterionic)	Affects cell wall	Sodium dodecyl sulfate and sodium deoxycholate (ionic)
				Triton X-100 (nonionic)
		Acids or bases	Affects cell wall	Acids—peracetic acid, deoxycholic acid, acetic acid (acids)
				Sodium hydroxide and calcium hydroxide (bases)
		Hypotonic or hypertonic solutions	Lyses cells by differences in osmotic pressure	Distilled water (hypotonic)
				1 N NaCl (hypertonic)
		Alcohol or other solvents	Lyses by dehydration of cells	Ethanol, isopropanol
	Biological methods	Chelating agents	Removes essential metal ions from access to cells	Ethylenediaminetetraacetic acid (EDTA)
		Enzymes	Targeted chain scission in protein backbone	Trypsin, ficin, proteases

	Process and Methods		Mode of Action	Examples
Crosslinking	Physical methods	Irradiation	Induces free radicals with attendant amino acid crosslinks	Gamma (Cobalt 60 isotope) or electron beam (Van DeGraaf Generator)
		Ultraviolet light	Low-energy photons create crosslinks	Black light lamps
			May be associated with oxidation	
	Chemical methods	Aldehydes, isocyanates, and carbodiimides	Chemical agents interact with device material to form covalent bonds	Formaldehyde, gluteraldehyde, genepin, diisocyanate, 1-Ethyl-3-(3-dimethylaminopropyl)carbodiimide
Preservation	Physical methods	Freezing at −4 °C to −10 °C	Slows any enzymatic or metabolic degradation of extracellular matrices or collagen	Commercial freezers
		Cryopreservation −80 °C to −196 °C		Liquid nitrogen in insulated flasks (dewar)
		Lyophilization	Removes water by application of high vacuum to frozen tissues	Freeze drier equipment
	Chemical methods	Wet preservation	Inhibition of hydrolysis	Storage of tissues in aqueous alcohols or antioxidants such as sodium benzoate

Continued

TABLE 1 Processing Methods of Acellular Dermal Matrices—cont'd

		Process and Methods		Mode of Action	Examples
Sterility assurance	Terminal sterilization	Physical methods	Gamma radiation	High-energy (gamma ray) destruction of cell DNA and cell wall integrity	Cobalt 60 isotope
			E-beam radiation	High energy (electron, beta ray) destruction of DNA and cell wall integrity	Van DeGraaf Generator
			Ethylene oxide	Chemical modification of cell proteins by ethoxylation and destroys cell metabolism	Commercial ethylene oxide gas
		Chemical methods	Alcohols, hydrogen peroxide, peracetic acid	Lyses by dehydration of cells Peroxide oxidizes cell walls	Commercial reagents, sterile
	Aseptic processing	Aseptic technique	Processes outlined in current good tissue practice and additional requirements for manufacturers of HCT/Ps	Prevent contamination and cross-contamination including control to prevent the introduction, transmission, or spread of communicable agents	Processing in clean rooms class 10 with chemical sterilization procedures

produced ADMs that, when implanted, induced immune responses and resulted in poor graft survival [55,56]. Although conceptually these attempts followed the principle of disrupting the cell membrane and rinsing away the cellular content, more effective and controlled processing techniques have been developed to improve upon these methods to remove antigenic materials.

Physical Methods

Physical methods rely on the access of processing chemicals to the ECM. Materials that are less dense with more open pore structure are easily penetrated by cleaning agents. Thicker and denser materials, such as dermis, require greater agitation and processing times to allow the reagents to penetrate and then have the solubilized materials diffuse out of the matrix [4,18,19].

Methods such as pressure, agitation, sonication, and thermal shock have been used to facilitate decellularization of tissues and organs [4,43]. Rapid controlled rate freezing protocols have been used for the decellularization of soft tissues such as ligaments and tendons, as well as nerves. Freezing and thermal shock techniques induce cell lysis, resulting in the exposure of the cell nuclear material to the chemical agents. Pressurizing methods have been found to be effective on tissues characterized with more open network and ECM, such as the lungs and the liver [43].

Sonication and agitation methods are most commonly used in conjunction with enzymatic and chemical methods in protocols for decellularization of commercial ADMs [4,43]. Examples of agitation methods include the use of an orbital shaker, roller, or stir plate. Alternatively, an ultrasonic cleaner can be used to disrupt cells. The rate and duration of agitation or sonication frequency are selected and optimized based on the tissue composition, structure, and quantity, as well as reagent used [43].

Chemical Methods

Chemical methods used in decellularization protocols include alkaline or acid treatment, detergents, including ionic and nonionic, as well as Zwitterionic detergents, hypotonic or hypertonic (hypertonic saline, 1% aqueous NaCl) treatment, and chelating agents, such as ethylenediaminetetraacetic acid (EDTA). Chemical methods are often used in combination to increase decellularization efficacy [57].

Detergents

Detergents effectively solubilize the phospholipid cell membrane, resulting in cell lysis. Detergents also serve to lyse the cell wall of the present bacteria. Saline (1 N) or pure water will also lyse cells [58]. Detergents, in the form of surfactants, are commonly used. Nonionic, ionic, and Zwitterionic detergents work by solubilizing cell membranes and disruption of nuclear material [4,18,43].

Triton X-100 and N-lauroylsarcosinate are the most commonly used nonionic detergents for decellularization and sodium dodecyl sulfate and sodium deoxycholate are the most widely used ionic detergents [55,59].

Analysis of DNA content post-processing of ADMs that utilize nondenaturing detergents demonstrate 97% reduction in DNA content as compared to normal dermis. Implantation studies also demonstrate the ability of fibroblasts to infiltrate the entire surface of the dermis with minimal inflammatory cells present [59]. In a study evaluating the use of different detergents on decellularization, scaffold composition, mechanical properties, and cytocompatibility, it was found that varying detergents were successful at decellularization of dermal tissue and had similar retained amounts of growth factors and glycosaminoglycan components and mechanical strength in response to multiaxial loading remained consistent to unprocessed dermal tissue. However, differences in the ability of the matrix to support cell growth did exist between the two treatments (trypsin/SDS/TritonX-100 and trypsin/TritonX-100) [60].

Acids or Bases

Acid or base treatments are used to solubilize the cell membrane and nuclear components. Examples of acids used in decellularization include acetic acid, peracetic acid, and deoxycholic acid. Examples of bases include ammonium hydroxide, sodium hydroxide, and calcium hydroxide. pH extremes in either the acid or alkaline range are also known to inactivate viruses, a necessary step in processing mammalian dermal tissues [61]. Incubation times in acidic or basic treatments are critical to determine, as low incubation times can fail in the complete removal of cells causing an induction in an inflammatory infiltrate after implantation, and longer periods of incubation time can cause visible damage to the ECM components [62]. Thus, the differences in incubation time can drastically affect the biological response postimplantation.

Hypotonic or Hypertonic Treatments

Hypotonic or hypertonic treatments are used to facilitate cell plasmolysis or cytolysis; however, they are typically used in combination with other chemical treatments in order to effectively remove DNA and cellular debris [58].

Alcohol and Other Solvents

Alcohols, such as glycerol, can be used to dehydrate cells to facilitate cell lysis [4,43]. Other alcohols are used in protocols for lipid removal, and include isopropanol, ethanol, and methanol [4,43]. Acetone is another solvent used for lipid removal as part of decellularization. Typical commercial processes use detergents or an alcohol or acetone solution first, in a strategy to remove the viscous and hydrophobic lipid components. With the lipids reduced or eliminated, the cells are more susceptible to lysis and physical removal by saline or successive water washes.

Biologic Methods

Chelating agents, such as EDTA and ethylene glycol tetraacetic acid (EGTA), dissociate cells from the ECM by binding to metal ions that play a role in cell adhesion [4,27,43]. Enzymes used in decellularization include trypsin, nucleases, collagenase, lipase, dispase, thermolysin, and α-galactosidase [4]. In decellularization of dermis, collagenases should not be used because the enzyme will break the peptide bonds in collagen destroying the intact collagen structure. Although enzymes are effective due to their high specificity for targeting cell proteins and components; they should be used in conjunction with other decellularization treatments to optimize cell removal [57].

Decellularization Criteria

Although no formal criteria have been established by the FDA to classify a tissue as acellular, methods evaluating the presence of nuclear material are used to determine if cell residues exist within the ECM. A common method is evaluating tissue sections stained with hematoxylin and eosin (H&E) for the lack of visible nuclear material [63]. Based upon findings of studies in which adverse cell and host responses have been avoided, a minimal criteria to satisfy the intent of decellularization has been proposed and used when evaluating new decellularization techniques. This criteria includes H&E or 4′6-diamidino-2-phenylindole (DAPI) tissue section staining, in addition to evaluating the dsDNA per mg ECM dry weight (<50 ng acceptable) and DNA fragment length (<200 bp DNA fragment length acceptable) [4,60] (Figure 1).

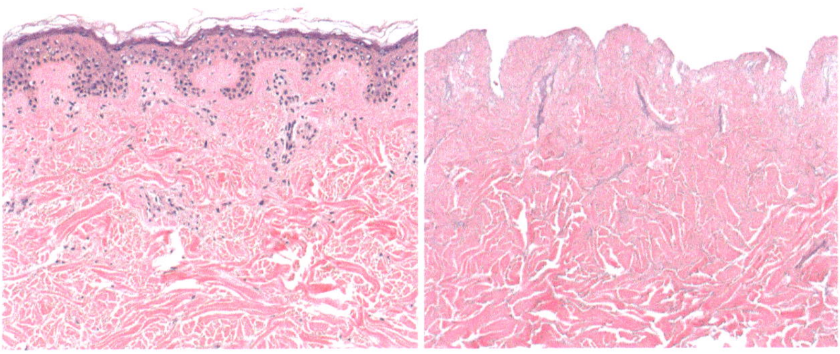

FIGURE 1 **Histological analysis of pre and postdecellularized human skin.** H&E staining was performed on pre and postprocessed human skin to demonstrate decellularization efficacy. Hematoxylin (blue) was used to stain for cellular nuclei and eosin to stain for collagen fibers. This figure is an example of human skin from the same donor preprocessed (left) and postprocessed (right) resulting in a commercially available allogeneic ADM. *Images courtesy of MTF: FlexHD® Acellular Dermal Matrix.*

Crosslinking

Crosslinking methods are often employed in the processing of ADMs, with the goal of improving degradation rates and mechanical properties for the scaffold materials. Crosslinking can be achieved using methods such as chemical reagents, high temperature, radiation, or ultraviolet light [64,65]. Regardless of the modality, the process of crosslinking adds nonnative crosslinks between polymer chains in the ADM which reduces the susceptibility of the collagen molecule to enzymatic degradation; affects growth factor binding to collagen, cellular attachment, and migration; and also can cause the implant to be recognized as a foreign body by the host species [42,66,67]. Additionally, crosslinking increases the tensile strength and stiffens the matrix (increases the modulus) and decreases the porosity of the matrix, while increasing the bulk density.

Although crosslinking increases the persistence of the material postimplantation, it decreases the ability of the matrix to be penetrated by cells and be revascularized, versus ADMs that are noncrosslinked [67–70]. Hence, crosslinking creates adverse effects to the detriment of the physical and biological performance of ADMs. Consequently, no commercially available ADM for dermal replacement utilizes crosslinking methodologies [66].

Preservation of ADMs

ADMs are preserved to facilitate shipping and storage. Preservation is required to reduce the activity of any residual endogenous microbiological materials present before sterilization and to slow the reactivity of enzymes or other endogenous active agents that may still be present in the ADMs after final packaging. Preservation methods include cryopreservation, wet freezing, dry freezing (lyophilization), and hydration with preservative agents. Depending on methodology, shelf life of an ADM can range from 18 months to 5 years. Products that are stored at room temperature offer an off-the-shelf advantage, which includes the avoidance of complex storage conditions as well decreased preparation time prior to implantation [1,71].

Wet Preservation

Wet preservation is accomplished for ADMs by hydrating the final decellularized tissue using a water and alcohol solution. The tissue is hydrated and is at equilibrium with the water component, rendering the tissue pliable. The alcohol component inhibits hydrolysis by the water, preventing the initiation of collagen degradation by any residual endogenous enzymes. The alcohol will also inhibit growth of any trace microorganisms present in the final tissue [72].

Cryopreservation

Cryopreservation refers to slowly freezing an ADM along with a cell preservative, such as dimethylsulfoxide (DMSO). The rate of freezing is specific, typically 1–2 °C per minute to a final temperature of −80 °C or to liquid nitrogen conditions (−196 °C). Such low temperatures greatly reduce the reactivity of any degradative enzymes and stabilize the material [73]. Clinically, no difference has been reported in the host-response postimplantation between cryopreserved and prehydrated human ADMs [74].

Freezing

Freezing typically achieves temperatures of about −4 to −10 °C. ADMs frozen with liquid nitrogen achieve a longer shelf life than products frozen using conditions of −4 °C to −10 °C or −80 °C. Both freezing and cryopreservation require special facilities, such as freezers or insulated Dewars to store the ADMs [75].

Lyophilization

Lyophilization, or freeze drying, is a method to achieve tissue preservation allowing product shipping and storage at room temperature. The tissue is held in a water vapor permeable container and frozen to about −40 °C. Vacuum is then applied and the water, in the form of ice is sublimated out of the matrix of the ADM. By controlling freeze temperatures, pore morphology can be controlled [76]. Advantages of lyophilization include longer shelf life and reduced loss of intrinsic growth factors; however, disadvantages include the reduction in the thickness of the scaffold and slight compaction of the collagen fibers.

Sterilization

Sterilization is defined as the process of killing all microorganisms, including viruses. A sterility-assurance level (SAL) is defined as the probability of an item being nonsterile after it has been exposed to a validated terminal sterilization process. Implantable medical devices are terminally sterilized to achieve a SAL of 10^{-6}, which implies a probability of finding not more than one viable microorganism in one million sterilized items of the final product [77]. Additionally, sterility assurance must be supported by sterility testing using validated and verified test methods, such as those found in the United States Pharmacopeia (USP) Chapter 71. An ADM labeled as "sterile" meets these requirements and is labeled with the SAL [77].

Terminal Sterilization

Radiation

Radiation is the most effective and widely validated sterilization process in industry [65]. The sterilization energy operates by either disrupting the cell wall of bacteria and viruses or destroying the nuclear DNA of the microbiological organism. Penetrating radiation can have the negative effect of breaking collagen bonds, while simultaneously crosslinking the collagen chain [78]. Radiation types include gamma ray or electron beam (beta ray). Either type is performed at a radiation dose level demonstrated to penetrate the sealed, final package containing the ADM. Currently, typical doses of sufficient irradiation for collagen range from 5 to 25 kGy, with lower doses less likely to induce damage to the material [65,78].

Ethylene Oxide

Ethylene oxide (EtO) gaseous treatment is accomplished at elevated humidity to reactivate any spore form of the microorganisms. The hydrated organisms react with the EtO to disrupt the cell wall, destroying the organism. Ethylene oxide gas will also react with the collagen molecules of the ADM creating a significant change in surface chemistry. A major advantage of EtO treatment over irradiation is that it can be performed at relatively low temperatures, which is important in avoiding denaturation of collagen [65,78].

Terminal Sterilization Optimization

Although sterilization procedures free ADMs of viable microorganisms, the effects of terminal sterilization can affect biomechanical properties and material stability [78]. However, no consensus exists on the definitive effects of sterilization methods, since testing is performed on various collagen constructs using a multitude of sterilization conditions including radiation doses and environmental settings [78]. Additionally, the tests used to evaluate the effects are often in vitro enzymatic digestions, with the results needing to be interpreted carefully since they cannot reflect the in vivo situation completely [65]. Optimization of sterilization conditions for ADMs can be performed to achieve the appropriate SAL and limit or minimize the negative effects, including collagen denaturation or foreign body reaction postimplantation.

Aseptic Processing

To avoid potential negative effects and lengthy studies to optimize terminal sterilization, an alternative to terminal sterilization is aseptic processing of the dermal tissue. Aseptic processing occurs in clean room facilities with controlled procedures and environmental conditions. ADMs processed aseptically must meet the USP standards for sterility assurance; however, while these products are in fact sterile, as mandated by the FDA, they cannot be labeled as sterile, but only as aseptically processed and sterility tested [54,79].

Viral Inactivation

The methods of EtO and radiation, when combined with chemical decellularization, are effective in devitalizing bacteria, spores, and fungi. However, natural dermal tissues from either human or animal sources will contain a range of viruses that must also be inactivated. Two broad classes of virus are distinguished by the presence or absence of a viral jacket. This barrier is usually composed of a phospholipid. Such lipid barriers are readily susceptible to disruption by simple detergent treatments and are removed to levels greater than 10^{-6}. However, many pathogenic viruses do not have the jacket structure and are more difficult to remove. Chemical treatments such as hydrogen peroxide and sodium hydroxide are effective, as is gamma treatment at doses above 25 kGy. Viral reduction procedures may degrade the underlying matrix chemistry, as noted. Successful ADMs must set a careful balance between microbiological (bacterial and viral) reduction and the protection of the matrix properties [80].

BIOLOGICAL RESPONSES TO ADMs

The intended purpose of an ADM in a full-thickness wound is to provide an immunologically tolerated dermal ECM that is capable of being repopulated and revascularized by the host, resulting in the generation of dermal tissue. The implantation of an ADM can alter the default mammalian healing response to tissue injury by modulating the reparative phases of healing including hemostasis, inflammation, granulation tissue formation, and remodeling [81,82]. In order for the response to be altered and transitioned toward a more regenerative response, the processing techniques used to produce the ADM must render the dermis free of cellular material and retain the biological integrity of the native dermal ECM components [3,5,16,24,27,37,40] (Figure 2).

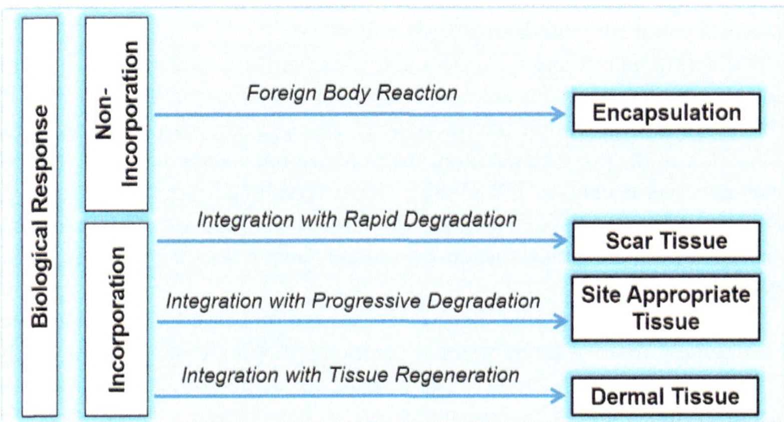

FIGURE 2 **Biological responses postimplantation of an ADM.** The processing techniques used to produce an ADM ultimately dictate the response postimplantation. These biological responses can be categorized as nonincorporation or incorporation. A nonincorporation response is one in which cells are unable to penetrate the material and detect the implant as a foreign object. This leads to a heavy collection of inflammatory cells around the implant periphery and the deposition of a layer of dense connective tissue encapsulating the implant. An incorporation response is one in which cells can penetrate the matrix and based on levels of inflammatory response and rate of matrix degradation can lead to scar tissue formation, site appropriate tissue deposition, or dermal tissue generation.

ADMs that undergo processing including crosslinking, poor-preservation techniques, and nonoptimal sterilization procedures which change the ultrastructure, composition, and surface topology of the ECM components have been found to elicit an intensified fibrotic response [37,83]. Following implantation, host cells recognize the material as foreign, and a collection of macrophages and neutrophils surround the periphery of the implant, without penetration into the material. Over time, the persistent inflammatory stimuli lead to a chronic inflammatory response, characterized by the presence of multinucleated foreign body giant cells and a dense connective tissue encapsulating the implant [69,82,84].

In contrast to encapsulation of the implant without cellular penetration, ADMs can also exhibit a fundamentally different set of host–implant interactions, where the material is penetrated by cells, the ADM is degraded, and ECM proteins are deposited to replace the resorbed implant [3,37]. The main determinants of the degradation profile of an implanted ECM material include the host inflammatory response to remaining cellular components and/or altered or denatured proteins, growth factors, or glycosaminoglycans.

Materials that can be penetrated by cells and can be incorporated result in various tissue responses, including the formation of scar tissue, site appropriate tissue, or histologically normal dermal tissue [3,24,37]. Scar tissue is the result of implant degradation with no divergence from the normal wound healing repair mechanism. Postimplantation, cells rapidly penetrate and populate the matrix and break down the material. The implant is resorbed and the body replaces the implant with disorganized scar tissue.

Site appropriate tissue forms through constructive remodeling of the scaffold. This process includes cellular infiltration of the matrix accompanied by ECM scaffold degradation, recruitment of endogenous progenitor cells, and modulation of the pro-inflammatory macrophage phenotype to a tissue remodeling and repair phenotype [37,85–87]. The material that is degraded is replaced with host connective tissue. Although the formation of site appropriate tissue is a departure from the typical response to tissue injury, it does not represent tissue regeneration [37,88].

It has been demonstrated that ADMs that are processed in such a manner to adequately remove immunogenic components without destruction of the ECM components are capable of generating tissue histologically similiar to normal dermis [3,89–92]. Postimplantation, the ADM undergoes progressive in vivo remodeling characterized by a transient infiltrate of inflammatory cells, followed by mesenchymal cells, and vascular development within the implanted collagen matrix [89]. Further investigation is warranted to evaluate the long-term characteristics of the generated tissue including the persistence of the collagen dermal fibers found in the implanted ADM (Figure 3).

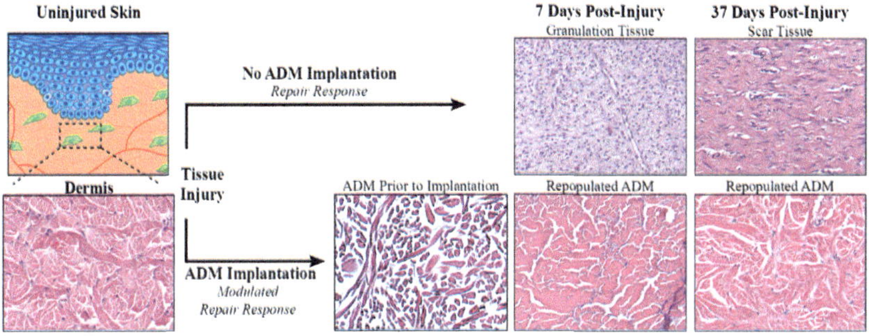

FIGURE 3 **Ability of an ADM to modulate the repair response generating a tissue with features consistent with uninjured dermis.** A full-thickness porcine wound model was used to compare the normal healing response to the biological response of an implanted fetal bovine ADM (PriMatrix, TEI Biosciences Inc.). Biopsies at 7 and 37 days postinjury were evaluated using H&E staining. At 7 days, granulation tissue consisting of cells, blood vessels, and loose connective tissue was observed in the wound beds left to heal by secondary intent. In contrast, a mature dermal collagen fiber architecture repopulated with cells was observed at the ADM-treated sites. Thirty-seven days postimplantation the granulation tissue was remodeled into scar tissue characterized by small bundles of collagen aligned in a parallel fashion to the epidermis. In the ADM-treated wounds, the dermal collagen fiber architecture of the fetal bovine ADM implant persisted, and the fibers were reconstituted with a cellular and vascular density consistent with uninjured dermis. Comparison of the tissue observed in the control group versus the tissue-generated post-ADM implantation suggests the ability of the ADM to modulate the repair response. *Figure courtesy of TEI Biosciences Inc.: PriMatrix.*

CLINICAL USE

In 1995, the first successful implantation of a processed acellular, structurally intact, dermal matrix in a full-thickness pig wound was demonstrated, suggesting the potential for ADMs as a permanent dermal substitute to augment the performance of an STSG in the closure of full-thickness wounds [33]. Since then, ADMs have been used to reconstruct full-thickness wounds in burn surgery, demonstrating the ability to reduce the thickness of donor skin required for STSGs and improve donor site healing time [8,15,93], promote cultured epithelial autograft cell take [94,95], and generate a tissue bed for STSG application in a second surgical procedure over exposed tendon and/or bone [96]. Biopsies demonstrate that the acellular dermis supports fibroblast infiltration, neovascularization, and epithelialization in the absence of an inflammatory response [8,90] (Table 2).

Long-term follow-up studies evaluating skin quality one year post-ADM and delayed STSG versus immediate STSG-treated wounds indicated that wounds grafted with the ADM and delayed STSG were significantly more elastic and had better scar quality than the wounds treated with STSG alone [97]. Range of motion of jointed areas treated with ADM showed restoration of joint movement ranges over 75% for 30 of 31 treated areas [95]. Studies have also extended the use of ADMs in the reconstruction of released scar contractures,

TABLE 2 Commercially Available ADMs

Trade Name and Manufacturer	Source Tissue	Processing			Product Features			
		Decellularization	Preservation	Sterilization	Storage	Application	Regulations	Indications
PriMatrix® TEI Biosciences Inc., Boston, MA	Fetal bovine skin	Proprietary technology	Lyophilized	Ethylene oxide Sterile to a SAL of 10^{-6}	Room temperature for 5 years	Nonsided product Intraoperative rehydration (1–2 min) with Sterile saline Available solid, fenestrated, and 2:1 meshed	Regulated as a medical device by CDRH (FDA) 510(K)	Management of wounds that include • Partial and full-thickness wounds • Surgical wounds • Trauma wounds • Tunneled/ Undermined wounds • Pressure, diabetic, and venous ulcers
PriMatrix Ag® TEI Biosciences Inc., Boston, MA					Room temperature for 34 months			
AlloDerm® Regenerative Tissue Matrix LifeCell Corporation Branchburg, NJ	Donated allograft adult human skin	Proprietary technology that includes the treatment with buffered salt solution followed by a series of washes with nondenaturing detergent solutions. Antibiotics and Polysorbate 20 are additionally used in processing the tissue	Stored in cryoprotective solution and lyophilized	No—aseptically processed	1 °C–10 °C for 2 years following date of processing	Sided product Intraoperative rehydration (10–40 min)	Regulated as human tissue for transplantation by CBER (FDA) Processed and marketed in accordance with the FDA's requirement for banked human tissue (21 CFR, Part 1270 and Part 1271) and standards for tissue Banking of the AATB	To be used for repair or replacement of damaged or inadequate integumental tissue or for other homologous uses of human integument
AlloDerm® Regenerative Tissue Matrix Ready to Use LifeCell Corporation Branchburg, NJ			Stored in preservative solution	Electron beam irradiated Sterile to a SAL of 10^{-3}	Room temperature for 2 years following date of processing	Sided product Intraoperative rinses are required		
AlloPatch™ Pliable Allograft Dermal Matrix Musculoskeletal Transplant Foundation Edison, NJ		Proprietary technology utilizing antibiotics	Packaged in ethanol solution	No—aseptically processed	Room temperature for 2 years following date of processing	Sided product Intraoperative rinses in a sterile solution are required		
BellaDerm® Acellular Dermis Musculoskeletal Transplant Foundation Edison, NJ								

Trade Name and Manufacturer	Source Tissue	Processing			Product Features			
		Decellularization	Preservation	Sterilization	Storage	Application	Regulations	Indications
DermACELL LifeNet Health, Virginia Beach, VA		Proprietary technology utilizing antibiotics, detergent, and endonuclease	Glycerol	Gamma irradiation Sterile to a SAL of 10^{-6}	Room temperature for 2 years following date of processing	Sided product Intraoperative rinses in a sterile isotonic solution is optional		
DermaPure™ Tissue Regenix Wound Care Inc., San Antonio, TX		dCELL® technology utilizing antibiotics and processing reagents including SDS, DNase, RNase, and glycerol	Glycerol	Gamma irradiated Sterile to a SAL of 10^{-6}	Room temperature for 2 years following date of processing	Sided product Intraoperative rinses in a sterile isotonic solution are required (10–15 min)		
Graftjacket® Regenerative Tissue Matrix LifeCell Corporation Branchburg, NJ		Proprietary processing utilizing antibiotics and Polysorbate 20	Lyophilized	No—aseptically processed	Room temperature for 2 years	Sided product Available 1:1 meshed Intraoperative rinses in a sterile isotonic solution are required (10–15 min)		
Epiflex® DIZG Berlin, Germany	Donated allograft adult human skin	Proprietary technology utilizing anionic detergent solution	Lyophilized	No–aseptically processed	Room temperature for 5 years following date of processing	Sided product Intraoperative rehydration with isotonic saline	Licensed pharmaceutical according to the German Pharmaceuticals Act	Soft tissue reconstruction
Glyaderm® Euro Skin Bank Huidbank, Beverwijk, The Netherlands		Proprietary technology utilizing low concentrations of NaOH without freezing	85% Glycerol	No—aseptically processed	2 °C–8 °C for 36 months	Sided product Intraoperative rinses in sterile water or saline are required (minimum of 10 min)	Regulated in the EU by the directives on human tissue and cells (2004/23/EG, 2006/17/EG, and 2006/86/EG)	Deep burns and reconstruction of scars, chronic wounds

demonstrating that improved range of motion can be achieved when ADMs are used with an STSG [90,98,99].

In addition to use in burn surgery, ADMs have recently been incorporated in the reconstructive ladder, with their use as a strategy to close a variety of complex defects [100]. Wounds resulting from avulsive trauma with exposed critical structures, such as exposed bone without intact periosteum and exposed tendon without intact paratenon, typically require free tissue transfer, local flap coverage, or limb amputation. ADMs coupled with a delayed STSG have been used to close these wounds, helping the surgeon simplify these reconstructions and avoid associated morbidity and salvage limbs marked for amputation [96,100,101].

ADMs have also been shown to be effective in managing chronic wounds, with excellent healing outcomes in both prospective and retrospective studies. These wounds are precluded from healing, due to biochemical imbalances detaining the wound typically in the inflammatory phase of healing [102]. Post-debridement of the wound, the ADM can assist in tissue formation and provide the base for reepithelialization from the healthy wound margin. In a prospective randomized control trial evaluating the use of an ADM versus moist wound therapy to heal chronic diabetic foot ulcers, it was found that 70% of the ADM-treated subjects healed in 12 weeks versus 46% healing of the control group [103]. A prospective study evaluating similar wounds reported the percent healing of chronic diabetic foot ulcers as high as 76% in 12 weeks [104]. It has also been demonstrated that the use of an ADM in other chronic wounds, including venous stasis and pressure ulcers, can help to achieve successful wound closure despite complicating patient factors such as advanced age, obesity, peripheral vascular disease, and concomitant use of anticoagulants [105].

SUMMARY

The recognition of the importance of reconstructing full-thickness wounds with dermal tissue leads to the development of ADMs. ADMs offer an advantage over other materials and tissues for use as dermal substitutes, due to the intrinsic biologic and structural organization of the dermal ECM. However, because of the allogeneic or xenogeneic nature of the starting tissue, all cellular components must be removed in order to avoid rejection of the material.

Over the past two decades, research has been conducted evaluating processing techniques and the effects on decellularization and retention of the features of the dermis. It should be appreciated that ADMs are not equivalent, and that remaining cells and damage to the structure and composition of the ADM directly impact the biological response postimplantation and resulting clinical outcome.

The use of ADMs in burn surgery has demonstrated improved functional and cosmetic results over that of standard autografting procedures. Furthermore, ADMs have been used to simplify complex reconstructions, salvage limbs marked for amputation, as well as aid in the closure of chronic wounds. Interdisciplinary cooperation between researchers, clinicians, and surgeons have identified that ADMs are an excellent platform technology, and further investigation of modification strategies to incorporate and deliver specific peptides, growth factors, or antimicrobial molecules [106], as well as combination with living cell technologies is warranted. Development of these novel functionalized matrices, designed to promote specified healing responses, could expand the utility of ADMs and further improve clinical outcomes.

REFERENCES

[1] International Consensus. Acellular matrices for the treatment of wounds. An expert working group review. London: Wounds International; 2010.

[2] Uitto J, Olsen DR, Fazio MJ. Extracellular matrix of the skin: 50 years of progress. J Invest Dermatol 1989:61–77.

[3] Cornwell KG, Landsman A, James KS. Extracellular matrix biomaterials for soft tissue repair. Clin Podiatr Med Surg October 2009;26(4):507–23.

[4] Crapo PM, Gilbert TW, Badylak SF. An overview of tissue and whole organ decellularization processes. Biomaterials 2011;32(12):3233–43. Elsevier Ltd.

[5] Chan BP, Leong KW. Scaffolding in tissue engineering: general approaches and tissue-specific considerations. Eur Spine J 2008;17(Suppl. 4):467–79.

[6] Singer A, Clark R. Cutaneous wound healing. N Engl J Med 1999;341(10):738–46.

[7] Clark RA. Wound repair: overview and general considerations. The molecular and cellular biology of wound repair. 2nd ed. 1995. p. 3–50.

[8] Wainwright D. Use of an acellular allograft dermal matrix (AlloDerm) in the management of full-thickness burns. Burns 1995;21(4):243–8.

[9] Guo S, Dipietro LA. Factors affecting wound healing. J Dent Res 2010;89(3):219–29.

[10] Hinshaw JR, Miller ER. Histology healing split-thickness, full-thickness autogenous skin grafts and donor sites. Arch Surg 1965.

[11] Corps B. The effect of graft thickness, donor site and graft bed on graft shrinkage in the hooded rat. Br J Plast Surg 1969;22(2):125–33.

[12] Kraemer M, Jones T, Deitch E. Burn contractures: incidence, predisposing factors, and results of surgical therapy. J Burn Care Rehabil 1988;9(3):261–5.

[13] Rudolph R, Suzuki M, Guber S. Control of contractile fibroblasts by skin grafts. Surg Forum 1977;28:524.

[14] El Hadidy M, Tesauro P, Cavallini M, Colonna M, Rizzo F, Signorini M. Contraction and growth of deep burn wounds covered by non-meshed and meshed split thickness skin grafts in humans. Burns 1994;20(3):226–8.

[15] El-hadidy MR, El-hadidy AR. Clinical and histological evaluation of acellular allograft dermal matrix in full-thickness burns. Egypt J Plast Reconstr Surg 2003;27(1):61–72.

[16] van der Veen V, van der Wal M, van Leeuwen M, Ulrich M, Middelkoop E. Biological background of dermal substitutes. Burns 2010;36(3):305–21.

[17] Arenas-Herrera JE, Ko IK, Atala A, Yoo JJ. Decellularization for whole organ bioengineering. Biomed Mater 2013;8(1):014106.

[18] Song J, Ott H. Organ engineering based on decellularized matrix scaffolds. Trends Mol Med 2011;17(8):424–32.

[19] Fu R, Wang Y, Liu S, Shih T, Lin H, Chen Y, et al. Decellularization and recellularization technologies in tissue engineering. Cell Transplant 2014;23(2):621–30.

[20] Zhong S, Zhang Y, Lim C. Tissue scaffolds for skin wound healing and dermal reconstruction. WIREs Nanomed Nanobiotechnol September/October 2010;2:510–25.
[21] Bosman FT, Stamenkovic I. Functional structure and composition of the extracellular matrix. J Pathol 2003;200(4):423–8.
[22] Badylak SF, Weiss DJ, Caplan A, Macchiarini P. Engineered whole organs and complex tissues. Lancet 2012;379(9819):943–52.
[23] Bissell MJ, Hall HG, Parry G. How does the extracellular matrix direct gene expression? J Theor Biol November 1982;99(1):31–68.
[24] Carey LE, Dearth CL, Johnson SA, Londono R, Medberry CJ, Daly KA, et al. In vivo degradation of 14C-labeled porcine dermis biologic scaffold. Biomaterials 2014:1–8. Elsevier Ltd.
[25] Badylak S. The extracellular matrix as a biologic scaffold material. Biomaterials 2007;28: 3587–93.
[26] Frantz C, Stewart KM, Weaver VM. The extracellular matrix at a glance. J Cell Sci 2010;123(Pt 24):4195–200.
[27] Badylak SF, Freytes DO, Gilbert TW. Extracellular matrix as a biological scaffold material: structure and function. Acta Biomater 2009;5(1):1–13.
[28] Gibson T. The physical properties of skin. 2nd ed. Philadelphia: WB Saunders; 1990.
[29] van Zuijlen P, Ruurda J, van Veen H, van Marle J, van Trier A, Groeneveit F, et al. Collagen morphology in human skin and scar tissue: no adaptations in response to mechanical loading at joints. Burns 2003;29(5):423–31.
[30] Osman OS, Selway JL, Harikumar PE, Stocker CJ, Wargent ET, Cawthorne MA, et al. A novel method to assess collagen architecture in skin. BMC Bioinformatics 2013;14(1):260.
[31] Coolen NA, Schouten KCWM, Middelkoop E, Ulrich MMW. Comparison between human fetal and adult skin. Arch Dermatol Res 2010;302:47–55.
[32] Krieg T, Aumailley M. The extracellular matrix of the dermis: flexible structures with dynamic functions. Exp Dermatol 2011;20(7):689–95.
[33] Livesey S, Herndon D, Hollyoak M, Atkinson Y, Nag A. Transplanted acellular allograft dermal matrix. Transplantation 1995;60(1):1–9.
[34] Yukna R, Turner D, Robinson L. Variable antigenicity of lyophilized allogeneic and lyophilized xenogeneic skin in guinea pigs. J Periodontal Res 1977;12:197–201.
[35] Lynn A, Yannas I, Bonfield W. Antigenicity and immunogenicity of collagen. J Biomed Mater Res B Appl Biomater 2004;71(2):343–54.
[36] Hutter H, Vogel BE, Plenefisch JD, Norris CR, Proenca RB, Spieth J, et al. Cell adhesion and extracellular matrix genes. Science February 2000;287:989–94.
[37] Badylak S. Decellularized allogeneic and xenogeneic tissue as a bioscaffold for regenerative medicine – factors that influence the host response. Ann Biomed Eng 2014; 42(7):1517–27.
[38] Castagnoli C, Stella M, Magliacani G, TeichAlasia S, Richiardi P. Anomalous expression of HLA class II molecules on keratinocytes and fibroblasts in hypertrophic scars consequent to thermal injury. Clin Exp Immunol 1990;82:350–4.
[39] Sedmak D, Orosz C. The role of vascular endothelial cells in transplantation. Arch Pathol Lab Med 1991;115:260–5.
[40] Patnaik SS, Wang B, Weed B. Decellularized scaffolds: concepts, methodologies, and applications in cardiac tissue engineering and whole-organ regeneration. In: Liu Q, Wung H, editors. Tissue regeneration: where nano-structure meets biology. 2nd ed. World Scientific Publishing Company; 2013. p. 1–48.
[41] Watt FM, Fujiwara H. Cell-extracellular matrix interactions in normal and diseased skin. In: Hynes R, Yamada K, editors. Cold Spring Harb Perspect Biol 2011. p. 1–14.

[42] Cornwell K, Downing B, Pins G. Characterizing fibroblast migration on discrete collagen threads for applications in tissue regeneration. J Biomed Mater Res 2004;71A(1):55–62.

[43] Gilbert TW. Strategies for tissue and organ decellularization. J Cell Biochem March 2012;113(7):2217–22.

[44] Achilli M, Mantovani D. Tailoring mechanical properties of collagen-based scaffolds for vascular tissue engineering: the effects of pH, temperature and ionic strength on gelation. Polymers (Basel) 2010;2:664–80.

[45] Miller A, Karmas E, Lu M. Age-related changes in the collagen of bovine corium: studies on extractability, solubility and molecular size distribution. J Food Sci 1983;48(3):681–885.

[46] Larson BJ, Longaker MT, Lorenz HP. Scarless fetal wound healing: a basic science review. Plast Reconstr Surg 2010;126:1172–80.

[47] Epstein EH. Alpha1(III)3 human skin collagen, release by pepsin digestion and preponderance in fetal life. J Biol Chem 1974;249:3225–31.

[48] Smith LT. Patterns of type VI collagen compared to types I, III and V collagen in human embryonic and fetal skin and in fetal skin-derived cell cultures. Matrix Biol 1994;14(2):159–70.

[49] Meigel W, Gay S, Weber L. Dermal architecture and collagen type distribution. Arch Derm Res 1977;259:1–8.

[50] Liu X, Wu H, Byrne M, Krane S, Jaenisch R. Type III collagen is crucial for collagen I fibrillogenesis and for normal cardiovascular development. Proc Natl Acad Sci USA March 1997;94:1852–6.

[51] Smith L, Holbrook K, Madri J. Collagen types I, III, and V in human embryonic and fetal skin. Am J Anat 1986;175(4):507–21.

[52] AATB. 13th Edition of the standards for tissue banking. 2014.

[53] FDA. FDA Guidance for industry-eligibility determination for donors of human cells, tissues, and cellular and tissue-based products (HCT/Ps). 2007.

[54] FDA. FDA guidance for industry-current good tissue practice (CGTP) and additional requirements for manufacturers of human cells, tissues, and cellular and tissue-based products (HCT/Ps). 2011.

[55] Walter R, Matsuda T, Reyes H, Walter J, Hanumadass M. Characterization of acellular dermal matrices (ADMs) prepared by two different methods. Burns 1998;24(2):104–13.

[56] Truong AN, Kowal-Vern A, Latenser BA, Wiley DE, Walter RJ. Comparison of dermal substitutes in wound healing utilizing a nude mouse model. J Burns Wounds 2005;4:72–82.

[57] Hrebikova H, Diaz D, Mokry J. Chemical decellularization: a promising approach for preparation of extracellular matrix. Biomed Pap Med Fac Univ Palacky Olomouc Czech Repub 2013;157:1–6.

[58] Ngo MD, Aberman HM, Hawes ML, Choi B, Gertzman AA. Evaluation of human acellular dermis versus porcine acellular dermis in an in vivo model for incisional hernia repair. Cell Tissue Bank 2011;12:135–45.

[59] Moore M, Samsel B, Wallis G, Triplett S, Chen S, Linthurst Jones A, et al. Decellularization of human dermis using non-denaturing anionic detergent and endonuclease: a review. Cell Tissue Bank 2015;16(2):249–59.

[60] Reing J, Brown B, Daly K, Freund J, Gilbert T, Hsiong S, et al. The effects of processing methods upon mechanical and biologic properties of porcine dermal extracellular matrix scaffolds. Biomaterials 2011;31(33):8626–33.

[61] Nishide M, Tsujimoto K, Uozaki M, Ikeda K, Yamasaki H, Koyama AH, et al. Effects of electrolytes on virus inactivation by acidic solutions. Int J Mol Med 2011;27:803–9.

[62] Richters CD, Pirayesh A, Hoeksema H, Kamperdijk EWA, Kreis RW, Dutrieux RP, et al. Development of a dermal matrix from glycerol preserved allogeneic skin. Cell Tissue Bank 2008;9:309–15.

[63] USP 32 Official Monographs. Bovine acellulardermal matrix. United States Pharmacopeia 2011;32:1701–4.
[64] Rault I, Frei R, Herbage D, Abdul-Malak N, Huc A. Evaluation of different chemical methods for cross-linking collagen gel, films, and sponges. J Mater Sci Mater Med 1996;7:215–21.
[65] Friess W. Collagen–biomaterial for drug delivery. Eur J Pharm Biopharm 1998;45:113–36.
[66] Charulatha V, Rajaram A. Influence of different crosslinking treatments on the physical properties of collagen membranes. Biomaterials 2003;24(5):759–67.
[67] Cornwell K, Lei P, Andreadis S, Pins G. Crosslinking of discrete self-assembled collagen threads: effects on mechanical strength and cell-matrix interactions. J Biomed Mater Res Part A 2007;80(2):362–71.
[68] Pieper J, Oosterhof A, Dijkstra P, Veerkamp J, van Kuppervelt T. Preparation and characterization of porous crosslinked collagenous matrices containing bioavailable chondroitin sulphate. Biomaterials 1999;20(9):847–58.
[69] Carlson T, Lee K, Pierce L. Effect of cross-linked and non-cross-linked acellulardermal matrices on the expression of mediators involved in wound healing and matrix remodeling. Plast Reconstr Surg 2013;131(697):697–705.
[70] Cavallo J, Greco S, Liu J, Frsella M, Deeken C, Matthews B. Remodeling characteristics and biomechanical properties of a crosslinked versus a non-crosslinked porcine dermis scaffolds in a porcine model of ventral hernia repair. Hernia 2015;19(2):207–18.
[71] Atala A, Lanza R, Nerem R, Thomson J. Principles of regenerative medicine. Elsevier Inc. 2008.
[72] Rutala W, Weber D, HICPAC. Center for Disease Control. Guideline for disinfection and sterilization in healthcare facilities. 2008. p. 1–158.
[73] Pegg D. Principles of cryopreservation. Methods Mol Biol 2007;368:39–57.
[74] Seth A, Persing S, Connor C, Davila A, Hirsch E, Fine N, et al. A comparative analysis of cryopreserved versus prehydrated human acellular dermal matrices in tissue expander breast reconstruction. Ann Plast Surg 2013;70(6):632–5.
[75] Kuleshova L, Gouk S, Hutmacher D. Vitrification as a prospect for cryopreservation of tissue-engineered constructs. Biomaterials 2007;28(9):1585–96.
[76] Qian L, Zhang H. Controlled freezing and freeze drying: a versatile route for porous and micro-/nano-structured materials. J Chem Technol Biotechnol 2011;86(2):172–84.
[77] FDA. FDA draft guidance – submission and review of sterility information in premarket notification (510(k)) submissions for devices labeled as sterile. 2008.
[78] Faraj KA, Brouwer KM, Geutjes PJ, Versteeg EM, Wismans RG, Deprest JA, et al. The effect of ethylene oxide sterilisation, beta irradiation and gamma irradiation on collagen fibril-based scaffolds. Tissue Eng Regen Med 2011;8(5):460–70.
[79] Code of Federal Regulations 21CFR1271.230; 2015. Title 21, vol. 8.
[80] WHO. Guidelines on viral inactivation and removal procedures intended to assure the viral safety of human blood plasma products. WHO Tech Rep 2004;Annex 4(Series No. 924):150–224.
[81] Anderson JM. Biological responses to materials. Annu Rev Mater Res 2001;31:81–110.
[82] Anderson J, Rodriguez A, Change D. Foreign body reaction to biomaterials. Semin Immunol 2008;20(2):86–100.
[83] Wagner C, Owens R, Harper J, McQuillan D. Human-derived acellular matrices for dermal replacement. Biomater Treat Skin Loss 2009:142–73.
[84] Sandor M, Singh D, Ronald B, Silverman P, Xu H, De Deyne P. Comparative host response of 2 human acellulardermal matrices in a primate implant model. Eplasty 2014;14(e7):52–64.

[85] Badylak S, Valentin J, Ravindra A, McCabe G, Stewart-Akers A. Macrophage phenotype as a determinant of biologic scaffold remodeling. Tissue Eng 2008;14(11):1835–42.
[86] Gilbert T, Stewar-Akers A, Badylak S. A quantitative method for evaluating the degradation of biologic scaffold materials. Biomaterials 2007;28(2):147–50.
[87] Sicari B, Turner N, Badylak S. An in vivo model system for evaluation of the host response to biomaterials. In: Gourdie R, Myers T, editors. Wound regeneration and repair: methods and protocols, methods in molecular biology. New York: Springer Science; 2013. p. 3–25.
[88] Agrawal V, Johnson SA, Reing J, Zhang L, Tottey S, Wang G, et al. Epimorphic regeneration approach to tissue replacement in adult mammals. Proc Natl Acad Sci USA 2010;107(8):3351–5.
[89] Rennert RC, Sorkin M, Garg RK, Januszyk M, Gurtner GC. Cellular response to a novel fetal acellular collagen matrix: implications for tissue regeneration. Int J Biomater 2013;2013 (Article ID 527957).
[90] Neill J, James K, Lineaweaver W. Utilizing biologic assimilation of bovine fetal collagen in staged skin grafting. Ann Plast Surg 2012;68:451–6.
[91] Strong A, Bennett D, Spreen E, Adhvaryu D, Littleton J, Mencer E. Fetal bovine collagen matrix in the treatment of a full thickness burn wound: a case report with long-term follow-up. J Burn Care Res 2014:1–6.
[92] Lineaweaver W, Bush K, James K. Suppression of α smooth muscle actin accumulation by bovine fetal dermal collagen matrix in full thickness skin wounds. Ann Plast Surg 2015:1–4.
[93] Wainwright D, Madden M, Luterman A, Hunt J, Monafo W, Heimbach D, et al. Clinical evaluation of an acellular allograft dermal matrix in full-thickness burns. J Burn Care Rehabil 1996;17:124–36.
[94] Compton C, Hickerson W, Nadire K, Press W. Acceleration of skin regeneration from cultured epithelial autografts by transplantation to homograft dermis. J Burn Care Res 1993;14(6):653–62.
[95] Yim H, Cho YS, Seo CH, Lee BC, Ko JH, Kim D, et al. The use of AlloDerm on major burn patients: AlloDerm prevents post-burn joint contracture. Burns 2010;36(3):322–8. Elsevier Ltd and International Society of Burns Injuries.
[96] Hayn E. Successful treatment of complex traumatic and surgical wounds with a foetal bovine dermal matrix. Int Wound J March 4, 2013:1–6.
[97] Pirayesh A, Hoeksema H, Richters C, Verbelen J, Monstrey S. Glyaderm(®) dermal substitute: clinical application and long-term results in 55 patients. Burns 2015;41(1):132–44.
[98] Askari M, Cohen M, Grossman P, Kulber D. The use of acellular dermal matrix in release of burn contracture scars in the hand. Plast Reconstr Surg 2011;127(4):1593–9.
[99] Juhasz I, Kiss B, Lukacs L, Erdei I, Peter Z, Remenyik E. Long-term followup of dermal substitution with acellular dermal implant in burns and postburn scar corrections. Dermatol Res Pract 2010;2010:1–7.
[100] Janis JE, Kwon RK, Attinger CE. The new reconstructive ladder: modifications to the traditional model. Plast Reconstr Surg 2011;127(Suppl. 1):205S–12S.
[101] Thomas S, Yen P, Melton A, Ravindra-Patil A, Bush K, Rue L. Reconstruction of complex wounds with poor vascularity utilizing fetal bovine acellular dermal matrix: an effective alternative to flaps and amputation. Society of Advanced Wound Care; 2013.
[102] Tarnuzzer R, Schultz G. Biochemical analysis of acute and chronic wound environments. Wound Repair Regen 1996;4:321–5.

[103] Reyzelman A, Crews RT, Moore JC, Moore L, Mukker JS, Offutt S, et al. Clinical effectiveness of an acellular dermal regenerative tissue matrix compared to standard wound management in healing diabetic foot ulcers: a prospective, randomised, multicentre study. Int Wound J June 2009;6(3):196–208.

[104] Kavros S, Dutra T, Marcus B. Clinical effectiveness of PriMatrix, a fetal bovine acellular dermal matrix, in healing chronic diabetic foot ulcers: a multicenter prospective study. Adv Skin Wound Care 2014;27(8):356–62.

[105] Lullove EJ. Acellular fetal bovine dermal matrix in the treatment of nonhealing wounds in patients with complex comorbidities. J Am Podiatr Med Assoc 2012;102(3):233–9.

[106] McCurdy B. The top ten innovations in podiatry. Pod Today 2014;27(7).

Chapter 6

Clinical Applications of Acellular Dermal Matrices in Reconstructive Surgery

Anders H. Carlsson[1,2], Edward M. Gronet[3], Lloyd F. Rose[1,2], Rodney Chan[1,4]

[1]*Quality Skin Collaborative for Advanced Reconstruction and Regeneration (Q-SCARRTM), United States Army Institute of Surgical Research, San Antonio, TX, USA;* [2]*Dental and Craniofacial Trauma Research and Tissue Regeneration, San Antonio, TX, USA;* [3]*Division of Plastic Surgery, Baylor Scott and White, Temple, TX, USA;* [4]*Plastic and Reconstructive Surgery, Clinical Division and Burn Center, United States Army Institute of Surgical Research, San Antonio, TX, USA*

Chapter Outline

INTRODUCTION

Understanding the body's response to insult and injury improves our ability to adapt interventions that optimize long-term functional and esthetic outcomes, while limiting morbidity. Pursuant to this goal, one of the more exciting advancements in plastic and reconstructive surgery has been the implementation and successful application of acellular dermal matrices (ADMs). ADM describes a class of scaffold that is used to augment and replace deficient or missing skin and soft tissue. As our understanding of the biological characteristics of this material increases, the indications for its use evolve and expand. ADMs have proven utility and reliability in several reconstructive applications and have gradually become an indispensable tool in the treatment of these complex problems. A plethora of ADM and ADM-derived products are available and that number continues to rise [1]. While there are data supporting their

Skin Tissue Engineering and Regenerative Medicine. http://dx.doi.org/10.1016/B978-0-12-801654-1.00006-1

use for various types of soft tissue reconstructions, there is a paucity of data comparing their relative efficacy, cost-effectiveness, and optimal-use conditions.

ADMs are divided into three classes: biological, synthetic, and composite [2]. Biological matrices are derived from human, animal, or plant origin. Synthetic ADMs are derived from manmade materials, and the composites are derived from a mix of biologic and synthetic materials. This chapter reviews the history of ADMs, and the preclinical data supporting their use, as well as the current clinical indications for wound healing, reconstructive surgery, and the yet unmet gaps that may prompt next-generation development of ADMs.

The European Union regulatory agency classifies most ADM products as Class III medical devices and requires the CE-mark. Human ADM tissue is regarded as minimally manipulated and not significantly changed in structure from the natural material, classifying it as banked human tissue not requiring approval by the U.S. Food and Drug Administration (FDA) [3]. Allograft ADM products derived from donated human tissue are supplied by tissue banks complying with guidelines provided by the American Association of Tissue Banks and the FDA. Animal ADM products are usually classified as medical devices and require, at a minimum, FDA 510(k) clearance [3]. The flexibility and fast-track FDA regulatory pathway afforded human ADMs expanded their clinical adoption over that of animal ADMs. However, clinical investigation and utilization of animal ADMs is increasing, due to their relatively low cost, availability, and biological and mechanical properties comparable to human ADMs.

The utility of ADMs continues to evolve. While originally developed as a dermal replacement scaffold, minimizing the morbidity of skin harvests, new generations of ADMs are now used to replace and augment missing and deficient soft tissue in the reconstruction of the head and neck, breast, and abdominal wall. Different preparations are finding specific clinical niches. Historically, most ADMs have been harvested from the superficial (papillary) dermis just beneath the epidermis. The collagen arrangement here is denser and more organized when compared to the deeper (reticular) dermis. Several newer ADMs are specifically sectioned in this deeper layer, as it may be more supple than the superficial dermis and potentially more conducive to ingrowth [4]. Further clinical study and experience is likely to refine specific ADMs, with their uses tailored to specific applications.

ANIMAL DATA

Assessments of the value of ADMs in wound healing animal models focus on two major criteria: successful integration and effect on long-term scarring. The primary factor governing successful integration of an ADM is whether or not the matrix vascularizes [5]. This is especially important if the ADM will be overlaid with a split-thickness skin graft. Although manufacturers typically recommend allowing a period of revascularization prior to overlay with skin graft, numerous studies report successful application of ADM and skin graft in single-stage procedures in porcine models [6]. Similar findings have been reported in humans [7]. A study in rats indicates that vascularization can be hastened by closer proximity of the ADM to vascularized muscle, resulting in enhanced induction of angiogenic factors [8]. These studies point out the importance of the wound bed for matrix integration. Grafting on highly vascularized tissues is more likely to result in positive outcomes; if failure occurs, it is often related to insufficient oxygenation or infection [9]. Because the status of the wound bed is not always clear, efforts have been directed at enhancing the ADM so that it has a better chance of integrating in the face of negative, undesirable, or suboptimal wound bed conditions. One major avenue of study involves the addition of stem cells. In in vitro and preclinical studies, stem-cell additions to ADMs suggest beneficial results, with improving angiogenesis, cellular infiltration, and tissue incorporation [5,10,11]. However, while the study of stem cells has grown exponentially over the past two decades, the use of stem-cell therapies as a standard procedure in the clinic may still be decades away [12].

The second major consideration in the value of ADMs is their effect on long-term remodeling and scarring. The assessment of an ADM can be measured with respect to the degree to which it is remodeled into tissue that structurally and functionally resembles the tissue it is intended to replace. In a porcine model of hernia repair, the use of ADM was compared with primary closure [13]. The study found that mechanical remodeling of the ADM was observed over time, with elastic modulus and tensile stress approaching that of fascia by 6 weeks. The biology of the ADM more closely resembled native abdominal wall, with less scarring and bony hyperplasia along the incision line. In a primate model of rotator cuff repair, ADM was observed to remodel into tendon-like architecture with homogenous distribution of cells and parallel collagen alignment [14]. However, success in animal models can be dependent on the health of the animal. When used for bladder augmentation, ADM works successfully in healthy pigs; however, it fails to properly integrate into diseased tissue [15]. These results support the contention that matrices designed for human bladder augmentation should be tested in diseased animals before use in humans and suggests the potential for more global caution in that ADMs might work best in tissues that are capable of properly directing their integration. Diseased tissue, or tissue surrounding diseased tissue, may not generate appropriate signals for angiogenesis and cell migration into the ADM.

CLINICAL APPLICATIONS

In order for ADMs to be successfully applied in clinical applications, they must allow host vascular ingrowth through angiogenesis, where host blood vessels grow into empty spaces left after ADM manufacture. With vascular ingrowth the ADM is said to "integrate" into the patient and become live connective tissue. Initial uses for ADMs were to augment skin replacement therapies in burn patients. Soon after, the usage of ADMs evolved into other, varied applications.

Skin/Burn

Skin is the largest and most esthetically defining organ of the body. Restoration after severe trauma not only depends on restoring skin continuity but also its natural elasticity, contour, and texture. When the dermis is sufficiently damaged, a skin graft consisting of epidermis and a thin layer of dermis is required to replace and restore this tissue to achieve both epithelial coverage and durability. Ideally, any soft tissue repair has texture and color that resemble the surrounding tissue, adheres to the remaining tissue bed, provides mechanical strength and elasticity, promotes cellular infiltration, as well as being able to fully incorporate and remodel over time. An ideal skin substitute used in this setting should not induce an immune response, be simple to apply and maintain, have a long shelf-life and expedite healing.

Acellular dermal grafts were initially developed for use in the treatment of full-thickness burns. Management of the severely burned is complicated by a lack of available donor skin. Furthermore, contraction of split-thickness skin grafts and mesh pattern scars are additional downsides of using thin, highly expanded skin. In order for skin grafts to "take" they must be applied onto clean, healthy, and vascularized wound beds. Such ideal conditions do not often exist; these are the situations where ADMs must survive in order to be useful in skin reconstruction. When an ADM is applied onto a full-thickness defect with an ultrathin, split-thickness skin graft overlaid, cellular migration and ingrowth into the ADM is seen from the skin graft as well as from the underlying tissue and peripheral skin edge [16]. The retained basement membrane provides laminin needed for adhesion and type IV collagen necessary for epithelialization. Processing of allogenic grafts results in a framework that retains normal collagen organization, as well as acellular vascular channels [17]. Under optimal conditions these channels become repopulated by the host within a week, and endothelium is restored [18]. The collagen scaffold initially contains empty spaces created by vacated fibroblasts, which repopulate with host fibroblasts over time, making the ADM indistinguishable from host tissue after about 4 weeks.

In theory, ADMs eliminate the need to harvest grafts containing thick autogenous dermis and ultimately allow a much thinner graft to be harvested, reducing donor site scarring seen with harvests of thick skin grafts [19]. Autologous, full-thickness skin grafts sometimes fail given the metabolic demands of living cellular material, while ADMs can be more forgiving in this scenario. It is not until the ADM is integrated and transformed into living tissue that the metabolic demands match that of natural tissue. ADMs have also been reported to resist encapsulation, leading to reduced inflammatory cytokine aggregation, likely because they lack antigenic signals (this may explain the decreased capsular contracture seen in breast surgeries that incorporate ADMs [20,21]).

Practically, ADMs are not widely used for the coverage of large areas of skin loss and large surface area burns. The conditions required for ADMs to properly incorporate, including a healthy vascular wound bed devoid of any necrotic tissue and infectious nidus, rarely exist in these instances. The presence of these conditions negates many of the beneficial effect of ADMs and actually may make it more difficult for an overlying skin graft to "take". Furthermore, while single-stage reconstruction has been described, ADMs are frequently used in two stages to maximize vascularity. The increased time and operative workload required for a two-stage reconstruction further discourages the use of ADMs in the acute setting. Nevertheless, ADMs have found their place in the coverage of smaller wounds and in late-stage reconstructions, where it is easier to achieve an optimal wound bed [22] (Figure 1).

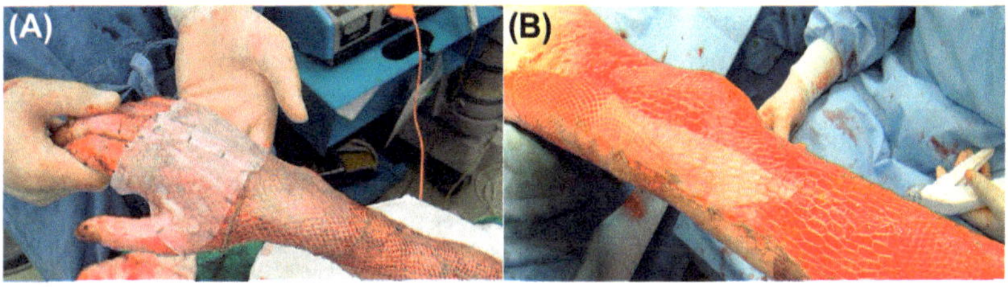

FIGURE 1 ADM use in burns. Panel A shows an example of AlloDerm used together with 1:3 meshed autograft on the arm of a burn patient. Panel B shows use of AlloDerm with 1:4–6 meshed autograft and a suspension of cultured epithelial autograft on the leg of a burn patient. *Used with permission from Elsevier.*

Head and Neck

The head and neck contains an array of specialized tissue types including skin, gingiva, and keratinized and nonkeratinized mucosa that provide many of the most important functions for daily living. The head and neck surgeon faces unique challenges, because choices of donor tissue with acceptable color and texture match are limited and local flaps may result in further esthetic compromise. The face and oral cavity in particular are fortunately receptive to grafting and resistant to infection, perhaps due to their immense blood supply. Several examples of ADM use in this context have been described.

Human ADMs have been shown to be useful in repairing oral mucosal and nasal septal defects, manifesting only mild contraction and no infection six months postoperatively [23–25]. ADMs offer a straightforward means to correct nasal septal defects, and have been shown to close defects less than 2 cm in diameter as well as multiple, small septal defects.

Temporomandibular joint ankylosis, a rare but debilitating problem, has historically been treated with temporalis muscle and fascia flaps used as an interpositional graft [26,27]. ADMs provide an alternative that results in equal functional improvement and reduced morbidity [28]. Human ADMs used as an interpositional graft have also been effective in preventing Frey's syndrome following parotidectomy; however, there have been reports of increased rates of seroma formation and infection compared to traditional superficial parotidectomy [29–32]. Similar to the application of skin coverage, ADM use elsewhere still requires the same mechanism of incorporation and is thus susceptible to similar problems.

Cleft palate repair is another application where ADMs have gained increasing acceptance. While tension-free closure using adjacent soft tissue is ideal in the repair of a cleft palate, this may be difficult to obtain, especially if at the hard and soft palatal junction. Additionally, lack of a muscular layer along the hard palate midline and the constant motion of the soft palate against the hard palate create a hostile healing environment. Consequently, wound breakdown and palatal fistulas may develop at the anterior most part of a complete palatoplasty near the region of the incisive foramen [17]. Fistulas permit regurgitation of oral contents into the nasal cavity leading to mucosal inflammation, infection, and halitosis. Local flaps used to salvage a palatal fistula, such as the buccal fat pad flap, FAMM flap, and tongue flaps are popular, but they are bulky and often require a second- or third-stage surgery. Amaratunga et al. reported a 21% fistula rate in 346 cleft palates repaired via the von Langenback method [33]. Techniques to alleviate palatal flap tension at closure include wide undermining of soft tissue, skeletonization of the greater palatine neurovascular bundle, and modification of the greater palatine foramen, or fracturing of the hamulus. However, despite these interventions, breakdowns can still occur. A 2006 study reviewed the use of Alloderm in combination with primary palatoplasty repair using the Furlow technique [34]. The ADM was placed over the nasal closure at the junction of the hard and soft palate, with the basement membrane side of the graft toward the nasal mucosa. Fistulas developed in only two of the 31 cases reviewed. Losee et al. report their main indication for ADM in primary palatoplasty is when it is difficult to obtain a two-layer closure resulting in a defect of the nasal mucosal flap [35]. Furlow double opposing Z-plasty is the most commonly used procedure for cleft palate repair in that study. In larger clefts, it is not uncommon for the anterior left-sided nasal mucosal flap to not reach right-sided mucosa at the posterior edge of the hard palate. In this case, there can be a defect of the nasal lining to which ADM can be sewn where the tensor aponeurosis would normally lie. When used in primary and secondary palatoplasty, no negative consequences on speech were found, as the ADM does not seem to limit palate mobility [35]. By laying AlloDerm over the bony palate defect, then sewing it to the undersurface mucoperiosteal flap, Steele et al. decreased the failure rate of cleft palate revision surgery to zero [36].

According to some reports, ADMs appear to be appropriate for all cases of primary cleft palate repair failure that result in oronasal fistula formation, because ADM grafting leads to a lower failure rate of secondary palatoplasty and supports spontaneous healing when revisionary procedures fail [17,34,37,38]. ADM use in primary palatoplasty has shown a lower fistula rate than more traditional techniques using autogenous local and free flaps; however, there is insufficient evidence to recommend its routine use in this setting.

A systematic review of cleft palate repair indicated that the addition of ADM to both primary cleft palate repair and secondary repair of fistulas lowered the rate of fistula formation [39]. Using a weighted average of their reviewed studies, they found a 5.4% fistula rate in the ADM group and 10.6% in non-ADM historical control groups. In revisionary reports, they found a weighted average for the recurrence of fistulas after secondary repair to be 8.1% when ADM was used at secondary repair compared to a 12.9% recurrence of fistulas in patients without ADM at secondary repair. Although the addition of ADM to both primary and secondary repair of cleft palate is advantageous, the data are retrospective and thus its routine use cannot be recommended without higher-level data.

Another use of ADMs in the treatment of a cleft palate patient is to protect cancellous bone grafted into alveolar ridge defects [40]. Several authors have reported the use of ADMs to cover grafted bone to prevent its exposure. There were no positive or negative effects on mucosal wound healing, yet there was significantly less exposure of the grafted bone. Further, the ADM did not impede canine eruption when compared to controls.

Tympanic membranes have traditionally been repaired with temporalis fascia but have now been successfully repaired with human ADMs, particularly Alloderm [41–43]. A study comparing more traditional techniques versus human ADM showed 40% reduction in operative time, 50% reduction in postoperative pain, and improved graft success rates associated with ADM use [44–47]. Though many of these studies have touted favorable outcomes, little long-term data exist.

Full-thickness injuries to the eyelids pose a particularly challenging problem. The reconstructed eyelid must be pliable yet protective and have a smooth, moist posterior surface to allow for retraction and protraction over the globe of the eye. ADMs have been used extensively in eyelid reconstruction, as they can add length and support to either eyelid and not require harvest of autogenous tissue. ADMs can lengthen the eyelid and serve as a spacer graft, over which conjunctiva can reepithelialize [48] or buccal mucosa graft can be placed [49], thus creating length and a smooth gliding surface. ADM has also been used to cover eye implants after enucleation, add volume to the sulci, and address contour defects [48]. ADMs have been used to reconstruct the tarsal plates of both upper and lower lids [50] to provide protection, as well as a connection point to the muscles that enable eyelid excursion.

ADM has also found cosmetic applications in lip augmentation, as opposed to traditional bovine collagen sources. Injectable ADMs may have more prolonged effects when compared to injectable collagen for lip augmentation [51] (Figure 2).

Breast

There is significant data supporting the use of ADM in breast reconstruction, becoming the mainstay in providing inferior pole support after mastectomy. Breast cancer is the most common form of cancer and the second leading cause

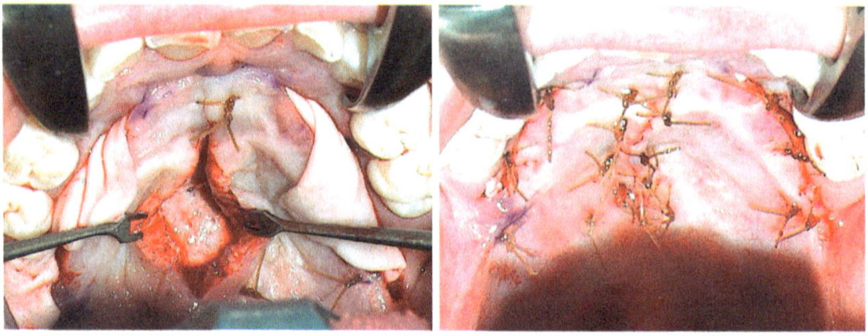

FIGURE 2 ADM use in oronasal fistulae repair. The left panel shows use of AlloDerm placed as an interposition graft beneath the oral mucoperiosteum. The right panel shows closure of the bilateral mucoperiosteal incisions and the oral mucosa over the ADM. *Used with permission from Wolters Kluwer.*

of cancer-related death among women [52]. When given the reconstructive options after mastectomy, women most frequently choose implant-based reconstructions over autologous tissue transfer due to harvest site morbidity [53]. Decisions regarding implant-based reconstructions may be guided by many factors: (1) recovery time, (2) lack of adequate donor tissue in slender women, and (3) surgeon unwillingness to harvest autologous tissue in obese women due to higher donor-site wound healing problems. Implant-based reconstruction is classically a staged procedure. At the time of mastectomy, a tissue expander is placed in a subpectoral pocket, serial expansions are conducted until the soft tissues are of adequate size to accept a permanent prosthetic implant, and a second surgery for tissue expander to permanent implant swap is conducted. One issue with the subpectoral placement of a tissue expander (or permanent prosthesis) is the lack of prosthetic coverage at the inferolateral pole. Historically this was addressed with elevation of serratus anterior fascia and rectus abdominus fascia which were then advanced to meet the lateral border of the pectoralis. However, this technique may result in tissue expanders that are too high and poorly expanded at the inferolateral pocket secondary to restrictive serratus muscle fascia.

The use of ADMs in breast surgery began in 2001, when ADMs were reported to reduce visible implant rippling in primary augmentation revision with capsulectomy and ADM placement [54]. The first published application of ADM use in breast reconstruction came in 2005 from Breuing and Warren who documented 10 cases in which patients underwent bilateral mastectomy with immediate (single-stage) reconstruction and an inferior AlloDerm sling [21]. These initial publications opened the door to ADM use in breast reconstruction, making it one of the most popular applications.

Most breast implants are placed below the pectoralis major muscle, which is dissected off the chest wall, creating a pocket in a lateral to supero-medial direction. At this lateral border of the pectoralis muscle, implants have less tissue coverage. It is at this location where ADMs are used instead of raising local serratus/rectus fascia defined by the lateral extent of the pocket. The ADM can allow the surgeon to actively sculpt the inferior pole, including the opportunity to establish the critically important infra-mammary fold at an ideal location.

ADMs have made single-stage breast reconstructions possible by re-creating the inferior pole, while taking advantage of the excess available skin after a mastectomy. Single-stage reconstruction not only eliminates the cost of a staged procedure including the tissue expander, multiple clinic visits, and a second operation. The established negative of single-stage reconstruction is difficulty with achieving symmetry, an issue that ADMs also have helped decrease. The benefits of ADMs in two-stage breast reconstruction include decreased postoperative pain, expedited expansion, and elimination of elevation of the serratus anterior fascia [20]. Although one study showed that expedited expansion can be achieved by the ADM, enabling a larger pocket to be created at the first stage, thereby allowing more initial fill volume in the expander—as high as 85% of final volume [55]. Level 1 evidence found no difference in expansion times in non- vs ADM tissue expander reconstruction [56].

ADMs are also used when there are viable but compromised or thin post-mastectomy skin flaps that are at risk of dehiscence or necrosis, as well as when the infra-mammary fold and lateral mammary folds have been undermined during mastectomy. According to some authors, ADMs also have the added benefit of minimizing capsular contracture rates and correcting surface irregularities [54,57]. However, postoperative pain and expansion rates remain unchanged [56,58].

In decreasing rates of capsular contracture, a histological comparison study of ADMs revealed less fibroblastic activity, B cells, T cells, macrophages, myofibroblasts, and type I and III collagens than in native breast capsules from the same patient [59]. This suggests an attenuation of the inflammatory response in the presence of ADMs. Less capsular contracture with ADMs are also found in primate studies [60].

Capsular contracture is a common phenomenon among women undergoing radiation therapy. Current evidence suggests that capsular contracture rates are similar among women undergoing radiation therapy with ADM-based reconstruction [55]. The effect of radiation on ADM shows that radiation does not significantly change the architecture or cellular components of the ADM, and there is a similar amount of neoangiogenesis regardless of whether the ADM was irradiated or not. When comparing periprosthetic capsules in irradiated breasts with ADM vs irradiated breasts without ADM, the irradiated capsule with ADM has less cellular infiltrate, elastin, and nonvascular alpha-smooth muscle actin than irradiated capsule without ADM.

The use of implantable ADMs is not without complications. Reported complications of breast reconstruction with the addition of ADMs include increased rates of seromas, infections, and skin flap necrosis [61]. When breast reconstructions with ADMs are subjected to external beam radiation treatment, the overall incidence of complications does not increase, but the incidence of seromas and infections do increase [55]. Seroma is the most common complication in ADM-based reconstruction. Newer versions of ADM are being manufactured with small fenestrations thought to facilitate fluid movement and thus decrease seroma [62].

Early results from the latest prospective randomized controlled trial for breast reconstruction with ADMs show a complication rate of 36%, about half being minor, and half requiring some form of intervention [63]. Analysis showed that poor ADM integration increased the rate of infection. Poor integration is seen with increasing breast size, expander size, initial tissue-expander fill volume, as well as with obesity and prolonged drain time [63]. The surgeon should be aware that extended drain output may be a clue to poor integration of the ADM.

Besides breast reconstruction, ADMs have also gained increased acceptance for breast cosmetic revision surgery. Allergan reported a reoperation rate of 30% for primary augmentation patients and 41% for revision-augmentation patients [64]. An infrequent but dreaded complication of cosmetic breast augmentation is overdissection of the subpectoral pocket medially, resulting in a continuous implant pocket extending from one side of the chest to the pocket on the contralateral side. Revision surgery for this complication is unlikely to eliminate the defect, but ADMs can be used to create a medial barrier to decrease implant migration in revisional symmastia surgeries. Much is yet to be determined about the long-term merits of ADMs in breast surgery, but there is little doubt, with current data and investigations, that ADMs will continue to serve as a tissue substitute for breast reconstruction (Figure 3).

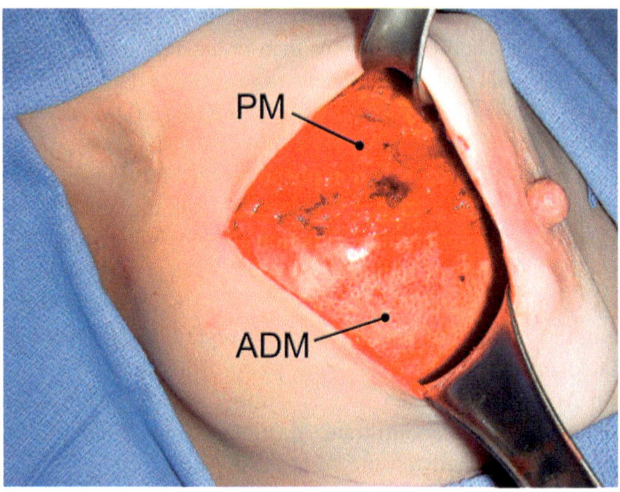

FIGURE 3 ADM use in breast reconstruction. Photograph shows the right breast after a total skin-sparing mastectomy performed through a lateral breast incision and placement of a tissue-expander with ADM sutured to the pectoralis major (PM) and infra-mammary fold. *Used with permission from Wolters Kluwer.*

Abdominal/Hernia

Closure of large abdominal and chest wall defects remains a challenge to the surgical community, and incisional hernias are by far the most common complication after laparotomy [65]. Factors such as defect size, infection, obtaining adequate skin coverage, among other concerns pose significant challenges. Historically, biologic and synthetic mesh has been the "gold standard" for complex abdominal and chest wall reconstruction, in an attempt to provide tension-free closure. However, the ideal method of implantation remains controversial, even though different types of mesh are used with good results. Retromuscular hernia repairs have been shown to be the technique of choice for open ventral hernia repair by the American Hernia Society [66].

The use of mesh materials in infected wounds results in a high incidence of mesh infection, fibrosis, fistulas, and/or adhesion formation. In an attempt to avoid these complications, biologic tissue grafts derived from animal or human sources are now in use [67]. ADMs have been touted as an alternative to synthetic mesh, with the goal of improving integration with lower rates of infection, breakdown, extrusion, adhesion formation, and rejection compared with synthetic materials [67,68]. However, publications on long- and short-term complications, such as hernia recurrence and infection, are lacking.

Strength of hernia repair is important to a successful outcome, longevity, and hernia recurrence. An ex vivo study has shown that when bovine-derived ADMs are compared to porcine-derived products, the bovine materials are stronger with regard to tensile strength, suture retention, and tear resistance [69]. Further, bovine ADMs tear in a predictable manner (in-line with applied force), while porcine ADMs tear in directions not parallel with the force vector. Most ADM repairs fail to do so at the musculo–fascial junction. Definitive comparative studies are however still lacking (Figure 4).

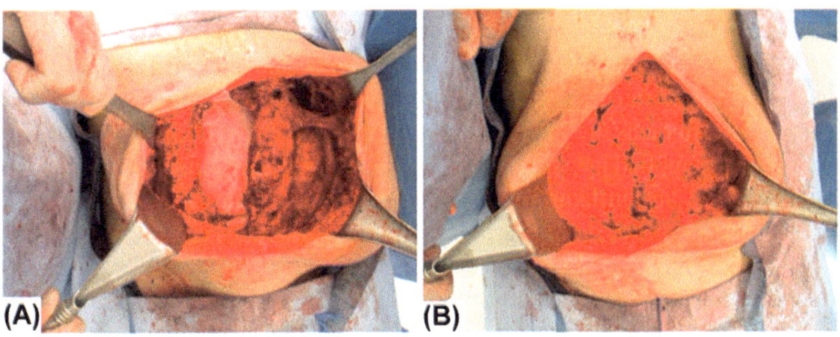

FIGURE 4 ADM use in abdominal wall hernias. Panel A shows use of porcine-derived ADM placement in an abdominal wall hernia. Panel B shows the completion of midline fascial closure over the ADM. *Used with permission from Springer.*

CONCLUSION/FUTURE

The search for a rapid, renewable, and cost-effective means to restore tissue to its preinjury state continues. ADMs provide multiple indispensable alternatives to reestablish functional and esthetic tissues. Though ADMs in many applications continue to be debated due to conflicting data, there remain many indications for which they seem to be the best option. As the process of tissue healing becomes more fully understood, research has revealed many new sources for providing extracellular matrices. ADMs reliably promote wound healing through fostering an environment that facilitates the body's regenerative processes and promotes angiogenesis. Many variations exist and are under development. However, no single product or surgical technique has proven most versatile or cost-effective. Efforts to create synthetic ADMs have been encouraging and products made from equine collagen or bovine tendon collagen and chondroitin-6 sulfate from porcine collagen have shown excellent results.

One future avenue of employing ADMs for reconstructive surgery may involve the addition of exogenous cells, and/or growth factors to enhance performance. The addition of adipose-derived stem cells (ASC) has shown some promise. Alharbi et al. showed that human adipocytes, harvested by standard liposuction procedures, can be seeded onto collagen and elastin matrices in as little as 1 h, a time interval feasibly achieved during operative procedures [70]. ASCs represent a potentially potent armament for soft-tissue reconstruction, and support of these cells with a matrix may increase survival of ASCs, as well as safely guide the cells into recipient sites.

Viable alternatives to ADMs increasingly hold great promise. Decellularized adipose tissue (DAT) provides a complex microenvironment, which mediates tissue-specific cellular responses such as proliferation, migration, and differentiation that facilitate wound repair and regeneration [71]. DAT is rich in basement membrane and may be processed from tissue blocks or lipoaspirate. Further, decellularized human adipose tissue extracellular matrix serves as an excellent substrate for ASCs, promoting cell adhesion, proliferation, and in vivo integration [72]. Numerous laboratories are investigating the use of these novel scaffolds for site- and tissue-specific applications.

Another potential source of exogenous cells and growth factors include the use of amniotic membrane and umbilical cord-derived products. Human amniotic membranes are currently used in ophthalmic surgeries [73] and have recently been applied in cutaneous wound healing, with results showing reduced pain, scarring, and inflammation while increasing neovascularization [74,75]. One of the benefits of decellularized amniotic membranes are that they contain a vast array of growth factors and cytokines, which are believed to contribute to the therapeutic efficacy seen in wound healing [76,77].

Since most clinics are under financial restraints, new applications not only need to provide better results, but also must be cost-effective. As the cost for ADMs decreases and implementation of stem cells and growth factors in transplants increases, the use of ADMs will likely increase. Future generations of ADMs will likely have very specific properties optimizing the intended function in each specific tissue.

REFERENCES

[1] Shevchenko RV, James SL, James SE. A review of tissue-engineered skin bioconstructs available for skin reconstruction. J R Soc Interface 2010;7(43):229–58.

[2] International consensus. Acellular matrices for the treatment of wounds. An expert working group review. London: Wounds International; 2010.

[3] Derwin KA, Badylak SF, Steinmann SP, Iannotti JP. Extracellular matrix scaffold devices for rotator cuff repair. J Shoulder Elbow Surg 2010;19(3):467–76.

[4] Musculoskeletal Transplant Foundation Research and Development Department. Development of a novel acellular dermal matrix (ADM) designed specifically for breast reconstruction. 125 May Street, Edison, NJ 08837.

[5] Garcia Jr O, Scott JR. Analysis of acellular dermal matrix integration and revascularization following tissue expander breast reconstruction in a clinically relevant large-animal model. Plast Reconstr Surg 2013;131(5):741e–51e.

[6] Bannasch H, Unterberg T, Fohn M, Weyand B, Horch RE, Stark GB. Cultured keratinocytes in fibrin with decellularised dermis close porcine full-thickness wounds in a single step. Burns 2008;34(7):1015–21.

[7] Sohn WI, Han SH, Jung SN. One-stage skin grafting of the exposed skull with artificial dermis after cancer removal: long-term experiences. Head Neck Oncol 2012;4:73.

[8] Sahin I, Ozturk S, Deveci M, Ural AU, Onguru O, Isik S. Experimental assessment of the neo-vascularisation of acellular dermal matrix in the wound bed pretreated with mesenchymal stem cell under subatmospheric pressure. J Plast Reconstr Aesthet Surg 2014;67(1):107–14.

[9] Unal S, Ersoz G, Demirkan F, Arslan E, Tutuncu N, Sari A. Analysis of skin-graft loss due to infection: infection-related graft loss. Ann Plast Surg 2005;55(1):102–6.

[10] Komatsu I, Yang J, Zhang Y, et al. Interstitial engraftment of adipose-derived stem cells into an acellular dermal matrix results in improved inward angiogenesis and tissue incorporation. J Biomed Mater Res A 2013;101(10):2939–47.

[11] Huang SP, Hsu CC, Chang SC, et al. Adipose-derived stem cells seeded on acellular dermal matrix grafts enhance wound healing in a murine model of a full-thickness defect. Ann Plast Surg 2012;69(6):656–62.

[12] Daley GQ. The promise and perils of stem cell therapeutics. Cell Stem Cell 2012;10(6):740–9.

[13] Monteiro GA, Delossantos AI, Rodriguez NL, Patel P, Franz MG, Wagner CT. Porcine incisional hernia model: evaluation of biologically derived intact extracellular matrix repairs. Journal Tissue Eng 2013;4:2041731413508771.

[14] Xu H, Sandor M, Qi S, et al. Implantation of a porcine acellular dermal graft in a primate model of rotator cuff repair. J Shoulder Elbow Surg 2012;21(5):580–8.

[15] Akbal C, Lee SD, Packer SC, Davis MM, Rink RC, Kaefer M. Bladder augmentation with acellular dermal biomatrix in a diseased animal model. J Urol 2006;176(4 Pt 2):1706–11.

[16] Livesey SA, Herndon DN, Hollyoak MA, Atkinson YH, Nag A. Transplanted acellular allograft dermal matrix. Potential as a template for the reconstruction of viable dermis. Transplantation 1995;60(1):1–9.

[17] Clark JM, Saffold SH, Israel JM. Decellularized dermal grafting in cleft palate repair. Arch Facial Plast Surg 2003;5(1):40–4; discussion 5.

[18] Wainwright D, Madden M, Luterman A, et al. Clinical evaluation of an acellular allograft dermal matrix in full-thickness burns. J Burn Care Rehabil 1996;17(2):124–36.

[19] Schwanholt C, Greenhalgh DG, Warden GD. A comparison of full-thickness versus split-thickness autografts for the coverage of deep palm burns in the very young pediatric patient. J Burn Care Rehabil 1993;14(1):29–33.

[20] Bindingnavele V, Gaon M, Ota KS, Kulber DA, Lee DJ. Use of acellular cadaveric dermis and tissue expansion in postmastectomy breast reconstruction. J Plast Reconstr Aesthet Surg 2007;60(11):1214–8.

[21] Breuing KH, Warren SM. Immediate bilateral breast reconstruction with implants and inferolateral AlloDerm slings. Ann Plast Surg 2005;55(3):232–9.

[22] Debels H, Hamdi M, Abberton K, Morrison W. Dermal matrices and bioengineered skin substitutes: a critical review of current options. Plast Reconstr Surg Glob Open 2015;3(1):e284.

[23] Shi LJ, Wang Y, Yang C, Jiang WW. Application of acellular dermal matrix in reconstruction of oral mucosal defects in 36 cases. Journal Oral Maxillofac Surg 2012;70(11):e586–91.

[24] Ayshford CA, Shykhon M, Uppal HS, Wake M. Endoscopic repair of nasal septal perforation with acellular human dermal allograft and an inferior turbinate flap. Clin Otolaryngol Allied Sci 2003;28(1):29–33.

[25] Chhabra N, Houser SM. Endonasal repair of septal perforations using a rotational mucosal flap and acellular dermal interposition graft. Int Forum allergy Rhinol 2012;2(5):392–6.

[26] Bayat M, Badri A, Moharamnejad N. Treatment of temporomandibular joint ankylosis: gap and interpositional arthroplasty with temporalis muscle flap. Oral Maxillofac Surg 2009;13(4):207–12.

[27] Su-Gwan K. Treatment of temporomandibular joint ankylosis with temporalis muscle and fascia flap. Int J Oral Maxillofac Surg 2001;30(3):189–93.

[28] Yazdani J, Ali Ghavimi M, Pourshahidi S, Ebrahimi H. Comparison of clinical efficacy of temporalis myofascial flap and dermal graft as interpositional material in treatment of temporomandibular joint ankylosis. J Craniofac Surg 2010;21(4):1218–20.

[29] Sinha UK, Saadat D, Doherty CM, Rice DH. Use of AlloDerm implant to prevent frey syndrome after parotidectomy. Arch Facial Plast Surg 2003;5(1):109–12.

[30] Ye WM, Zhu HG, Zheng JW, et al. Use of allogenic acellular dermal matrix in prevention of Frey's syndrome after parotidectomy. Br J Oral Maxillofac Surg 2008;46(8):649–52.

[31] Athavale SM, Phillips S, Mangus B, et al. Complications of alloderm and dermamatrix for parotidectomy reconstruction. Head Neck 2012;34(1):88–93.

[32] Govindaraj S, Cohen M, Genden EM, Costantino PD, Urken ML. The use of acellular dermis in the prevention of Frey's syndrome. Laryngoscope 2001;111(11 Pt 1):1993–8.

[33] Amaratunga NA. Occurrence of oronasal fistulas in operated cleft palate patients. J Oral Maxillofac Surg 1988;46(10):834–8.

[34] Helling ER, Dev VR, Garza J, Barone C, Nelluri P, Wang PT. Low fistula rate in palatal clefts closed with the Furlow technique using decellularized dermis. Plast Reconstr Surg 2006;117(7):2361–5.

[35] Losee JE, Smith DM. Acellular dermal matrix in palatoplasty. Aesthet Surg J 2011;31(7 Suppl.):108S–15S.

[36] Steele MH, Seagle MB. Palatal fistula repair using acellular dermal matrix: the University of Florida experience. Ann Plast Surg 2006;56(1):50–3. discussion 3.

[37] Kirschner RE, Cabiling DS, Slemp AE, Siddiqi F, LaRossa DD, Losee JE. Repair of oronasal fistulae with acellular dermal matrices. Plast Reconstr Surg 2006;118(6):1431–40.

[38] Cole P, Horn TW, Thaller S. The use of decellularized dermal grafting (AlloDerm) in persistent oro-nasal fistulas after tertiary cleft palate repair. J Craniofac Surg 2006;17(4):636–41.

[39] Shararah AW, Aldekhayel S, Williams HB. Iatrogenic arteriovenous fistula in the hand: a case report. J Plast Reconstr Aesthet Surg 2012;65(12):e351–353.

[40] Clavijo-Alvarez JA, Vecchione L, DeCesare G, et al. Autologous bone grafting with adjunctive use of acellular dermal matrix for alveolar cleft defects: early outcomes. Cleft Palate Craniofac J 2010;47(2):116–21.

[41] Haynes DS, Vos JD, Labadie RF. Acellular allograft dermal matrix for tympanoplasty. Curr Opin Otolaryngol Head Neck Surg 2005;13(5):283–6.

[42] Downey TJ, Champeaux AL, Silva AB. AlloDerm tympanoplasty of tympanic membrane perforations. Am J Otolaryngol 2003;24(1):6–13.

[43] Laidlaw DW, Costantino PD, Govindaraj S, Hiltzik DH, Catalano PJ. Tympanic membrane repair with a dermal allograft. Laryngoscope 2001;111(4 Pt 1):702–7.

[44] Raj A, Sayal A, Rathore PK, Meher R. Sutureless tympanoplasty using acellular dermis. Am J Otolaryngol 2011;32(2):96–9.

[45] Fishman AJ, Marrinan MS, Huang TC, Kanowitz SJ. Total tympanic membrane reconstruction: AlloDerm versus temporalis fascia. Otolaryngol Head Neck Surg 2005;132(6): 906–15.

[46] Lai P, Propst EJ, Papsin BC. Lateral graft type 1 tympanoplasty using AlloDerm for tympanic membrane reconstruction in children. Int J Pediatr Otorhinolaryngol 2006;70(8):1423–9.

[47] Fayad JN, Baino T, Parisier SC. Preliminary results with the use of AlloDerm in chronic otitis media. Laryngoscope 2003;113(7):1228–30.

[48] Rubin PA, Fay AM, Remulla HD, Maus M. Ophthalmic plastic applications of acellular dermal allografts. Ophthalmology 1999;106(11):2091–7.

[49] Gu J, Zhai J, Chen J. The use of acellular human dermis composite graft for upper eyelid reconstruction in ocular injury. J Trauma Acute Care Surg 2012;72(1):288–92.

[50] Jiaqi C, Zheng W, Jianjun G. Eyelid reconstruction with acellular human dermal allograft after chemical and thermal burns. Burns 2006;32(2):208–11.

[51] Sclafani AP, Romo 3rd T, Jacono AA. Rejuvenation of the aging lip with an injectable acellular dermal graft (Cymetra). Arch Facial Plast Surg 2002;4(4):252–7.

[52] Freitas-Silva R, Conde DM, de Freitas-Junior R, Martinez EZ. Comparison of quality of life, satisfaction with surgery and shoulder-arm morbidity in breast cancer survivors submitted to breast-conserving therapy or mastectomy followed by immediate breast reconstruction. Clinics 2010;65(8):781–7.

[53] Rawlani V, Buck 2nd DW, Johnson SA, Heyer KS, Kim JY. Tissue expander breast reconstruction using prehydrated human acellular dermis. Ann Plast Surg 2011;66(6):593–7.

[54] Duncan DI. Correction of implant rippling using allograft dermis. Aesthet Surg J 2001;21(1):81–4.

[55] Clemens MW, Kronowitz SJ. Acellular dermal matrix in irradiated tissue expander/implant-based breast reconstruction: evidence-based review. Plast Reconstr Surg 2012;130(5 Suppl. 2): 27S–34S.

[56] McCarthy CM, Lee CN, Halvorson EG, et al. The use of acellular dermal matrices in two-stage expander/implant reconstruction: a multicenter, blinded, randomized controlled trial. Plast Reconstr Surg 2012;130(5 Suppl. 2):57S–66S.

[57] Hartzell TL, Taghinia AH, Chang J, Lin SJ, Slavin SA. The use of human acellular dermal matrix for the correction of secondary deformities after breast augmentation: results and costs. Plast Reconstr Surg 2010;126(5):1711–20.

[58] Davila AA, Seth AK, Wang E, et al. Human acellular dermis versus submuscular tissue expander breast reconstruction: a multivariate analysis of short-term complications. Archives of plastic surgery 2013;40(1):19–27.

[59] Leong M, Basu CB, Hicks MJ. Further evidence that human acellular dermal matrix decreases inflammatory markers of capsule formation in implant-based breast reconstruction. Aesthet Surg J 2015;35(1):40–7.

[60] Stump A, Holton 3rd LH, Connor J, Harper JR, Slezak S, Silverman RP. The use of acellular dermal matrix to prevent capsule formation around implants in a primate model. Plast Reconstr Surg 2009;124(1):82–91.

[61] Chun YS, Verma K, Rosen H, et al. Implant-based breast reconstruction using acellular dermal matrix and the risk of postoperative complications. Plast Reconstr Surg 2010;125(2):429–36.

[62] Rosenberg MH, Palaia DA, Cahan AC, Arthur KS, DeLuca-Pytell DM, Bonanno PC. Breast reconstruction with or without human acellular dermal matrices: a single-clinic, review of esthetic outcomes and risk factors for complications. Am J Cosmet Surg 2014;31(1):7–17.

[63] Mendenhall SD, Anderson LA, Ying J, et al. The BREASTrial: stage I. Outcomes from the time of tissue expander and acellular dermal matrix placement to definitive reconstruction. Plast Reconstr Surg 2015;135(1):29e–42e.

[64] Allergan. Company document. Important information for women about breast augmentation with Natrelle® Gel-filled breast implants. 2015. http://www.allergan.com/assets/pdf/ca_natrelle_gel_aug_info_en.pdf.

[65] Zhu KQ, Carrougher GJ, Gibran NS, Isik FF, Engrav LH. Review of the female Duroc/Yorkshire pig model of human fibroproliferative scarring. Wound Repair Regen 2007;15(Suppl. 1):S32–9.

[66] Pauli EM, Rosen MJ. Open ventral hernia repair with component separation. Surg Clin North Am 2013;93(5):1111–33.

[67] Garvey PB, Martinez RA, Baumann DP, Liu J, Butler CE. Outcomes of abdominal wall reconstruction with acellular dermal matrix are not affected by wound contamination. J Am Coll Surg 2014;219(5):853–64.

[68] Guerra O. Noncrosslinked porcine-derived acellular dermal matrix for single-stage complex abdominal wall herniorrhaphy after removal of infected synthetic mesh: a retrospective review. Am Surg 2014;80(5):489–95.

[69] Adelman DM, Selber JC, Butler CE. Bovine versus porcine acellular dermal matrix: a comparison of mechanical properties. Plast Reconstr Surg Glob Open 2014;2(5):e155.

[70] Alharbi Z, Oplander C, Almakadi S, Fritz A, Vogt M, Pallua N. Conventional vs. micro-fat harvesting: how fat harvesting technique affects tissue-engineering approaches using adipose tissue-derived stem/stromal cells. J Plast Reconstr Aesthet Surg 2013;66(9):1271–8.

[71] Yu C, Bianco J, Brown C, et al. Porous decellularized adipose tissue foams for soft tissue regeneration. Biomaterials 2013;34(13):3290–302.

[72] Wang L, Johnson JA, Zhang Q, Beahm EK. Combining decellularized human adipose tissue extracellular matrix and adipose-derived stem cells for adipose tissue engineering. Acta Biomater 2013;9(11):8921–31.

[73] Aldekhayel SA, Sinno H, Gilardino MS. Acellular dermal matrix in cleft palate repair: an evidence-based review. Plast Reconstr Surg 2012;130(1):177–82.

[74] Niknejad H, Peirovi H, Jorjani M, Ahmadiani A, Ghanavi J, Seifalian AM. Properties of the amniotic membrane for potential use in tissue engineering. Eur Cell Mater 2008;15:88–99.

[75] Maan ZN, Rennert RC, Koob TJ, Januszyk M, Li WW, Gurtner GC. Cell recruitment by amnion chorion grafts promotes neovascularization. J Surg Res 2015;193(2):953–62.

[76] Koob TJ, Lim JJ, Zabek N, Massee M. Cytokines in single layer amnion allografts compared to multilayer amnion/chorion allografts for wound healing. J Biomed Mater Res B Appl Biomater 2014;103(5):1133–40.

[77] Koob TJ, Lim JJ, Massee M, Zabek N, Denoziere G. Properties of dehydrated human amnion/chorion composite grafts: implications for wound repair and soft tissue regeneration. J Biomed Mater Res B Appl Biomater 2014;102(6):1353–62.

Chapter 7

Advances in Acellular Extracellular Matrices (ECM) for Wound Healing

Rhiannon Harries[1], Jared Torkington[2], Keith Harding[3]
[1]*School of Medicine, Cardiff University, Cardiff, UK;* [2]*Cardiff and Vale Health Board, Cardiff, UK;* [3]*Cardiff University, Medical Director of Welsh Wound Innovation Initiative, Cardiff, UK*

Chapter Outline

INTRODUCTION

The majority of wounds heal uneventfully; however, a small proportion fail to progress through an orderly sequence of repair in an appropriate time and become chronic and nonhealing. The common types of wounds can be classified as seen in Figure 1.

It is estimated that around 1% of the general population suffer from chronic wounds (of more than 4 weeks duration) that have failed the standard treatment of care [1–4]; this burden will inevitably grow with an increasing population. The management of chronic wounds is a large financial drain on healthcare resources. In 1998, the estimated annual cost of wounds to the National Healthcare System in the UK was £1 billion [5], and since, in 2005–2006, it was estimated that chronic wounds cost £2.5–3 million per 100,000 population [6]. These financial costs do not take into account the personal costs to the patient, such as a reduction in a patient's ability to work and reduction in quality of life. The management of chronic wounds represents a difficult dilemma for clinicians.

Skin Tissue Engineering and Regenerative Medicine. http://dx.doi.org/10.1016/B978-0-12-801654-1.00007-3

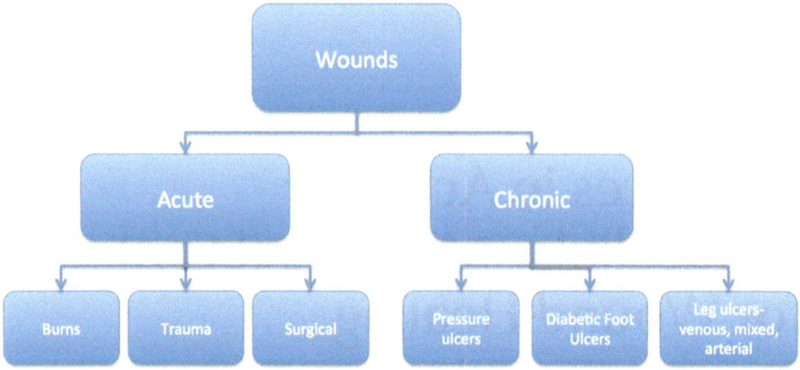

FIGURE 1 Suggested classification of wounds.

While chronic wounds may be classified by etiology, they are often challenging to treat, present at differing times after their onset and are associated with various comorbidities, thus making clinical trials (in human subjects) of treatments rather difficult [7]. Animal models are currently used in research, by subjecting an acute wound to a range of clinical causes of chronic wounds such as ischemia, diabetes, pressure, and reperfusion damage; however, these have limitations including lack of genetic tractability, high costs, technical difficulties, and the inability to replicate aging or microbiotica [8].

Wound healing takes place when a series of complex interactions between epidermal cells, dermal cells, the extracellular matrix (ECM), and plasma-derived proteins are controlled by cytokines and growth factors. These are described in a series of stages, which often overlap [9]:

1. Hemostasis: Interaction between endothelial cells, platelets, and coagulation factors in order to achieve hemostasis after tissue wounding. Trapped cells within the clot, predominantly platelets, trigger an inflammatory response by the release of vasodilators and chemoattractants and activation of the complement cascade.
2. Inflammation: Firstly, phagocytotic neutrophils predominate, removing bacteria and other foreign material from the wound. Subsequently, neutrophils are replaced by macrophages, which coordinate transition from inflammation to proliferation by releasing soluble mediators including growth factors.
3. Proliferation: Angiogenesis takes place. Fibroblasts contribute to the production of the new ECM, while existing matrix is degraded by enzymes (regulated by tissue inhibitors) such as matrix metalloproteinases (MMPs) and plasminogen activators. Keratinocytes at the wound edge migrate across the wound bed, resulting in reepithelialization.
4. Remodeling: Following wound closure, remodeling of the resulting scar takes place over months or years, with a reduction of both cell content and blood flow in the scar tissue.

There are a range of factors that can contribute to a failure for wound healing, either local, including inadequate tissue oxygenation, infection, maceration, and presence of foreign bodies, or patient-related factors, including malnutrition, diabetes, renal disease, immunosuppression, and advanced age [9]. In addition, both local and patient-related factors, aberrant biological responses including inadequate fibroblast function, a reduction in tissue growth factors, excessive or uncontrolled protease activity can all lead to a dysfunctional ECM [10–14].

The ECM is a gel-like matrix that is an integral component of the dermal skin layer, providing a structural, biochemical, and biomechanical support for the cellular constituents in order for tissue morphogenesis, differentiation, homeostasis, and eventual regeneration to take place [15]. It is comprised of water and two main classes of macromolecules: fibrous proteins and polysaccharides. Fibroblasts and epidermal cells secrete these ECM macromolecules [16].

Fibrous proteins include collagen, fibronectin, elastin, and laminin and give the ECM strength and resilience [17]. Fibrillar collagens are the most commonly found fibrous protein within the dermal ECM, consisting of primarily Type I (80–85%) and Type III (8–11%) collagens [15]. They provide structural support to resident cells, tensile strength, regulate cell adhesions, and support migration. Fibroblasts are responsible for the secretion of collagen and organize it into sheets and cables. Fibronectin, a dimeric molecule that exists in both soluble and tissue isoforms, is the second most abundant component in the dermal ECM. It has many types of binding sites for collagen and fibrin molecules and proteoglycans, allowing it to link together different types of ECM molecules and facilitates cell adhesion [18]. Elastin fibers provide recoil or elasticity to tissues that undergo repeated stretch, consist of elastin molecules forming a core of elastin fibers, and are covered by glycoprotein microfibrillins [19]. Laminin helps to bind ECM components [17].

Polysaccharides in the ECM consist of glycosaminoglycans and proteoglycans [20]. Glycosaminoglycans are composed of polysaccharide chains and are strongly hydrophilic; they attract osmotically active Na^+, causing large amounts of water to be drawn in. Hyaluronic acid is one of the largest glycosaminoglycans but, unlike other glycosaminoglycans, is not found in proteoglycans. Hyaluronic acid regulates cell behavior by stimulating macrophage cytokine production, and thus promoting angiogenesis [20]. Glycosaminoglycans attach to ECM proteins to form proteoglycans, which retain water, form a gel-like substance through which ions, hormones, and nutrients can move freely and allow storage of growth factors [20]. There are an extensive number of growth factors naturally found within the ECM, including vascular endothelial cell growth factor (VEGF), fibroblast growth factor (FGF), epithelial cell growth factor (EGF), platelet-derived growth factor (PDGF), transforming growth factor β (TGF-β), and keratinocyte growth factor (KGF) [21–25]. These bind to some components of the ECM and create a reservoir of active molecules that can be rapidly mobilized following tissue injury in order to stimulate cell proliferation and migration.

Traditionally, split-thickness skin grafts (STSG) have been the "gold standard" for wound coverage in skin defects requiring surgery, burns wounds, and chronic wounds. However, this can be technically difficult to perform (particularly, if there is a shortage of available autologous donor sites; i.e., in patients with significant burns), painful, result in scarring or failure of graft take, and have associated donor-site complications (Figure 2).

Advances in the understanding of chronic wound biology have led to the development of tissue-engineered products, targeting the proliferation stage of wound healing, which can be applied to the wound bed. These contain a collagen matrix or scaffold that replaces or stimulates the ECM by facilitating attachment, migration, proliferation, differentiation, and three-dimensional spatial organization of the cell population required for tissue regeneration [26,27].

These tissue-engineered products can be either acellular (biologically inert) or cellular (containing living cells). In this chapter, we discuss acellular ECM products for use in chronic wound healing.

ACELLULAR MATRICES

Acellular matrices have living cells removed in order to prevent an inflammatory or immunogenic response [28], and can be derived from:

- Biological tissue (synthesized by nature, with all cells removed during manufacture)
    - Animal
    - Human

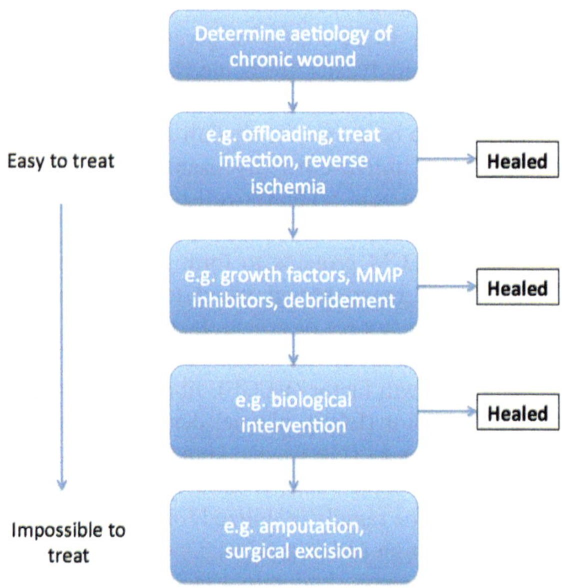

FIGURE 2 Suggested flow diagram illustrating differential management of chronic wounds.

- Synthetic (derived from manmade products)
- Composite (derived from a mixture of various materials)

Products derived from animal are referred to as xenografts. They are developed from harvested living tissue, such as dermis, urinary bladder, pericardium, amniotic membrane, small bowel, or stomach. They can be developed from various donor animals (porcine, equine, and bovine). These matrices undergo a process called decellularization (removal of cells) in order to leave the remaining collagen matrix. Many animal-derived products can consist of a tissue scaffold only or can be combined with synthetic materials to create a composite product. Products derived from human sources, usually from screened donated human cadaver skin, are referred to as allografts. They undergo processes to remove living cells and destroy pathogens.

These products are produced in a variety of delivery systems: temporary skin substitute (e.g., Biobrane®), implant (e.g., OASIS™), a staged procedure (e.g., Integra®), or result in a biodegradable gel (e.g., Promogran Prisma®). They contain active products such as collagen, oxidized regenerated cellulose, vitronectin, and/or hyaluronic acid. These are discussed in detail later in this chapter.

MANUFACTURING PROCESS

When manufacturing acellular matrices, the aim is to remove all cellular components, using nondamaging processes, but maintain the structure and function of the source tissue, thereby preventing rapid degradation of tissue and minimizing inflammatory or immunologic reactions [29]. If degradation of the source tissue by manufacturing processes occurs, it can lead to reabsorption of the matrix by the host and scar tissue formation [29]. Adverse reactions can occur as a result of inflammatory or immunologic responses to cellular components, with accumulation of cells around the matrix preventing vascular infiltration.

Manufacturing of these acellular ECM can involve a number of different processes.

Cross-linking

Cross-linking is the process of stabilizing collagen by creating new links between strands of collagen; this process inhibits degradation of the collagen by proteases and prolongs its presence in the wound [28]. The type of cross-linking process that takes place directly affects the durability of acellular matrices. Methods of cross-linking involving chemical (e.g., aldehydes) or mechanical processes (e.g., heat or radiation) may result in little control over the degree of cross-linking and produce bonds between collagen strands that are short and inflexible [28]. These shorter inflexible links can result in inhibition of cell migration and vascular regeneration, and residual cross-linking agents in the product may result in cytotoxicity or an inflammatory response, causing the matrix to be rapidly reabsorbed by the host tissue, and may result in scar tissue formation [28]. Newer processes of cross-linking produce more elastic and flexible cross-links between strands of collagen and are less prone to enzymatic breakdown [27]. These more flexible bonds produced by newer processes are thought to allow cells to migrate and proliferate in an organized manner similar to normal tissue repair [30].

Sterilization

Animal-derived products undergo sterilization in order to prevent disease transmission. This process may result in either an inflammatory reaction within the host tissue attributed to residual chemicals used during the process or the radiation may cause significant damage to the matrix [28]. Most human-derived acellular products are aseptically processed and not terminally sterile.

Preservation and Shelf-Life

The preservation media used will affect product stability and overall shelf-life. Product shelf-life is variable, depending on the product, and ranges up to 5 years. It is sensible to perform a saline wash prior to implantation to remove any residual chemicals that are used for product preservation.

MODE OF ACTION

The matrix or scaffold functions by binding to the host and permitting matrix–cell interaction. The matrix contains ECM proteins such as collagen, hyaluronic acid, or fibronectin, thus ensuring biocompatibility [15]. Following matrix implantation, it is thought that a series of events take place:

1. Imbibition
   Wound fluid containing fibrin and red blood cells permeates the matrix, due to its porous nature. Fibrin causes the matrix to adhere to the wound [31].
2. Fibroblast migration
   The matrix can contain virus vectors or plasmids, which can transcribe and translate the inbuilt DNA leading to secretion of specific growth factors [21,23–25]. These growth factors carry out their specific functions by stimulating host cells. Mononuclear cells (macrophages and mast cells) infiltrate the matrix [32], which secrete cytokines and other signaling factors for fibroblast proliferation. Fibroblasts migrate into the matrix using the collagen fibers as a scaffold.
3. Neovascularization
   Roughly 3 weeks after matrix application, endothelial cells migrate and lumen formation takes place, in order to support newly settled fibroblasts. Neovascularization is established by week 4 [31].

4. Matrix remodeling and maturation
   The final stage of matrix integration consists of remodeling with new collagen deposition and tissue regeneration [31]. It is at this stage that either the implanted matrix is often resorbed or biodegraded by the host or the temporary dressing is removed. This stage may be affected by the rate at which the matrix is degraded. This stage should occur without a significant reaction since it may lead to increased scarring [28].

The skin substitute can work either as:

- A temporary skin substitute (acting like a dressing), which provides transient physiologic wound closure, protects against trauma, acts as a barrier to prevent infection, and maintains a moist wound environment until wound healing has taken place. Examples include:
  - Biobrane® skin substitute (Bertek Pharmaceuticals, Morgantown, West Virginia, USA) in use since the 1970s for treatment of partial-thickness burn wounds. It is a synthetic dressing composed of a collagen peptide-coated nylon mesh, which acts as a "dermis," and a semipermeable silicon membrane, which acts as an "epidermis," that are embedded in porcine collagen and incorporated by chemical linkage to enhance its bond to the wound bed [33,34]. It provides a flexible and adherent surface for wound coverage, with the silicone surface controlling water vapor loss from the wound. Biobrane is intended as a temporary wound dressing and is removed either when the wound is healed or when autograft skin is available.
  - Endofoam™ Dermal Template (Mesynthes, Wellington, New Zealand) is sourced from propria–submucosa layers of ovine forestomach and sterilized by ethylene oxide. It is used as a temporary dressing in wound care.
  - EZDerm® (Molnlycke, Gothenburg, Sweden) is an aldehyde cross-linked porcine dermis-derived product, and is intended as a temporary coverage of wounds prior to split-thickness skin grafting.
  - Hyalomatrix PA (Fidia Advanced Biopolymers, AbanoTerme, Italy) is a synthetic bilayered product consisting of a thin moisture vapor–permeable silicone sheet adhered to a nonwoven pad composed entirely of the benzyl ester of hyaluronic acid.
  - Puracol® Plus Microscaffold Collagen (Dr. Suwelack Skin & Health Care AG, Germany) is derived from bovine collagen, and is available with impregnated silver.
  - Suprathel® (BioMed Sciences, Allenstown, Pennsylvania, USA) is composed of Lacto-capromer and is gamma irradiated.
- A permanent skin substitute (acting like an implant) provides a permanent wound closure, and replaces lost skin components. Examples include:
  - AlloDerm® Regenerative Tissue Matrix (LifeCell Corporation, Branchburg, New Jersey, USA) is an allograft created from cadaveric human dermis and is aseptically processed to remove any cellular matter, then freeze dried to

render it immunologically inert, but its basement membrane remains intact [33,34]. It is currently available in a ready-to-use product that is stored at room temperature.

- AlloMax™ Surgical Graft (Bard Davol Inc., New Jersey, USA) is a noncross-linked human dermis allograft processed using Tutoplast® sterilization process and low-dose gamma irradiation to remove all noncollagenous cellular components.
- AminoFix® (MiMedix, Marietta, Georgia, USA) is a composite dehydrated amnion tissue membrane sterilized using the PURION® process.
- Biodesign® (Cook Medical, Bloomington, Indiana, USA) is a cross-linked xenograft derived from porcine small intestine submucosa.
- BIOPAD® Collagen Wound Dressing (Euroresearch, Milan, Italy) is derived from equine flexor tendon and sterilized using gamma irradiation.
- CollaMend Implant (Bard Davol Inc., New Jersey, USA) is an EDC cross-linked xenograft sourced from porcine dermis. It is sterilized using ethylene oxide.
- DermaMatrix™ (Synthes, Solothurn, Switzerland) is an allograft derived from donated human skin tissue. DermaMatrix Acellular Dermis is processed by the Musculoskeletal Transplant Foundation®.
- FlexHD® (Musculoskeletal Transplant Foundation, Edison, New Jersey, USA) is an acellular dermal matrix derived from donated human allograft skin. The tissue is prehydrated and ready to use.
- GraftJacket® Regenerative Tissue Matrix (Wright Medical Technology Inc., Memphis, Tennessee, USA) is a regenerative tissue matrix that has been aseptically processed from screened donated human skin supplied from US tissue banks.
- OASIS™ Wound Matrix (Cook Medical, Bloomington, Indiana, USA) is a xenogeneic collagen scaffold derived from porcine small intestinal mucosa. It is sterilized by ethylene oxide.
- Matriderm® (Dr. Suwelack Skin & Health Care AG, Germany) is xenograft derived from bovine dermis used in combination with an STSG and sterilized using gamma irradiation.
- Matristem™ Wound Care Matrix (Acell Inc., Columbia, Maryland, USA) is a xenograft product derived from porcine urinary bladder matrix sterilized by electron beam irradiation.
- Neoform™ (Mentor, Irving, Texas, USA) is derived from human dermis and sterilized using Tutoplast® process and low-dose gamma irradiation.
- Permacol® (Tissue Science Laboratories Inc., Andover, Maryland, USA) is an HDMI cross-linked xenograft derived from porcine dermis which is sterilized using gamma irradiation.
- PriMatrix™ (TEI Biosciences, Boston, Massachusetts, USA) is a xenogeneic matrix from fetal bovine dermis that is processed using ethylene oxide.

 - Strattice™ Reconstructive Tissue Matrix (LifeCell Corporation, Branchburg, New Jersey, USA) is derived from porcine dermis and processed using electron beam irradiation.
 - SureDerm (Hans Biomed, South Korea) is an allogenic acellular human dermis product from donated screened skin.
 - Unite® Biomatrix Collagen Wound Dressing (Synovis Orthopedic and Woundcare, Irvine, California, USA) is derived from equine pericardium and EDC processed.
 - Veritas® Collagen Matrix (Synovis Orthopedic and Woundcare, Irvine, California, USA) is derived from bovine pericardium and sterilized by sodium hydroxide.
 - XenMatrix™ Surgical Graft (Bard Davol Inc., New Jersey, USA) is derived from bovine dermis and is processed with electron beam irradiation.
- A mixture of both temporary and permanent skin substitute. An example:
 - Integra® Dermal Regeneration Template and Integra® Bilayer Matrix Wound Dressing (Integra LifeSciences, Plainsboro, New Jersey, USA) are synthetic bilayered bovine collagen/shark chondroitin-6-sulphate glycosaminoglycan dermal replacements covered by a silicone temporary epidermal substitute for barrier replacement [34,35]. The dermal substitute becomes populated with cells from the wound bed, and wound revascularization takes place over 2–3 weeks following application, thus incorporating it into the wound. After this time, the superficial silicone layer can be removed and an STSG can be applied.
- A product that results in a biodegradable gel. Examples include:
 - Promogran Prisma® (Systagenix, Gatwick, UK) is a sterile, freeze-dried matrix comprised of collagen, oxidized regenerated cellulose, and silver. In the presence of wound exudate, the matrix absorbs liquid and forms a biodegradable gel.
 - Excellagen® (Taxus Cardium Pharmaceuticals Group, San Diego, California, USA) is a formulated fibrillar type I bovine collagen gel.
 - Biopolymer composite (a combination of atelocollagen and chitosan) is a biodegradable molecular matrix in gel form as a biomimetic surface for cell attachment [36].

MATRIX APPLICATION

Prior to application of the acellular matrix, it is important to assess the patient for suitability and compliance, to establish a diagnosis for the wound and to exclude ischemia, infection, and allergy [37]. Adequate wound bed preparation should be performed consisting of tissue management (e.g., debridement of nonviable tissue), inflammation and infection control, moisture balance, and epithelial advancement [38]. It is recommended that the manufacturer's instructions be followed, as each product will vary. It may be necessary to fenestrate the matrix, if there is concern about heavy exudate from the wound, to ensure the matrix is in complete contact with the wound bed and that the exudate is removed. Once the matrix is in place, it should be secured with steri-strips, staples, or sutures, depending on clinician preference and wound edge appearance. A nonadherent primary dressing should be placed over the matrix, with a secondary dressing to hold this in place.

It is important that the matrix should not be disturbed for a minimum of 7 days, in order to avoid displacement [37]. If it is displaced, it is advisable to replace with a new matrix. The attachment method should be removed at a maximum of 1 week for staples and 2 weeks for sutures or steri-strips. The wound should be monitored for signs of complications (i.e., infection, allergic reaction, and so on). Evidence suggests that a reduction in the size of the wound bed within the first 4 weeks following matrix application is predictive of complete healing at 12 weeks duration [39]. STSG may be applied, if necessary. Large wound areas may require repeated application of the matrix (Table 1).

EVIDENCE OF ACELLULAR MATRICES USE IN OTHER CONDITIONS

Biobrane use started in the late 1970s and has been used as a skin substitute in superficial or mid-dermal partial-thickness burns since then [40–42]. There is also a plethora of evidence on the use of other acellular matrices in the management of burns, such as Integra [43,44], Alloderm [45,46], and Matriderm [47,48].

Acellular dermal matrices have also been described for use in complex hernia repair and abdominal wall reconstruction [49], breast reconstruction [50], perineal defect repair following abdominoperineal resection of the rectum [51], soft tissue repairs [52], axillary reconstruction following hidradenitis suppurativa excision [53], and facial reconstructions [54].

CONCLUSIONS

Chronic wounds are relatively common and have huge financial implications. The etiology of chronic wounds can be diverse, and the healing process is affected by patients' comorbidities, as well as local wound factors. There is an urgent need for improved treatment options for the management of chronic wounds; however, wound biology is complex, and there are limitations to research in this area. Tissue-engineered products have been developed that are directed at the proliferation stage of wound healing and aim to replicate or replace the ECM. When considering the use of acellular dermal matrices, it is important to select patients appropriately, target the treatment to the clinical assessment, and integrate with clinical care.

TABLE 1 Evidence of Acellular Matrices Use in Chronic Wounds

Publication	Type of Study	Number of Patients	Type of Wound	Product Used	Results
Veves et al. [75], Arch Surg 2002; 137(7):822–827.	Randomized control trial	276	Diabetic foot ulcer	Promogran® versus standard of care	37.0% of the Promogran®-treated patients had complete wound closure, compared to 28.3% of control at 12-week follow-up
Brigido et al. [55], Orthopedics 2004; 27(1 Suppl):s145–9.	Prospective randomized single blind pilot	40	Diabetic foot ulcer	GraftJacket® versus hydrogel wound dressing (Curasol®)	85% in the GraftJacket® group had healed compared to 5% in the control group at 12-week follow-up
Demling et al. [60], wounds 2004; 16(1):18–22	Prospective multicentre randomized control trial	84	Venous leg ulcer	Oasis® and compression versus compression alone	71% in the Oasis® group had healed compared to 46% in the control group at 12-week follow-up
Niezgoda et al. [71], Adv skin wound care 2005; 18(5):258–66.	Prospective multicentre randomized control trial	73	Diabetic foot ulcer	Oasis® versus becaplermin wound gel (Regranex®)	49% in the Oasis® group had healed compared to 28% in the Regranex® group at 12-week follow-up
Mostow et al. [70], J Vasc Surg 2005; 41(5):837–43.	Prospective multicentre randomized control trial	120	Venous leg ulcer	Oasis® and compression versus compression alone	55% in the Oasis® group had healed compared to 34% in the control group at 12-week follow-up

Continued

TABLE 1 Evidence of Acellular Matrices Use in Chronic Wounds—cont'd

Publication	Type of Study	Number of Patients	Type of Wound	Product used	Results
Campitiello et al. [57], Acta BioMed 2005; 76(suppl 1):69–71.	Retrospective case series	7	Chronic wounds	Integra™ and split-thickness skin grafts	100% were ready for split skin graft at 15 days
Silverstein et al. [73], Foot Ankle Surg 2006; 45(1):28–33.	Retrospective case series	5	Diabetic foot ulcer	Integra™ and split-thickness skin grafts	Despite two grafts failing all wounds healed
Brigido et al. [56], int wound J 2006; 3(3):181–87.	Prospective randomized control trial	28	Diabetic foot ulcer	GraftJacket® versus hydrogel wound dressing (Curasol®)	85.7% in the GraftJacket® group had healed compared to 28.6% in the control group at 16-week follow-up
Helgeson et al. [63], J Orthop trauma 2007; 21:394–399.	Retrospective case series	16	Traumatic extremity wound	Integra™	94% healed at completion of follow-up
Romanelli et al. [78], Int Wound J 2007; 4(1):3–7.	Prospective randomized control trial	54	Mixed arterial/ venous ulcers	Oasis® versus hyaluronic acid (HA) biomaterial (Hyaloskin)	82% of wounds achieved complete healing at 16 weeks
Winters et al. [77], Adv skin wound care 2008; 21(8):375–81.	Retrospective multicentre case series	75 patients with 100 wounds	Diabetic foot ulcer	GraftJacket®	91% of wounds achieved complete healing
Brigido et al. [79], Foot Ankle Spec 2009; 2(2):67–72.	Retrospective case series	12	Diabetic foot ulcer	GraftJacket® Xpress scaffold	10 of 12 patients achieved complete healing at 12-week follow-up; the average time to complete healing was 8.5 weeks
Reyzelman et al. [72], Int Wound J 2009; 6(3):196–208.	Prospective multicentre randomized control	86	Diabetic foot ulcer	GraftJacket® versus standard of care	69.6% were completely healed at 12 weeks in the GraftJacket® compared with 46.2% in the

Lower Extrem Wounds 2009; 8(4):209–12.			ulcer		follow-up
Fleischli et al. [61], J Am Pod med 2009; 99(4):301–05.	Prospective pilot case study	23 patients with 34 wounds	Diabetic foot ulcer	Unite®	94% of wounds had improved on removal of Unite®; at 12 weeks, 47% of wounds had healed
Cervelli et al. [58], Int Wound J 2010; 7(4):291–96.	Prospective case report	1	Diabetic foot ulcer	Matriderm® with split skin graft	At 15 days, reduced pain, exudate secretion, and size of ulcer noted
Taras et al. [74], J Hand Surg Am 2010; 35:415–421.	Retrospective case series	17	Traumatic extremity wound	Integra™	80% healed at 12-month follow-up
Clerici et al. [59], Int Wound J 2010; 7:176–183.	Retrospective case series	30	Diabetic foot ulcer	Integra™	87% healed at average follow-up of 74 days
Karr [67], Adv Skin Wound Care 2011; 24(3):119–25.	Retrospective cohort study	68	Diabetic foot and venous leg ulcers	Primatrix™ versus Apligraf (a cellular dermal matrix)	Healing times were quicker in both diabetic foot and venous leg ulcers for Primatrix compared to Apligraf
Iorio et al. [64], Plast Reconstr Surg 2011; 127(1):260–7.	Retrospective review	105 patients with 121 wounds	Diabetic foot ulcer	Integra™	Use of a collagen bilayer matrix appears to be a viable option for reconstruction and stable closure in the diabetic patient at low risk for amputation
Weigert et al. [76], J Hand Surg Eur Vol 2011; 36:185–193.	Retrospective case series	15	Traumatic extremity wound	Integra™	87% healed at 37-month follow-up
Khan et al. [66], J Burn Care Res 2011; 32:124–8.	Retrospective case series	10	Chronic wounds of various etiologies	Integra™	87% healed at 9-month follow-up. 87% had take of split skin graft at 21 days

Continued

TABLE 1 Evidence of Acellular Matrices Use in Chronic Wounds—cont'd

Publication	Type of Study	Number of Patients	Type of Wound	Product used	Results
Fraccalvieri et al. [62], int Wound J 2012; 9:214–220.	Retrospective case series	7	Diabetic foot ulcers with associated osteomyelitis	Integra™	94% healed at 36-month follow-up
Lullove [69], J Am Podiatr Med Assoc 2012; 102(3):233–9.	Retrospective case series	29 patients with 34 wounds	Chronic wounds of various etiologies	PriMatrix™	30 of 34 wounds healed; mean time for healing was 105 days for diabetic foot ulcers and 82 days for venous, traumatic, and other chronic wounds.
Jeon et al. [65], Arch Plast Surg 2013; 40(4):403–08.	Prospective	60	Diabetic foot ulcer	Matriderm® and split-thickness skin graft versus split-thickness skin graft only	A statistical significance in duration for complete healing was observed $p<0.05$; 8.61 weeks in the Matriderm® and split-thickness skin graft group compared to 12.94 weeks in the split-thickness skin graft only group
Harding et al. [12], Int Wound J 2014; 112:129–37.	Multicentre cohort study	45	Venous leg ulcer	Synthetic extracellular matrix	Healing was achieved in 35.6% and wound size decreased in 93.9% of patients
Kavros et al. [68], Adv Kin Wound Care 2014; 27(8):356–62.	Prospective multicentre case study	55	Diabetic foot ulcer	PriMatrix™	76% of ulcers had healed at 12 weeks

LIST OF ACRONYMS AND ABBREVIATIONS

AATB American Association of Tissue Banks
DNA Deoxyribonucleic acid
ECM Extracellular matrix
EGF Epithelial cell growth factor
FDA Food and Drug Administration
FGF Fibroblast growth factor
KGF Keratinocyte growth factor
MMPs Matrix metalloproteinases
NICE National Institute of Health and Care Excellence
PDGF Platelet-derived growth factor
STSG Split-thickness skin graft
TGF-β Transforming growth factor β
VEGF Vascular endothelial cell growth factor

REFERENCES

[1] Collins L, Seraj S. Diagnostic and treatment of venous ulcers. Am Fam Physician 2010;81:989–96.
[2] Hansson C, Andersson E, Swanbeck G. Leg ulcer epidemiology in Gothenburg. Acta Chir Scand Suppl 1988;544:12–6.
[3] Sen CK, Gordillo GM, Roy S, Kirsner R, Lambert L, Hunt TK, et al. Human skin wounds: a major and snowballing threat to public health and the economy. Wound Repair Regen 2009;17:763–71.
[4] Skene AI, Smith JM, Dore CJ, Charlett A, Lewis JD. Venous leg ulcers: a prognostic index to predict time to healing. BMJ 1992;305:1119–21.
[5] Harding KG. The future of wound healing. In: Leaper DJ, Harding KJ, editors. Wounds: biology and management. Oxford: Oxford University Press; 1998. p. 191.
[6] Drew P, Posnett J, Rusling L. Wound Care Audit Team. The cost of wound care for a local population in England. Int Wound J 2007;4:149–55.
[7] Gottrup F, Apelqvist J. Present and new techniques and devices in the treatment of DFU: a critical review of evidence. Diabetes Metab Res Rev 2012;28(Suppl. 1):64–71.
[8] Nunan R, Harding KG, Martin P. Clinical challenges of chronic wounds: searching for an optimal animal model to recapitulate their complexity. Dis Model Mech 2014;7:1205–13.
[9] Harding KG, Morris HL, Patel GK. Healing chronic wounds. BMJ 2002;324:160–3.
[10] Gibson D, Cullen B, Legerstee R, Harding KG, Schultz G. MMPs made easy. Wounds International 2009;1(1).
[11] Harding KG, Morre K, Phillips TJ. Wound chronicity and fibroblast senescence- implications for treatment. Int Wound J 2005;2:364–8.
[12] Harding KG, Aldons P, Edwards H, Stacey M, Finlayson K, Gibb M, et al. Effectiveness of an acellular synthetic matrix in the treatment of hard-to-heal leg ulcers. Int Wound J 2014;112:129–37.
[13] Hodde JP, Johnson CE. Extracellular matrix as a strategy for treating chronic wounds. Am J Clin Dermatol 2007;8:61–6.
[14] Lukashev ME, Werb Z. ECM signalling: orchestrating cell behaviour and misbehaviour. Trends Cell Biol 1998;8(11):437–41.
[15] Frantz C, Stewart KM, Weaver VM. The extracellular matrix at a glance. J Cell Sci 2010;123:4195–200.

[16] Järveläinen H, Sainio A, Koulu M, Wight TN, Penttinen R. Extracellular matrix molecules: potential targets in pharmacotherapy. Pharmacol Rev 2009;61(2):198–223.

[17] Alberts B, Johnson A, Lewis J, Raff M, Roberts K, Walter P. Molecular biology of the cell. London: Garland Science; 2007.

[18] Smith ML, Gourdon D, Little WC, Kubow KE, Eguiluz RA, Luna-Morris S, et al. Force-induced unfolding of fibronectin in the extracellular matrix of living cells. PLoS Biol 2007;5(10):e268.

[19] Wise SG, Weiss AS. Tropoelastin. Int J Biochem Cell Biol 2009;41(3):494–7.

[20] Schaefer L, Schaefer RM. Proteoglycans: from structural compounds to signaling molecules. Cell Tissue Res 2010;339(1):237–46.

[21] Bonadio J, Smiley E, Patil P, Goldstein S. Localized, direct plasmid gene delivery in vivo: prolonged therapy results in reproducible tissue regeneration. Nat Med 1999;5(7):753–9.

[22] Boudreau N, Bissell MJ. Extracellular matrix signalling: integration of form and function in normal and malignant cells. Curr Opin Cell Biol 1998;10(5):640–6.

[23] Luu YK, Kim K, Hsiao BS, Chu B, Hadjiargyrou M. Development of a nanostructured DNA delivery scaffold via electrospinning of PLGA and PLA-PEG block copolymers. J Control Release 2003;89(2):341–53.

[24] Murphy WL, Mooney DJ. Controlled delivery of inductive proteins, plasmid DNA and cells from tissue engineering matrices. J Periodontal Res 1999;34(7):25.

[25] Shea LD, Smiley E, Bonadio J, Mooney DJ. DNA delivery from polymer matrices for tissue engineering. Nat Biotechnol 1999;17(6):551–4. 413-9.

[26] Caplan AI. Design parameters for functional tissue engineering. In: Guilak F, Butler DL, Goldstein SA, Mooney DJ, editors. Functional tissue engineering. Springer-Verlag; 2003. p. 129–30.

[27] Zhong SP, Zhang YZ, Lim CT. Tissue scaffolds for skin wound healing and dermal reconstruction. WIREs Nanomed Nanobiotchnol 2010;2:510–25.

[28] Nataraj C, Ritter G, Dumas S. Extracellular wound matrices: novel stabilisation and sterilisation method for collagen-based biologic wound dressings. Wounds International 2007;19(6):148–56.

[29] Cornwall KG, Landsman A, James KS. Extracellular matrix biomaterials for soft tissue repair. Clin Podiatr Med Surg 2009;26:507–23.

[30] Powell HM, Boyce ST. EDC cross-linking improves skin substitute strength andd stabilty. Biomaterials 2006;27(34):5821–7.

[31] Moiemen NS, Staiano JJ, Ojeh NO, Thway Y, Frame JD. Reconstructive surgery with a dermal regeneration template: clinical and histologic study. Plast Reconstr Surg 2001;108:93–103.

[32] Badylak SF, Valentin JE, Ravindra AK, McCabe GP, Stewart-Akers AM. Macrophage phenotype as a determinant of biologic scaffold remodeling. Tissue Eng Part A 2008;14(11): 1835–42.

[33] Horch RE, Jeschke MG, Spilker G, Herndon DN, Kopp J. Treatment of second-degree facial burns with allografts-preliminary results. Burns 2005;31:597–602.

[34] Supp DM, Boyce ST. Engineered skin substitutes: practices and potentials. Clin Dermatol 2005;23:403–12.

[35] Bello YM, Falabella AF, Eaglstein WH. Tissue-engineered skin. Current status in wound healing. Am J Clin Dermatol 2001;2:305–13.

[36] Judith R, Nithya M, Rose C, Mandal AB. Biopolymer gel matrix as acellular scaffold for enhanced dermal tissue regeneration. Biologicals 2012;40(4):231–9.

[37] International consensus. Acellular matrices for the treatment of wounds. An expert working group review. Wounds International; 2011.

[38] Falanga V. Wound bed preparation: science applied to pratice. In: European wound management association (EWMA). Position document. Wound bed preparation in pratice. London: MEP Ltd; 2004. p. 2–5.

[39] Sheehan P, Jones P, Giurini JM, Caselli A, Veves A. Percent change in wound area of diabetic foot ulcers over a 4-week period is a robust predictor of complete healing in a 12-week prospective trial. Plast Reconstr Surg 2006;117(7 Suppl.):239S–44S.

[40] Barret JP, Dziewulski P, Ramzy PI, Wolf SE, Desai MH, Herndon DN. Biobrane versus 1% silver sulfadiazine in second-degree paediatric burns. Plast Reconstr Surg 2000;105(1):62–5.

[41] Hansbrough JF, Zapata-Sirvent R, Carroll WJ, Dominic WJ, Wang XW, Wakimoto A. Clinical experience with biobrane biosynthetic dressing in the treatment of partial thickness burns. Burns Incl Therm Inj 1984;10(6):415–9.

[42] Lesher AP, Curry RH, Evans J, Smith VA, Fitzgerald MT, Cina RA, et al. Effectiveness of biobrane for treatment of partial-thickness burns in children. J Pediatr Surg 2011;46(9):1759–63.

[43] Branski LK, Herndon DN, Pereira C, Mlcak RP, Celis MM, Lee JO, et al. Longitudinal assessment of Integra in primary burn management: a randomized pediatric clinical trial. Critical Care Medicine 2007;35(11):2615–23.

[44] Nguyen DQ, Potokar TS, Price P. An objective long-term eveluation of Integra (a dermal skin substitute) and split thickness skin graft, in acute burns and reconstructive surgery. Burns 2010;36(1):23–8.

[45] Sheridan RL, Choucair RJ, Donelan M, Lydon RJ, Petras L, Tompkins R. Acellular allogenic dermis in burn surgery: 1 year results of pilot trial. J Burn Care Rehabil 1998;19:528–46.

[46] Wainwright D, Madden M, Luterman A, Hunt J, Monafo W, Heimbach D, et al. Clinical evaluation of an acellular allograft dermal matrix in full-thickness burns. J Burn Care Rehabil 1996;17(2):124–36.

[47] Haslik W, Kamolz LP, Manna F, Hladik M, Rath T, Frey M. Management of full-thickness skin defects in the hand and wrist region: first long-term experiences with the dermal matrix Matriderm. J Plast Reconstr Aesthet Surg 2010;63(2):360–4.

[48] Ryssel H, Germann G, Kloeters O, Gazyakan E, Radu CA. Dermal substitution with Matriderm in burns on the dorsum of the hand. Burns 2010;36(8):1248–53.

[49] Patton Jr JH, Berry S, Kralovich KA. Use of human acellular dermal matrix in complex and contaminated abdominal wall reconstructions. Am J Surg 2007;193(3):360–3.

[50] Spear SL, Sinkin JC, Al-Attar A. Porcine acellular dermal matrix (strattice) in primary and revision cosmetic breast surgery. Plast Reconstr Surg 2013;131(5):1140–8.

[51] Harries RL, Luhmann A, Harris DA, Shami JA, Appleton BN. Prone extralevator abdominoperineal excision of the rectum with porcine collagen perineal reconstruction (Permacol™): high primary perineal wound healing rates. Int J Colorectal Dis 2014;29(9):1125–30.

[52] Barber FA, Burns JP, Deutsch A, Labbe MR, Litchfield RB. A prospective, randomized evaluation of acellular human dermal matrix augmentation for arthroscopic rotator cuff repair. Arthroscopy 2012;28(1):8–15.

[53] Melkun ET, Few JW. The use of biosynthetic skin substitute (Biobrane) for axillary reconstruction after surgical excision for hidradenitis suppurativa. Plast Reconstr Surg 2005;115(5):1385–8.

[54] Phillips J, Riley KO, Woodworth BA. Porcine small intestine submucosal grafts for post-tumor resection orbital reconstruction. Laryngoscope 2014;124(6):E219–23.

[55] Brigido SA, Boc SF, Lopez RC. Effective management of major lower extremity wounds using an acellular regenerative tissue matrix: a pilot study. Orthopedics 2004;27(Suppl. 1):s145–9.

[56] Brigido SA. The use of an acellular dermal regenerative tissue matrix in the treatment of lower extremity wounds: a prospective 16-week pilot study. Int Wound J 2006;3:181–7.

[57] Campitiello F, Della Corte A, Fattopace A, D'Acunzi D, Canonico S. The use of artificial dermis in the treatment of chronic and acute wounds: regeneration of dermis and wound healing. Acta BioMed 2005;76(Suppl. 1):69–71.
[58] Cervelli V, Lucarini L, Cerretani C, Spallone D, Palla L, Brinci L, et al. The use of Matriderm and autologous skin grafting in the treatment of diabetic ulcers: a case report. Int Wound J 2010;7(4):291–6.
[59] Clerici G, Caminiti M, Curci V, Quarantiello A, Faglia E. The use of a dermal substitute to preserve maximal foot length in diabetic foot wounds with tendon and bone exposure following urgent surgical debridement for acute infection. Int Wound J 2010;7:176–83.
[60] Demling RH, Niezgoda JA, Haraway GD, Mostow EN. Small intestinal submucsa wound matrix and full-thickness venous ulcers: preliminary results. Wounds 2004;16(1):18–22.
[61] Fleischli JG, Laughlin TJ, Fleischli JW. Equine pericardium collagen wound dressing in the treatment of the neuropathic diabetic foot wound: a pilot study. J Am Pod Med 2009;99(4): 301–5.
[62] Fraccalvieri M, Pristerà G, Zingarelli E, Ruka E, Bruschi S. Treatment of chronic heel osteomyelitis in vasculopathic patients: can the combined use of Integra, skin graft and negative pressure wound therapy be considered a valid therapeutic approach after partial tangential calcanectomy? Int Wound J 2012;9:214–20.
[63] Helgeson MD, Potter BK, Evans KN, Shawen SB. Bioartificial dermal substitute: a preliminary report on its use for the management of complex combat-related soft. J Orthop Trauma 2007;21:394–9.
[64] Iorio ML, Goldstein J, Adams M, Steinberg J, Attinger C. Functional limb salvage in the diabetic patient: the use of a collagen bilayer matrix and risk factors for amputation. Plast Reconstr Surg 2011;127:260–7.
[65] Jeon H, Kim J, Yeo H, Jeong H, Son D, Han K. Treatment of diabetic foot ulcer using matriderm in comparison with a skin graft. Arch Plast Surg 2013;40(4):403–8.
[66] Kahn SA, Beers RJ, Lentz CW. Use of acellular dermal replacement in reconstruction of nonhealing lower extremity wounds. J Burn Care Res 2011;32:124–8.
[67] Karr JC. Retrospective comparison of diabetic foot ulcer and venous stasis ulcer healing outcome between a dermal repair scaffold (PriMatrix) and a bilayered living cell therapy (Apligraf). Adv Skin Wound Care 2011;24(3):119–25.
[68] Kavros SJ, Dutra T, Gonzalez-Cruz R, Liden B, Marcus B, McGuire J, et al. The use of PriMatrix, a fetal bovine acellular dermal matrix, in healing chronic diabetic foot ulcers: a prospective multicentre study. Adv Skin Wound Care 2014;27(8):356–62.
[69] Lullove E. Acellular fetal bovine dermal matrix in the treatment of nonhealing wounds in patients wit complex comorbidities. J Am Podiatr Med Assoc 2012;102(3):233–9.
[70] Mostow EN, Haraway GD, Dalsing M, Hodde JP, King D. OASIS Venus Ulcer Study Group. Effectiveness of an extracellular matrix graft (OASIS Wound Matrix) in the treatment of chronic leg ulcers: a randomized clinical trial. J Vasc Surg 2005;41(5):837–43.
[71] Niezgoda JA, Van Gils CC, Frykberg RG, Hodde JP. Randomised clinical trial comparing OASIS Wound Matrix to Regranex Gel for diabetic ulcers. Adv Skin Wound Care 2005;18(5):258–66.
[72] Reyzelman A, Crews RT, Moore JC, Moore L, Mukker JS, Offutt S, et al. Clinical effectiveness of an acellular dermal regenerative tissue matrix compared to standard wound management in healing diabetic foot ulcers: a prospective, randomised, multicentre study. Int Wound J 2009;6:196–208.
[73] Silverstein G. Dermal regeneration template in the surgical management of diabetic foot ulcers: a series of five cases. J Foot Ankle Surg 2006;45(1):28–33.

[74] Taras JS, Sapienza A, Roach JB, Taras JP. Acellular dermal regeneration template for soft tissue reconstruction of the digits. J Hand Surg Am 2010;35:415–21.

[75] Veves A, Sheehan P, Pham HT. A randomized, controlled trial of Promogran (a collagen/oxidized regenerated cellulose dressing) vs. standard treatment in the management of diabetic foot ulcers. Arch Surg 2002;137(7):822–7.

[76] Weigert R, Choughri H, Casoli V. Management of severe hand wounds with Integra dermal regeneration template. J Hand Surg Eur Vol 2011;36:185–93.

[77] Winters CL, Brigido SA, Liden BA, Simmons M, Hartman JF, Wright ML. A multicenter study involving the use of a human acellular dermal regenerative tissue matrix for the treatment of diabetic lower extremity wounds. Adv Skin Wound Care 2008;21:375–8.

[78] Romanelli M, Dini V, Bertone M, Barbanera S, Brilli C. OASIS wound matrix versus Hyaloskin in the treatment of difficult-to-heal wounds of mixed arterial/venous aetiology. Int Wound J 2007;4(1):3–7.

[79] Brigido SA, Schwartz E, McCaroll R, Hardin-Young J. Use of an acellular flowable dermal replacement scaffold on lower extremity sinus tract wounds: a retrospective series. Foot Ankle Spec 2009;2(2):67–72.

[80] Clerici G, Caminiti M, Curci V, Quarantiello A, Faglia E. The use of a dermal substitute (Integra) to preserve maximal foot length in a diabetic foot wound with bone and tendon exposure following urgent surgical debridement for an acute infection. Int J Low Extrem Wounds 2009;8(4):209–12.

Chapter 8

Natural Biomaterials for Skin Tissue Engineering

Mihail Climov[1], Tripp Leavitt[1], Joseph Molnar[2], Dennis Orgill[1]
[1]*Tissue Engineering and Wound Healing Laboratory, Division of Plastic Surgery, Department of Surgery, Brigham and Women's Hospital, Harvard Medical School, Boston, MA, USA;* [2]*Department of Plastic Surgery, Wake Forest University School of Medicine, Winston-Salem, NC, USA*

Chapter Outline

INTRODUCTION

There have been significant advances in the field of biomaterials over the past 50 years. More recently, investigators have begun combining cellular materials with biomaterials to develop an area that many have referred to as tissue engineering (TE). Skin, being the largest organ of the body and reasonably easy to study, has been the genesis of many TE constructs. Our goal in this chapter is to explore in detail natural biomaterials that have been used for skin-replacement therapies. The initial approach to skin tissue engineering (STE) is to mimic the natural components of the skin including collagens, glycosaminoglycans (GAGs), and elastins. Some of the first commercially available semisynthetic biologic dressings include products like EZ-Derm®, aldehyde cross-linked porcine dermal collagen (Mölnlycke Health Care, Gothenburg, Sweden); Biobrane®, porcine collagen chemically bound to silicon/nylon membrane (UDL

Skin Tissue Engineering and Regenerative Medicine. http://dx.doi.org/10.1016/B978-0-12-801654-1.00008-5

Biomaterials for Skin Tissue Engineering			
Natural			Synthetic
ECM-derived		Source other than ECM	Biopolymers
Decellularized matrix	Isolated biomolecules	Fibrin Silk Chitosan Alginate Dextran Agarose Cellulose Glycogen	Polyesters polyglycolide polylactic acid polycaprolactone
Decellularized dermal matrix Decellularized adipose matrix	Collagen Fibronectin Laminin Elastin Glycosaminoglycans hyaluronan chondroitin sulfate		

FIGURE 1 Classification of biomaterials for STE.

Laboratories, Inc., Sugarland, TX); as well as first semisynthetic scaffolds such as Integra®, bovine type 1 collagen and shark condroitin-6-sulphate covered with a silicone membrane (Plainsboro, NJ) or Matriderm®, bovine collagen types 1, 3, 5, and elastin (Billerbeck, Germany). These natural components are assembled as porous scaffolds and can provide support for an overlying epidermis, as well as an internal architecture capable of allowing new blood vessels to grow within the scaffold. These materials are biodegradable, with the time constant of biodegradation being approximately equal to the time constant of synthesis of new biological material within the scaffold. Just as in the construction of buildings, the scaffold provides the architecture for the subsequent structure. With natural biomaterials, this structure is ideally replaced by the synthesis of new tissue—differing from scar. Design characteristics for these scaffolds include color match, match of elasticity and strength, and the capacity to renew following injury.

Natural biomaterials can be altered in several of their physicochemical properties, including porosity, cross-link density, the degree of crystallinity, and the presence of copolymers. Fundamentally, one can obtain these constructs through either a decellularization process of an existing anatomic structure, such as dermis, or a more purifying process where individual chemical components are isolated and then subsequently reassembled into a matrix. Such biomaterials could come from human or animal origins. The most used biomaterials for the generation of skin substitutes, to date, belong to the category of extracellular matrix (ECM) proteins (Figure 1).

Dermal ECM is a dynamic network of fibrous adhesion proteins, proteoglycans, and GAGs [1] that interact in order to give skin its function (Figure 2). Reducing this structure to only some of its components and incorporating them into a scaffold cannot replace the entirety; however, it can promote regeneration clinically. Use of natural biomaterials despite the clinical success has some

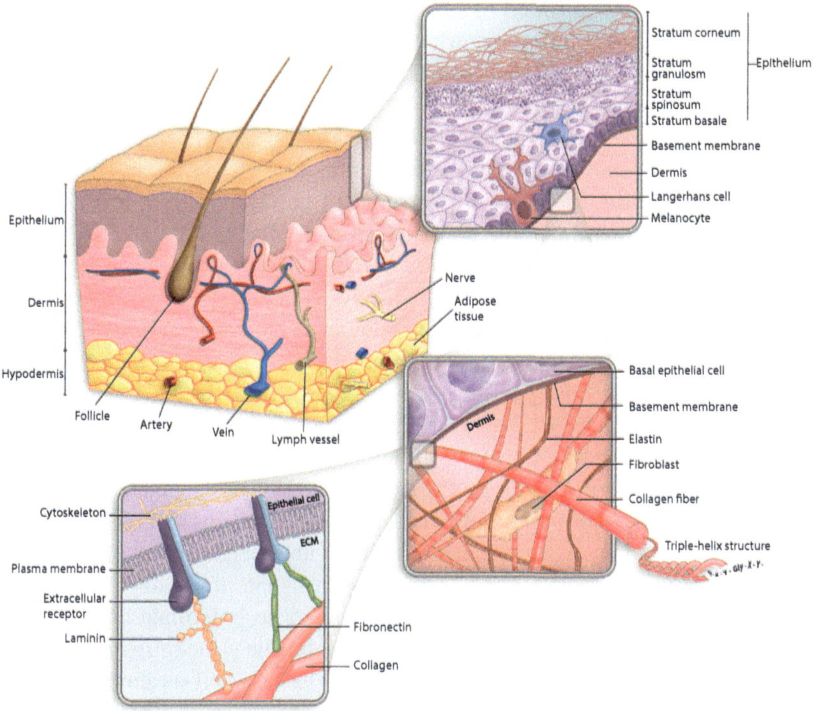

FIGURE 2 Schematic diagram of the macro- and microstructure of skin and its natural ECM-derived biomaterials.

disadvantages including the risk of disease transmission. This risk has been mitigated by using disease-free animal herds and by rigorous screening for infection in cadaveric donors. Not all natural biomaterials, or their derivatives, have optimal biocompatibility and very few can prevent wound contraction [2]. Despite our ability to control certain parameters in semisynthetic biomaterials, the natural dermis is very difficult to replicate mainly because of the anisotropic microanatomy [3]. These properties are the result of an elaborate process that involves the processing of the messenger RNA, as well as the enzymatic trimming within the extracellular space where the collagen fibers assemble. Despite their semisynthetic nature, it was shown clinically that interactions between scaffolds and host can induce regeneration [4].

NATURAL BIOMATERIALS FOUND IN ECM

Collagen

Collagen is the major protein of the human body and comprises 75% of total skin protein. There are at least 29 types of collagens that have been described, with 11 of them being found in skin [1,5]. The most important role of collagen is to maintain structural integrity [6]. Collagens are synthesized by fibroblasts and myofibroblasts and can be extracted from bone, cartilage, skin, ligament, and tendon. Biomaterials based on collagen can originate from two different techniques: the decellularization of dermis or the extraction of collagen. The resulting products can be subjected to various cross-linking methods, if needed [7]. From all the types of collagen, only a few are used for STE, and among them, the gold standard is collagen type 1 [7]. The skin is composed primarily of type 1 and 3 collagens, but also 4, 5, 6, and 12 [1]. Most of the collagen polypeptide chains are formed from a repetitive amino acid sequence, in which every *third* residue is *Glycine*; it can be described as Gly-X-Y (Figure 2), where X and Y are proline and hydroxyproline, respectively. Three collagen polypeptide sequences make a left-handed *superhelix*, approximately 300 nm long with a diameter of 1.4 nm, comprised of 1000 amino acid residues [8]. Chemical structure and 3-dimensional arrangement play a decisive role in the stability of the protein [5]. Collagen exists in either fibrillar or nonfibrillar forms. The fibril-forming collagens self-assemble into cross-striated fibrils with a characteristic 67 nm repeat. The collagens are synthesized in a precursor form—procollagen—that, after enzymatic cleavage, gets assembled in the extracellular space. The procollagen is excreted outside the cell, and both ends are trimmed by the *procollagen peptidase* to form the tropocollagen. Tropocollagen molecules self-assemble into the collagen microfibril, which subsequently undergoes cross-linking with the help of *lysyl oxidase* and forms the collagen fiber (Figure 2). Fibril-forming collagens provide a structural framework and include types 1, 2, 3, 5, and 11 [9]. Some collagens have been known to form mesh-like networks; for example, collagen type 4 forms the basal membrane along with other proteins. Other collagens that are known to form networks are types 8 and 10. There are collagens that associate with fibril surfaces (types 6, 9, 12, and 14), transmembrane proteins (types 13 and 18), or form periodically beaded structures (type 6) [6]. Collagen structures contain motifs—ligands that cells can recognize. A ligand is a particular amino acid motif in the protein that is located on the matrix surface, where cell adhesion molecules bind [10]. Using collagen in STE has many advantages: (1) low immunogenicity; (2) high mechanical strength; (3) good biocompatibility; and (4) the potential to tailor cross-linking for the desired mechanical, degradation and water uptake properties. Disadvantages include difficulties in processing, sterilization, and controlling the rate of degradation.

For commercial use, collagen is isolated from bovine, porcine, or chicken sources [8] but can also be obtained through recombinant technology [11]. The recombinant version has the advantage of having the same amino acid sequence as human collagen without the problems of extracted collagen (risk of infection, purity) [12]. However, the majority of existing biomaterials in clinical use are based on natural or extracted collagen [2].

Gelatin is a denatured collagen. Gelatin can support cell attachment, migration, and proliferation and was tested in different experimental matrices [11]. Cell-matrix interactions are of primary importance in STE. These can be categorized as direct or indirect interactions between cells and collagen.

Fibronectin

Fibronectin is a high-molecular weight glycoprotein (440 kDa) also known as fibrin pectin. It is a member of the adhesive protein family and has an important role in ECM. Within the ECM, it binds integrins and other components like collagen, fibrin, and heparin, playing a crucial role in indirect cell–collagen contact [7]. Fibronectin also plays an important role in wound healing, especially in the process of migration and growth of cells during the recovery process [13]. Fibronectin exists as a protein dimer formed of two monomers linked by a pair of disulfide bonds. Soluble and insoluble types of fibronectin exist. The soluble form is found in the blood, participates in wound healing processes, and is produced by hepatocytes. The insoluble fibronectin is produced by various cells, mostly fibroblasts that contribute to ECM. These two forms are structurally and biologically similar but not identical [14,15]. Fortunately, the plasma form can be insolubilized in vivo and in vitro [14,16], allowing for its use in regenerative matrices or coating other biomaterials to enhance cell to collagen interactions. The primary sources of fibronectin are bovine and human plasma.

Laminin

Laminin is a widely present protein within the ECM and one of the main components of the basal membrane [6]. The basal membrane is a key structure within all epithelial and endothelial tissues, such as the skin, and the lining of the digestive, respiratory, and vascular systems. Laminin is produced by various cells and plays a role in cell adhesion, migration, differentiation, embryonic development, and angiogenesis during wound healing [6,17,18]. Currently, 12 laminin isoforms have been isolated. Structurally, laminin is composed of three disulfide-linked chains (a combination of five α chains, three β chains, and three γ chains) [6]. All laminins share common elements, such as small globular domains, involved in chain polymerization the epidermal growth factor-like repeats, which indirectly mediate the connection to collagen type 4 within the basal membrane. Laminin 1 was the first to be isolated and is considered the representative for the class; it has a cross shape as depicted in Figure 2. Laminin 1 is present in the basal membrane of epithelial cells and is structurally composed of $\alpha 1$, $\beta 3$, and $\gamma 2$ subunits. The majority of cells are in contact with various basal membranes, making laminin of interest for STE given the possibility to improve the interface between host and implant [6]. By incorporating laminin into a skin-engineered construct, a formerly nonadherent surface can be augmented in terms of cell adhesion, potentially demonstrating improved healing and regeneration [19].

Defects in laminin expression can be associated with muscular dystrophy and skin blistering diseases [17] and may have a role in cancer and metastases [6].

Elastin

Elastin is another structural protein found in the ECM of mammals. Within tissues, it forms a complex three-dimensional (3D) structure that helps bind the tissues, represents the dominant part of the elastic fiber, and confers the resilience and elasticity to tissues [20]. Elasticity is the mechanical property that allows solid materials to return to their original structure following deformation; an essential attribute of tissues subjected to stretch forces. It thus follows that tissues such as blood vessels, which stretch and relax more than a billion times over a lifetime, are, therefore, abundant with elastin, as are ligaments, lungs, and skin [21]. Another important property of tropoelastin (the precursor of elastin) and elastin-like peptides is their ability to self-assemble under physiological conditions [21], which is the basis of the process called coacervation. Coacervation consists of the alignment of molecules prior to cross-linking. The half-life of elastin is 70 years, and it is the most stable known protein [22]. Elastin is of importance for STE given its properties of elasticity, self-assembly, long-term stability, and biological activity [21]. Elastin can be applied in STE in various forms such as insoluble elastin fibers, hydrolyzed soluble elastin, recombinant tropoelastin (fragments), repeats of synthetic peptide, block copolymers of elastin, and possibly in combination with other biopolymers [21]. Elastin is isolated by treating the elastin-rich tissues with sodium hydroxide or guanidine hydrochloride; this mix is then either simultaneously or subsequently heated to remove other proteins and cellular material [20]. The difficulty of manipulating elastin in vitro, due to extensive cross-linking and associated insolubility, makes elastin less attractive for TE [20,21]. However, the elastic properties of elastic tissues have been replicated with elastin–collagen matrices synthesized with freeze–drying processes [20]. Given that collagen–elastin matrices can support fibroblasts and keratinocytes, these heterogeneous matrices have been explored as dermal regeneration templates [20,21]. Matrices composed of elastin with fibrin have also been synthesized; however, they have not yet been tested in vivo [20].

Glycosaminoglycans

GAGs, also termed mucopolysaccharides, are a category of large linear polysaccharides in which the repeating disaccharide is composed of one amine sugar (*N*-acetylglucosamine or *N*-acetylgalactosamine) and an uronic acid (glucuronic acid or iduronic acid). GAGs are highly polar, attracting and binding water molecules, filling gaps between collagen fibrils within the ECM and thereby facilitating their role as both a lubricant and a shock absorber [23]. Virtually all cells produce GAGs, which are then secreted, making them exceptionally prevalent within the ECM. The most important for STE are hyaluronan and chondroitin sulfate. GAGs are introduced into the composition of the porous scaffolds, in combination with collagen, chitosan, or other natural biomaterials, to modulate specific properties, such as the availability of binding sites. For example, condroitin-6-sulfate is introduced in Integra Dermal Regeneration Template® to confer it the increased bioavailability.

OTHER NATURAL BIOMATERIALS

Chitosan

Chitosan is another biomaterial with extensive use in STE. It is derived from chitin, which is a polysaccharide found in the exoskeleton of shrimp and other sea crustaceans [24]. Chitin is the most abundant sugar found in nature, making the chitosan source an almost unending reservoir, and it is relatively inexpensive to extract. This polysaccharide is unique and has structural characteristics similar to GAGs; moreover, it mimics their functional behavior [24]. It is advantageous to use because of a minimal foreign body reaction, does not require aggressive chemicals to be dissolved, can be incorporated into matrices, and allows the control of porosity. Chitosan can promote cell adhesion and act as a modulator of cell morphology, differentiation, movements, synthesis, and function [25]. It was shown that chitosan can also stimulate fibroblasts to release interleukin-8, which itself is involved in migration and proliferation of fibroblasts. Chitosan can be combined in porous matrices with other natural biomaterials to enhance their properties, for example, gelatin. Such a combination increases cell proliferation, reduces apoptosis, and also shows improved cell mobility, migration, and multiplication [25,26].

Fibrin

Fibrin is a versatile biopolymer, which shows excellent potential for tissue regeneration and wound healing [11]. Chemically it is a fibrous glycoprotein that can be found in the plasma fraction of blood. It results from thrombin-mediated cleavage of fibrinogen as part of the coagulation cascade. Fibrin serves as a transient wound healing matrix that forms after injury. Fibrin also supports keratinocyte and fibroblast growth to enhance motility of the cells in wound healing [13]. It represents the major component of the blood clot that is of crucial importance for preventing hemorrhage and enhancing wound healing. Fibrin has excellent biocompatibility, because fibrin matrices contain sites for cellular adhesion [1]. Fibrin alone, or in combination with other biomaterials, was employed as a biological scaffold to promote stem or primary cells to regenerate. Currently, fibrin has a variety of clinical applications, including its use as a medical sealant or "tissue glue." Gels and sprays can also be synthesized from a patient's blood for use as an autologous matrix, augmenting/improving homeostasis, tissue handling, and accelerating the wound healing [13]. The fibrin matrix properties could be modified according to the amount of fibrin introduced [11]. Fibrin is considered a potential biomaterial for STE due to its availability, autologous nature, and presence of growth factors that favor wound healing, skin cell proliferation, and migration [13,27]. Fibrin is naturally degraded without generating any toxic substances or inflammation. Following enzymatic polymerization of fibrinogen incubated with thrombin at room temperature, fibrin will aggregate to form a gel [28]. However, fibrin gels are limited in mechanical strength, preventing their use alone for STE.

Silk

Silk is a natural biomaterial widely used in surgery as suture material. It is composed of a filament protein core—fibroin and a glue-like coating—sericin protein. The most recognized silk is naturally made by the mulberry silkworm (*Bombyx mori*). However, other insects, like spiders and crickets, have been reported to be able to produce silk. It has relatively good biocompatibility, slow degradability, and desirable mechanical properties if used within scaffold materials [29]. Traditionally, silk has led to an inflammatory response in adjacent native tissue, but modern techniques of modifying its chemistry allows for modulation of the inflammatory response.

Alginate

Alginate is a naturally occurring polysaccharide that can be extracted from brown seaweed and bacteria. It is the most abundant biosynthetic biomaterial known [30]. Alginate is composed of blocks of (1–4)-linked β-D-mannuronic acid and α-L-guluronic acid monomers. It is used in a broad range of applications; however, the most important role is to provide a supporting matrix and delivery system for wound repair and regeneration. It has outstanding biocompatibility, biodegradability, nonantigenicity, and chelating ability. The most exploited forms of alginate are hydrogels, microspheres, porous scaffolds, and fibers. It can be easily processed in water and is nonadherent [31]. The porosity of alginate can be controlled, and other biologically active molecules could be added [32]. Matrices made out of the alginate can be used for cell seeding [32]. Porous scaffolds can be applied as 3D structures for a temporary skeleton to accommodate and stimulate tissue regeneration [30]. The major drawbacks of alginates are their weak mechanical properties and poor cell adhesion. Some of these issues can be addressed by combining alginate with other natural biomaterials like chitosan or agarose [33,34].

In addition to the aforementioned natural biomaterials, several others, such as cellulose, glycogen, agarose, and dextran have been tested for STE [35,36].

APPLICATIONS OF NATURAL BIOMATERIALS IN STE

One of the distinctive characteristics of collagen is its biological activity [7]. It is characterized by superior biocompatibility, biodegradability, and low immunogenicity [7]. These qualities make it the protein of choice for medical applications [7]. Wound dressings containing collagen have been used for decades in burns and chronic wounds [13]. The uses of collagen in TE, and especially in STE, are numerous. Collagen can be used as a scaffold, matrix or gel, acting as an adhesive for cells within regenerated organs [7]. However, due to its specificity, collagen has been shown to be useful in organ regeneration [37]. Thus, the molecular architecture of collagen, as well as the other proteins used within the scaffold, is crucial for proper integration and the ability to be infiltrated by cells, due to particular cell adhesion and migration. Only very few porous scaffolds can bind the contractile fibroblasts via integrin ligand-binding and, by doing so, block the deformation generated by the contractile forces produced over the course of normal wound healing [10]. Yannas et al. investigated the beneficial effect of GAGs within an optimized collagen–GAG (CG) scaffold [11,38]. This resulted in a particular class of scaffolds that has been used more than others in the clinical setting. They are analogous to the ECM and are synthesized as graft copolymers of type 1 collagen and GAGs [10]. A commercial product, Integra Dermal Regeneration Template® (IntegraLifeScience Inc. Plainsboro, NJ) has been designed based on these principles. A major component of such a dermal substitute is processed collagen, in which the telopeptides have been removed (Figure 3).

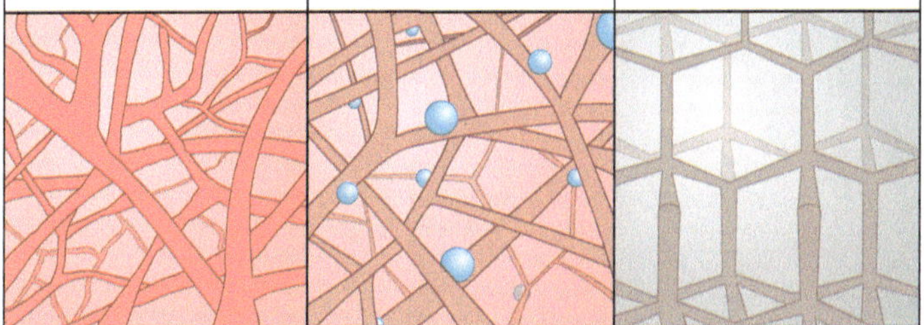

	Decellularized	Semi-synthetic	Synthetic
Advantages	• Native architecture – derived from slightly modified natural tissue	• Some control of matrix structural characteristics; Reproducible results through careful control of processing variables • Utilizes rich chemistry of biomolecules: Abundance of binding sites to cell integrin receptors Co-polymerization process can incorporate multiple biomolecules	• Ultimate control of structure • High consistency of synthesis process and product • Easily sterilizable • No risk of disease transmission • Potential for cost savings. • Complete homogeneity
Disadvantages	• Source variability • Potential disease transmission • Some are not sterile	• Some source variability • Risk of prion transmission (theoretical)	• Current synthetic chemistry technology is unable to mimic biological interaction characteristics comparable to semi-synthetics • Complete homogeneity
Processing Considerations	• Scaffold harvest: Thickness Anatomic source (papillary vs. reticular dermis) • Necessitates chemical decellularization • Freeze-drying • Cross-linking	• Crystallinity • Cross-link density • Orientation of polymer • Pore size • Composition/Geometry • Thickness • Co-polymerization biomolecules (e.g. GAGs, Hyaluronan)	• Complete control of structure Most reproducible Most adaptable

FIGURE 3 Advantages, disadvantages, and processing considerations of scaffold types.

Despite the possibility of immunogenicity of the telopeptides [13,39], it was shown that such substitutes do not undergo rejection, a critical factor for clinical applications [13,40]. Multiple clinical studies of Integra have been performed, and together with a thin skin graft in a two-stage procedure, it produces excellent cosmetic and functional results (Figure 4) [41–43].

Collagen can be used in the form of a gel to deliver cells for wound treatment. For example, Kim et al. used collagen as a vehicle to deliver adipose-derived stem cells (ASCs) to cutaneous wounds [1,44]. Alternatively, Organogenesis Inc. (Canton, MA) developed a bilayer collagen gel, seeded with human

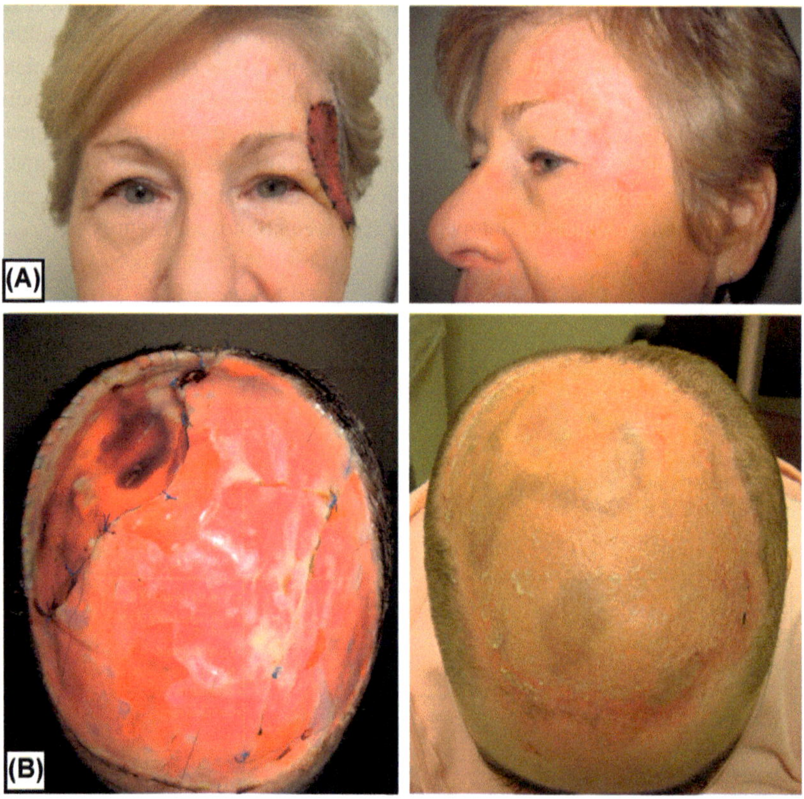

FIGURE 4 Clinical demonstration of wound treatment using Integra Dermal Regeneration Template® in a two-stage approach. (A) A 65-year-old woman with a melanoma on the left temple area was initially treated by resection and application of a dermal regeneration template. At 1 month, the silicone was removed and a thick skin graft was taken from the upper arm. During this period, the specimen was cleared on permanent sections to show that there was complete resection of the tumor. At 1 year, there is an excellent color match of the skin. (B) A 38-year-old woman with a squamous cell carcinoma of the scalp. She required three excisions and burning down of the calvaria to clear the tumor. She was treated with dermal regeneration template followed by a skin graft 6 weeks later. She subsequently had this area irradiated with 3500 cGy. The photograph was obtained 3 months after completion of the irradiation. *Reproduced with permission from [42]*

fibroblasts in the lower component and human keratinocytes in the upper part, to be used as the "dermal" matrix termed Apligraf® [11]. To date, these complex skin substitutes have been expensive to produce. The work is ongoing in order to show that anatomically correct composite skin substitutes can be engineered in a cost-effective manner. Boyce et al. have pioneered one approach [45–48] referred to as Permaderm®/NovaDerm®. It consists of a CG scaffold seeded with autologous fibroblasts and an autologous epidermal layer of differentiated keratinocytes. It showed relatively good engraftment, and it can help in burns covering more than 60% of the total body surface area [49]. However, it requires a skin biopsy in advance and takes several weeks for preparation, which includes ex vivo culture and expansion. Another approach is microfabrication by 3D printing described by Chin et al. [11,50].

In attempt to explore other properties of combined scaffolds, chitosan was added to CG scaffold by Auxenfans et al. to create a matrix that acts as an attractive scaffold for fibroblasts; this biomaterial potentially could be used as a dermal scaffold for a future composite skin substitute [11,51]. Another promising, although expensive, approach is to rely on molecular properties of self-assembly to generate a dermal scaffold. This approach has the natural advantage of producing all of the components of a decellularized dermis (collagen, fibronectin, elastin) and maintaining the physiologic microstructure and ligand density, without the need for sterilization. This process is possible due to the ability of fibroblasts to generate a collagen-based ECM when cultured under special conditions in vitro [52,53]. Climov et al. demonstrated in swine that an autologous dermal–epidermal skin generated by self-assembly could engraft and heal full- and partial-thickness wounds with a cosmetic result similar to skin grafts [54]. Another regenerated self-assembled autologous skin substitute proposed by Boa et al. showed clinical use in healing of chronic venous leg ulcers [55]. Multiple other semi-natural skin substitutes, which are based on collagen as the primary biomaterial, are available at this time for dermal regeneration, such as Matriderm® (collagen–elastin polymer), Alloderm® and Repliform® (decellularized matrices) [13].

Gelatin-based constructs offer flexibility in STE applications in that they can be produced in a variety of conformations by modulating variables such as temperature, pH, and solvent used in the manufacturing process. They have been shown to support cell adhesion, migration, and proliferation similarly to collagen, elastin, and recombinant human tropoelastin [11,56]. A bilayered gelato-chondroitin 6 sulfate–hyaluronic acid membrane with variable pore size developed by Huang et al. was tested to encapsulate cells or microparticles of bioactive substances for delivery, and as a biodegradable membrane to be used in STE [11,57].

Fibronectin has a major effect on wound healing, especially on migration and cell growth. It was used by several researchers alone, or in combination with collagen, within a scaffold as a dermal substitute to treat burns and diabetic ulcers [13]. Fibronectin alone can be aggregated in the form of mats; these mats contain pores oriented in the same direction that can guide regeneration of nerves and other cells by providing cell adhesion sites [11]. The philosophy behind a functional skin substitute is the ability to act as a bridge between the cytoskeleton and integrins of fibroblasts and their respective ECM ligands. Fibronectin plays a significant role in mediating this connection. To date, there are only a limited number of available fibronectin-containing biomaterials, most likely due to the failure to promote vascularization and because fibronectin function is inhibited by mechanical stretching [13]. Very few relatively recent studies have demonstrated that silk can promote fibroblast growth and attachment in vivo [58,59]. Attempts to combine silk with other natural biomaterials have been made, such that in combination with chitosan a new biomaterial termed Shrilk was developed.

Fibrin gels can be generated from a patient's own blood and used as autologous scaffolds. This approach can avoid the problem of blood-borne infections and immunogenicity. Fibrin is used as glue in such products as Tisseel/Artiss (Baxter), Beriplast (CSL Behring), Evicel (Ethicon), and others. It is used extensively in surgery due to its excellent biocompatibility. Fibrin scaffolds have been used for cell transplantation (mesenchymal stem cells, bone, cartilage), and notably with skin cells, to develop the autologous product named MyDerm[1,11]. MyDerm is a fully autologous skin substitute that contains keratinocytes and fibroblasts in a fibrin scaffold. A new generation of skin substitutes that learned from the mistakes of Apligraf, Dermagraft, and TransCyte is the ICX-PRO/ICX-SKN [60,61]. It is designed to be inexpensive, uses collagen-coated biomaterials and allogeneic fibroblasts, has low immunogenicity, and was tested in a clinical trial [13], however for unknown reasons failed to pass the phase of clinical trials.

Not only have GAGs been used with other molecules such as collagen to produce effective biosynthetic constructs for STE, esters of hyaluronic acid can be used alone to create useful products. Most of the experience has been with a bilaminate product termed Hylomatrix® PA (Anika Therapeutics, Srl, Abano, Terme, Italy) that, similar to Integra®, has an outer silicone layer. Hyaluronic acid products have been used in burns, reconstructive surgery, and chronic wounds alone or with cultured cell technology [62–65]. Based on current studies, it appears to aid in epthithelialization and angiogenesis. In a comparison to Integra®, long-term results did not show favorable physical, mechanical, and optical properties [66].

GENERAL CONSIDERATIONS ON THE APPLICABILITY OF BIOMATERIALS IN STE

We would like to emphasize some critical points that guide the actual trends in STE. Success in STE occurs when the product promotes regeneration without inducing an immune response and causing scar formation. To date, the best clinical success has come from CG dermal regeneration scaffolds. This success is defined by safety, efficiency, ease of production, ease of use, and cost. The architecture of the scaffold is necessary for success [67]. Yannas et al. established that the structural features that are required for regenerative activity include *pore structure*, *degradation rate*, and the *surface chemistry* [37]. Proper porosity is needed to ensure cellular infiltration, as well as an adequate infusion of nutrients to the cell and effusion of waste products back to the ECM. This is a major issue of actual constructs; most of them do not allow this bidirectional communication by not allowing vascularization due to the inadequate porosity [67]. Along with the appropriate number of pores, the mean pore size is critically important for the regenerative capacity of the scaffold. Small pores do not allow cellular infiltration, or, in contrast, excessively large pores do not maintain the proper cellular interactions [67]. The primary interaction between cells and scaffold is realized by ligands on the biomaterial surface and integrins on the cell surface. Natural materials, like collagen, possess such ligands in the form of Arg-Gly-Asp (RGD) binding sequences, as opposed to the synthetic biomaterials that may require incorporating them. The density of such ligands is critical for the bioavailability of a biomaterial [67]. An optimal biomaterial should be degraded in an appropriate amount of time, not only to prevent encapsulation due to longer degradation but also from being degraded too quickly causing scar formation. Other important criteria identified by stakeholders (patient, insurer, surgeon, hospital, manufacturer) are shown in Figure 5. The common qualities

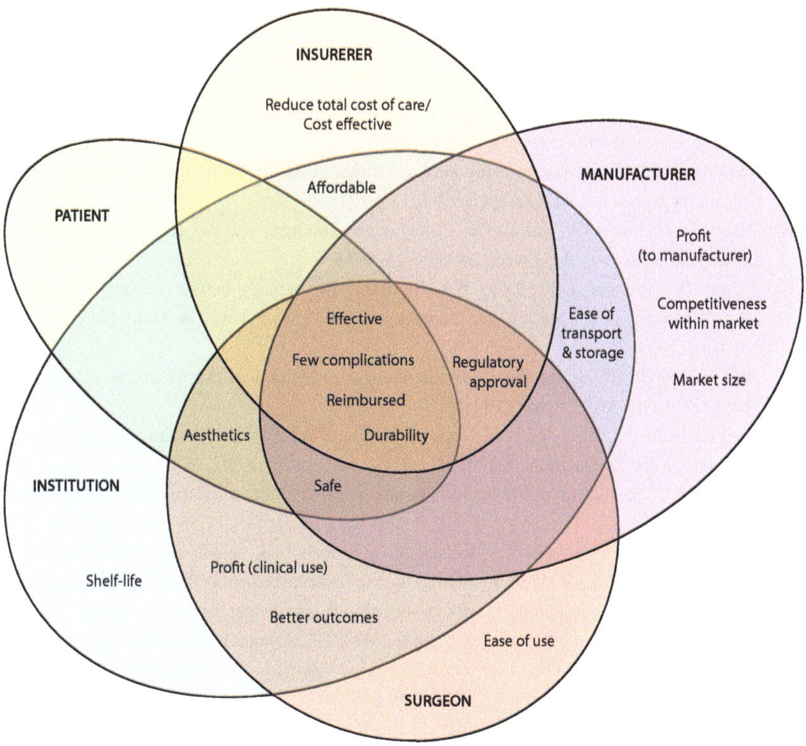

FIGURE 5 Stakeholder needs in the US.

identified are the efficiency of the biomaterial, low complication rate, durability, and reimbursement. However, the parameters should be judged in their entirety when considering the development of a novel biomaterial for STE.

LIST OF ACRONYMS AND ABBREVIATIONS

CG Collagen-glycosaminoglycan
ECM Extracellular matrix
GAG Glycosaminoglycan
STE Skin tissue engineering
TE Tissue engineering

REFERENCES

[1] Hu MS, Maan ZN, Wu J-C, et al. Tissue engineering and regenerative repair in wound healing. Ann Biomed Eng 2014;42:1494–507.
[2] MacNeil S. Biomaterials for tissue engineering of skin. Mater Today 2008;11:26–35.
[3] Bose S, Bandyopadhyay A. Introduction to biomaterials. Charact Biomater 2013:1–9.
[4] Orgill D, Straus F, Lee R. The use of collagen–GAG membranes in reconstructive surgery. Ann NY Acad Sci 1999;888:233–48.
[5] Shoulders MD, Raines RT. Collagen structure and stability. Annu Rev Biochem 2009;78: 929–58.
[6] Bosman FT, Stamenkovic I. Functional structure and composition of the extracellular matrix. J Pathol 2003;200:423–8.
[7] Parenteau-Bareil R, Gauvin R, Berthod F. Collagen-based biomaterials for tissue engineering applications. Materials (Basel) 2010;3:1863–87.
[8] Aberoumand A. Comparative study between different methods of collagen extraction from fish and its properties. World Appl Sci J 2012;16:316–9.
[9] Silver FH, Freeman JW, Seehra GP. Collagen self-assembly and the development of tendon mechanical properties. J Biomech 2003;36:1529–53.
[10] Yannas IV, Tzeranis DS, Harley BA, So PTC. Biologically active collagen-based scaffolds: advances in processing and characterization. Philos Trans A Math Phys Eng Sci 2010;368:2123–39.
[11] Patel H, Bonde M, Srinivasan G. Biodegradable polymer scaffold for tissue engineering. Trends Biomater Artif Organs 2011;25:20–9.
[12] Yang C, Hillas PJ, Báez JA, et al. The application of recombinant human collagen in tissue engineering. BioDrugs 2004;18:103–19.
[13] Morshed M. The current available biomaterials being used for skin tissue engineering, vol. 3. 2014. p. 17–22.
[14] McKeown-Longo P, Mosher D. Mechanism of formation of disulfide-bonded multimers of plasma fibronectin in cell layers of cultured human fibroblasts. J Biol Chem 1984;259:12210–5.
[15] Yamada K, Kennedy D. Fibroblast cellular and plasma fibronectins are similar but not identical. J Cell Biol 1979. http://jcb.rupress.org/content/80/2/492.abstract [accessed 09.12.14].
[16] Hayman E, Ruoslahti E. Distribution of fetal bovine serum fibronectin and endogenous rat cell fibronectin in extracellular matrix. J Cell Biol 1979;83:255–9.
[17] Mercurio A. Laminin receptors: achieving specificity through cooperation. Trends Cell Biol 1995:419–23.

[18] Li J, Zhang Y-P, Kirsner RS. Angiogenesis in wound repair: angiogenic growth factors and the extracellular matrix. Microsc Res Tech 2003;60:107–14.
[19] Ratner BD, Hoffman AS, Schoen FJ, Lemons JE. Biomaterials science: an introduction to materials in medicine. 2001.
[20] Nivison-Smith L, Weiss A. Elastin based constructs. In: Eberli D, editor. Regenerative medicine and tissue engineering-cells and biomaterials. InTech; 1997. p. 323–40.
[21] Daamen W, Veerkamp J. Elastin as a biomaterial for tissue engineering. Biomaterials 2007. http://www.sciencedirect.com/science/article/pii/S0142961207005030 [accessed 10.12.14].
[22] Mecham RP, Broekelmann TJ, Fliszar CJ, Shapiro SD, Welgus HG, Senior RM. Elastin degradation by matrix metalloproteinases: cleavage site specificity and mechanisms of elastolysis. J Biol Chem 1997;272:18071–6.
[23] Esko JD, Kimata K, Lindahl U. Chapter 16 proteoglycans and sulfated glycosaminoglycans. In: Varki A, Cummings RD, Esko JD, Freeze HH, Stanley P, Bertozzi CR, et al., editors. Essentials of glycobiology. Cold Spring Harbor Laboratory Press; 2009. p. 784.
[24] Chandy T, Sharma CP. Chitosan–as a biomaterial. Biomater Artif Cells Artif Organs 1990;18:1–24.
[25] Gorgieva S, Kokol V. Collagen- vs. gelatine-based biomaterials and their biocompatibility: review and perspectives. In: Pignatello R, editor. Biomaterials applications for nanomedicine. InTech; 2011. http://dx.doi.org/10.5772/1957.
[26] Mao JS, Cui YL, Wang XH, et al. A preliminary study on chitosan and gelatin polyelectrolyte complex cytocompatibility by cell cycle and apoptosis analysis. Biomaterials 2004;25: 3973–81.
[27] Ruszymah BH. Autologous human fibrin as the biomaterial for tissue engineering. Med J Malaysia 2004;59(Suppl. B):30–1.
[28] Lee KY, Mooney DJ. Hydrogels for tissue engineering. Chem Rev 2001;101:1869–80.
[29] Khang G. Handbook of intelligent scaffold for tissue engineering and regenerative medicine. Pan Stanford; 2012. http://books.google.com/books?id=PID9d-1ppEMC.
[30] Sun J, Tan H. Alginate-based biomaterials for regenerative medicine applications. Materials (Basel) 2013;6:1285–309.
[31] Glicklis R, Shapiro L. Hepatocyte behavior within three-dimensional porous alginate scaffolds. Biotechnol Bioeng 2000;67.
[32] Rowley JA, Mooney DJ. Alginate type and RGD density control myoblast phenotype. J Biomed Mater Res 2002;60:217–23.
[33] Chung TWW, Yang J, Akaike T, et al. Preparation of alginate/galactosylated chitosan scaffold for hepatocyte attachment. Biomaterials 2002;23:2827–34.
[34] Orive G, Hernandez RM, Gascon AR. Survival of different cell lines in alginate-agarose microcapsules. Eur J Pharm Sci 2003;18:23–30.
[35] Bačáková L, Novotná K, Pařízek M. Polysaccharides as cell carriers for tissue engineering: the use of cellulose in vascular wall reconstruction. Physiol Res 2014;63. http://biomed.cas.cz/physiolres/pdf/63Suppl 1/63_S29.pdf [accessed 11.12.14].
[36] Lévesque SG, Shoichet MS. Synthesis of cell-adhesive dextran hydrogels and macroporous scaffolds. Biomaterials 2006;27:5277–85.
[37] Yannas IV. Emerging rules for inducing organ regeneration. Biomaterials 2013;34:321–30.
[38] Yannas IV, Burke J. Design of an artificial skin. II. Control of chemical composition. J Biomed Mater Res 1980;14:107–31.
[39] DeLustro F, Dasch J, Keefe J, Ellingsworth L. Immune responses to allogeneic and xenogeneic implants of collagen and collagen derivatives. Clin Orthop Relat Res 1990:263–79.

[40] Lynn AK, Yannas IV, Bonfield W. Antigenicity and immunogenicity of collagen. J Biomed Mater Res B Appl Biomater 2004;71:343–54.
[41] Orgill DP, Butler C, Regan JF, Barlow MS, Yannas IV, Compton CC. Vascularized collagen-glycosaminoglycan matrix provides a dermal substrate and improves take of cultured epithelial autografts. Plast Reconstr Surg 1998;102:423–9.
[42] Yannas IV, Orgill DP, Burke JF. Template for skin regeneration. Plast Reconstr Surg 2011;127(Suppl. 1):60S–70S.
[43] Heimbach DM, Warden GD, Luterman A, et al. Multicenter postapproval clinical trial of Integra dermal regeneration template for burn treatment. J Burn Care Rehabil 2003;24:42–8.
[44] Kim W-S, Park B-S, Sung J-H, et al. Wound healing effect of adipose-derived stem cells: a critical role of secretory factors on human dermal fibroblasts. J Dermatol Sci 2007;48:15–24.
[45] Boyce ST, Kagan RJ, Yakuboff KP, et al. Cultured skin substitutes reduce donor skin harvesting for closure of excised, full-thickness burns. Ann Surg 2002;235:269–79.
[46] Boyce ST, Kagan RJ, Meyer NA, Yakuboff KP, Warden GD. The 1999 clinical research award cultured skin substitutes combined with Integra artificial skin to replace native skin autograft and allograft for the closure of excised full-thickness burns. J Burn Care Rehabil 1999;20:453–61.
[47] Boyce ST, Kagan RJ, Greenhalgh DG, et al. Cultured skin substitutes reduce requirements for harvesting of skin autograft for closure of excised, full-thickness burns. J Trauma 2006;60:821–9.
[48] Boyce ST, Goretsky MJ, Greenhalgh DG, Kagan RJ, Rieman MT, Warden GD. Comparative assessment of cultured skin substitutes and native skin autograft for treatment of full-thickness burns. Ann Surg 1995;222:743–52.
[49] Kamel RA, Ong JF, Eriksson E, Junker JPE, Caterson EJ. Tissue engineering of skin. J Am Coll Surg 2013;217:533–55.
[50] Chin CD, Khanna K, Sia SK. A microfabricated porous collagen-based scaffold as prototype for skin substitutes. Biomed Microdevices 2008;10:459–67.
[51] Auxenfans C, Builles N, Andre V, et al. Porous matrix and primary-cell culture: a shared concept for skin and cornea tissue engineering. Pathol Biol (Paris) 2009;57:290–8.
[52] Pouyani T, Ronfard V, Scott PG, et al. De novo synthesis of human dermis in vitro in the absence of a three-dimensional scaffold. In Vitro Cell Dev Biol Anim 2009;45:430–41.
[53] Bourget J-M, Gauvin R, Larouche D, et al. Human fibroblast-derived ECM as a scaffold for vascular tissue engineering. Biomaterials 2012;33:9205–13.
[54] Climov M, Matar AJ, Medeiros E, et al. A novel tissue engineered construct for wound treatment and skin immunity research. Plast Reconstr Surg 2013;132:56–7.
[55] Boa O, Cloutier CB, Genest H, et al. Prospective study on the treatment of lower-extremity chronic venous and mixed ulcers using tissue-engineered skin substitute made by the self-assembly approach. Adv Skin Wound Care 2013;26:400–9.
[56] Li M, Mondrinos MJ, Gandhi MR, Ko FK, Weiss AS, Lelkes PI. Electrospun protein fibers as matrices for tissue engineering. Biomaterials 2005;26:5999–6008.
[57] Huang S, Deng T, Wang Y, et al. Multifunctional implantable particles for skin tissue regeneration: preparation, characterization, in vitro and in vivo studies. Acta Biomater 2008;4:1057–66.
[58] Minoura N, Aiba S, Higuchi M, Gotoh Y, Tsukada M, Imai Y. Attachment and growth of fibroblast cells on silk fibroin. Biochem Biophys Res Commun 1995;208:511–6.
[59] Sofia S, McCarthy MBM, Gronowicz G, Kaplan DL. Functionalized silk-based biomaterials for bone formation. J Biomed Mater Res 2001;54:139–48.

[60] Boyd M, Flasza M, Johnson PA, Roberts JSC, Kemp P. Integration and persistence of an investigational human living skin equivalent (ICX-SKN) in human surgical wounds. Regen Med 2007;2:369–76.

[61] Hodgkinson T, Bayat A. Skin substitutes. In: Maibach HI, Gorouhi F, editors. Evidence based dermatology. 2nd ed. 2011. p. 997–1018.

[62] Gravante G, Gelogu D, Giordan N, Morano G, Montone A, Esposito G. The use of hyalomatrix PA in the treatment of deep partial-thickness burns. J Burn Care Res 2007;28:269–74.

[63] Gravante G, Sorge R, Merone A, Tamisano A, DiLonardo A, Scalise A, et al. Hylomatrix PA in burn care practice: results from a national retrospective survey 2005 to 2006. Ann Plast Surg 2010;64:69–79.

[64] Caravaggi C, Grioletto F, Scuderi N. Wound bed preparation with a dermal substitute (hylomatrix® PA) facilitates re-epithelialization and healing: results of a multicenter prospective, observational study on complex chronic ulcers (The FAST study). Wounds 2011;23:228–35.

[65] Scuderi N, Onesti MG, Bistoni G, Ceccarelli S, Rotolo S, Angeloni A, et al. Biomaterials 2008;29:1620–9.

[66] Nicoletti G, Brenta F, Pellegatta T, Malovini A, Paga A, Perugini P. Long-term in vivo assessment of bioengineered skin substitutes; a clinical study. J Tissue Eng Regen Med 2014;10:1002–111.

[67] O'Brien FJ. Biomaterials and scaffolds for tissue engineering. Mater Today 2011;14:88–95.

Chapter 9

Synthetic Biomaterials for Skin Tissue Engineering

Mohammed Hussein Kailani[1,2], Hanan Jafar[1], Abdalla Awidi[1]
[1]*Cell Therapy Center, University of Jordan, Amman, Jordan;* [2]*Department of Chemistry, Faculty of Science, University of Jordan, Amman, Jordan*

Chapter Outline

INTRODUCTION

Historically, efforts in the medical and biomedical engineering fields have been directed not only at preserving the health and integrity of existing tissues and organs, but also toward regenerating and replacing the damaged ones. Therefore, tissue engineering emerged as an interdisciplinary field that merges the best of two worlds: the biological aspect of replacing lost cells and the physical/chemical one of fabricating artificial extracellular matrices that would support such cells [1]. The anatomy of the skin, organized into two distinct layers, as well as its vital role in maintaining homeostasis, offered a very tempting

Skin Tissue Engineering and Regenerative Medicine. http://dx.doi.org/10.1016/B978-0-12-801654-1.00009-7

opportunity for scientists to pursue developing biosynthetic skin substitutes. As a matter of fact, skin substitutes are among the first and most versatile products of tissue engineering to become commercially available [2].

Different materials were tried as possible skin-replacement therapies. First, completely biologic biomaterials were extracted and treated to act as scaffolds for cultured cells, and later, various synthetic and semisynthetic materials were used either to support existing tissues and protect them or to replace them all together [3]. This presented an exciting opportunity for scientists to experiment with novel materials or produce some modifications on existing ones and resulted in a surge of patents, many of which were further developed into commercial products. That said, one should keep in mind that we are still far from the "perfect skin substitute," and this indeed is a field in which material sciences have a lot to offer.

This chapter deals with the developments in artificial skin products that contain fully synthetic polymeric materials. It is not intended to give a comprehensive review of the literature but to shed light on a selective perspective that, the authors hope, will be useful for future improvements in this field. Specifically, we focus on issued patents on the topic, as we feel this is a rather neglected area in published literature reviews elsewhere.

CHARACTERISTICS OF THE IDEAL SKIN SUBSTITUTE

To date, the "ideal" bioengineered skin model does not exist. Such a model should completely replicate the biology, anatomy, and physiology of natural skin. In that regard, skin substitutes are required to possess some essential properties and meet certain requirements. Tissue-engineered skin substitutes, whether biologic, synthetic, or semisynthetic must adhere to many requirements, three of which are regarded as most essential [2,3]. These are (a) safety, (b) efficacy, and (c) convenience. In general, biomaterials should be nontoxic, biocompatible, nonimmunogenic, and pathogen free. More specifically, synthetic biomaterials used for skin-replacement therapies must degrade in a controlled manner in vivo, enhance normal healing of tissues, prevent wound desiccation, and protect the wound from infection. Additionally, biomaterials are preferred to be easily manufactured, readily available, inexpensive, durable, and have a long shelf-life. Fundamentally, the goal of skin tissue–engineering scientists is to fulfill as many of these criteria as feasible, when constructing novel skin-replacement products.

The "ideal artificial skin," as a final product, needs to provide the barrier function in a similar fashion to the epidermal barrier, adhere securely to the underlying connective tissue, show or support a good "take" demonstrated by vascularization, and display mechanical properties comparable or superior to uninjured skin.

THE NEED FOR SYNTHETIC MATERIALS IN SKIN SUBSTITUTES

The split-thickness graft remains the "gold standard" in the treatment of skin defects, such as burns. However, as discussed extensively in the literature, they have the disadvantages of donor site morbidity, extended healing periods, and limited availability (especially in defects affecting large body surface area). Moreover, although life-saving in many situations, their outcomes are far from being aesthetically acceptable and often produce disfiguring or disabling scarring. Alternatively, cadaveric donor skin has been extensively used in clinical settings and in a variety of formats such as de-epithelized, de-cellularized, frozen viable, and frozen nonviable. All of these are considered temporary biological dressings and are associated with risks of antigenicity and disease transmissibility.

Almost as soon as the drawbacks associated with autologous and allogenic grafts were identified, the search began for alternatives. The most obvious, and indeed the most commonly employed, was collagen-based skin-replacement therapies [2]. Other examples of natural biomaterials used in skin substitutes include hyaluronan, glycosaminoglycans (GAGs), chitosan, fibronectin, and alginate. These materials offer the advantages of low toxicity, reduced immunogenicity, and produce limited inflammatory responses. However, the above-mentioned materials, which work well as scaffolds, could hold great risk if not properly handled. The extraction process and subsequent purification and sterilization should be meticulously planned and executed to maximize safety for patients. For example, the Food and Drug Administration (FDA) approved bovine-derived collagen as "low risk," but this is dependent on whether or not the herds were exposed to the agent of bovine spongiform encephalopathy [4].

Synthetic biomaterials hold great promise for providing solutions to most of the problems associated with autografts and allografts. Most notably, the risk of disease transmissibility from xenogenic collagen, for example, can be reduced or even eliminated through using recombinant produced collagen or synthetic collagen-mimetic products [4]. In addition, they would possibly provide superior traits such as inherent antimicrobial activity and controlled biodegradability. It is also important to note that their properties can essentially be modified both on the chemical and physical levels. For example, chemical properties such as side chains of polymers and composition could be tailored so as to fit specific needs. On the other hand, one could manipulate the physical characteristics of these materials such as the structure, shape, and consistency. Indeed, a wide range of products exists on the market, each dedicated for one or more prespecified uses. These products are found in the form of films, gels, foams, sponges, fibers, multilayer composite membranes, and some products also incorporate living cells or cell products. As such, synthetic skin substitutes may potentially provide the overall needed properties for fast recovery with reduced pain, hospitalization, cost, scarring, and loss of function.

DEVELOPMENT OF SYNTHETIC SKIN SUBSTITUTES OVER TIME

Although not well documented in the literature, the first endeavor of a "skin substitute" graft in the United States was done by Dr. E.W. Lee in 1880 [4a]. He used sheep skin to treat an extensive burn in a young girl. Since then, several attempts were performed using heterografts and synthetic substitutes. These were classified according to their form/shape. Different macroshapes have been used for the treatment of burn patients and other skin injuries, mostly in the form of films, gel dressings, spray-on films, foams, and sponges.

1. **Films**

   Five types of films have been introduced.

   a. **Methylcellulose films.** In 1942, the first films were made with methylcellulose and were used as a wound dressing [4b]. These films were short lived and disintegrated in a short period of time (1 day if covered with skin flap and up to 5 days if exposed to air).
   b. **Animal collagen films.** In 1959, films made from animal collagen were reported [4c]. The properties of these films depended on the preparation procedures, with emphasis on the fact that these films should cause no inflammation or rejection reactions [5–8].

      Collagen films can be made with antimicrobial agents [5] and with reinforcing materials to enhance tensile strength [9]. The outer surface of collagen film had to be covered with water evaporation retarding agents in order to prevent drying and cracking [10]. Collagen films had better adherence at both 5 and 72 h after application than autograft, homograft, heterograft, and silicon membrane, on partial and full-thickness wounds [11,12].
   c. **Hydron film (poly (2-hydroxyethyl methacrylate) polymerized with polyethylene glycol 400).** In 1966 and in the late 1970s, Hydron film was prepared, which had many favorable characteristics [12a,13–22]. The advantages of Hydron dressing were greater patient comfort, equivalent control of bacterial growth, less frequent dressing changes, and faster coverage by epithelial cells. Disadvantages include difficulty of application, poor adherence during the first 24 h after application, the frequent need to repair cracks in the film, and the inability to monitor the wound for microorganisms.

      Hydron may be useful in the treatment of a partial-thickness skin loss, burns covering less than 20 per cent of the body surface, but should not be used for full-thickness skin injuries. The poor tensile strength of the film has been improved by grafting a thin film of polybutadiene to the outer side of the polymethacrylate film [23].
   d. **Opsite (elastomeric polyurethane).** In 1975, a film made from elastomeric polyurethane (Opsite, ~28 μm thick) was recommended as a dressing for skin graft donor sites and partial-thickness burns but not for full-thickness burns [24–26]. The film has the advantages of reducing pain, promoting earlier wound coverage by epithelial cells (in 10 ± 5 days compared with 14 ± 7 days beneath the dressing Bactigras), and lack of bulky dressings. Prolonged contact between the film and tissues did not produce an allergic reaction [27]. The main disadvantage of the film was

its poor water permeability (less than 800 g/m^2/24 h), which allowed the accumulation of fluid under the film and resulted in microorganism growth, in the early period of treatment after injury.

e. **Plasticized polyvinylchloride (PVC) and polyacrylonitrile.** In 1976, two different membrane materials were introduced: one based on plasticized polyvinylchloride (PVC) and the second on polyacrylonitrile [28,29].

PVC films containing plasticizers, available commercially as Vitafilm, Clingfilm, or Stretch'n Seal, have been used as a dressing for partial-thickness burns and skin graft donor sites [28]. These films were very thin (12 μm), had water vapor permeability (up to 800 g/m^2/24 h), and had appreciable permeability to both oxygen and carbon dioxide. When used as dressing for partial-thickness burns and donor sites, the film reduced pain and was less expensive and quicker to change than conventional dressings. However, in the presence of virulent bacteria, the films led to infectious complications, including destruction of epithelial cell remnants and conversion of partial to full-thickness skin loss [29]. The water impermeability of the film allowed the accumulation of exudate between the tissues and the film. Bacteria grew profusely in this fluid and invaded the tissues, producing between 10^5 and 10^8 *Pseudomonas aeruginosa* organisms per gram of biopsy tissue 3 days after application.

2. **Gel dressings**

In 1952, gel dressings were introduced based on polymerized hydrolyzed casein [30]. In 1972, gels based on block copolymers, derived from oxides of propylene and ethylene, were introduced. These gels were unable to provide mechanical protection to injured skin.

3. **Spray-on films**

In the 1950s, spray-on films were introduced using polyvinylchloride (PVC) dissolved in ethylacetate and propanone. Application of these sprays caused stinging pain and the films, after solvent evaporation, prevented external bacterial contamination. These spray-on films had good adhesion but the film water permeability was insufficient to handle the large volumes of exudates, usually produced after burn injuries. The spray-on method was later used with an acrylic polymer, an incorporated antimicrobial agent [31] and poly-ε-caprolactone, dissolved in tetrahydrofuran [32].

4. **Foams and sponges**

In the 1960s, foams and sponges were introduced to the market. The advantages of hydrophilic sponges were their absorbency, thickness providing protection against external mechanical forces, low cost, and ease of application compared to the standard mixture of gauze, cotton wool, and bandages [33]. The first material to be used in the shape of a sponge was poly(vinyl alcohol). This sponge adhered well to tissues and changed from a soft pliable and elastic material to a rigid sponge when dry, but the wounds scarred badly when the sponge fragmented with age and left fragments in the healing tissue. Also microorganisms tended to proliferate in the spaces within the sponge and the loss of water was not controlled.

Modified versions of this sponge—by making a bilaminate with a thin silicon rubber top layer or tightly compressed cells in the top layer—did not confer a clear beneficial effect [33–39]. In 1966, silicon foam (1–2 mm thick) and a terpolymer foam made of acrylate (4 mm thick) [10] were introduced but were not successful due to microorganism growth and uncontrolled water evaporation.

In 1966, an important work was published on the use of a reconstituted bovine collagen sponge [10]. The porosity of the sponge allowed fibroblast colonization, and adhesion of this sponge was greater than porcine xenograft when applied to full-thickness burns in an animal model. Wound contraction, in the absence of microorganism contamination, was only 17% compared to pigskin (27%) and a polyurethane sponge (44%). The main disadvantage of this material was bacterial growth underneath the sponge. This could be reduced by soaking it with 3.2% gentamicin.

In 1973, polyurethane foam was introduced [40] and was reported to have good adhesion to both burns and grafting sites, irrespective of wound contamination with bacteria. Periodic removal of the foam caused a large amount of pain to the patient, but provided a clean surface for autografting.

In 1975, polyether foams such as Lyofoam (Molnlycke Health Care, USA) were introduced [41]. They have similar adhesive properties to polyurethane foams, with an initial healing rate two to three times more rapid than wounds covered with traditional tulle gras, gauze, cotton, wool, and crepe bandage but at a lower cost than the latter.

SELECTED PATENTS ON "ARTIFICIAL SKIN"

Historically, working with composite films started in 1966, with an outer layer composed of silicon rubber (silastic) film [17,42–44]. Later different materials were used for the outer layer, such as polyvinyl fluoride [43], polyvinyl chloride, polyurethane film [45,46], polypeptide film [44,47–49], polytetrafluoroethylene [50,51], and poly-ε-caprolactone films [32]. The primary functions of these layers were to provide mechanical strength, while being elastic and durable, to protect the injured tissue from bacterial contamination, to provide an egress for exudate from the wound area and to be water vapor permeable.

The major disadvantage of all the different types of films mentioned above was the accumulation of fluid beneath them in the dead spaces between the tissues and the films, since they failed to adhere firmly to the underlying tissue. The film should have a high surface area structure, in order to obtain a good adhesion between the film and the tissue within few hours after application [32].

Subsequently, relevant US patents and all commercial synthetic skin materials were based on composite/hybrid macrostructures (laminate composite) of a thin outer film covering of microporous plastic film and an adhesion promoting porous inner layer that can have different form(s)/shape(s)/structure(s) (knitted fabric, foam, mat). The porous inner layer functions to produce quick and firm adherence to the wound surface to eliminate voids between the underlying tissue and the composite membrane where fluid can accumulate and microorganisms can grow. It is critical for the substitute skin membranes to have strong adherence to the wound surface to prevent its detachment from the high flux of fluid produced in the early stages of the burn/trauma.

At least Ten "artificial skin" patents have been issued since the 1970s from five different countries (summarized in Table 1). Following is a short description for selected US patents:

US Pat 7815931 (Woodroof E.A. and Enright M.K., 2010)

This describes a method to produce artificial skin, which is very similar to that of Biobrane (discussed below) but with much higher porosity in the silicon rubber sheet (up to 12%), using a vacuum technique that does not damage the nylon weave used as the inner layer. The inventor of this product is also the coinventor of Biobrane (patent number US Pat 4725279 (1988)).

The porosity in the top silicon layer was increased to 6%, 11%, and 12% depending on the type. These values are much higher than that of Biobrane with a porosity of 1.2%. This high porosity in the top silicon sheet is reported to better handle the large volume of exudate released in the early stages of a burn injury.

US Pat 6383220 (Van Blitterswijk et al., 2002)

A significant improvement in the morphology of formed neodermis was reported in pigs grafted with polyethylene glycol terephthalate/polybutylene terephthalate copolymer (PEGT/PBT copolymer) containing fibroblasts (autologous or allogenic) compared with the same product lacking fibroblasts, both in collagen appearance and distribution and also in terms of synthetic skin degradation in which the degradation of the fibroblast-populated product was low compared to the high degradation of the cell-free product. Vascularization took place after 4 days following application in both cell-free and fibroblast populated artificial skin.

This invention is feasible for culturing fibroblasts, due to its porous characteristic. The fibroblasts can grow both on the upper and on the lower side of the skin, as well as within the skin, while the keratinocytes will be normally provided on the upper layers of the skin.

US Pat 4882162 (Ikada et al., 1989)

This invention is composed of silicon rubber top sheet, surfaced by a fibrous wound contacting layer of polyglycolic acid and/or copolymers, which will be subsequently degraded to a form having an optimum bioabsorption, in a period of 2–4 weeks. Thus, damage to the newly formed, fragile epidermis is minimized, when the silicon top sheet is removed as it is no longer needed.

TABLE 1 Some Selected US Patents Since 1970 on Artificial Skin

	Year Issued	Ref./Pat. Number	Country of Inventors	Synthetic Polymer Used/Covered
1	2010	7815931	USA	Silicone elastomer
2	2004	6733530	Hong Kong	Silicone elastomer
3	2002	6383220	Netherlands	Polyalkylene glycol and an aromatic polyester, polyalkylene glycol comprises one or more of polyethylene glycol, polypropylene glycol, or polybutylene glycol. Aromatic polyester is one or more of poly(ethyleneterephthalate), poly(propyleneterephthalate), and poly(butyleneterephthalate). Copolymer of polyethylene glycol/poly(butyleneterephthalate) copolymer (PEGT/PBS copolymers).
4	1997	5658331	Italy	*Biocompatible materials*: Polylactic acid, polyglycolic acid, and their copolymer and their derivatives, polydioxanoes, polyphoshpazenes, polysulphones, and polyurethane. *Other synthetic materials*: Silicone, silane, and silane rubbers, fluoropolymers, polystyrene, vinyl polychloride, polyarcylate, and derivatives thereof, polyhydroxyacrylate, polyhydroxymethacrylate, carboxyvinyl polymers, and their derivatives, polyvinylchloride, polyvinylalchol and its derivatives, polyethylene and polypropylene. Fluoropolymers, such as polyfluoroethylene, polyfluoropropylene, and polyfluoroethers.
5	1997	5650164	Italy	Same as in number 4 above
6	1994	5326356	Italy	Polylactic acid, polyglycolic acid, and their copolymer and their derivatives, polydioxanoes, polyphoshpazenes, polysulphones, and polyurethane

7	1992	5147401	The Netherlands	Nonporous top layer of a segmented thermoplastic copolyester made from polyoxyethylene/polybutylene terephthalate (PEO/PBT)
8	1991	4996292	USA	Thermally polymerized copolyamino acids used to make self-sealing skin;
9	1991	4985036	The Netherlands	nonbiodegradable elastomer, polyurethane having elastomeric properties is used as elastomer. The elastic polyurethane is made from polyether urethane or a polyester urethane which is based on the elements chosen from the group comprising polytetramethylene adipate, polyethylene glycol adipate, polytetramethylene oxide, polydiethylene glycol adipate, 3,4′- and 4,4′-diphenylmethane diisocyanate, toluene diisocyanate, and hexamethylene diisocyanate.
10	1989	4882162	Japan	Fibrous material prepared from a bioabsorbable substance selected from polyglycolic acid, polylactic acid, and copolymers thereof and polyglycolic acid; polymer support film is prepared from silicone elastomer.

US Pat 5147401 (Bakker D. et al., 1992)

This artificial skin is composed of (a) a nonporous top layer of segmented thermoplastic copolyester that is water-vapor permeable and (b) a macroporous lower layer of a degradable, biocompatible (co)polymer having a thickness of 30–300 μm, a pore diameter of 30–300 μm, and a macroporosity of 30–80%. Porosity was generated by salt-leaching technique.

US Pat 4985036 (Lommen, E.J.C.M.P. et al., 1991)

A material that transmits wound moisture and is comprised of two layers of thermoplastic elastomer: (a) an upper layer with a thickness of 0.01–0.2 mm, a pore diameter not exceeding 0.7 μm, and a pore density of 1–75%, (b) and a lower layer having a thickness of 0.05–1 mm and a pore diameter of 50–200 μm. The artificial skin immediately and firmly adheres to the surface of the wound until the wound healing process is complete. It reduces the production of wound moisture.

The drainage capacity of the top layer is very large and normally adequate for the quantity of exudate originating from skin lesions, and the micropores in the walls play an important role in exerting a strong capillary suction force, which causes immediate adherence to the wet wound bed and also promotes the transport of exudates.

SELECTED EXAMPLES OF COMMERCIALLY AVAILABLE SKIN SUBSTITUTES

Massive skin defects, most commonly those resulting from burns or trauma, require immediate coverage. As mentioned above and elsewhere in this book, the "gold standard" for such treatment is the split-thickness autograft. Full-thickness grafts can also be used, and even though they heal with less contracture and scarring, their take is not as optimum, and they produce higher donor site morbidity. Skin substitutes can range from being completely biologic, such as an acellular dermal matrix (Alloderm Lifecell Corporation, Branchburg, NJ), to being almost completely synthetic, such as the bilayered Biobrane® (discussed below). Other useful products on the market include Integra, Dermagraft, Apligraf, Epicel, and others. This section only addresses products that have synthetic components (Additional products under development are summarized in Table (2). For a comprehensive review of commercially available products, the reader is directed to Supp and Boyce [52], Ehrenreich and Ruszczak [53], and Alrubaiy and Al-Rubaiy [54]).

Biobrane® (Smith & Nephew, USA) The use of this skin substitute started in the late 1970s, and it has been widely utilized as a temporary coverage in acute cases. It is comprised of two synthetic layers mimicking the layers of natural skin. The outer layer, which acts as an "epidermis," is made of a silicon membrane, while the inner layer, which acts as a "dermis," is a nylon mesh.

TABLE 2 Skin Substitute Products Currently Under Development

Product	Company	Incorporated Cells	Scaffold Type	Synthetic Material
Polycaprolactone collagen nanofibrous membrane	Nanoscience and Nanotechnology Initiative, Division of Bioengineering, National University of Singapore, Singapore	Cultured dermal fibroblasts	Synth	Polycaprolactone-blended collagen electrospun nanofibrous membrane
Tegaderm-nanofibre construct	Nanoscience and Nanotechnology Initiative, Division of Bioengineering, National University of Singapore, Singapore	Cultured dermal fibroblasts	Xeno + synth	Poly(ε-caprolactone)/gelatin nanofibrous scaffold electrospun on polyurethane dressing
Hybrid nanofibrous PLGA/chitosan membrane	School of Materials Science and Engineering, Tianjin University, Tianjin, China	—	Synth	PLGA/chitosan hybrid electrospun nanofibrous membrane
Biodegradable polyurethane microfibres	Department of Materials Science and Engineering, University of Delaware, Newark, DE, USA	—	Synth	Biodegradable polyurethane microfibres
Silk fibroin and alginate	Department of Veterinary Physiology, College of Veterinary Medicine and School of Agricultural Biotechnology, Seoul National University, Seoul, South Korea	—	Xeno + synth	Silk fibroin/alginate-blended sponge
Polyvinyl alcohol/chitosan/fibroin-blended sponge	Department of Sericulture and Entomology, National Institute of Agriculture and Technology, Suwon, Korea	—	Xeno + synth	Polyvinyl alcohol/chitosan/fibroin-blended sponge
Composite nano-titanium oxide–chitosan artificial skin (NTCAS)	Department of Nursing, Cardinal Tien College of Healthcare and Management, Taipei County, Taiwan	—	Allo + recomb	Composite nano-titanium oxide–chitosan with gelatin and hyaluronic acid

Both layers are embedded in porcine collagen and chemically linked, in order to enhance the construct's adherence to the wound bed [52]. The main indication for its use is superficial or partial-thickness wounds, burns, donor sites, and congenital diseases such as epidermolysis bullosa [55]. Commonly advertised advantages of Biobrane® include availability, reduced pain, shorter hospital stays, and accelerated healing. However, some authors reported adverse events resulting from its use, such as toxic shock syndrome due to the accumulation of fluid beneath it [56].

Integra® (Integra Life Sciences, USA) This skin substitute is considered a milestone in skin reconstruction therapies. It is a bilayered product composed of a silicone membrane acting as an epidermal layer and a dermal layer made of bovine collagen and shark chondroitin-6-sulphate glycosaminoglycan [52]. The construct is impermeable to water and therefore protects the wound from desiccation and infection. The dermal part will be vascularized in 2–3 weeks, after which the superficial layer is removed and replaced by a very thin split-thickness autograft. Advantages of Integra are ready availability and acceptable cosmetic results. On the other hand, it requires a two-step procedure and is expensive. Being impermeable may cause exudate to accumulate underneath it, which could lead to infection [52,57].

Epigard® (Medisafe GmbH, Germany) This product is marketed as a synthetic skin replacement and temporary wound dressing. Epigard is composed of a nontextile, double-ply synthetic material. The top surface consists of a thin, microporous film permeable to air and steam, which would prevent exudate build-up while at the same time providing protection to the wound. The underside forms an open matrix made of pliable, soft polyurethane. The wound exudate adheres to the fine porous structure and is removed together with the necrotic tissue when the bandage is changed. Fibroblasts and endothelial cells migrate into the wound area and the open polyurethane foam during the mechanical wound cleansing phase, which conditions the wound for subsequent transplantation. Two types of Epigard are available for the outer microporous film, one is made from polytetrafluoroethylene [50,51], while the other is made from polyurethane [46]. The porous layer (inner layer) is made from reticulated polyurethane foam 1 and 2 mm thick with pores of 50 μm or more in diameter, which accommodates fibroblast cell invasion. Due to the polyurethane foam used, a foreign body reaction is produced in 14 days. Antimicrobial agents such as silver sulphadiazine and gentamicin can be added to the inner layer foam to control microbial growth in contaminated wounds/burns.

Transcyte® (TransCyte Inc., USA) It is a tissue-engineered skin-replacement therapy composed of a nylon mesh and a silastic, semipermeable, biocompatible layer. Allogenic neonatal fibroblasts are embedded in the mesh and maintained for 3–6 weeks to produce a collagenous cellular matrix [55]. The construct is left in place until it spontaneously detaches, which indicates wound bed healing. Alternatively, it can be utilized for wound bed conditioning for subsequent surgery [58]. It has been approved by the FDA for use in burns.

Dermagraft® (Organogenesis Inc., USA) This product is similar to Transcyte®, but it lacks the silicone layer. Instead, it is produced by mixing viable neonatal foreskin fibroblasts with a biodegradable mesh of polyglycolic acid (PGA) in a bag with circulating nutrients. The fibroblasts are then cryopreserved at −80 °C to maintain their viability. After application to the wound, the cells proliferate and produce a variety of growth factors, as well as extracellular collagen matrix components while the PGA mesh is absorbed within 3–4 weeks [59].

LIMITATIONS OF AVAILABLE SKIN SUBSTITUTES (PROBLEMS WITH "OFF-THE-SHELF" SKIN SUBSTITUTES)

Despite the pressing need for commercially available skin substitutes that mimic the normal skin for cases of burn, skin injury, and different skin diseases, the available substitutes are far from being optimal [60]. The main problems with the limited "off-the-shelf" skin grafts and challenges facing their improvement are:

1. Short-term survival and inability to "take": Most commercially available skin substitutes do not allow rapid angiogenesis to occur. As a consequence, the grafted cells in the substitute die, resulting in a failure to "take." Understanding the revascularization process would lead to improvement of skin substitutes and incorporating prefabricated vessels to increase the speed of "take" could be the solution.
2. Scarring: In few cases where the skin substitutes do "take," the margins are often different from the original skin, both functionally and aesthetically. Hair follicles and sweat glands are absent within the scar tissue, which is also less resistant to ultraviolet radiation [61].
3. Lack of functional skin characteristics: Successful grafted skin provides a barrier, like normal skin does, but lacks temperature control due to the absence of sweat glands. Immune regulation provided by Langerhans cells is also absent. The incorporation of melanocytes, Langerhans cells, and sweat glands, or induction of their differentiation, would be a major clinical development in skin-replacement therapies.
4. Slow autologous cell expansion protocols: Skin cells taken from a patient (autologous) take 2–3 weeks to expand in sufficient numbers to form dermis and epidermis substitutes for grafting purposes. This delay is critical in severely injured patients, and the development of rapid expansion protocols would be essential in these cases.
5. Safety of allogenic cells in skin substitutes: Large surface area burns and skin disorders necessitate the use of epidermal and dermal cells isolated from donor skin (allogenic). Immediate skin replacement can also be achieved using an "off-the-shelf" skin product that is made by donor skin. The long-term safety has not yet been demonstrated in terms of the lack of subsequent transformation of the grafted allogenic cells [62]. The bioengineered skin products need to be examined for undesirable local or systemic inflammation or rejection responses.

6. Persistence of allogeneic skin grafts: Commercially available skin products incorporating cells, such as Apligraf (Organogenesis Inc., USA), have shown that the presence of skin cells assisted in wound healing [63]. Cell survival in such grafts may be affected by the age of cells used. Neonatal skin cells and fetal fibroblasts are among the best sources and were shown to persist for up to 2.5 years.
7. Biocompatibility and mechanical properties: The materials used in the skin graft need to mimic the normal skin composition and support all the cellular activities that ensure regeneration. In addition, the physical properties of the skin substitutes should allow the clinicians to handle them with ease in surgical settings [64].
8. Development and product costs: The research and development of a skin substitute followed by the safety assessment and clinical trials that lead to its approval are costly. In addition, the cost of production, storage, and marketing, all make the task of having a ready to use shelf product a hard one.

SELECTED MATERIALS AND FABRICATION METHODS

The current focus of skin substitute research is aimed at developing a tissue-engineered skin equivalent that combines both living cells and natural or synthetic cellular components [65]. Engineering skin substitutes requires planned design and fabrication consistent with prespecified functional requirements [66]. Efforts have been directed at the development of synthetic scaffolds, which can support the growth of cells. Accordingly, the development of synthetic scaffolds is a major objective for the skin tissue engineer. Some of the earliest materials to be utilized, and approved by the FDA for clinical use, are poly-L-lactic acid (PLLA), polycaprolactone (PCL), and polyglycolic acid (PGA) [67,68]. Other examples of synthetic materials include polytetrafluoroethylene and polyethylene terephthalate [69].

In general, polymers employed for tissue engineering are usually resorbed or degraded in vivo. Importantly, any polymers, as well as their degradation products, should be nontoxic and nonimmunogenic upon implantation and degradation. Moreover, there are two major challenges to overcome: (1) limitations in mechanical properties and (2) fabrication into functionally useful three-dimensional scaffolds [70,71]. Improvements in mechanical properties could be achieved through developing new classes of biopolymers with superior properties, or alternatively, modifications of previously used polymers or the use of different ratios of these polymers have been attempted.

Fabrication methods can vary according to many factors, such as the material used, purpose of the scaffold, availability, and others. Some methods have attracted more attention than others over the past years, namely three-dimensional printing, plasma polymerization, and electrospinning. There is much interest in producing scaffolds by electrospinning [2]. Using this technique, one is able to produce three-dimensional porous structures that mimic the structure of fibrous dermis.

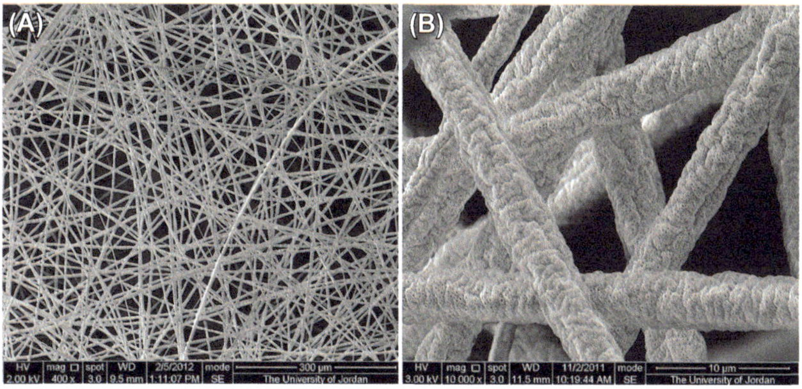

FIGURE 1 Scanning Electron Micrograph of Nonwoven Electrospun PLGA fibers (A), showing corrugated surface at higher magnification (B).

Of special importance in the manufacturing process is the control of properties of material for successful use in tissue engineering. The macrostructure (macroshape) of polymeric materials plays a very important role in determining the overall properties. With the advent of new techniques, opportunities are emerging for the production of controlled macroshapes that are more suitable to serve the emerging field of tissue engineering [72]. Figure (1) shows PLGA electrospun fibers, fabricated into a membrane to be used as a scaffold for cultured skin cells. A greater magnification in Figure 1(B) shows the corrugated surface of the fibers, which enhances cell attachment to the scaffold (our unpublished data). It is noteworthy that these fibers may be loaded with active chemicals, like antimicrobial agents, to further control the local environment and positively enhance healing.

FUTURE PROSPECTS AND CONCLUDING REMARKS

The ultimate goal of skin tissue engineering is to rapidly produce skin constructs that offer the complete regeneration of functional skin including all layers of the skin (epidermis, dermis, and fatty subcutus) and all the skin appendages (hair follicles, sweat glands, and sensory organs) [73–76]. In addition, the engineered constructs should be rapidly vascularized and produce scar-free integration with the surrounding host tissue, including the establishment of a functional vascular and nerve network [77]. Such constructs should allow the skin to fulfill its many normal functions, such as barrier formation, pigmentation defense against UV irradiation, thermoregulation, and mechanical and aesthetic functions [78]. Some, but not all, of the functions can be restored with existing skin substitutes. Cultured keratinocytes, applied as a cell layer in fibrin matrix or as cells in spray techniques, can rapidly restore barrier function on a burn or major trauma [79], while cultured keratinocytes on a dermal substitute such as collagen/fibronectin mixture can partially restore some mechanical function [80].

Despite the many potential advantages of autologous cultured cells, the potentiality of their rapid availability as an off-the-shelf item for the treatment of burns and trauma is considered one of their major drawbacks. On the other hand, heterologous cells such as poorly differentiated or stem cells, which are not immunologically rejected and are genetically stable to prevent subsequent tumor formation, would appear to hold the promise as an off-the-shelf widely available therapy.

Understanding how to engineer such cells using lessons learned from embryonic development, stem cell biology, and biomaterial engineering will be a key to developing the next generation of skin substitutes [81].On one hand, these substitutes can be engineered to be highly sophisticated, for example, to release in a time-dependent fashion various signaling molecules, differentiation factors, and protein domains [82]. Such substitutes may be further engineered to produce various three-dimensional patterns and densities. On the other hand, the substitutes may be minimally engineered, with the aim of producing substitutes that work as an easily handled carrier (e.g., thermo-/hydrogel) [83] for appropriate cells, which when the carrier disappears in the body (by noninflammatory dissociation) allows the cells to interact to regenerate all of the skin structures.

Tissue engineering of skin is based on more than 25 years of research and relies on a strong background of material technologies and cell and molecular biology. The challenge that still remains is the generation of a complex dermo-epidermal substitute that can be securely and conveniently transplanted with minimal scarring in one single surgical intervention [84,85]. Investigating new means to speed up vascularization in such complex grafts, seeding endothelial cells directly into the matrix, and engineering the vasculature directly into the tissue may well contribute to reach this goal.

The development of novel ECM matrices and scaffolds will still be essential in engineering optimized skin grafts [86]. Several attempts to develop novel skin substitutes are still (and will be in the future) based on purified ECM components such as collagens, fibrin, and hyaluronic acid. In addition, growth factor administration can dramatically influence development of a successful skin substitute [87,88]. It is promising that the outcome of growth factor administration can be improved enormously with the employment of relatively simple slow-release schemes. However, it also needs to be taken into account that an epithelium and its mesenchyme support each other in an equilibrated and complete manner. Thus, the growth factors and matrix components released by different, but interacting, cell types may be even more beneficial than slowly released factors and matrices. It remains to be investigated whether purified factors alone or distinct cell types in concert, or combinations, will provide the "skin engineering" concept of the future [89].

The field of stem cell biology must also be integrated in this future concept. The potential for using human-originated adult stem cells as a source of in vitro generation of skin is huge [90–93]. Human keratinocytes can be derived from different sources, such as epidermal stratum basale, hair follicles, and from eccrine sweat glands. These cells can be considered sources that allow scaling up the self-renewing keratinocytes fraction in engineered human skin grafts [94].

Over 25 years of experience has identified that, in order to achieve the production of tissue-engineered materials, the focus must move more quickly from laboratory to clinical use. Yet safety precautions must be addressed. Production costs must be carefully considered, so that reimbursement produces sufficient income to sustain a new product. Of importance also is that new products have to be rapidly adapted to a changing regulatory environment.

Up-to-date, tissue-engineered skin has already delivered considerable benefits to patients with burns, trauma, infections, and chronic wounds. "Skin engineering" has enormous potential that has just begun to be realized.

ACKNOWLEDGMENTS

This work was partially supported by the Research Support fund of the Ministry of Higher Education/Jordan. Grant no. 2009/01/1 year 2009.h. The authors wish to thank Fatima Jamali, Dana Hattab, Sophia Odwan, Nazneen Aslam, and Maram Hasan for critically reviewing the manuscript.

REFERENCES

[1] Bottcher-Haberzeth S, Biedermann T, Reichmann E. Tissue engineering of skin. Burns 2010;36(4):450–60.

[2] MacNeil S. Progress and opportunities for tissue-engineered skin. Nature 2007;445(7130): 874–80.

[3] Shevchenko RV, James SL, James SE. A review of tissue-engineered skin bioconstructs available for skin reconstruction. J R Soc Interface 2010;7(43):229–58.

[4] [a] MacNeil S. Biomaterials for tissue engineering in skin. Materialstoday 2008;11(5):26–35.
[b] Bromberg B, Song I, Mohn M. The use of pig skin as a temporary biological dressing. Plast Reconstr Surg 1965;36(1):80–90.
[c] Pickrell K. A sulfonamide film for use as surgical dressing: preliminary report. Bull Johns Hopk Hosp 1942;71:304.
[d] Pappas A, Hyatt GW. The evaluation of collagen film applied to skin defects in mice. Surg Forum 1959;10:844.

[5] Hait MR, Battista OA, Stark RB, McCord CW. Microcrystalline collagen as a biologic dressing, vascular prosthesis, and hemostatic agent. Surg Forum 1969;20:51–3.

[6] Lorenzetti OJ, Fortenberry B, Busby E. Influence of microcrystalline collagen in wound healing. II. Comparison of several collagen dressings on excised wounds of pigs and rabbits. Res Commun Chem Pathol Pharmacol 1973;5(2):431–40.

[7] Wilkinson TS, Tenery JH, Zufi D. The skin graft donor site as a model for evaluation of hemostatic agents. Plast Reconstr Surg 1973;51(5):541–4.

[8] Tavis MJ, Harney JH, Thornton JW, Bartlett RH. Modified collagen membrane as a skin substitute: preliminary studies. J Biomed Mat Res 1975;9(3):285–301.

[9] Song IC, Bromberg BE, Mohn MP, Koehnlein E. Heterografts as biological dressings for large skin wounds. Surgery 1966;59(4):576–83.

[10] Davies JW. Synthetic materials for covering burn wounds: progress towards perfection. Part I. Short term dressing materials. Burns Incl Therm Inj 1983;10(2):94–103.
[11] Tavis MJ, Thornton JW, Harney JH, Woodroof EA, Bartlett RH. Graft adherence to de-epithelialized surfaces: a comparative study. Ann Surg 1976;184(5):594–600.
[12] [a] Tavis MJ, Thornton JW, Harney JH, Danet RT, Woodroof EA, Bartlett RH. Mechanism of skin graft adherence: collagen, elastin, and fibrin interactions. Surg Forum 1977;28: 522–4.
[b] Harrison H, Lindberg R, Mason A. Management of the burn wound: bacteriological and histological response of excised rat wounds to synthetic grafts. In: Wallace A, Wilkinson A, editors. Res Bums. Edinburgh: Livingstone; 1966. p. 327–43.
[13] Saymen DG, Nathan P, Holder IA, Hill EO, Macmillan BG. Control of surface wound infection: skin versus synthetic grafts. Appl Microbiol 1973;25(6):921–34.
[14] Nathan P, Macmillan BG, Holder IA. Effect of a synthetic dressing formed on a burn wound in rats: a comparison of allografts, collagen sheets, and polyhydroxyethylmethacrylate in the control of wound infection. Appl Microbiol 1974;28(3):465–8.
[15] Nathan P, MacMillan BG, Holder IA. In situ production of a synthetic barrier dressing for burn wounds in rats. Infect Immun 1975;12(2):257–60.
[16] Nathan P, Law EJ, MacMillan BG, et al. A new biomaterial for the control of infection in the burn wound. Trans Am Soc Artif Intern Organs 1976;22:30–41.
[17] Nathan P, Robb EC, Dressler D, MacMillan BG. A silicone-nylon laminated dressing (IP-758) for closure of excised or debrided burn wounds. Burns Incl Therm Inj 1982;8(5):328–32.
[18] Tavis MJ, Thornton J, Danet R, Bartlett RH. Current status of skin substitutes. Surg Clin North Am 1978;58(6):1233–48.
[19] Dressler DP, Barbee WK, Sprenger R. The effect of Hydron burn wound dressing on burned rat and rabbit ear wound healing. J Trauma 1980;20(12):1024–8.
[20] Robb EC, Fitz DG, Nathan P. Delivery of the topical antimicrobial agents silver sulfadiazine, gentamicin and nystatin to infected burn wounds in rats from preloaded synthetic dressings. Trans Am Soc Artif Intern Organs 1980;26:533–6.
[21] Brown AS. Hydron for burns. Plast Reconstr Surg 1981;67(6):810–1.
[22] Robb EC, Nathan P. Control of experimental burn wound infections: comparative delivery of the antimicrobial agent (silver sulfadiazine) either from a cream base or from a solid synthetic dressing. J Trauma 1981;21(10):889–93.
[23] Migliaresi C, Carfagna C, Nicolais L. Laminates of poly(2-hydroxyethyl methacrylate) and polybutadiene as potential burn covering. Biomaterials 1980;1(4):205–8.
[24] James JH, Watson AC. The use of Opsite, a vapour permeable dressing, on skin graft donor sites. Br J Plast Surg 1975;28(2):107–10.
[25] Conkle W. Op-Site dressing: new approach to burn care. J Emerg Nurs 1981;7(4):148–52.
[26] Neal DE, Whalley PC, Flowers MW, Wilson DH. The effects of an adherent polyurethane film and conventional absorbent dressing in patients with small partial-thickness burns. Br J Clin Pract 1981;35(7–8):254–7.
[27] Cavlak Y. Application of Opsite at donor sites, burns and postoperative wounds (author's transl). Aktuelle Traumatologie 1980;10(6):311–6.
[28] Lendrum J, Bowen-jones E. A new dressing for burns : enclosure in a plasticized polyvinyl chloride sheet. Burns 1976;2(2):86–9.
[29] Harris DR, Filarski Jr SA, Hector RE. The effect of silastic sheet dressings on the healing of split-skin graft donor sites. Plast Reconstr Surg 1973;52(2):189–90.
[30] Spangler PE. A new local treatment for burns. US Armed Forces Med J 1952;3(1):105–14.
[31] Branemark PI, Birch J. Microvascular damage produced by a plastic film wound dressing. Scand J of Plast Reconstr Surg 1968;2(2):71–6.

[32] Schwope AD, Wise DL, Sell KW, Skornik WA, Dressler DP. Development of a synthetic burn covering. Trans Am Soc Artif Intern Organs 1974;20A:103–7.

[33] Chardack WM. The use of synthetic substitutes for skin in the treatment of burns. Mil Med 1962;127:335–44.

[34] Blumberg JB, Griffith PC, Merendino KA. The effect of specific compression on soft-tissue response to formalinized polyvinyl alcohol (ivalon) sponge: a critical evaluation. Ann Surg 1960;151:409–18.

[35] Chardack WM, Brueske DA, Santomauro AP, Fazekas G. Experimental studies on synthetic substitutes for skin and their use in the treatment of burns. Ann Surg 1962;155:127–39.

[36] Chardack WM, Martin MM, Jewett TC, Boyer BE. Synthetic substitutes for skin. Clinical experience with their use in the treatment of burns. Plast Reconstr Surg Transplant Bull 1962;30:554–67.

[37] Jewett Jr TC, Chardack WM. Temporary closure of a granulating burn wound with a synthetic sponge. Am J Surg 1963;106:24–8.

[38] Taylor PH, Moncrief JA, Switzer WE, Rose LR, Pugsley LQ. Ivalon as a cover for excised burn wounds. An evaluation. Arch Surg 1963;86:250–1.

[39] Mutschler W, Burri C, Plank E. Experimental and clinical experiences with the synthetic skin cover of polyvinylalcohol-formaldehyde foam (PVA). Helv Chir Acta 1980;47(1–2):163–6.

[40] Guldalian J, Jelenko 3rd C, Callaway D, McKnight JT. A comparative study of synthetic and biological materials for wound dressings. J Trauma 1973;13(1):32–5.

[41] Winter GD. Epidermal wound healing under a new polyurethane foam dressing (Lyofoam). Plast Reconstr Surg 1975;56(5):531–7.

[42] Hall CW, Liotta D, De Bakey ME. Artificial skin. Trans Am Soc Artif Intern Organs 1966;12:340–5.

[43] Hall CW, Liotta D, Ghidoni JJ, De Bakey ME, Dressler DP. Velour fabrics applied to medicine. J Biomed Mater Res 1967;1(2):179–96.

[44] Spira M, Fissette J, Hall W, Hardy SB, Gerow FJ. Evaluation of synthetic fabrics as artificial skin grafts to experimental burn wounds. J Biomed Mater Res 1969;3(2):213–34.

[45] Skornik WA, Dressler DP, Rhicard JK. Adherence of prosthetic skin. J Biomed Mater Res 1968;2(4):447–56.

[46] Silverstein P, McDonough RJ. Laboratory evaluation of a new synthetic bilaminate physiologic dressing. Surg Forum 1977;28:538–9.

[47] Walder AI, May PD, Bingham CP, Wright JR. Evaluation of synthetic films as wound covers. Trans Am Soc Artif Intern Organs 1969;15:29–32.

[48] Hall CW, Spira M, Gerow F, Adams L, Martin E, Hardy SB. Evaluation of artificial skin models: presentation of three clinical cases. Trans Am Soc Artif Intern Organs 1970;16:12–6.

[49] Buckley CJ, Chambers CE, Kemmerer WT, Rawlings CA, Casey HW, Hall CW. Evaluation of a synthetic bioadherent dressing as a temporary skin substitute. Trans Am Soc Artif Intern Organs 1971;17:416–20.

[50] O'Neill Jr JA. Comparison of xenograft and prosthesis for burn wound care. J Pediatr Surg 1973;8(5):705–9.

[51] Kleine-Natrop HE, Sebastian G. The use of synthetic material in preparing old leg ulcers and fresh burn injuries for transplantation (author's transl). Dermatol Monatschr 1979;165(5):331–6.

[52] Supp DM, Boyce ST. Engineered skin substitutes: practices and potentials. Clin Dermatol 2005;23(4):403–12.

[53] Ehrenreich M, Ruszczak Z. Update on tissue-engineered biological dressings. Tissue Eng 2006;12(9):2407–24.

[54] Alrubaiy L, Al-Rubaiy KK. Skin substitutes: a brief review of types and clinical applications. Oman Med J 2009;24(1):4–6.

[55] Shakespeare PG. The role of skin substitutes in the treatment of burn injuries. Clin Dermatol 2005;23(4):413–8.

[56] Shakespeare P, Shakespeare V. Survey: use of skin substitute materials in UK burn treatment centres. Burns 2002;28(4):295–7.

[57] Lee KH. Tissue-engineered human living skin substitutes: development and clinical application. Yonsei Med J 2000;41(6):774–9.

[58] Bello YM, Falabella AF, Eaglstein WH. Tissue-engineered skin. Current status in wound healing. Am J Clin Dermatol 2001;2(5):305–13.

[59] Raguse JD, Gath HJ. The buccal fad pad lined with a metabolic active dermal replacement (Dermagraft) for treatment of defects of the buccal plane. Br J Plast Surg 2004;57(8):764–8.

[60] Seet WT, Manira M, Khairul Anuar K, et al. Shelf-life evaluation of bilayered human skin equivalent, MyDerm. PloS one 2012;7(8):e40978.

[61] Markeson D, Pleat JM, Sharpe JR, Harris AL, Seifalian AM, Watt SM. Scarring, stem cells, scaffolds and skin repair. Journal of tissue engineering and regenerative medicine 2013.

[62] Han SK, Kim HS, Kim WK. Efficacy and safety of fresh fibroblast allografts in the treatment of diabetic foot ulcers. Dermatol Surg 2009;35(9):1342–8.

[63] Falanga V, Sabolinski M. A bilayered living skin construct (APLIGRAF) accelerates complete closure of hard-to-heal venous ulcers. Wound Repair Regeneration 1999;7(4):201–7.

[64] Wang X, You C, Hu X, et al. The roles of knitted mesh-reinforced collagen-chitosan hybrid scaffold in the one-step repair of full-thickness skin defects in rats. Acta Biomater 2013;9(8):7822–32.

[65] Auger FA, Berthod F, Moulin V, Pouliot R, Germain L. Tissue-engineered skin substitutes: from in vitro constructs to in vivo applications. Biotechnol Appl Biochem 2004 ;39(Pt 3):263–75.

[66] Boyce ST, Warden GD. Principles and practices for treatment of cutaneous wounds with cultured skin substitutes. Am J Surg 2002;183(4):445–56.

[67] Hernon CA, Dawson RA, Freedlander E, et al. Clinical experience using cultured epithelial autografts leads to an alternative methodology for transferring skin cells from the laboratory to the patient. Regen Med 2006;1(6):809–21.

[68] Moustafa M, Simpson C, Glover M, et al. A new autologous keratinocyte dressing treatment for non-healing diabetic neuropathic foot ulcers. Diabetic Med 2004;21(7):786–9.

[69] Vats A, Tolley NS, Polak JM, Gough JE. Scaffolds and biomaterials for tissue engineering: a review of clinical applications. Clin Otolaryngol Allied Sci 2003;28(3):165–72.

[70] Griffith LG. Emerging design principles in biomaterials and scaffolds for tissue engineering. Ann N Y Acad Sci 2002;961:83–95.

[71] Griffith LG, Swartz MA. Capturing complex 3D tissue physiology in vitro. Nat Rev Mol Cell Biol 2006;7(3):211–24.

[72] Metcalfe AD, Ferguson MW. Harnessing wound healing and regeneration for tissue engineering. Biochem Soc Trans 2005;33(Pt 2):413–7.

[73] Wendt H, Hillmer A, Reimers K, et al. Artificial skin–culturing of different skin cell lines for generating an artificial skin substitute on cross-weaved spider silk fibres. PloS one 2011;6(7):e21833.

[74] Bellas E, Seiberg M, Garlick J, Kaplan DL. In vitro 3D full-thickness skin-equivalent tissue model using silk and collagen biomaterials. Macromol Biosci 2012;12(12):1627–36.

[75] Nathoo R, Howe N, Cohen G. Skin substitutes: an overview of the key players in wound management. J Clin Aesthet Dermatol 2014;7(10):44–8.

[76] Shahrokhi S, Arno A, Jeschke MG. The use of dermal substitutes in burn surgery: acute phase. Wound Repair Regen 2014;22(1):14–22.

[77] Schenck TL, Chavez MN, Condurache AP, et al. A full skin defect model to evaluate vascularization of biomaterials in vivo. J Visualized Exp 2014;90.
[78] Fernandez TL, Van Lonkhuyzen DR, Dawson RA, Kimlin MG, Upton Z. Characterization of a human skin equivalent model to study the effects of ultraviolet B radiation on keratinocytes. Tissue Eng Part C Methods 2014;20(7):588–98.
[79] Wojtowicz AM, Oliveira S, Carlson MW, Zawadzka A, Rousseau CF, Baksh D. The importance of both fibroblasts and keratinocytes in a bilayered living cellular construct used in wound healing. Wound Repair Regen 2014;22(2):246–55.
[80] Reinertsen E, Skinner M, Wu B, Tawil B. Concentration of fibrin and presence of plasminogen affect proliferation, fibrinolytic activity, and morphology of human fibroblasts and keratinocytes in 3D fibrin constructs. Tissue Eng Part A 2014;20(21–22):2860–9.
[81] Chlapanidas T, Tosca MC, Farago S, et al. Formulation and characterization of silk fibroin films as a scaffold for adipose-derived stem cells in skin tissue engineering. Int J Immunopathol Pharmacol 2013;26(Suppl. 1):43–9.
[82] Greaves NS, Ashcroft KJ, Baguneid M, Bayat A. Current understanding of molecular and cellular mechanisms in fibroplasia and angiogenesis during acute wound healing. J Dermatol Sci 2013;72(3):206–17.
[83] Zhu J, Marchant RE. Design properties of hydrogel tissue-engineering scaffolds. Expert Rev Med Devices 2011;8(5):607–26.
[84] Clement AL, Moutinho Jr TJ, Pins GD. Micropatterned dermal-epidermal regeneration matrices create functional niches that enhance epidermal morphogenesis. Acta Biomater 2013;9(12):9474–84.
[85] Rehim SA, Singhal M, Chung KC. Dermal skin substitutes for upper limb reconstruction: current status, indications, and contraindications. Hand Clin 2014;30(2):239–52. vii.
[86] Baiguera S, Urbani L, Del Gaudio C. Tissue engineered scaffolds for an effective healing and regeneration: reviewing orthotopic studies. Biomed Res Int 2014;2014:398069.
[87] Lee K, Silva EA, Mooney DJ. Growth factor delivery-based tissue engineering: general approaches and a review of recent developments. J R Soc Interface 2011;8(55):153–70.
[88] Choi JS, Kim JD, Yoon HS, Cho YW. Full-thickness skin wound healing using human placenta-derived extracellular matrix containing bioactive molecules. Tissue Eng Part A 2013;19(3–4):329–39.
[89] Correia SI, Pereira H, Silva-Correia J, et al. Current concepts: tissue engineering and regenerative medicine applications in the ankle joint. J R Soc Interface 2014;11(92):20130784.
[90] Tan KK, Salgado G, Connolly JE, Chan JK, Lane EB. Characterization of fetal keratinocytes, showing enhanced stem cell-like properties: a potential source of cells for skin reconstruction. Stem Cell Rep 2014;3(2):324–38.
[91] Aguilera V, Briceno L, Contreras H, et al. Endothelium trans differentiated from Wharton's jelly mesenchymal cells promote tissue regeneration: potential role of soluble pro-angiogenic factors. PloS one 2014;9(11):e111025.
[92] Sivan U, Jayakumar K, Krishnan LK. Constitution of fibrin-based niche for in vitro differentiation of adipose-derived mesenchymal stem cells to keratinocytes. Biores Open Access 2014;3(6):339–47.
[93] Leiros GJ, Kusinsky AG, Drago H, et al. Dermal papilla cells improve the wound healing process and generate hair bud-like structures in grafted skin substitutes using hair follicle stem cells. Stem Cells Transl Med 2014;3(10):1209–19.
[94] Mazzone L, Pontiggia L, Reichmann E, Ochsenbein-Kolble N, Moehrlen U, Meuli M. Experimental tissue engineering of fetal skin. Pediatr Surg Int 2014;30(12):1241–7.

Chapter 10

Hybrid Biomaterials for Skin Tissue Engineering

John E. Greenwood
Royal Adelaide Hospital, Adelaide, SA, Australia

Chapter Outline

INTRODUCTION

The origin of the word hybrid is Latin (hybrida) and pertains to what we understand today as "biological hybrids." These can be defined as "the offspring of two plants, or animals, of different species or varieties, such as a mule—a hybrid of a donkey and a horse." A broader definition is "a thing made by combining two different elements; a mixture."

This chapter focuses on total split-thickness skin graft replacement, where the "hybrid" element involves a two-stage strategy utilizing a synthetic/synthetic in vivo sealed dermal scaffold, followed by application of a composite cultured skin (CCS), produced in vitro and consisting of a similar synthetic

Skin Tissue Engineering and Regenerative Medicine. http://dx.doi.org/10.1016/B978-0-12-801654-1.00010-3

scaffold, seeded with autologous dermal and epidermal cells. This is prefaced by an elucidation of other hybrid in vivo scaffolds, and in vitro composites, which will illuminate their shortcomings and illustrate the thought processes behind the synthetic solution.

Already an important distinction has been introduced, that of in vivo versus in vitro manipulation. In vivo manipulation occurs when materials that do not contain cells are implanted *into the wound*, and then encourage, facilitate, or guide cellular and vascular elements from the recipient wound bed to populate the structure, naturally realigning to produce a design outcome (physiological postimplantation manipulation). Such outcomes might include the coverage of avascular structures such as bone, cartilage, or tendon in the bed of complex wounds with vascularized "tissue" capable of allowing subsequent skin graft survival, *or* otherwise "temporizing" the wound, resisting infection, minimizing contraction, and so on until definitive wound closure is possible later. In vitro manipulation involves the culture and incorporation of desirable cell lines within three-dimensional matrices, or suspension within alternative delivery systems, *before* the material is implanted into the wound, requiring less physiological postimplantation manipulation to be undertaken by the wound bed.

In order to understand the development of all materials used and designed for skin tissue engineering, it is important to appreciate the structure and function of normal skin and the mechanisms for its repair after injury. This has been ably elucidated in other chapters, but a quick précis is offered here.

SKIN: STRUCTURE AND FUNCTION

The structure and functions of skin are intimately related. Although skin is often considered as one organ, it is actually two—each completely different from the other, even arising from different embryological germ lines (epidermis from ectoderm, dermis from mesoderm). The epidermis is cellular. When it is lost, it can be regenerated by cellular proliferation, differentiation, and migration without scar formation. The dermis is molecular, consisting mainly of collagen and elastic proteins, supported by glycosaminoglycan (GAG) ground substance. It makes up 95% of the thickness of skin and is the key to skin suppleness and mobility. When damaged or lost, it can only be repaired. This occurs by the rapid laying down of collagen, which subsequently undergoes remodeling to result in what we call a scar. Whenever injury involves the dermis, scar is inevitable. The deeper the dermal injury, the more significant the scar will be in terms of poor function and appearance.

SKIN INJURY

A proportion of the injuries that we see as surgeons are sufficiently deep and extensive that they, untreated, would normally undergo prolonged healing by secondary intention (granulation tissue formation and wound contraction), often complicated by recurrent episodes of wound infection, resulting in "pathological" scarring, which reduces, impairs, or abolishes function, is hypertrophic, dysaesthetic, and often symptomatic (causing itch and pain). Examples are those burn injuries described clinically as deep dermal (deep second-degree) and full thickness (third degree). Loss of all adenexal structures (and thus nests of potential epidermal cells within the wound) and the poor quality of the deep (reticular) dermis ensure that epithelialization can only progress from keratinocyte migration from the wound edge. Since the reticular dermis or underlying fat will not allow keratinocyte survival, let alone their migration, a vascular tissue of blood vessels and loose collagen (granulation tissue) from the wound bed creates a layer across which keratinocytes could migrate. Even then, keratinocyte migration only proceeds over 5 mm from the wound edge. Wound closure is thus necessarily facilitated by the differentiation of fibroblasts into myofibroblasts that have the capacity to contract, pulling on collagen strands as they do so, resulting in wound contraction. The outcome is a significantly smaller wound (thus a shorter distance across which epithelialization has to occur). In humans, if such healed wounds reduce or abolish movement, or cause facial disfigurement, they may prevent or reduce capacity to work and earn, or result in social withdrawal, depressive illness, and suicide. In the case of burn injury, the retention of large volumes of burn eschar allows release and absorption of lipid–protein complexes, resulting in deleterious effects on the immune system and on general organ function, leading to multiorgan failure and death. Once burn is removed, unless the wound can be closed definitively, it will also tend to heal spontaneously by granulation tissue formation and contraction.

HOW CAN WE "CLOSE" AN EXTENSIVE CUTANEOUS WOUND?

The split-thickness skin graft remains the mainstay of extensive wound closure. Its survival as a technique for ~140 years suggests that it is highly effective in closing large wounds. The characteristics and properties that make it successful can be summed up quite succinctly. It is "self" (autologous), has an inosculatory capacity for rapid take and contains some dermal supporting elements, a basement membrane and an intact epidermis, thus definitively closing the wound. It is fiscally "free," and the donor site heals spontaneously, apparently allowing a deep wound to be closed without physiological cost.

WHY DO WE NEED ALTERNATIVES TO THE SKIN GRAFT?

Although useful, the split-thickness skin graft has several disadvantages. As wound surface area increases, available donor site decreases. Eventually, even with maximal (reasonable) mechanical expansion, sufficient skin cannot be harvested to cover all excised burn wounds and still achieve an acceptable functional result and cosmetic appearance. The number of re-harvests from a donor area is finite, because the dermal component repairs (scars) rather than regenerates and it is seldom, if ever, that re-harvesting from a donor site can be contemplated within 10 days. The dermal component of the graft means that a donor site scar is always created, since the dermis is not capable of regaining its pre-injury architecture. In the very large burn, donor sites for graft harvest are simply unavailable. With the patient already susceptible to infection due to loss of mechanical restriction to pathogen ingress and general immune compromise, the donor site extends the area of skin loss and inflicts further physiological insult.

The pain generated by the donor site usually creates more patient complaint than the burn wound and can be sufficiently severe to delay mobilization and therapy, compounded by opiate administration. The pain of the donor site in the elderly can "keep them in bed," which can result in pneumonia and pressure injury, and the dermis is thinner, so that the donor site frequently struggles to heal, causing pain with recurrent breakdown and repeated infection. Thin grafts used to replace deep or full-thickness skin loss cannot confer the same degree of robustness to future injury, which thus requires less insult. A reduced thickness of elastic dermis often mars complete restitution of the supple envelope that uninjured skin provides to facilitate joint mobility and range.

WHAT NEEDS TO BE CONSIDERED IN TREATING DEEP SKIN INJURY/LOSS?

Burn eschar must be removed, and the resulting wound must be quickly closed to reduce inflammation, minimize wound contraction/risk of infection, and improve the scarring outcome. In major burn injury, "staged" closure is necessary. Several variations of staged closure exist. In the developed world, burn excision is most often followed by application of cadaver allograft to the excised wound bed. In the immune-compromised burn patient, this can "take," and as long as the epidermal components are removed before the immune system recovers, the dermal component can remain permanently, forming a base for definitive closure with widely meshed autograft. More frequently, the cadaver tissue causes alterations in the quality of the burn bed before being rejected, thus temporizing the wound as well as improving it for graft application and closure.

An alternative is the use of a bioabsorbable "temporizing" dermal matrix, which "physiologically," but temporarily, closes the excised burn wound by virtue of a pseudo-epidermal component, while the dermal component integrates and provides "dermal" structure in anticipation of later (delayed) application of split-thickness skin graft, cultured epithelial cells, or (eventually) cultured composite skin. Effective materials in this group must resist wound contraction, inhibit infection prior to overgrafting, reduce graft contraction, and improve functional and cosmetic outcomes after split-thickness skin grafting. This is particularly important where burn excision has extended deeply into fat, fascia, or muscle. A marked cosmetic improvement is observed when these materials subsequently receive widely meshed graft application over the integrated matrix (compared to the unsightly, wide "lizard-skin" appearance otherwise achieved by applying widely meshed graft on fat). Again, split-thickness skin graft donor site is needed to effect permanent closure; however, this can be acquired serially. Such technologies have existed since 1980.

More recently, the need for a split-thickness skin graft has been questioned by the creation of CCSs. These use a three-dimensional matrix as a scaffold for the serial culture of fibroblasts and keratinocytes, resulting in a bilayer product containing both a dermis and an epidermis. It takes about 3–4 weeks to create a CCS, if autologous cellular material is used to create both dermis and epidermis; thus, the excised wound needs to be stabilized to limit contraction and infection. Nonautologous (allogeneic) composites have been created, but remain prohibitively costly.

Research in the field of skin tissue engineering has accelerated in the past 5 years, with over 500 articles published on a range of materials, processes, and techniques.

AN INTRODUCTION TO BIODEGRADABLE TEMPORIZING MATRICES

Five biosynthetic materials are commercially available for in vivo reconstruction of the dermis. These materials are characterized by having an integrating dermal component and a temporary epidermal component. Since none of these materials afford definitive closure and require secondary procedures to create an epidermal component (skin grafting, CCS application, or keratinocyte seeding), their function is one of "active temporization." This implies that they do not merely buy time for surgeon and patient by mimicking physiological wound closure, when early definitive closure is prevented by skin graft donor site shortage (temporization), but also modify the quality of the subsequent wound bed to accept that definitive closure (active, rather than passive, temporization). All allow invasion of the dermal component by fibroblasts, other host cells, and blood vessels (neovascularization), which allows autologous neodermis formation. The biological support subsequently degrades and is absorbed over time. The function of the pseudo-epidermis (silicone in all five materials) is to minimize evaporative water loss, one of the key physiological stimuli in recognizing the presence of a wound. They have common disadvantages, which have reduced their impact on burn care globally. These materials are summarized briefly here.

Integra® Dermal Regeneration Template

Designed by John F. Burke and Ionnas Yannas in Boston, Massachusetts, and introduced in 1980 [1–3], Integra® Dermal Regeneration Template consists of two components. An "integrating" porous dermal component composed of cross-linked bovine type I collagen fibrils in a GAG ground substance (chondroitin-6-sulphate derived from shark cartilage) is bonded to a "pseudo-epidermis" of reinforced silicone [1]. The integration and neovascularization of Integra® varies according to structures in the wound bed and the physiology of the injury and can be complete in a "hot" acute burn situation by 21 days, but may take 4–5 weeks in a "cold" reconstructive situation. As a dermal template, it still requires definitive wound closure, often with a split-thickness skin graft. The silicone layer is peeled off once the dermal component is integrated to expose the vascularized neodermis below.

Integra® has become easier to use quite recently with simpler packaging, storage, and preparation being introduced. It has a 2-year shelf life, is relatively simple to handle and fix, and proffers low immunogenicity. The main downside is cost [4]. Meticulous surgical excision and hemostasis of the wound bed is required, fastidious monitoring and frequent dressing changes help guarantee a good take of Integra®, but a high rate of material infection has been reported [5,6].

Pelnac™ Artificial Dermal Template

Pelnac™ has been in use in Asia since the mid-1990s and consists of a superficial silicone film layer and porcine type I collagen sponge layer derived from pig tendon (atelocollagen), produced by Gunze Ltd, Tokyo, Japan. At about 3 mm thick, it is considerably thicker than Integra (about 1 mm). Three "two-stage" variants are produced: a "Standard type" (as above), a fortified variety, which has reinforcement of the silicone film with a silicone gauze, and a "Fenestrated form" to allow wound fluid egress. The collagen is freeze dried, and undergoes a cross-linking process stimulated by glutamic acid. The acid is washed out before the cross-linked collagen is refreeze dried and packaged. It is stored dry and has a 3-year shelf life.

Most of the initial work was documented in *Tissue Engineering* in 1996 [7] and, like Integra, it has been regulated for multiple uses, including deep excised burn injuries, traumatic skin loss, posttumour/neval excisional defects, and free-flap donor site reconstructions. Suzuki reported good to excellent long-term results in 90% of cases when Pelnac™ was used in combination with extra-thin split-thickness skin grafts to treat deep burns, full-thickness skin defects, and in the reconstruction after hypertrophic skin-graft donor-site scar excision [8]. Its profile (time to neovascularization and thus neodermal formation) appears similar to Integra® and Terudermis™ and, like those materials, a second-stage procedure is necessary to afford definitive closure.

Terudermis™

Terudermis™ is a lyophilized collagen sponge dermal component bonded to a silicone membrane epidermal component and manufactured by Olympus Terumo

Biomaterials Corp., Tokyo, Japan. It is regulated for use in Japan, Taiwan ROC, and Korea. The collagen is reconstituted from a mixture of fibrous and heat-denatured bovine collagen cross-linked by dehydrothermal treatment [9]. Terudermis™ has been designed for deep burn management, especially where the wound bed is complex (bone, muscle, or ligament exposed) [10]. It has also found utility in other reconstructive indications [11–15].

Hyalomatrix (PA)™

This material is produced by Anika Therapeutics Inc, Bedford, MA/Medline Industries Inc. It is a bilayered material with a dermal component derived from the GAG, hyaluronic acid (HYAFF). Like the other materials above, a pseudo-epidermis is provided by a silicone layer. The origin of the hyaluronic acid is not disclosed (can be sourced xenogenically from rooster combs, or produced by bacterial recombination). HYAFF is derived by esterification of the derived hyaluronic acid with benzyl alcohol [16,17]. Indicated primarily for chronic wound management, it has shown some conflicting results [18,19]. It appears to be efficacious in the management of deep partial-thickness burns [20].

Renoskin™

A bilayer, in vivo dermal matrix composed of purified bovine collagen bonded to a reinforced silicone pseudo-epidermis. Early animal work was reported in 2001 [21] and the material is produced by Perouse Plastie SAS, Bornell, France. Preclinical data were released in 2007 [22]. Renoskin™, Integra®, and Hyalomatrix™ were compared in 2012 in a porcine study which revealed that Hyalomatrix™ integrated most effectively, Renoskin™ was the most rapidly absorbed, but no long-term difference was reported in macro- or microscopic scar quality or wound contraction at 6 months postapplication [23].

NovoSorb™ BTM (Biodegradable Temporizing Matrix)

The high cost of biologically based dermal matrices have priced them out of the market for second and third world nations. The vulnerability to infection has limited their use in the first world to very extensive deep burn injuries and complex wound repair and burn reconstruction. The desire to produce an inexpensive dermal matrix, affordable to developing nations, while inedible by pathogenic organisms stimulated my exploration for a synthetic solution. Fiscal and supply arguments are not the only impetus for this exploration, as synthetic polymers have the added advantage of "customized" design to enable properties such as molecular elution or fluid absorption to effect better wound healing.

Biodegradable polyurethanes are of great interest for medical application because of their variable mechanical properties, biocompatibility, and structural versatility. NovoSorb™ represents a family of novel polyurethanes, produced by PolyNovo Biomaterials Pty Ltd, Port Melbourne, Victoria, Australia. A subclass of this family is biodegradable, designed, and constructed to degrade by hydrolysis to naturally occurring and biologically tolerated products.

The material is made rapidly and in large volumes from simple raw materials, and the process can be easily and greatly upscaled. The materials produced are dry packed, factory fenestrated, and gamma sterilized. Each piece is extremely light (1 oz for the 20×40 cm), can be transported by any means, stored without any special considerations, and used in austere environments. They are cut to size and shape with scissors and fixed with staples. The material's nonbiodegradable polyurethane seal has been designed to avoid spontaneous separation (delamination) and to date has remained in a human burn wound, resisting contraction, for 7 weeks before delamination and overgrafting. This negates some of the issues with the spontaneous delamination of many of the collagen/GAG products, which often occurs (when the materials are being used to temporize the wounds during serial grafting) before donor sites are ready for re-harvest.

A development program was commenced in 2004 to assess the potential of biodegradable polyurethane as both an in vivo template (BTM) and an in vitro scaffold for CCS production. BTM is designed to overcome most of the issues encountered during the use of dermal matrices of biologic origin. The BTM is a fully synthetic hybrid product, with a biodegradable polyurethane (NovoSorb™) foam dermal component bonded to a nonbiodegradable polyurethane pseudo-epidermis.

Several papers document the development of this product [24–31]. A pilot trial using BTM in the reconstruction of free-flap donor sites was completed in 2012 [30], and since then, the material has been used extensively for this indication under a Therapeutic Goods Administration (TGA) authorization for use of an unapproved product [31]. Figures 1–11 illustrate the process of integration of the BTM, including schematic explanation and histological confirmation, in a "cold" surgical wound (free radial forearm flap donor site). The schematics equally well illustrate the integration of collagen/silicone bilayer dermal matrices. Figures 12 and 13 demonstrate the BTM degradation process in the same wound. The BTM is currently being evaluated in significant full-thickness burn injuries in Australia, with recruitment of full-thickness burn injury covering between 20% and 50% total body surface area (TBSA). Figures 14 and 15 illustrate the utility of the BTM in actively temporizing a deep excised burn wound to create a useful, vascularized neodermis, which facilitates subsequent grafting at day 46 post-BTM implantation.

AN INTRODUCTION TO IN VITRO DERMO-EPIDERMAL (COMPOSITE) CULTURED SKINS

Irrespective of composition, any scaffold used to support and guide culture of autologous or allogeneic cells in the production of a CCS will form part of a long-term implantable device and thus must be composed of components that are

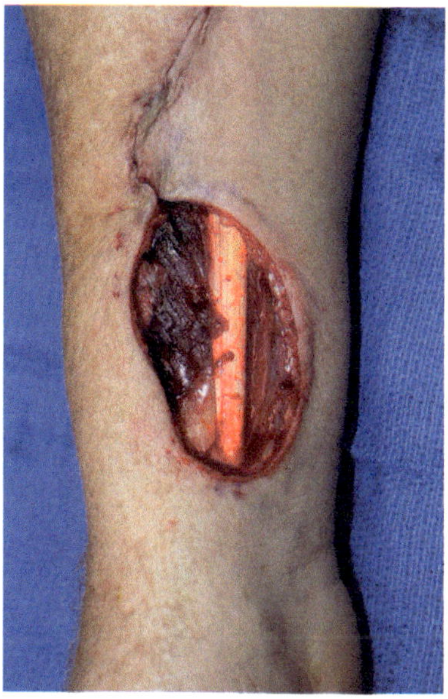

FIGURE 1 Day 0 (day of BTM insertion): radial forearm free-flap donor site just after flap harvest. The pedicle wound has been closed (proximal). The tendon of Flexor Carpi Radialis runs across the wound bed from proximally to distal. The wound area is 19.1 cm^2.

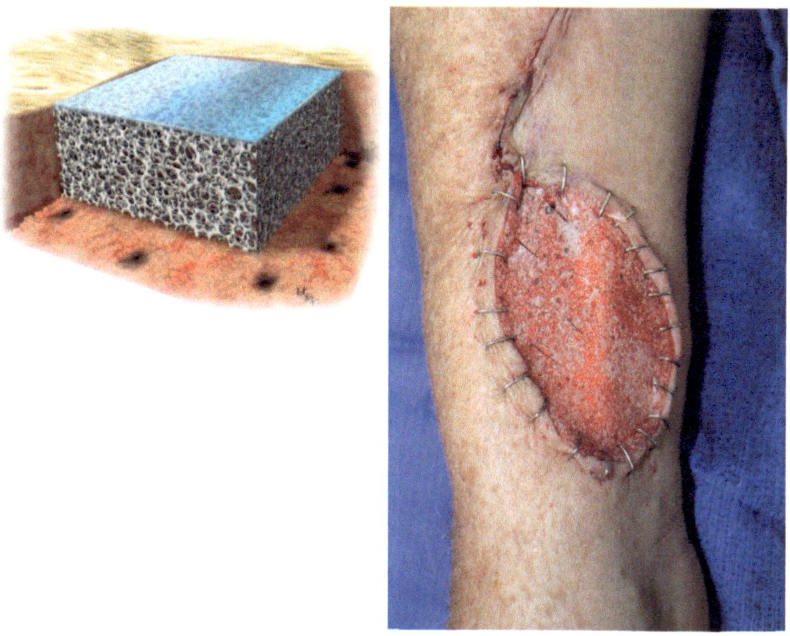

FIGURE 2 Day 0 (day of BTM insertion): with hemostasis in the wound bed secured, the fenestrated BTM is cut to shape, inserted into the wound, and affixed with staples (area = 19.1 cm^2).

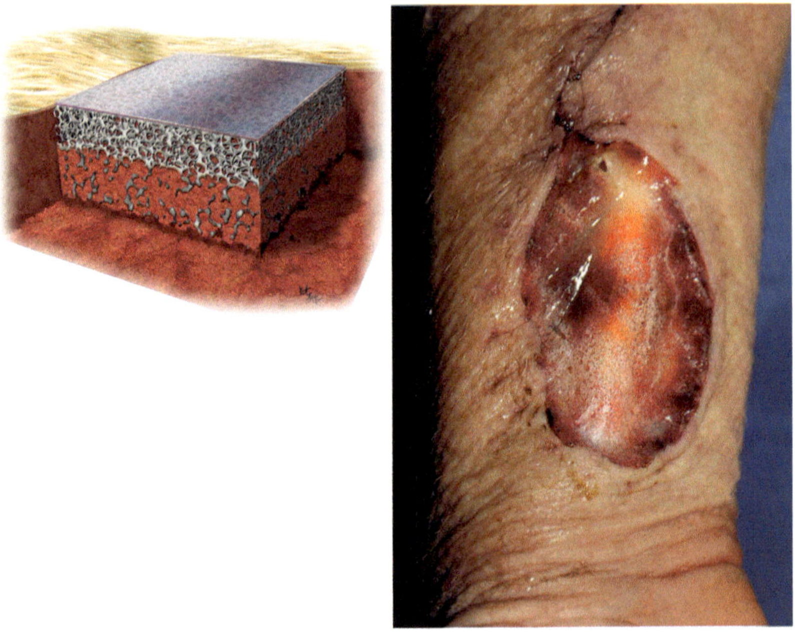

FIGURE 3 Day 15 post-BTM insertion: the foam structure is still visible beneath the seal but, deeply, tissue from the wound bed graft has begun to invade the foam matrix (area = 18.3 cm^2).

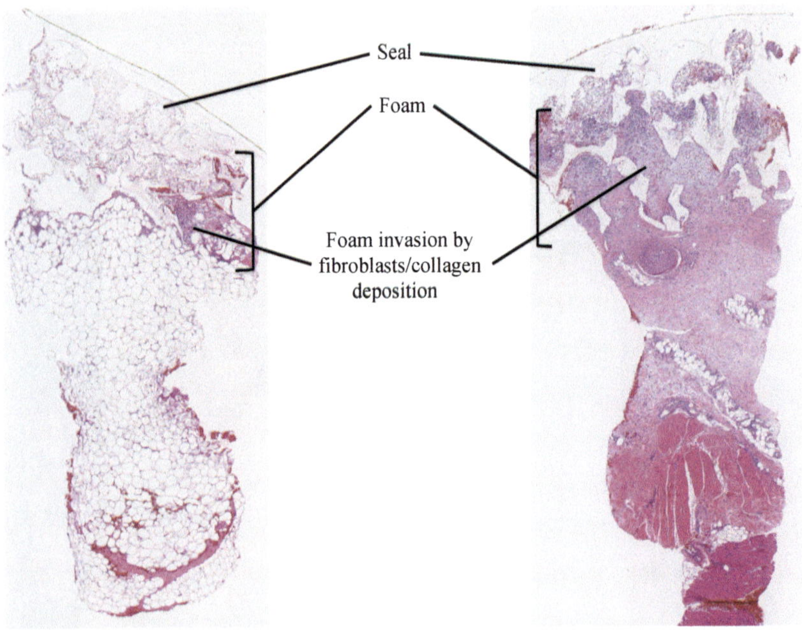

FIGURE 4 Integration progress: day 16 (left) demonstrates deep invasion of the polymer dermal matrix by fibroblasts laying down collagen. Day 23 (right) shows that the dermal invasion extends almost to the BTM seal.

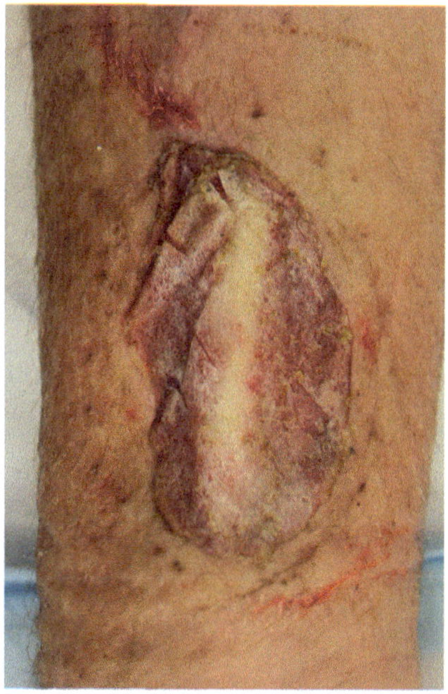

FIGURE 5 Day 36 post-BTM insertion (before seal removal): the seal is firmly adherent, despite some wrinkling secondary to edema from the original surgery settling while the BTM was integrating (area = 18.5 cm^2). The pallor over the tendon represents lesser thickness of foam integration.

biocompatible. The biocompatibility of a scaffold or matrix for a tissue-engineering product refers to the ability to perform as a substrate that will support the appropriate cellular activity, including the facilitation of molecular and mechanical signaling systems, in order to optimize tissue regeneration, without eliciting any undesirable effect in those cells, or inducing any undesirable local or systemic responses in the eventual host. The standard applicable to medical device biocompatibility is the ISO 10993, which entails multiple testing of the scaffold, including tests for genotoxicity, carcinogenicity, and reproductive toxicity; tests for interactions with blood; tests for in vitro cytotoxicity; tests for local effects after implantation, a framework for the identification and quantification of potential degradation products; tests for irritation and delayed-type hypersensitivity; tests for systemic toxicity, and so on. Obviously, the degradation products of a biodegradable matrix must also be assessed for biocompatibility. Biocompatibility thus covers the in vitro interaction with cultured cell lines, as well as any potential effect in vivo after implantation, including antigenicity.

Additionally, the matrix should be designed to resist infection, be able to withstand wound hypoxia, and not be reliant on rapid neovascularization (unless it is inherently capable of such). It should provide permanent wound coverage and offer long-term wound stability. It should be able to resist shear forces.

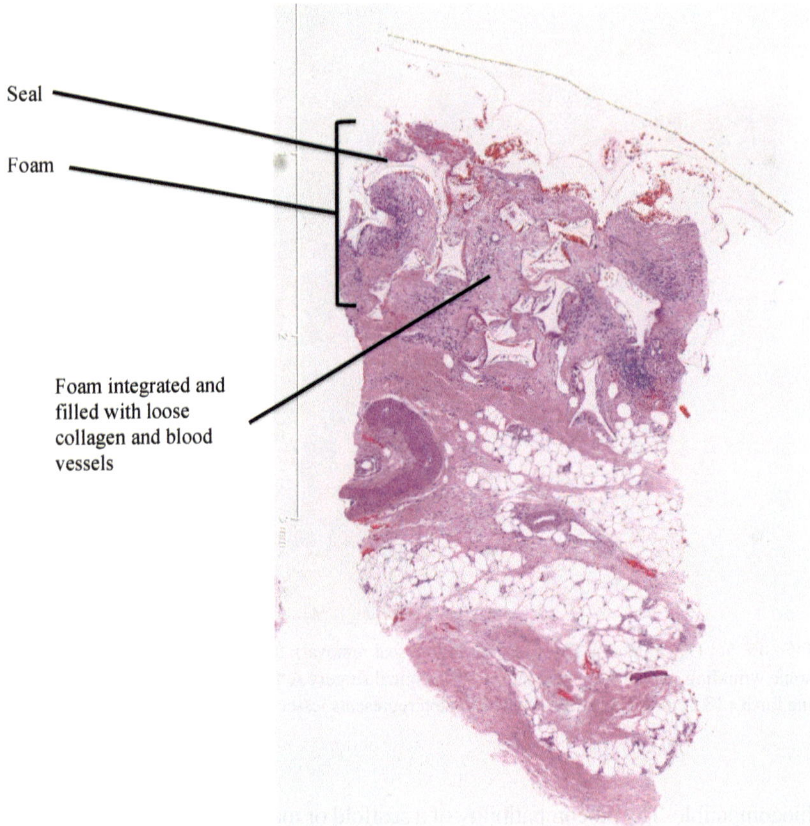

FIGURE 6 Completed integration: day 35—the BTM dermal foam is filled with fibroblasts and collagen. Blood vessels can also be seen.

Adherence to the above properties will ensure that the matrix will perform well in small and large animal models, and possibly have utility in restricted numbers of humans. There are many other characteristics that will determine whether a matrix will have the potential to be manufactured on a mass scale and be commercially marketable. These include the cost of raw materials, the cost of the processes to convert them into the finished product and the time taken for those processes (these will largely determine the cost of each piece of product to the end user), the ability to upscale production, the ease of sterilization, the ease of packaging and storage, the density (and thus the weight, for transportation), the ease of preparation before use, the ease of handling, flexibility, ease of cutting to size and fixation, and access to a storage and distribution network to ensure supply. Finally, the matrix must satisfy global regulatory authorities and, generally, the more components, the greater the complexity, the more difficult, lengthy, and costly will be the regulatory process.

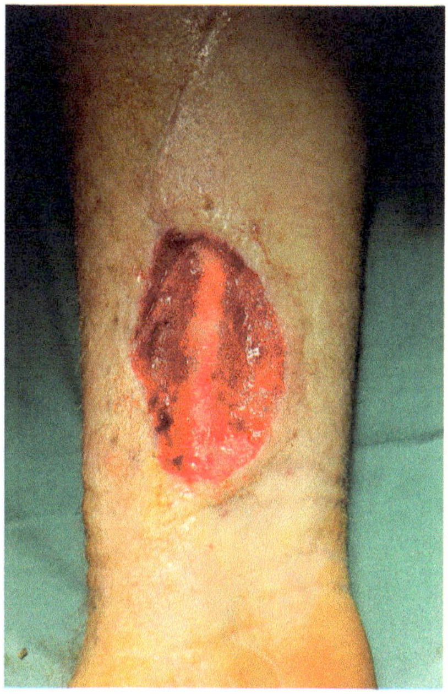

FIGURE 7 Day 36 post-BTM insertion (after seal removal): the seal of the BTM has been removed and the wound surface gently refreshed by dermabrasion. The tendon is patently covered by vascularized foam. The wound area is 18.5 cm^2, contraction of only 0.6 cm^2 in 5 weeks.

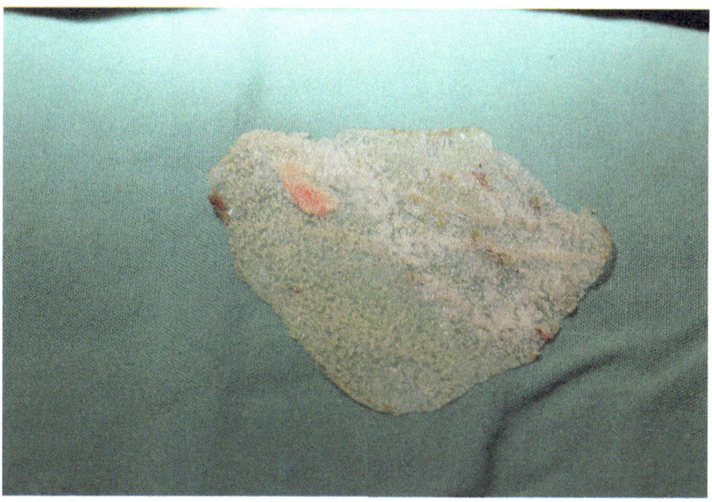

FIGURE 8 A typical seal having been removed. The under surface (deep side) is shown uppermost and the superficial parts of the foam (ripped off during delamination) can be seen bonded to the seal.

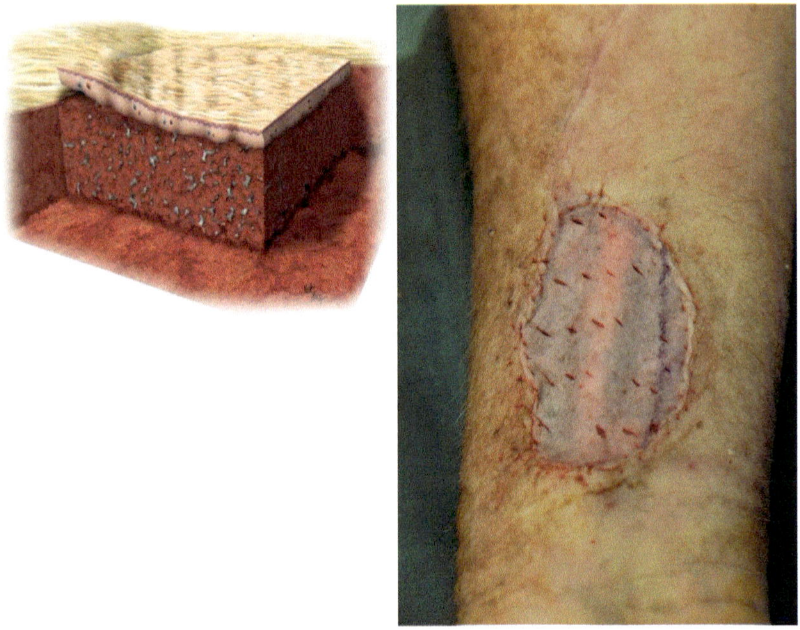

FIGURE 9 Day 36 post-BTM insertion (skin graft application): the graft has been applied and sutured with absorbable suture material (area = 18.5 cm^2).

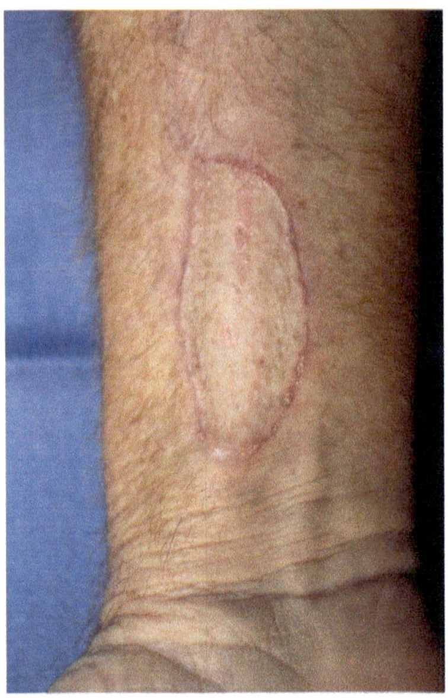

FIGURE 10 Day 75 post-BTM insertion (39 days post-skin graft application): the graft has contracted as thin grafts do (area = 10.6 cm^2). The visible scarring is around the periphery of the BTM/graft.

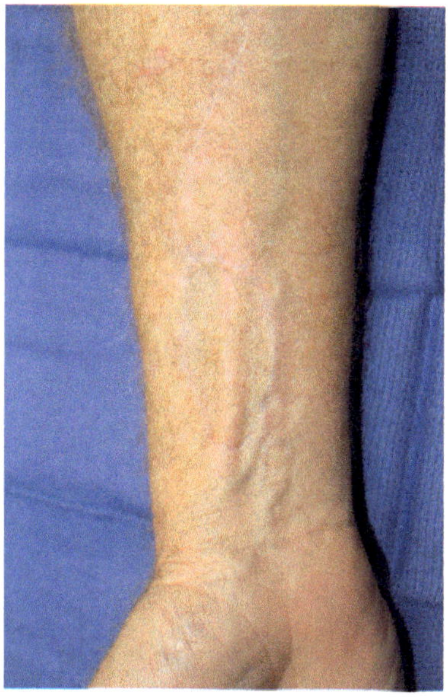

FIGURE 11 Day 271 (9 months) post-BTM insertion (day 235 post-skin graft application): the scar around the BTM/graft has softened and stretched (area = 13.5 cm^2) and is comparable to the scar further up the forearm. The proximal scar is where the vascular pedicle was dissected out, and this narrow wound was closed with sutures only. The BTM wound is flush with the surrounding skin and difficult to see if you are not looking specifically for it. The graft is mature, despite being only 9 months old.

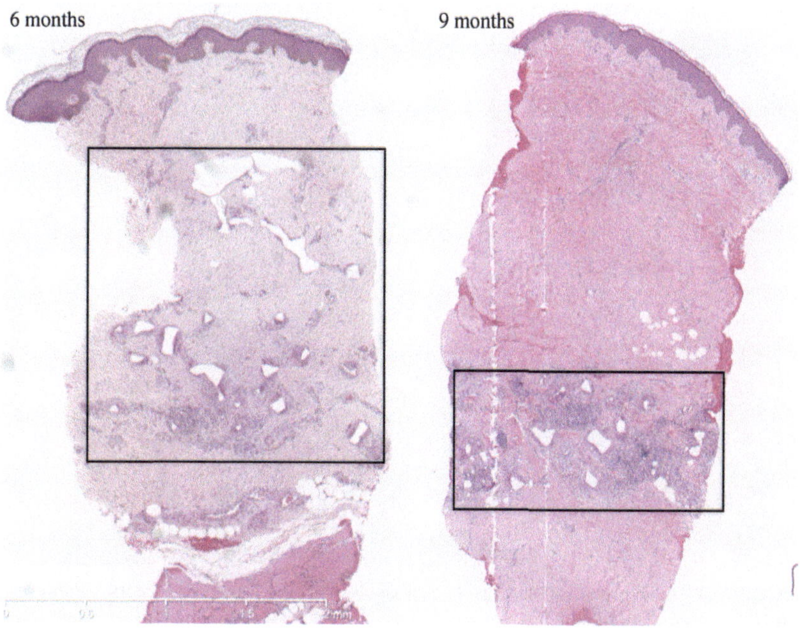

FIGURE 12 Degradation progress: (left) 6 months, polymer thickness undiminished but "sharp edges" (seen in earlier slides) rounded. (right) 9 months, polymer residue smaller, foam thickness visibly diminished.

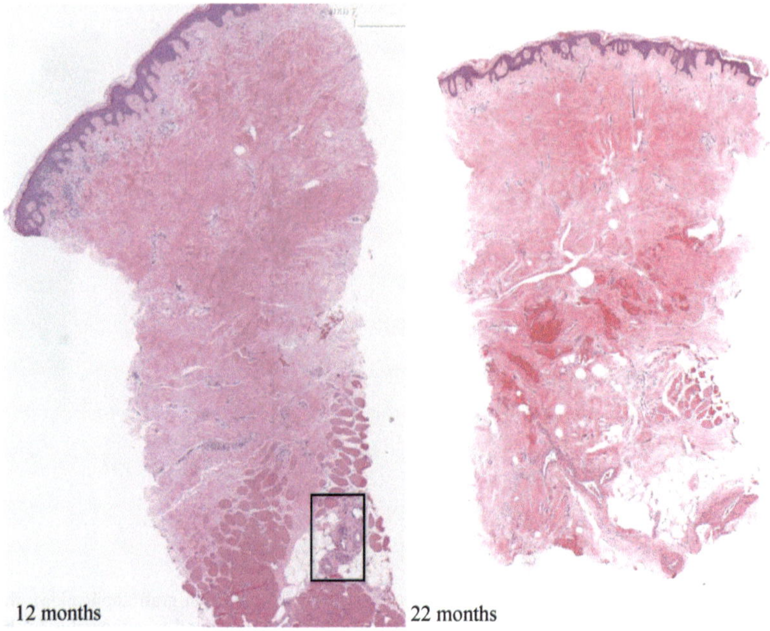

FIGURE 13 Degradation progress (continued): (left) 12 months, microscopic remnants of polymer being phagocytosed by giant cells (shown in box). (right) 22 months, no residual polymer.

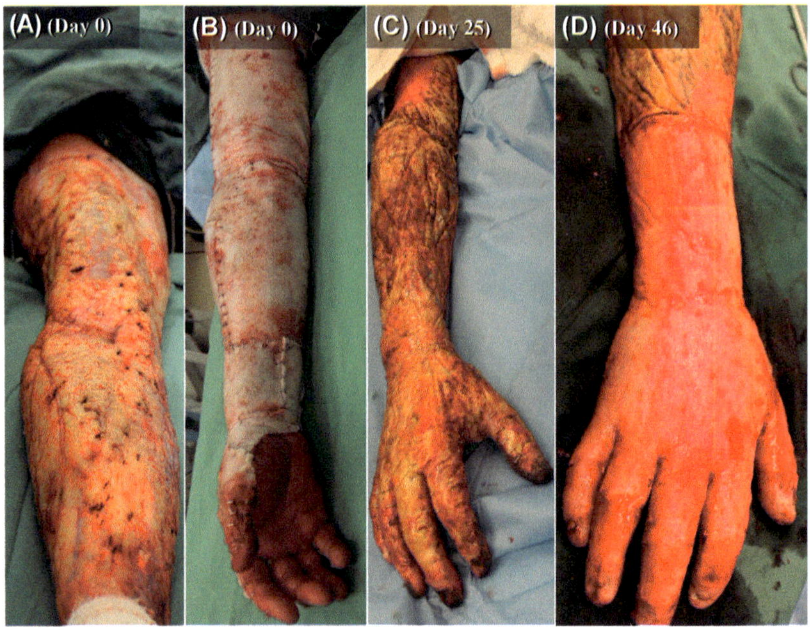

FIGURE 14 Example of BTM utility in first significant burn patient (73% TBSA burns, 48% TBSA full thickness). Right upper limb: (A) At day 0, the burn is excised to viable fat and meticulously hemostased. (B) The BTM is cut to shape and affixed into the excised wounds with staples. (C) At day 25, integration is well under way. (D) By day 46, split-thickness skin donor sites were available for re-harvest definitive cover with autograft, so the BTM was delaminated.

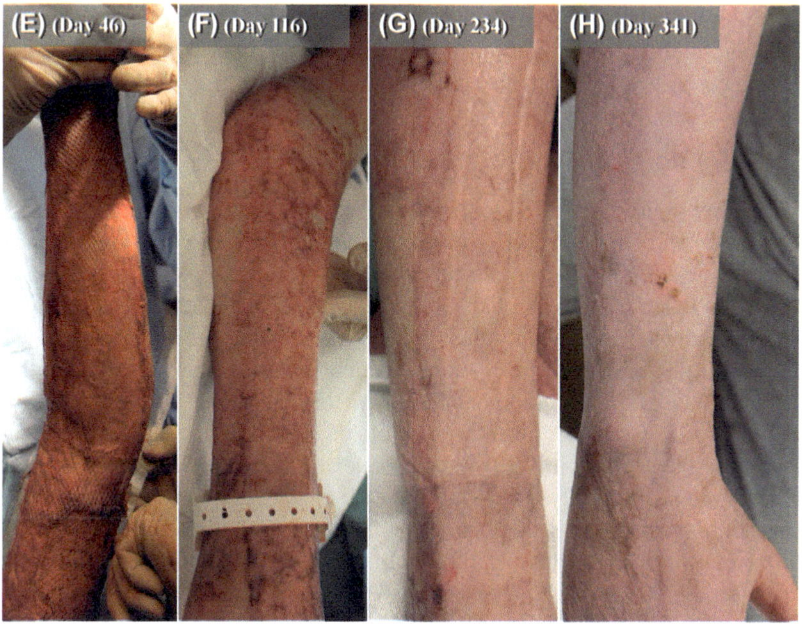

FIGURE 15 Example of BTM utility in first significant burn patient (73% TBSA burns, 48% TBSA full-thickness) (continued). (E) Meshed autograft applied. (F) Day 116 since the process began (day 50 post-autograft application). (G) Day 188 post-grafting. (H) 299 days since graft application and nearly 12 months post-injury.

While many of the materials described in the literature have potential utility in laboratory and small clinical situations, production on a global scale is likely to prove impossible. Electrospinning, for example, is costly and results in a dense matrix (using a large volume of material). The smaller the fibers, the more layers are necessary to achieve matrix thickness. This adds to the cost, since more raw materials are necessary, and also increases the volume of degradation products, with which the biological system into which they are implanted must cope. Density also reduces flexibility, especially at the likely matrix thicknesses required for composite skin production. Some authors have countered this with additional components. This adds to cost, complexity, and regulatory difficulty.

The most successful CCS to date is Permaderm™, developed by Steven Boyce at the University of Cincinnati, recently purchased by Regenicin, Inc., Little Falls, NJ. Each piece has been marketed previously at US \$28 per cm^2 (US \$14,000 for a piece 20 cm × 25 cm, equating to >US \$450,000 for a 100% TBSA full-thickness burn), which is prohibitively expensive outside (and, sometimes, even within) first world economies.

Again, many of these materials have formed the basis of other chapters within this book, so detailed description is not necessary here. However, to put the production of CCS based in a polyurethane foam into perspective, it is important to briefly revisit other materials, to highlight their downsides, and illustrate why a novel synthetic solution has been the subject of my studies.

Biologic/Biologic Hybrids

The ability to produce CCSs required two pivotal technological leaps. Firstly, Rheinwald and Green in 1975, developed a reliable technique to culture keratinocytes in vitro, enabling confluent sheets of epidermal cells to be grown on an irradiated mouse 3T3 fibroblast feeder layer. This heralded techniques to culture a wide variety of human cell lines, including other cells intimately involved in skin structure and function (fibroblasts, microvascular endothelial cells, melanocytes, and so on). Later, this direction would lead to the identification, isolation, and propagation of stem cell technologies, which would enable the in vitro growth of adenexal structures (sweat glands and hair follicles). Secondly, Jack Burke and Ionnas Yannas in 1980 revealed a three-dimensional "dermal regeneration template," composed of cross-linked collagen fibrils in a chondroitin-6-sulphate GAG ground substance. This structure, after wound excision and material application, would integrate by a mixture of neovascular and fibroblast invasion from the wound bed. The three-dimensional collagen/GAG would be found also to provide scaffolding support to fibroblasts and microvascular elements in vitro. These two innovations together gave the burn and wound scientific community the two building blocks required to create CCS.

There are several issues which mar the potential of biological materials for large-scale CCS production. The derivation of the biological material form, whatever starting point (bovine nuchal ligament, porcine atellocollagen, salmon, human cadaver tissue, and so on), is generally a time-consuming and expensive process. As such, in order to recoup cost of production, the cost of the end product is thought provoking. The speed of the process is invariable and the raw materials for production often sparse, meaning that upscaling to cope with surge demand during crises or after mass casualty incidents is effectively impossible. Most of these matrices require storage in a manner some potential users cannot employ, and the logistics of packaging and transport of "wet" biological material are difficult to comprehend, and thus additionally expensive. Concerns regarding immunogenicity and prion disease transmission of xenogenically derived collagen have been largely unfounded, although biological materials are digestible by bacteria, which puts a potentially very large CCS at risk of in vitro infection. Certain biologics, such as chitosan (a linear polysaccharide produced commercially by deacetylation of chitin, the structural element in the exoskeleton of crustaceans) and silk fibroin (an insoluble protein from the silk created by spiders, the larvae of *Bombyx mori*, and several moth genera), have been extensively investigated as less expensive substrates for culture. Cellulose (the most abundant organic polymer and the important structural component of the primary cell wall of green plants and many algae) has similarly been employed.

Biologic/Synthetic Hybrids

This group of hybrid biomaterials for cellular growth incorporates those where biological materials are augmented or modified by combination with synthetic components, or vice versa. Of the materials currently under investigation and in use, the biological component is almost invariably chitosan, collagen, or gelatin with the most popular synthetic components being polycaprolactone (PCL) or lactic acid/glycolic acid copolymers. The advent of the mixed biosynthetic matrix is relatively recent (2011), reflecting the relative youth of the synthetic technology.

The combination of chitosan with PCL was evaluated by three groups [32–34]. In each case, the chitosan–PCL combination was electrospun to give nanofiber mats. All three assessed cytotoxicity by successfully culturing fibroblasts of various origins into the mats. Gomes [35] compared chitosan, gelatin, and PCL scaffolds individually in their ability to permit co-culture of keratinocytes and fibroblasts in vitro, and subsequently heal wounds in vivo. Although the three materials appeared to proliferate in all three scaffolds, the gelatin scaffold did not allow deep penetration of the seeded cells into the matrix. The chitosan matrix was most effective in preventing wound contraction and enabling neodermis formation in vivo. The combination of poly(lactide-co-glycolide (PLGA)) copolymer with gelatin has allowed human fibroblast culture within the matrix [36].

Collagen/PGA matrices allowed cellular migration and facilitated vascular in-growth compared to collagen matrices alone [37]. Chitosan/PVA nanofiber mats demonstrated fibroblast adherence to the fibers [38], and chitosan–collagen matrices reinforced with poly((lactic acid)-co-((glycolic acid)-alt-(L-lysine))) (PLGL) mesh demonstrate good mechanical and handling properties compared to chitosan–collagen matrices alone, while resisting contraction, allowing cell infiltration and neovascularization in vivo [39,40].

Like the biologic/biologic hybrids, these materials are expensive to produce because of the biological component. Again, market forces in commercial production will mandate that development and production costs are passed on to the end user.

Synthetic/Synthetic Hybrids

With the ability to grow composite skins in biological scaffolds achieved, the disadvantages of biological matrices were revealed. The scaffolds themselves were laborious and costly to produce, and the defined structures of biopolymers were limited by their chemistry. Newer synthetic polymers offered cheap, mass-producible alternatives, with the potential to tailor physical properties and degradation rates, allow pharmaceutical, growth factor and antimicrobial elution in a time-release fashion as material degradation proceeded. However, the best part of a century of synthetic polymer chemistry had been focused on industry, civil engineering materials, furniture, electronic component production, and product packaging. Very few existing materials had any of the desirable properties to make them scaffold candidates for tissue engineering. The advent of bioabsorbable polymers, heralded by materials like absorbable sutures and the biodegradable plastics developed as a response to nonbiodegradable plastic pollution, changed this.

The task that has beset polymer engineers is the production of materials that are inert and biocompatible; biotolerated; have appropriate structural properties that are maintained until no longer required; degrade into neutral, nontoxic, and readily removed or metabolized degradation products; and which encourage, facilitate, or guide tissue in-growth from the wound bed (in the case of temporizing dermal matrices, like NovoSorb™ BTM) or allow cellular co-culture within them to produce composite skins in vitro. The task that besets clinicians is the serial and responsible assessment of these materials, and their degradation products, which do not enjoy the long clinical history of their biological counterparts.

Several polymers of monomeric components such as lactic and glycolic acids, vinyl alcohol, and caprolactone have been produced. These are simple and cheap to produce and, as such, have enjoyed extensive investigation by several groups. Hybrids have also been produced and used, such as PLGA. The acid release caused by the degradation of some of the polyacids (PLA and PGA) is problematic.

NovoSorb™ CCS Development

In developing the two-stage strategy to abolish the need for the split-thickness skin graft (described in the preamble to this chapter), it was necessary to begin by evaluating the novel biodegradable NovoSorb™ polyurethanes for their potential as scaffolds for CCS production.

Development began by culturing human cell lines (keratinocytes, fibroblasts, and human microvascular endothelial cells) both in the presence of biodegradable polyurethane fibers and on biodegradable polyurethane sheets. Cells in culture are fragile, vulnerable, and do not generally tolerate changes to the culture environment. In this case the cells of each line grew to confluence [24]. A three-dimensional mat was produced by "spinning" alternately perpendicular layers. This time serial co-culture demonstrated that fibroblasts moved within the scaffold by utilizing the polymer fibers and laid down sufficient collagen to fill the pore spaces between fibers. Keratinocytes added later formed a confluent layer over the fibroblast/collagen-containing matrix [24]. Endothelial cells were observed to surround the fibers. This raised the potential of creating a microvasculature based around a rapidly degrading polyurethane [24]. The simple, spun,

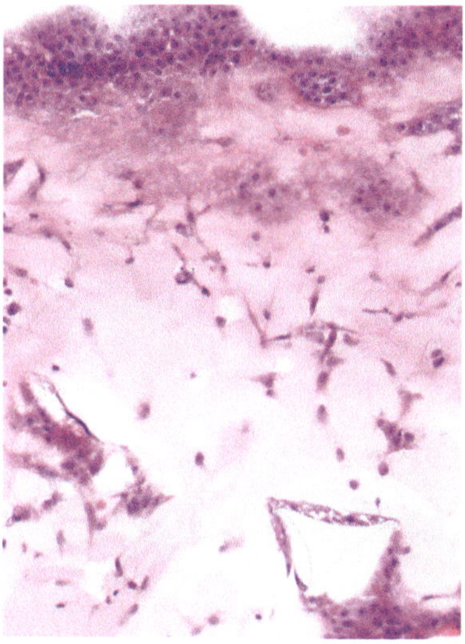

FIGURE 16 An example of CCS in vitro inside a polyurethane foam matrix. Foam struts are visible, surrounded by collagen and fibroblasts with a stratified epithelium of keratinocytes.

unsealed three-dimensional lattices were implanted into wound on the flanks of sheep and assessed in their ability to integrate while resisting contraction, favorably comparing against Integra®.

The issue with spinning fiber matrices is that a large volume of polymer is required. The "firing" of hot 50 μm fibers onto a roller leaving 200 μm between them is difficult. The first layer must then be cut, rotated 90°, and reapplied to the roller. A second layer is then laid down before cutting and rotating again. It becomes obvious that, having done this six times, the material is still only 300 μm thick! This process yields a dense and, thus, stiff material. The development of foam technology was mandated.

Foams generally contain "cells" of variable sizes and require surfactants and other chemicals in their manufacture, which may be deleterious to cell growth. The foam we developed utilized none of these chemicals and had pore sizes between 200 and 1000 μm (mean 500 μm). Studies were performed, which demonstrated that fibroblasts would invade and populate the foam, laying down collagen and that subsequently seeded keratinocytes formed a layer superficial to the fibroblasts within each pore, creating a bilayer skin in every foam cell (Figure 16) [28,29]. These composites were subsequently applied to porcine wounds, both fresh and those into which a NovoSorb™ BTM had already been integrated (Figures 17 and 18). Over integrated BTMs, the CCS created a stratified squamous cell epidermis with an intact basement membrane by 7 days after application (Figures 19(A–C)). The polymer scaffold in these cases was expelled from the surface leaving a healed wound behind [28].

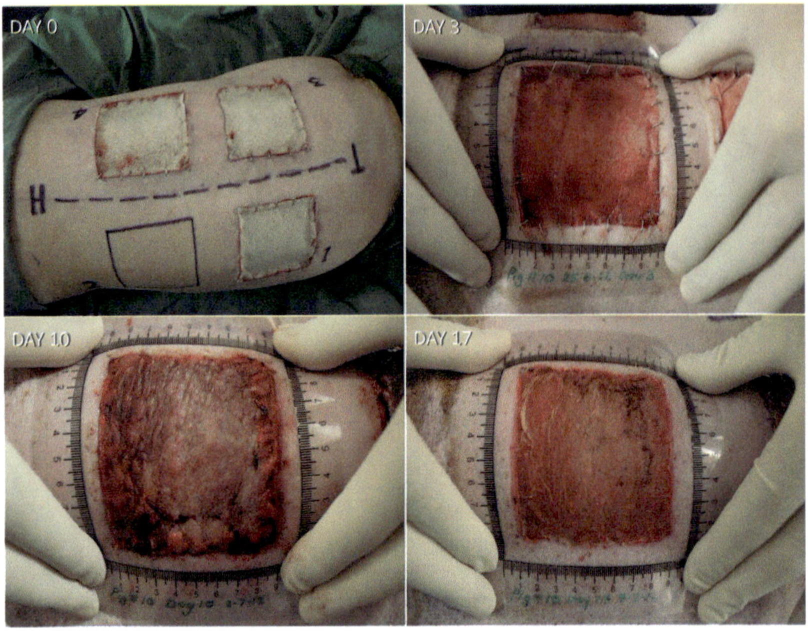

FIGURE 17 Integration of a BTM to day 17.

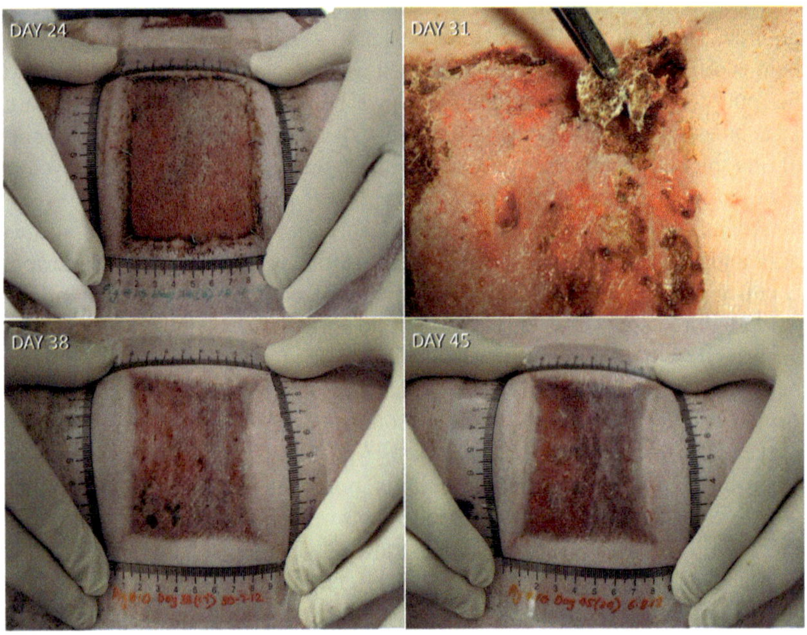

FIGURE 18 Once the CCS has been applied at day 21, this series shows reepithelialization to day 45. After 10 days (day 31), the polymer could be "picked off" revealing healing below which subsequently robusts.

(A)

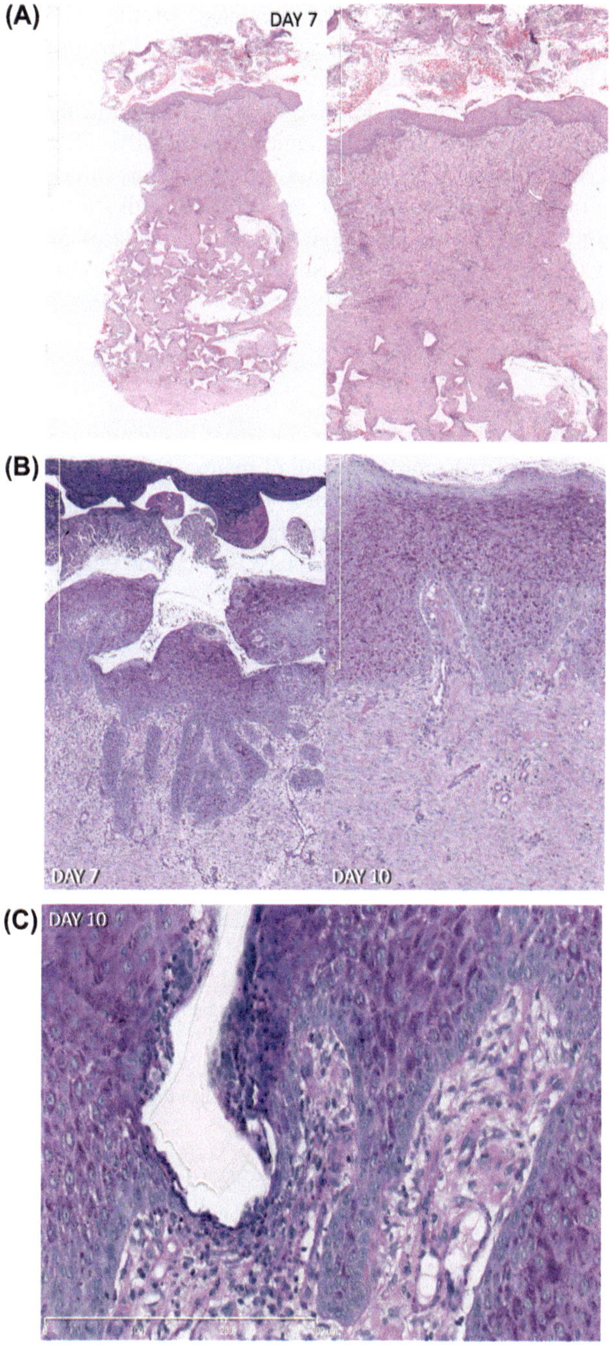

FIGURE 19 The CCS deposits its cellular contents, creating a stratified squamous epithelium and is subsequently "shed." (A) Left: at Day 7, CCS foam can be seen superficially (top of picture), with BTM foam deeply (bottom of picture); Right: higher magnification. (B) Left: at Day 7, as the CCS foam is extruded, a stratifies squamous epithelium is left behind; Right: 3 days later, the polymer scaffold is gone. (C) In a fresh wound, polymer remnants are seen within the epidermis at ten days postapplication.

In fresh wounds, the CCS scaffold integrated, forming a layer of vascular dermal tissue. By 10 days after application, islands of keratinocytes appeared, coalescing to generate a stratified squamous epidermis [28].

There is, however, a big difference in creating such healing in a wound 8 cm × 8 cm (64 cm^2), and doing the same in an 80–90% TBSA burn wound (~16,000 cm^2), and the tyranny of mass production has forced us and others to develop incubator-housed bioreactor facilities, with start/stop feed/drain or perfusion culture media exchange systems, capable of producing up to 2.5 m^2 of CCS in 28 days. These bespoke systems now exist and are under evaluation.

THE FUTURE

Newer technologies are already being employed in creating materials for skin engineering. Three-dimensional printing of matrices, the creation of microvascular structures within matrices, the use of stem cells, and the employment of gene modification make this an important and exciting field. However, the gulf between the laboratory and the marketplace is wide and deep and threatens to claim all those who cling to the "small picture."

REFERENCES

[1] Yannas IV, Burke JF. Design of an artificial skin. I Basic design principles. J Biomed Mater Res 1980;14:65–81.

[2] Yannas IV, Burke JF, Gordon PL, Huang C, Rubenstein RH. Design of an artificial skin. II Control of chemical composition. J Biomed Mater Res 1980;14:107–31.

[3] Dagalakis N, Flink J, Stasikelis P, Burke JF, Yannas IV. Design of an artificial skin. III Control of pore structure. J Biomed Mater Res 1980;14(4):511–28.

[4] Nguyen DQ, Potokar TS, Price P. An objective long-term evaluation of Integra (a dermal skin substitute) and split thickness skin grafts, in acute burns and reconstructive surgery. Burns 2010;36:23–8.

[5] Heimbach DM, Warden GD, Luterman A, Jordan MH, Ozobia N, Ryan CM, et al. Multicenter postapproval clinical trial of Integra dermal regeneration template for burn treatment. J Burn Care Rehabil 2003;24:42–8.

[6] Peck MD, Kessler M, Meyer AA, Bonham Morris PA. A trial of the effectiveness of artificial dermis in the treatment of patients with burns greater than 45% total body surface area. J Trauma 2002;52:971–8.

[7] Suzuki S, Matsuda K, Nishimura Y, Maraguchi Y, Maraguchi T, Ikada Y, et al. Review of acellular and cellular skins. Tissue Eng 1996;2(4):267–75.

[8] Suzuki S, Kawai K, Ashoori F, Morimoto N, Nishimura Y, Ikada Y. Long-term follow-up study of artificial dermis composed of outer silicone layer and inner collagen sponge. Br J Plast Surg 2000;53(8):659–66.

[9] Matsui R, Okura N, Osaki K-I, Konishi J, Ikegami K, Koide M. Histological evaluation of skin reconstruction using artificial dermis. Biomaterials 1996;17(10):995–1000.

[10] Choi MH, Yi SB, Hwang JW, Yang WS, Lee KK. Treatment of bone and tendon-exposed wounds using Terudermis. J Kor Soc Plast Reconstr Surg 1999;26:491–7.

[11] Lee JW, Jang YC, Oh ST. Use of the artificial dermis for free radial forearm flap donor site. Ann Plast Surg 2005;55(5):500–2.

[12] Yurugi S, Hatoku M, Kuwahara M, Tanaka A, Iioka H, Niitsuma K. Usefulness and limitations of artificial dermis implantation for posttraumatic deformity. Aesthetic Plast Surg 2002;26:360–4.

[13] Bessho K, Murakami K, Iizuka T. The use of a new bilayer artificial dermis for vestibular extension. Br J Oral Maxillofac Surg 1998;36:457–9.

[14] Soejima K, Nasaki M, Sasaki K, Takeudi M, Negishi N. Reconstruction of burn deformity using artificial dermis combined with thin split skin graft. Burns 1997;23(6):501–4.

[15] Ohno S, Hirano S, Tateya J, Kanemaru S, Umeda H, Suehiro A, et al. Atelocollagen sponge as a stem cell implanter scaffold for the treatment of scarred vocal folds. Ann Otol Rhinol Laryngol 2009;118(11):805.

[16] Price RD, Berry MG, Navsaria HA. Hyaluronic acid: the scientific and clinical evidence. J Plast Reconstr Aesthet Surg 2007;60:1110–9.

[17] Myers SR, Partha VN, Soranzo C, Price RD, Navasaria HA. Hyalomatrix: a temporary epidermal barrier, hyaluron delivery, and neodermis induction system for keratinocyte stem cell delivery. Tissue Eng 2007;13(11):2733–41.

[18] Caravaggi C, De Giglio R, Pritelli C, Sommaria M, Dalla Noce S, Faglia E, et al. HYAFF 11-based autologous dermal and epidermal grafts in the treatment of noninfected diabetic plantar and dorsal foot ulcers: a prospective, multicenter, controlled, randomized clinical trial. Diabetes Care 2003;26(10):2853–9.

[19] Giuggioli D, Sebastiani M, Cazzato M, Piaggesi A, Abatangelo G, Ferri C. Autologous skin grafting in the treatment of severe scleroderma cutaneous ulcers: a case report. Rheumatology (Oxford) 2003;42(5):694–6.

[20] Gravante G, Delogu D, Giordan N, Morano G, Montone A, Esposito G. The use of hyalomatrix PA in the treatment of deep partial-thickness burns. J Burn Care Res 2007;28:269–74.

[21] Braye FM, Stefani A, Venet E, Pieptu D, Tissot E, Damour O. Grafting of large pieces of human reconstructed skin in a porcine model. Br J Plast Surg 2001;54:532–8.

[22] Braye FM, Daniel H, Damour O, Michel M, Benjamin H, Hanane C. Development of a new bilayer dermal matrix, RENOSKIN: preclinical data. Burns 2007;33(1):S105.

[23] Sabatier F, Andrac-Meyer L, Philandrianos C, Mordon S, Feuerstein JM, Veran J. Comparison of five dermal substitutes in full thickness skin wound healing in a porcine model. Burns 2012;38:820–9.

[24] Li A, Dearman BL, Crompton KE, Moore TG, Greenwood JE. Evaluation of a novel biodegradable polymer for the generation of a dermal matrix. J Burn Care Res 2009;30(4):717–28.

[25] Greenwood JE, Dearman BL. Split-skin graft application over an integrating, biodegradable temporising polymer matrix: immediate and delayed. J Burn Care Res 2012;33(1):7–19.

[26] Greenwood JE, Dearman BL. Comparison of a sealed, polymer foam biodegradable temporising matrix against Integra™ dermal regeneration template in a porcine wound model. J Burn Care Res 2012;33(1):163–73.

[27] Wagstaff MJD, Driver S, Coghlan P, Greenwood JE. A randomised, controlled trial of negative pressure wound therapy via a biodegradable polyurethane foam in the management of decubitus ulceration. Wound Repair Regen 2014;22:205–11.

[28] Dearman BL, Li A, Greenwood JE. Optimisation of a polyurethane dermal matrix and experience with a polymer-based cultured composite skin. J Burn Care Res 2014;35(5):437–48.

[29] Dearman BL, Stefani K, Li A, Greenwood JE. Take of a polymer-based autologous cultured composite skin on an integrated temporising dermal matrix: proof of concept. J Burn Care Res 2013;34(1):151–60.

[30] Wagstaff MJD, Schmitt BJ, Coghlan P, Finkemeyer JP, Caplash Y, Greenwood JE. A biodegradable polyurethane dermal matrix in reconstruction of free flap donor sites: a pilot study. Eplasty 2015;15:102–18.

[31] Wagstaff MJD, Schmitt BJ, Caplash Y, Greenwood JE. Free flap donor site reconstruction: a prospective case series using an optimised polyurethane Biodegradable Temporising Matrix (BTM). ePlasty 2015;15.

[32] Chen H, Huang J, Yu J, Liu S, Gu P. Electrospun chitosan-graft-poly(epsilon-caprolactone)/poly(epsilon-caprolactone) cationic nanofibrous mats as potential scaffolds for skin tissue engineering. Int J Biol Macromol 2011;48(1):13–9.

[33] Shalumon KT, Anulekha KH, Chennazhi KP, Tamura H, Nair SV, Jayakumar R. Fabrication of chitosan/poly(caprolactone) nanofibrous scaffold for bone and skin tissue engineering. Int J Biol Macromol 2011;48(4):571–6.

[34] Lou T, Leung M, Wang X, Chang JY, Tsao CT, Sham JG, et al. Bilayer scaffold of chitosan/PCL-nanofibrous mat and PLLA-microporous disc for skin tissue engineering. J Biomed Nanotechnol 2014;10(6):1105–13.

[35] Gomes SR, Rodrigues G, Martins GG, Roberto MA, Mafra M, Henriques CM, et al. In vitro and in vivo evaluation of electrospun nanofibers of PCL, chitosan and gelatin: a comparative study. Mater Sci Eng C Mater Biol Appl January 2015;46:348–58. http://dx.doi.org/10.1016/j.msec.2014.10.051. [Epub 23.10.14].

[36] Norouzi M, Shabani I, Ahvaz HH, Soleimani M. PLGA/gelatin hybrid nanofibrous scaffolds encapsulating EGF for skin regeneration. J Biomed Mater Res A October 24, 2014;24. http://dx.doi.org/10.1002/jbm.a.35355. 35355.

[37] Sekiya N, Ichioka S, Terada D, Tsuchiya S, Kobayashi H. Efficacy of polyglycolic acid (PGA)/collagen composite nanofiber scaffold on cell migration and neovascularization in vivo skin defect model. J Plast Surg Hand Surg 2013;47(6):498–502.

[38] Sundaramurthi D, Vasanthan KS, Kuppan P, Krishnan UM, Sethuraman S. Electrospun nanostructured chitosan-poly(vinyl alcohol) scaffolds: a biomimetic extracellular matrix as dermal substitute. Biomed Mater August 2012;7(4):045005. http://dx.doi.org/10.1088/1748-6041/7/4/045005. [Epub 08.05.12].

[39] Wang X, Li Q, Hu X, Ma L, You C, Zheng Y, et al. Fabrication and characterization of poly(L-lactide-co-glycolide) knitted mesh-reinforced collagen-chitosan hybrid scaffolds for dermal tissue engineering. J Mech Behav Biomed Mater 2012;8:204–15.

[40] Wang X, You C, Hu X, Zheng Y, Li Q, Feng Z, et al. The roles of knitted mesh-reinforced collagen-chitosan hybrid scaffold in the one-step repair of full thickness defects in rats. Acta Biomater 2013;9(8):7822–32.

Chapter 11

Biologic Skin Substitutes

Steven Boyce[1], **Dorothy Supp**[1,2]

[1]*Department of Surgery, University of Cincinnati, Cincinnati, OH, USA;* [2]*Research Department, Shriners Hospitals for Children - Cincinnati, Cincinnati, OH, USA*

Chapter Outline

INTRODUCTION AND MEDICAL NEEDS

Wound coverage can be a limiting factor in several different medical conditions. In one extreme example, massive burn injuries can require replacement of skin covering nearly the entire body surface area (BSA). Burns remain an important medical problem in the United States and result in about 20,000 hospitalizations and 650 deaths annually [1]. Many advances in burn care have caused a sharp decline in burn mortality rates, including early excision, improved fluid resuscitation, infection control, nutritional support, and aggressive physical therapy [2]. Because most patients survive the initial resuscitation phase, even after very severe burns affecting a large percentage of total body surface area (TBSA), wound management is critical for recovery. Autografting with split-thickness skin, either meshed or unmeshed, has been considered the preferred treatment for coverage of excised burn wounds, but donor sites for autograft are limited in patients with very large burns. In these patients, wound coverage

Skin Tissue Engineering and Regenerative Medicine. http://dx.doi.org/10.1016/B978-0-12-801654-1.00011-5

requires repeated harvesting of available donor sites, which is associated with pain and scarring at the donor site and lengthy hospital stays.

Chronic wounds represent a different kind of challenge for wound healing. These wounds do not usually involve a large surface area, but they have a high incidence in the general population and thus have enormous medical and economic impacts. The most common chronic wounds include pressure ulcers and leg ulcers. In the United States alone, these wounds are estimated to affect more than two million people [3,4] with total treatment costs as high as $15 billion annually [5]. These figures may be expected to rise as the average age of the population increases. Historically, treatment of relatively small chronic wounds has included the use of topical agents and compression dressings, or grafting of split- or full-thickness skin. Skin grafts can provide timely wound coverage, but may lead to painful donor sites which are slow to heal and may be unsuccessful because of underlying deficiencies in wound healing.

Though they affect a relatively small fraction of the population, congenital skin conditions and diseases represent significant challenges for wound coverage. For example, giant congenital nevi can cover more than 50% TBSA and when untreated can significantly increase a patient's lifetime risk for development of melanoma [6,7]. Traditional treatment has involved serial excision and skin autografting, but can result in donor site morbidity, including hypertrophic scarring. Another example is epidermolysis bullosa (EB), an inherited mechanobullous disorder that is characterized by erosions and blistering of the epidermis [8]. The genetic causes are heterogeneous but primarily affect proteins of the basement membrane zone of the skin, leading to mechanical fragility. Wounds in patients with EB are difficult to avoid and tend to heal slowly or progress to chronic wounds. Because EB results from genetic mutations in proteins of the patient's own skin, autografting is not a viable treatment. Clinical management has focused on protection from damage and topical therapy, including sterile dressings, antibiotics, and analgesics [9]. Currently, treatment is supportive, and there is no known cure, but several approaches using gene therapy are underway, as described below.

There are clearly many medical needs for safe and effective therapeutic options for wound coverage in these diverse patient populations. Skin substitutes have been developed in response to these needs, providing new alternatives for temporary coverage and permanent wound closure with stable skin tissue.

BIOLOGICAL REQUIREMENTS AND CURRENT ALTERNATIVES

Skin provides a wide range of protective, perceptive, and regulatory functions for the human body. Protections include the epidermal barrier to fluid loss and infection; melanin to protect from mutagenesis caused by ultraviolet radiation; and immune surveillance and responses to foreign bodies, noxious chemicals, and agents; and invasion of virulent microorganisms. Neural networks extending from the central nervous system to the epidermis provide continuous perception of the environment and actuate involuntary and voluntary responses to temperature and pain at the local and systemic levels. Among the systemic regulatory functions that skin provides are thermoregulation (sweating, vasodilation, vasoconstriction) and moisture retention (stratum corneum, sebaceous secretion) to help maintain local and systemic homeostasis.

Full-thickness skin loss conditions, which may extend to the muscle fascia, can cause loss of some or all of the cutaneous anatomy and physiology. Therefore, a full-thickness, autologous skin flap is the closest substitute for uninjured skin [10,11]. For permanent closure of full-thickness wounds affecting deep dermis or subcutaneous fat, unmeshed, split-thickness skin autografts remain the prevailing standard of care, subject to availability [12,13]. For temporary wound coverage until autologous grafts are available, cadaveric split-thickness skin allograft and porcine xenograft remain effective and economical alternatives [14,15]. Allograft or xenograft [16] promotes development of granulation tissue that can facilitate subsequent engraftment of skin autograft; however, this can lead to scar. In addition, allograft or xenograft may carry transmissible pathogens and both are ultimately rejected.

Alternatives for tissue repair that reduce requirements for skin autograft include acellular dermal substitutes, temporary skin substitutes, and permanent skin substitutes (Table 1). Acellular dermal substitutes may be derived from decellularized human skin or purified polymers (i.e., collagen, glycosaminoglycans, derivatized hyaluronic acid, elastin), which can vascularize readily upon application to excised wounds [10,17,18]. Temporary skin substitutes may also include allogeneic cells, typically fibroblasts and/or keratinocytes, which are generally not detected within a few months after application [19,20]. Permanent skin substitutes often use similar compositions as temporary skin substitutes, consisting of autologous cell transplants as keratinocyte sheets, sprays of keratinocyte suspensions, or in combination with biopolymers and/or fibroblasts. The categorical distinction of permanent skin substitutes derives from the inclusion of autologous cells as an exclusive component or a chimeric fraction of the cellular composition. Table 1 contains only a few of the numerous examples of materials that follow these general designs and compositions from academic and industrial laboratories worldwide [21,22]. Figure 1 shows an example of a subject treated for 95% TBSA burns with an autologous skin substitute containing a biopolymer scaffold, dermal fibroblasts, and epidermal keratinocytes in comparison to meshed, split-thickness skin grafts. Autologous engineered skin substitutes (ESSs) have contributed to reduced morbidity from burns and chronic wounds [23,24], and are associated with reduced pain, and fewer needs for skilled nursing and social services [20,25]. These benefits represent not only prospective reductions in costs for inpatient care, but also significant improvements in recovery of activities of daily living, and long-term quality of life [26,27].

TABLE 1 Biologic Skin Substitutes; Commercially Available or in Clinical Trial

Acellular Skin Substitutes		
Model (References)	**Description**	**Indications for Use**
Integra dermal regeneration template [17,133]	Bovine collagen and chondroitin-sulfate coated with silicone	Burns, reconstructive surgery
AlloDerm [134,135]	Decellularized human dermis	Repair or replacement of damaged or inadequate integumental tissue
MatriDerm [136,137]	Bovine collagen and elastin	Burns, reconstructive surgery
Hyalomatrix [138,139]	Derivatized hyaluronic acid	Partial- and full-thickness wounds
Temporary Skin Substitutes (Dressings)		
Model (References)	**Description**	**Indications for Use**
Cadaver allograft [15]	Split-thickness skin from human donors; unfrozen or cryopreserved	Burns, reconstructive surgery
Porcine xenograft [16]	Split thickness porcine skin; cryopreserved or lyophilized	Burns, reconstructive surgery
Apligraf® [140,141]	Allo hF in collagen gel plus stratified allo hK	Diabetic foot ulcers
StrataGraft® [142,143]	Allo hF in collagen gel plus stratified all hK	Partial-thickness burns
DermaGraft® [144,145]	Allo hF on poly-galactin mesh	Diabetic foot ulcers
Permanent Skin Substitutes (Grafts)		
Model (References)	**Description**	**Indications for Use**
EpiCel® [32,146]	Cultured auto hK multilayer sheet	Full-thickness burns
ReCell® [147,148]	Uncultured suspension of auto hK, delivered as a spray	Partial-thickness burns
Reconstructed skin [24,78]	Auto hF on acellular scaffold of dermal ECM, plus stratified auto hK	Full-thickness burns; venous and mixed ulcers
Autologous engineered skin substitute [23,149]	Auto hF on a collagen-GAG scaffold, plus stratified auto hK	Full-thickness burns

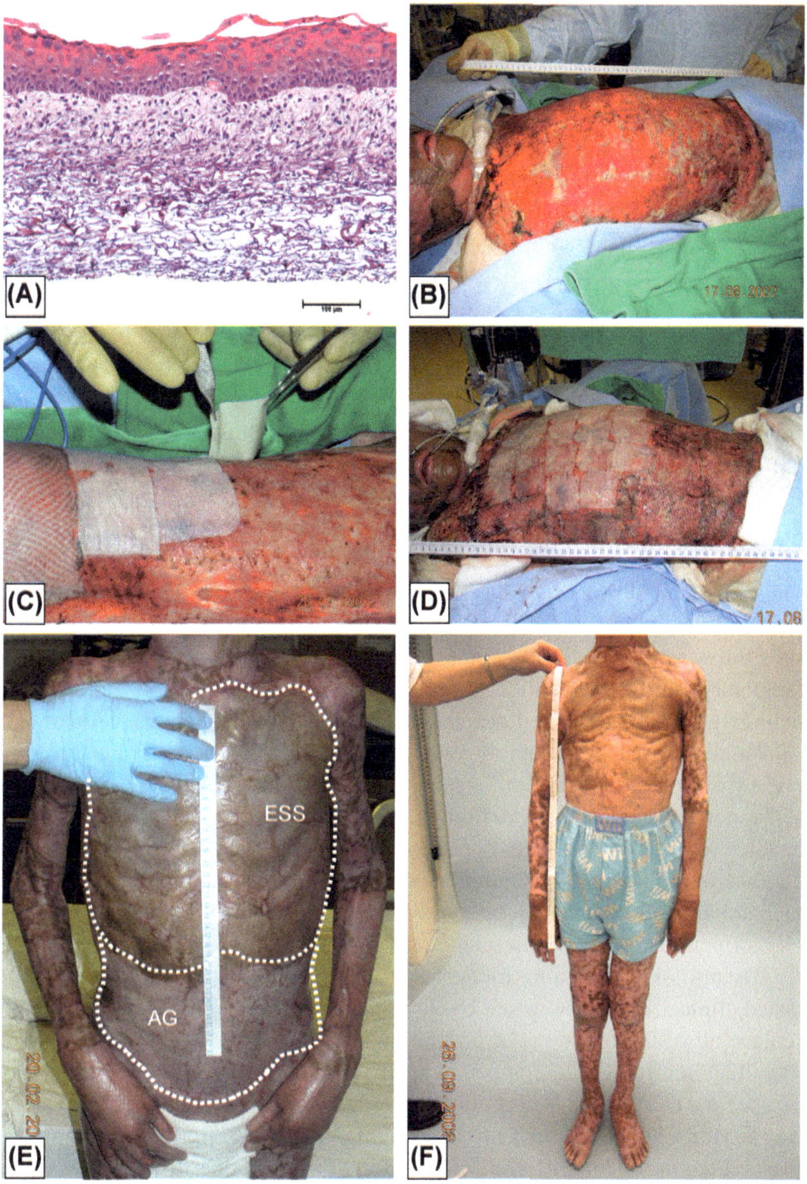

FIGURE 1 Treatment and outcome of 94% TBSA, full-thickness burn. (A) Autologous ESS consisting of a degradable biopolymer scaffold, dermal fibroblasts, and epidermal keratinocytes. Scale = 0.1 mm. (B) Subject with full-thickness wounds prepared for grafting. (C) Application of autologous ESS. (D) Engineered skin applied. (E) Comparative sites treated with autologous ESS and split-thickness skin autograft at discharge from hospital at post-burn day 227. (F) Clinical appearance at one year after first application of ESS. Scales in cm.

Despite reductions in morbidity, the current alternatives remain deficient in quality of outcome compared to sheet grafts of autologous skin. Among the anatomic deficiencies are hypopigmentation, absence of vasculature, hair, sebaceous glands and sweat glands, incomplete restoration of innervation, incomplete basement membrane, mechanical fragility, increased rates of infection, and repeated surgeries for graft failure and site reconstruction [21,28,29]. These constitute the challenges for future models of biologic skin substitutes.

DEFICIENCIES OF BIOLOGIC SKIN SUBSTITUTES

The most advanced biologic skin substitutes available currently exhibit physiologic deficiencies compared with native skin and healed split-thickness autograft. In general, deficiencies are attributable to the absence of biological complexity normally found in the native tissue, which is a common limitation of most engineered tissues.

Single or Multiple Cell Types

The most critical property of a skin substitute is permanent restoration of barrier function, which can only be achieved if engraftment is successful. Dermal substitutes can be used to facilitate healing in complex wounds, or to provide temporary barrier restoration, but an epidermal component is required for definitive and stable wound closure. Further, as described above, cellular skin substitutes must contain autologous cells for permanent wound closure. The presence of autologous keratinocytes competent to differentiate into a stratified, self-renewing epidermis, a definitive property of permanent skin substitutes, enables restoration of barrier function. However, certain deficiencies of some engineered skin models, including mechanical fragility, susceptibility to microbial contamination, and lack of a vascular plexus, can result in reduced rates of engraftment that compromise outcomes.

Because skin substitute models that are currently available, or have been tested clinically, are comprised of only one or two cell types, they cannot fully replace all of the functions of normal human skin. For example, cultured epithelial autografts (CEA) comprised solely of keratinocytes are limited by lack of a dermal component. This deficiency reduces engraftment due to mechanical fragility and increased susceptibility to infection [30,31]. In addition, blistering is a common complication of CEA grafting [32], due to the absence of a functional basement membrane [33]. The basement membrane at the dermal–epidermal junction of skin is required for anchoring of the epidermis to underlying connective tissue. Indeed, several human skin-blistering disorders, including EB, have been attributed to mutations in genes encoding vital basement membrane components [34]. Interactions with fibroblasts are central to deposition of basement membrane components by keratinocytes [35]. Thus, composite skin substitutes comprised of keratinocytes in direct contact with fibroblasts promote deposition of basement membrane components in vitro, which greatly reduces blistering after grafting [36,37]. Establishment of basement membrane in bilayered skin substitutes was found to stimulate stratification of the epidermal component, contributing to enhanced barrier function [36,38]. Composite skin substitutes are also thicker and exhibit greater mechanical strength than CEA [36,38], which facilitates handling and transplantation. Further, paracrine signaling between fibroblasts and keratinocytes in composite skin substitutes has a profound impact on gene expression, resulting in up-regulation of genes involved in wound healing [38,39]. Growth factors uniquely expressed by bilayer skin substitutes, compared with constructs containing only fibroblasts or keratinocytes, may serve to enhance recruitment of circulating cells and stimulate angiogenesis after transplantation, further enhancing wound healing beyond mere restoration of the epidermal barrier [38].

Anatomic Deficiencies

Even in composite skin substitutes that contain a fibroblast-populated dermal matrix in addition to epidermal keratinocytes, anatomic deficiencies can compromise engraftment or limit functional outcome after healing. A comparative analysis of the gene expression profiles of engineered skin and native skin revealed numerous differences, many of which could be attributed to the absence of specific cell types and structures in the engineered tissue [40]. Next generation skin substitutes currently in development include additional cell types and adnexal structures for increased homology with native human skin. To incorporate adnexal structures, a process of deconstruction and reconstruction is often required, involving isolation and expansion of relevant cells types, organization into complex three-dimensional structures that retain appropriate function, and incorporation into ESSs.

Among the most important deficiencies that compromise engraftment of engineered skin is the absence of a vascular plexus in vitro, which delays vascularization compared with split-thickness autograft, prolonging ischemia, nutrient deprivation, and microbial degradation, thereby reducing engraftment [41]. Preclinical studies investigated the use of genetically modified cells in skin substitutes to stimulate angiogenesis. For example, keratinocytes in composite skin substitutes were genetically modified by retroviral transduction to overexpress vascular endothelial growth factor (VEGF), which accelerated vascularization by host blood vessels and increased engraftment in athymic mice [42]. However, transplantation of genetically modified cells represents a potential safety risk that significantly increases the regulatory requirements for clinical translation [43]. A more direct approach involves incorporation of a vascular network into the engineered dermis in vitro, enabling vascularization to occur by inosculation in vivo, as with split-thickness autograft [44]. ESSs containing endothelial cells, in addition to fibroblasts and keratinocytes, have been described, in which endothelial cells self-assembled into organized networks and contributed to more rapid vascularization in mice [45,46]. Because all three cell types can be isolated from a single skin biopsy, clinical translation of this model is feasible using autologous cells.

Epidermal adnexal structures absent from current models of skin substitutes include hair follicles, sebaceous glands, and sweat glands. These structures are usually absent from split-thickness autograft as well, but this is dependent on the thickness of the harvested graft and the location of the donor site. Among these skin appendages, deficiency of eccrine sweat glands may have the greatest impact on survivors of large burns, as the ability to perspire is central to control of body temperature. Sweat glands are relatively simple, coiled tubular structures composed of a ductal portion, which opens at the epidermal surface, and a coiled secretory portion, deep within the dermis [47]. The secretory portion is composed of inner luminal cells, which are classified as clear (secretory) and dark (mucoid) cells, surrounded by outer myoepithelial cells [48]. Luminal cells transfer water and metabolites from the blood into sweat, which is discharged due to contractions of the myoepithelial cells. Adult sweat gland stem cells have been recently identified in mice, with three distinct populations described that each give rise to separate lineages. These include ductal progenitors, which could contribute to epidermal wound repair, and myoepithelial and luminal progenitors, which can give rise to myoepithelial cells and luminal cells, respectively, after injury [49]. Transplantation studies demonstrated that isolated populations of ductal and myoepithelial progenitors could give rise de novo to sweat glands, although their morphogenesis was incomplete and highly dependent on the microenvironment [49,50]. Human sweat gland cells, isolated from explants of glands microdissected from skin, showed a typical epithelial morphology when grown in two dimensions on plastic, but when three-dimensional cultures were grown in a mixture of extracellular matrix (ECM) components, tubular structures with coil-like shapes were formed [51,52]. In other studies, mesenchymal stem cells (MSCs) were induced to become sweat gland-like cells after co-culture with heat-inactivated sweat gland cells. After subcutaneous injection to burn wounds in rats, sweat gland-like structures were observed [53]. MSCs induced by co-culture with sweat gland cells were tested in a small clinical trial using autologous cells. The procedure for MSC grafting involved co-transplantation of perforated decellularized human allograft, which was covered with autograft and then human allograft on top as a protective layer. Sweat gland-like structures that perspired were observed in wounds treated with MSCs preexposed to sweat gland cells [53]. These studies suggest the feasibility of sweat gland reconstruction in engineered skin from autologous progenitor cells resident in the sweat gland and/or MSCs isolated from patient bone marrow.

De novo formation of pilo-sebaceous units, consisting of hair follicles, sebaceous glands, and erector pili muscles, has not yet been achieved in skin substitutes consisting of all adult, human cells. Mature follicles are relatively complex structures in which multiple types of cells interact to regulate postnatal cyclical hair growth. For hair follicle induction, both during embryogenesis and in adult skin during hair cycling, interactions are required between pluripotent epithelial cells and a specialized population of mesenchymal cells called dermal papilla cells. The dermal papilla of the postnatal hair follicle is found at the base of the follicle bulb. Epithelial matrix cells, in close association with dermal papilla cells, differentiate to form the cortex of the hair and the inner root sheath. These structures are surrounded by the outer root sheath, which is continuous with the interfollicular epidermis [54]. Also located at the base of the hair bulb are melanocytes, required for hair pigmentation. Near the upper portion of the follicle shaft is a small outcropping of cells called the "hair follicle bulge," which is a niche for epidermal stem cells. Adjacent to the bulge and associated with the hair follicle is the sebaceous gland, which helps to lubricate the outgrowth of the hair shaft and moisturize the interfollicular epidermis.

Much of what is known about hair follicle development and cycling has come from studies in mice, but hair morphogenesis assays have shown that human and mouse cells have different capacities for trichogenesis. Combinations of dermal and epidermal cells have been tested for hair follicle formation when grafted back to the skin in athymic mice. Cultured epidermal cells were able to contribute to follicle formation if combined with freshly isolated, but not cultured, dermal cells [55]. Dermal papillae isolated from rat or mouse vibrissae have been shown to induce follicle formation when combined with epidermal cells, suggesting the feasibility of maintaining inductive ability of isolated dermal papilla cells in culture. However, this has not yet been demonstrated with cells isolated and selectively cultured from adult human dermal papillae, indicating differences in signaling potential between rodent and human dermal papilla cells [56]. For example, in ESSs containing cultured fibroblasts, keratinocytes, and either mouse or human dermal papilla cells, neofollicles were observed in the skin substitutes containing murine, but not human, dermal papilla cells [56]. Chimeric hair follicles that formed from murine dermal papilla cells and human keratinocytes in skin substitutes did not contain sebaceous glands, which may have been required in that model for normal eruption of hair shafts [57]. Recently, reconstructed skin containing hair follicles of fully human origin have been demonstrated, but limitations have been encountered related to the age of the cell donor [58]. Epidermal and dermal cells were isolated from human fetal or adult scalp skin, and neonatal epidermal cells were isolated from foreskin. Cells were dissociated and expanded in culture, although individual cell types were not selectively isolated, and dermal and epidermal cells were combined and transplanted with a synthetic scaffold to wounds in immunodeficient mice. Hair follicle formation was optimal in this system when both dermal and epidermal cells were of fetal origin. Adult dermal cells could form some follicular structures when combined with fetal or newborn epidermal cells, but if both dermal and epidermal cells were of adult origin, no follicles were formed [58]. Identification of the basis for these age-related differences in inductive potency of dermal papilla cells is a major focus of current research that will be required for clinical translation of this technology.

Cellular Deficiencies

Although keratinocytes are responsible for the most important role of the epidermis, barrier function, other epidermal cells are critical for other important functions, including sensation, immune surveillance, and protection from solar radiation. The cells responsible for these functions, Merkel cells, Langerhans cells, and melanocytes, respectively, are currently absent from commercially available skin substitutes. Unlike the dermis, in which specialized cells are localized to specific adnexal structures, these different types of epidermal cells are distributed among keratinocytes in the basal layer (melanocytes, Merkel cells) or throughout the epidermis (Langerhans cells). Although skin biopsies used to establish primary cultures of keratinocytes contain these additional cell types, they are generally lost during selective culture in vitro under conditions that promote keratinocyte proliferation. This is particularly true for Langerhans cells and Merkel cells [50,59]. Langerhans cells, like other dendritic cells, arise from bone marrow-derived progenitors have been shown to repopulate skin substitutes after grafting [60]. Merkel cells are relatively rare touch-sensitive epidermal cells localized to the dermal–epidermal junction, in proximity to sensory neurons, which are integral to mechanosensory signal transduction [61]. A limited number of studies have investigated innervation of skin substitutes. In one clinical study of burn patients treated with split thickness autograft and CEA, with or without Integra® dermal scaffold, no differences in innervation were observed, but sensory function was significantly worse in the group treated with Integra® [62]. This suggests that innervation alone may be insufficient for complete restoration of sensation [62]. This emphasizes the need for further studies to determine if cellular deficiencies inherent to skin substitutes may be responsible for sensory defects after healing.

Currently available skin substitutes lack normal pigmentation, leaving the healed grafts vulnerable to ultraviolet light-induced damage and impaired cosmetic appearance [63]. Melanocytes usually persist at low levels during propagation of keratinocytes in culture. These cells have been termed "passenger melanocytes," and when present in sufficient numbers, they can result in foci of pigmentation after grafting [37]. This observation suggested that population of engineered skin grafts with greater numbers of melanocytes could enable uniform pigmentation. This was demonstrated in preclinical studies in mice, in which the area of pigmentation was proportional to the number of melanocytes added during skin substitute preparation [63]. Recent improvements in the techniques used for the culture of melanocytes enable relatively rapid expansion in vitro without tumorigenic transformation, which should enable population of engineered skin grafts with autologous melanocytes for timely treatment of acute burns [64].

Anatomic Site-Specific Phenotypes

In addition to lack of complexity due to the absence of multiple types of cells, skin substitutes are limited by the use of nonspecific populations of keratinocytes and fibroblasts. For treatment of severely burned patients, cell isolation would utilize whatever uninjured skin is available. However, keratinocytes and fibroblasts can exhibit significant heterogeneity, depending on their anatomical location. As well as site-specific differences, fibroblasts display substantial differences between different dermal regions at the same anatomical location. Fibroblasts from the papillary (superficial) and reticular (deep) layers of the dermis have very different gene expression profiles. Differences in expression of growth factors and production of ECM are retained, even when cells are propagated in culture [65]. Reticular fibroblasts exhibit relatively increased levels of genes involved in contraction and motility, whereas papillary fibroblasts display relatively higher proliferation rates and higher decorin levels [65]. Further, fibroblasts cultured from deep dermal layers were shown to produce more of the profibrotic cytokine TGF-β1, more alpha smooth muscle actin, and more collagen compared with fibroblasts isolated from superficial dermis [66]. Given the different profiles of ECM production, in particular, with a more fibrotic matrix produced by reticular fibroblasts, it is perhaps not surprising that deep dermal injuries that affect reticular fibroblasts are often accompanied by hypertrophic scar formation [67]. These differences in fibroblast gene expression can significantly affect the functioning of engineered skin prepared using fibroblasts from different dermal regions. When cultured in a collagen-based matrix, deep dermal fibroblasts caused greater matrix contraction than superficial fibroblasts [68]. Skin substitutes prepared with keratinocytes and superficial dermal fibroblasts were found to be superior to those made using deep dermal fibroblasts, with improved basement membrane formation and epidermal barrier function [69]. This may suggest an advantage for skin substitutes prepared using cells isolated from split-thickness skin biopsies, which would contain predominantly papillary fibroblasts.

In addition to distinct papillary or reticular phenotypes, dermal fibroblasts from different regions of the body exhibit distinct gene expression profiles, a phenomenon that has been termed *topographic differentiation* [70]. Relatively few genes are uniquely expressed in fibroblasts from any particular anatomical site. However, fibroblasts from different sites are defined by different combinatorial patterns of gene expression that are maintained when fibroblasts are cultured in vitro [70,71]. This is exemplified elegantly by the expression patterns of HOX family genes in fibroblasts. HOX genes are transcription factors that regulate pattern formation and positional identity along the anterior–posterior axis of the developing embryo. Vertebrate HOX genes are localized to four chromosomal clusters. An intriguing feature of HOX genes is that their order of expression along the anterior–posterior body axis is the same as their chromosomal order within each cluster. It has been suggested that HOX genes control the positional identity of skin in different anatomical regions [72]. It is well known that fibroblasts can affect gene expression in keratinocytes, and fibroblasts from different anatomical sites can regulate keratinocyte gene expression in a position-dependent manner. This is illustrated by co-culture studies involving palmoplantar keratinocytes, which have a specific profile of expressed keratin genes, including keratin 9, which is not normally expressed in nonpalmoplantar keratinocytes. Keratin 9 expression in nonpalmoplantar keratinocytes can be induced by co-culture with palmoplantar fibroblasts, and this induction is more efficient when the fibroblasts are from the papillary versus reticular palmoplantar dermis [73]. Recent studies have shown that HOXA13 in palmoplantar fibroblasts regulates WNT5A expression, which in turn induces expression of keratin 9 in keratinocytes [72]. Thus, HOX gene expression in fibroblasts helps to regulate epidermal cell identity. Interestingly, in other contexts, intrinsic regulation of keratin expression has been observed in keratinocytes. In burn patients grafted with CEA prepared using plantar keratinocytes, the cells in the healed CEA retained their phenotype, including thick stratum corneum and keratin 9 expression, after transplantation to non plantar sites [74]. This observation was attributed, in part, to an intrinsic capacity for site-specific gene expression. Taken together, these studies emphasize the complexity of keratinocyte–fibroblast interactions that determine site-specific differentiation of skin, with implications for tissue engineering. Thus, the source of keratinocytes and fibroblasts for fabrication of skin substitutes may have a tremendous impact on gene expression and subsequent function. This is particularly important for specialized skin regions, such as the eyelids, scalp, and nipples, as well as palmoplantar regions.

Roles of Biopolymer Scaffolds

In addition to the critical paracrine interactions between fibroblasts and keratinocytes that are required for stratification and differentiation of skin substitutes, it is becoming increasingly clear that the physical structure of the bioengineered scaffold plays an important role in epidermal function. There is a wide variety of natural and bioengineered scaffold materials that have been developed for skin tissue engineering, as described above. The scaffold functions as more than just a vehicle for transplantation of cells to a wound. It should provide the appropriate environment for cell localization and adhesion to support the expression of the natural cellular phenotypes [75]. In doing so, the scaffold guides or permits the development of new functional skin. It is generally assumed that the ideal scaffold should resemble the native extracellular environment [76,77]. Many of the composite skin substitutes tested clinically utilize animal-derived, collagen-based scaffolds as dermal matrices. To more closely replicate the natural microenvironment, with the added benefit of eliminating animal-derived materials, ECMs produced by human fibroblasts have been reported as scaffolds for skin substitutes [78]. To generate fibroblast-derived matrices, culture conditions are optimized to maximize cellular synthesis and accumulation of matrix. Human fibroblast-derived matrices populated with fibroblasts and keratinocytes were compared with similar constructs fabricated using rat collagen and were found to remain viable in culture for a significantly longer period, up to 20 weeks, compared with 8 weeks for tissue fabricated with rat collagen [79]. This was attributed to the biochemical composition of the fibroblast-derived matrices, which favored establishment of an appropriate balance of epidermal proliferation and differentiation as well as maintenance of an epidermal stem cell phenotype [79].

Reproducing the natural environment of fibroblasts and keratinocytes involves not only the biochemical environment, but the physical and biomechanical properties of the ECM as well [75]. Topographical features of natural skin, such as rete ridges, may affect keratinocyte function by compartmentalizing the cells into discreet populations. Rete ridges also serve to increase surface area of the dermal–epidermal junction, which may amplify paracrine interactions between cell layers. Most proliferating basal and suprabasal cells in the epidermis are localized to deeper rete ridges. Further, epidermal stem cells are believed to be localized to deep rete ridges, which provide a physically protected niche [80]. Recent developments in scaffold engineering have attempted to reproduce the function of rete ridges in skin. For example, Clement et al. fabricated a micro-patterned biopolymer matrix designed to duplicate different functional niches of keratinocytes [81]. Micro-patterned silicone molds were designed with channels of varying widths and were used to cast collagen gel matrices. Fibroblasts were populated on the nonpatterned side, and keratinocytes were populated on the micro-patterned surface. The investigators found that proliferative keratinocytes, as well as putative stem cell-like populations expressing integrin β1 and p63, were preferentially localized to the more narrow channels, similar to rete ridges, whereas keratinocytes in the wider channels expressed higher levels of the basement membrane component laminin-332 [81]. This study illustrates the regulatory activities of substrate topography in modulating keratinocyte physiology. By controlling the biochemical composition and physical structure of dermal scaffolds, tissue engineering enables fine-tuning of dermal–epidermal interactions and cell compartmentalization for optimal tissue development.

REGENERATIVE MEDICINE AND MECHANISMS OF DEVELOPMENTAL BIOLOGY

As advanced therapeutics have entered the market, prospective future products have begun to address the deficiencies of the current models of skin substitutes. As mentioned above, these deficiencies include epidermal appendages, basement membrane, pigmentation, nerve, vasculature, and subcutaneous fat. Each of these structures is formed by a specific process of morphogenesis that occurs during embryonic and fetal development (Table 2). Correction of current deficiencies will require an understanding of key developmental pathways and integration of essential regulatory mechanisms into tissue-engineered skin models.

Formation of an epidermal barrier results from a program of differentiation that is regulated by extracellular calcium and involves cross-linking of the cornified envelope by transglutaminases, and secretion of a liquid crystalline, multilamellar lipid matrix that seals the body against fluid loss and

TABLE 2 Developmental Pathways and Regulatory Factors for Cutaneous Phenotypes

Cutaneous Structures and Phenotypes	Regulatory Pathways and Factors
Epidermal barrier	Ca^{+2} [82] Transglutaminase, loricrin [83] Essential fatty acids, stratum corneum lipids [84]
Dermal–epidermal junction	Integrins [85] FAK–Ras–MapK [86,88] PKB/Akt–ERK1/2 [87]
Hair follicle genesis and cycling	Wnt/β-catenin/DKK4/BMPs [89,150] EDA/EDAR [90,151] Sox-2, -21 [91,152]
Sebaceous glands	Rac1–Sox9–Lrig1 [150] Wnt; Blimp1; IHH; c-myc [94] TGFβ-1 [95]
Sweat glands	Wnt/β-catenin [49] EDA/EDAR/NF-κB [97] DKK4, SHH [98]
Pigmentation	c-kit/SCF [99] ET-3/EDNRB2 [100] Sox-10/Mitf; Eph/EphR [101]
Sensory and motor innervation	c-jun [102] Sox-2,-10 [103] Oct-6; Krox-20; Pax 3 [104]
Cardiovascular system	Sox-7, -17, -18 [107] Mef2c/β-catenin [108] VEGF; HOXA9, VEZF1 [109]
Subcutaneous fat	PPARγ [113] Pref-1; Fabp4 [114] Myf5; Ebf2; Prdm16; Pgc-1α [115]

infection [82–84]. Basement membrane forms the critical boundary between the epidermis and dermis and regulates cell proliferation by integrin signaling through the focal adhesion kinase–MAP kinase pathway [85–88]. Morphogenesis and cycling of hair follicles has been studied extensively and is known to be regulated through the WNT–β-catenin pathway [89–91]. Taking advantage of this knowledge, modulation of WNT signaling has been used to promote trichogenic potential in cultured dermal papilla cells. For example, exposure of dermal papilla cells to Wnt10b maintained their trichogenic potential, and Wnt10b-treated dermal papilla cells combined with keratinocytes formed hairs after transplantation to athymic mice [92]. WNT proteins also regulate formation of sebaceous glands, which develop in the late stages of hair formation, and facilitate eruption of the hair through the stratum corneum [93–95]. Sweat glands, which may be associated with hair follicles (apocrine), or located in the interfollicular epidermis (eccrine), are also regulated by the WNT–β-catenin pathway. In addition, sweat gland morphogenesis is regulated by ectodysplasin (EDA) and its receptor (EDAR), DKK4 and SHH, congenital defects of which lead to absence of sweat glands and serious malformations of the nails, hair, and teeth [96–98].

Pigmentation of skin develops from melanocytes, which originate in the neural crest during embryogenesis and migrate to the epidermis where they reside on the basement membrane and express pigment [99]. The cell lineage and migration of melanocytes to the epidermis is regulated by Sox-10, microphthalamia transcription factor (MITF), and ephrin and its receptor [99–101]. After homing to the basal layer of the epidermis, melanocytes express the pigments eumelanin and pheomelanin, which are taken up by phagocytosis of melanocyte dendrites by keratinocytes. By this mechanism, pigment is distributed uniformly throughout the epidermis. The dendritic morphology of melanocytes resembles neurons and glial cells. This similarity may be due, in part, to the coordinated development of the epidermis and nervous systems, which both arise from the embryonic ectoderm. The sensory and motor innervation of skin is reported to be regulated by c-jun, Sox-2, Sox-10, Oct-6, Krox-20, and Pax-3 during murine fetal development [102–104]. Interestingly, expression of Sox-10 is also found during melanocyte migration, suggesting shared developmental pathways.

Direct connections between the skin and the central nervous system are established during development of the embryonic ectoderm, and in the absence of injury, remain connected throughout a lifetime. These connections provide continuous information from the external environment, through exquisitely sensitive mechano-transduction of the epidermis and its appendages to a variety of sensory nerve fibers, and ultimately to the brain. Because of this fundamental function of epidermal skin as a sensory organ, it is considered by some that the skin is the surface of the brain [105,106]. It is also clear that after full-thickness skin loss, such as in deep burns, the regrowth of nerves does not restore the complete neural physiology of uninjured skin. As in related areas of skin morphogenesis, it may be expected that greater understanding of the developmental mechanisms that connect the brain and the skin will lead eventually to improved restoration of sensory, neuro-exocrine, and motor functions after application of skin grafts and skin substitutes.

Sox-7, Sox-17, and Sox-18 are known to regulate early development of the vascular system and heart [107]. As vascular development proceeds, Mef2c activates β-catenin to promote formation of the peripheral vascular system [108]. Transcription factors HOXA9 and VEZF1 have been described as important regulatory factors during embryonic development of the vascular system [109].

HOXA9 acts to promote vasculogenesis through expression of target genes, endothelial nitric oxide synthase (eNOS), VEGF receptor 2 (VEGFR-2), and vascular endothelial cadherin (VE-cadherin). VEZF1 regulates angiogenesis by expression of endothelin 1(ET-1) and metallothionein 1 (MT1), which is associated with zinc-binding enzymes such as matrix metalloproteinases (MMPs). In postnatal wound healing, mesenchymal stem cells (MSCs) derived from the bone marrow have been shown to remain in healed tissue to support nascent blood vessels as pericytes [110].

The subcutaneous fat has been shown to be an important source of adipocyte stem cells that can reduce long-term scar after traumatic injury [111,112]. Fat development results, in part, from expression of PPARγ [113], and expression of characteristic markers, including Pref-1, fatty acid-binding protein 4 (Fabp4), Prdm 16, and Pgc-1α [114,115]. Sebaceous glands in skin also express some of these factors (i.e., PPARγ, Fabp4, Prdm 16). Notably, recent studies of hair follicle morphogenesis have found that these factors to be absent after transplantation of murine dermal papilla cells and human epidermal keratinocytes, which failed to form sebaceous glands [56,57]. These findings suggest that critical upstream regulatory genes are not expressed properly in this system, while a similar model using neonatal murine skin cells forms both hair shafts and complete sebaceous glands.

GENE THERAPY APPROACHES

The ability to genetically modify cells used to prepare skin substitutes enables ex vivo gene therapy approaches to treat cutaneous diseases and injuries. Nevertheless, gene therapy remains an unfulfilled promise of cell therapies with skin substitutes. However, selected trials of gene therapy in skin wounds have been performed in recent years [116–118]. Early models using overexpression of angiogenic growth factors with replication-incompetent retroviral vectors demonstrated feasibility for constitutive delivery of biologically active compounds, such as VEGF and PDGF, with potential for stimulation of wound healing in impaired wounds [119,120]. However, use of retroviral gene transfer for the treatment of hematopoietic disorders was associated with development of leukemia due to insertional mutagenesis, a finding that has essentially disqualified that approach from prospective consideration [43]. Similar risks have been identified with lentiviruses that have also reduced their prospective use in therapeutics. Alternatives to viral vectors have been developed that include plasmid transfection for expression of endogenous antimicrobial peptides, such as cathelicidin [121]. Other recently developed approaches, such as the clustered regularly interspaced short palindromic repeats (CRISPR) system, enable site-specific genome editing, greatly reducing the risk of insertional mutagenesis [122,123]. In addition, if gene therapies are administered in allogeneic cells that are eliminated immunologically after a limited time period, then risks may be minimized.

More recent studies have demonstrated feasibility for the delivery of structural proteins of the ECM to treat certain forms of EB [124,125]. Other technologies, such as induced pluripotent stem cells (iPSCs), can improve the efficiency of genome editing. The use of iPS cells involves reversion of donor cells to a pluripotent phenotype in vitro, expansion of cell numbers, and regulation of the cell population toward a differentiated phenotype of interest. The combination of genetic modification approaches, enabling addition of normal genes or correction of mutated genes, with iPSC technology provides the ability to correct inherited diseases, while extending the cellular lifespan significantly compared with primary skin cells. Approaches involving gene correction and/or iPSCs have been proposed recently to treat multiple forms of EB [126,127].

REGULATION OF SKIN SUBSTITUTES BY THE US FOOD AND DRUG ADMINISTRATION

Safety and efficacy of skin substitutes is regulated by the US Food and Drug Administration (FDA) in the United States. Biologic skin substitutes have increased in complexity from models that replace either dermis or epidermis, to dermal–epidermal models, to those that deliver combinations of biopolymer scaffolds, multiple cell types, or multiple cell sources, to those that express gene products for prospective stimulation of wound healing. This spectrum of unprecedented materials presented questions regarding the regulatory framework within which each model would be evaluated for consideration of permission to market. Traditionally, the FDA has consisted of three centers for evaluation of human therapeutics: the Center for Devices and Radiologic Health (CDRH), the Center for Biologics Evaluation and Research (CBER), and the Center for Drug Evaluation and Research (CDER). Availability of cadaveric allograft has been provided under regulations for tissue banking, which are administered by CBER. As the spectrum of research models of skin substitutes broadened during the 1980s and 1990s, several investigative therapies had components that required consideration by multiple centers at FDA. The agency responded proactively with two initiatives and with Guidances to Industry [128,129] that have contributed to greater clarity of the regulatory process on how to propose a path to market.

The first initiative was FDA's participation in the establishment of definitions and standards for Tissue Engineered Medical Products (TEMPs) via the formation of a subdivision of Committee F04 for medical devices through the American Society for Testing and Materials (ASTM) [130]. Beginning in 1997, this organization has had members from academics, government, and industry participating in a consensus process for composing definitions of materials and provision of methods for calibration and testing of the materials. With regard to skin substitutes, the ASTM process has resulted in a Standard Guide for Classification of Therapeutic Skin Substitutes [131], which provides consensus definitions and nomenclature. The second initiative was FDA's establishment in 2002 of the Office of Combination Products, by which investigative therapies are reviewed initially for their *primary mode of action* [132]. This Office confers with the centers for evaluation of human therapeutics to designate new therapies at a lead Center at FDA with participation from other Centers as appropriate. Together, these initiatives have added clarity to the assignment of novel therapeutics to a designated regulatory path.

CONCLUSIONS AND FUTURE DIRECTIONS

Future prospects for biologic skin substitutes are extensive and diverse. Advances in the use and regulation of stem cells in skin are quite likely to lead to autologous skin substitutes with greater homology to uninjured skin, by providing restoration of skin pigmentation, epidermal appendages (hair, sebaceous, and sweat glands), a vascular plexus, and subcutaneous tissues. Genetic modification of autologous cells opens tremendous opportunities for the regulation of wound closure, reductions in scar formation, and correction of congenital diseases. As these advances in biologic skin substitutes translate into clinical care, it can be predicted with confidence that reductions in morbidity from acquired and congenital skin diseases will also be realized.

LIST OF ACRONYMS AND ABBREVIATIONS

Allo Allogeneic
ASTM American Society for Testing and Materials
Auto Autologous
β-catenin Beta-catenin
Blimp1 B lymphocyte-induced maturation protein 1
BMPs Bone morphogenetic proteins
CBER Center for Biologics Evaluation and Research
CDER Center for Drug Evaluation and Research
CDRH Center for Devices and Radiologic Health
CEA Cultured epithelial autografts
DKK-1, -4 Dikkopf-1, -4
EB Epidermolysis bullosa
Ebf2 Early B cell factor 2
ECM Extracellular matrix
EDA/EDAR Ectodysplasin-A/Ectodysplasin-A receptor
EDNRB2 Endothelin receptor subtype B2
eNOS Endothelial nitric oxide synthase
ESSs Engineered skin substitutes
Eph/EphR Ephrin/Ephrin receptor
ET-1, -3 Endothelins-1, -3
Fabp4 Fatty acid-binding protein 4
FAK Focal adhesion kinase
FGF2 Fibroblast growth factor 2 (basic)
GAG Glycosaminoglycan

hF Human fibroblasts
hK Human keratinocytes
hM Human melanocytes
HOX genes Homeobox genes
IHH Indian hedgehog
KGF Keratinocyte growth factor
Krox-20 Kruppel-related box gene 20
Lrig1 Leucine-rich repeats and immunoglobulin-like domains 1
MapK Mitogen activated kinase protein
Mef2c Myocyte enhancer factor 2c
Mitf Microphthalmia-associated transcription factor
MMPs Matrix metalloproteinases
Myf5 Myogenic factor 5
NF-κB Nuclear factor of kappa light polypeptide gene enhancer in B cells
Oct-6 Organic cation/carnitine transporter 6
Pax-3 Paired box 3
PPARγ Peroxisome proliferator-activated receptor gamma
Pgc-1α Peroxisome proliferative activated receptor gamma co-activator 1 alpha
PRDM6 Promoter domain containing 6
Rac1 Ras-related C3 botulinum toxin substrate 1
Ras Rat sarcoma gene family
SCF Stem cell factor
SHH Sonic hedgehog
SOX genes Sex-determining region Y chromosome-related high mobility group box genes
TBSA Total body surface area
TEMPS Tissue engineered medical products
TGF-β1 Transforming growth factor-β1
US FDA United States Food and Drug Administration
VEGF Vascular endothelial growth factor
VEZF1 Vascular endothelial zinc finger-1 transcription factor
Wnt Wingless integration site of murine mammary tumor virus

GLOSSARY

Allograft Transplantation of skin, or any organs, tissues, or cells, from one individual to another individual of the same species.

Autograft Transplantation of skin or other tissues or cells from one part of the body to another in the same individual.

Fibroblasts The predominant parenchymal cell type found in connective tissue; in skin, the main cellular component of the dermis, responsible for synthesis of collagen and other ECM components.

Keratinocytes The predominant parenchymal cell type of the epidermis, responsible for keratin synthesis and epidermal barrier.

Melanocytes Pigment-producing cells of the epidermis.

Trichogenesis The process of generating hair.

Xenograft Transplantation of skin, or any organ, tissues or cells, from one individual to another individual of a different species.

REFERENCES

[1] American Burn Association. 2014 National burn repository. American Burn Association; 2014.

[2] Herndon DN. In: Phildelphia PA, editor. Total burn care. 4th ed. W.B. Saunders; 2012.

[3] Richmond NA, Lamel SA, Davidson JM, Martins-Green M, Sen CK, Tomic-Canic M, et al. US-National Institutes of Health-funded research for cutaneous wounds in 2012. Wound Repair Regen 2013;21(6):789–92.

[4] Markova A, Mostow EN. US skin disease assessment: ulcer and wound care. Dermatol Clin 2012;30(1):107–11, ix.

[5] Sen CK, Gordillo GM, Roy S, Kirsner R, Lambert L, Hunt TK, et al. Human skin wounds: a major and snowballing threat to public health and the economy. Wound Repair Regen 2009;17(6):763–71.

[6] Warner PM, Yakuboff KP, Kagan RJ, Boyce S, Warden GD. An 18-year experience in the management of congenital nevomelanocytic nevi. Ann Plast Surg 2008;60(3):283–7.

[7] Passaretti D, Billmire D, Kagan R, Corcoran J, Boyce S. Autologous cultured skin substitutes conserve donor autograft in elective treatment of congenital giant melanocytic nevus. Plast Reconstr Surg 2004;114(6):1523–8.

[8] Fine JD, Bruckner-Tuderman L, Eady RA, Bauer EA, Bauer JW, Has C, et al. Inherited epidermolysis bullosa: updated recommendations on diagnosis and classification. J Am Acad Dermatol 2014;70(6):1103–26.

[9] Goldschneider KR, Lucky AW. Pain management in epidermolysis bullosa. Dermatol Clin 2010;28(2):273–82, ix.

[10] Bottcher-Haberzeth S, Biedermann T, Schiestl C, Hartmann-Fritsch F, Schneider J, Reichmann E, et al. Matriderm(R) 1 mm versus Integra(R) single layer 1.3 mm for one-step closure of full thickness skin defects: a comparative experimental study in rats. Pediatr Surg Int 2012;28(2):171–7.

[11] Neuhaus K, Meuli M, Koenigs I, Schiestl C. Management of "difficult" wounds. Eur J Pediatr Surg 2013;23(5):365–74.

[12] Warden GD, Saffle JR, Kravitz M. A two-stage technique for excision and grafting of burn wounds. J Trauma 1982;22(2):98–103.

[13] Housinger TA, Warden GD, Lang D, Visse M, Pohlman S. The use of sheet skin grafts in pediatric burn patients. Proc Amer Burn Assn 1991;23:32.

[14] Kagan RJ. FDA regulation of human tissue: implications for skin banks and burn centers. J Burn Care Rehabil 2000;21(4):353–7.

[15] Kagan RJ, Robb EC, Plessinger RT. Human skin banking. Clin Lab Med 2005;25(3): 587–605.

[16] Hermans MH. Porcine xenografts vs. (cryopreserved) allografts in the management of partial thickness burns: is there a clinical difference? Burns 2014;40(3):408–15.

[17] Heimbach DM, Warden GD, Luterman A, Jordan MH, Ozobia N, Ryan CM, et al. Multicenter postapproval clinical trial of Integra dermal regeneration template for burn treatment. J Burn Care Rehabil 2003;24(1):42–8.

[18] Wainwright DJ. Use of an acellular allograft dermal matrix (AlloDerm) in the management of full-thickness burns. Burns 1995;21(4):243–8.

[19] Centanni JM, Straseski JA, Wicks A, Hank JA, Rasmussen CA, Lokuta MA, et al. StrataGraft skin substitute is well-tolerated and is not acutely immunogenic in patients with traumatic wounds: results from a prospective, randomized, controlled dose escalation trial. Ann Surg 2011;253(4):672–83.

[20] Falanga V, Margolis D, Alvarez O, Auletta M, Maggiacomo F, Altman M, et al. Rapid healing of venous ulcers and lack of clinical rejection with an allogeneic cultured human skin equivalent. Human Skin Equivalent Investigators Group. Arch Dermatol 1998;134(3):293–300.

[21] Shevchenko RV, James SL, James SE. A review of tissue-engineered skin bioconstructs available for skin reconstruction. J R Soc Interface 2010;7(43):229–58.

[22] Yildirimer L, Thanh NT, Seifalian AM. Skin regeneration scaffolds: a multimodal bottom-up approach. Trends Biotechnol 2012;30(12):638–48.

[23] Boyce ST, Kagan RJ, Greenhalgh DG, Warner P, Yakuboff KP, Palmieri T, et al. Cultured skin substitutes reduce requirements for harvesting of skin autograft for closure of excised, full-thickness burns. J Trauma 2006;60(4):821–9.

[24] Boa O, Cloutier CB, Genest H, Labbe R, Rodrigue B, Soucy J, et al. Prospective study on the treatment of lower-extremity chronic venous and mixed ulcers using tissue-engineered skin substitute made by the self-assembly approach. Adv Skin Wound Care 2013;26(9):400–9.

[25] Belatti DA, Phisitkul P. Declines in lower extremity amputation in the US Medicare population, 2000–2010. Foot Ankle Int 2013;34(7):923–31.

[26] Chou R, Dana T, Bougatsos C, Blazina I, Starmer A, Reitel K, et al. Pressure ulcer risk assessment and prevention: comparative effectiveness. AHRQ Comparative Effectiveness Reviews 2013.

[27] Saha S, Smith MEB, Totten A, Fu R, Wasson N, Rahman B, et al. Pressure ulcer treatment strategies: comparative effectiveness. AHRQ Comparative Effectiveness Reviews 2013.

[28] Lazarus G, Valle MF, Malas M, Qazi U, Maruthur NM, Doggett D, et al. Chronic venous leg ulcer treatment: future research needs. Wound Repair Regen 2014;22(1):34–42.

[29] Desai MH, Mlakar JM, McCauley RL, Abdullah KM, Rutan RL, Waymack JP, et al. Lack of long term durability of cultured keratinocyte burn wound coverage: a case report. J Burn Care Rehabil 1991;12:540–5.

[30] Wood FM, Kolybaba ML, Allen P. The use of cultured epithelial autograft in the treatment of major burn injuries: a critical review of the literature. Burns 2006;32(4):395–401.

[31] Matsumura H, Gondo M, Imai R, Shibata D, Watanabe K. Chronological histological findings of cultured epidermal autograft over bilayer artificial dermis. Burns 2013;39(4):705–13.

[32] Sood R, Roggy D, Zieger M, Balledux J, Chaudhari S, Koumanis DJ, et al. Cultured epithelial autografts for coverage of large burn wounds in eighty-eight patients: the Indiana University experience. J Burn Care Res 2010;31(4):559–68.

[33] Woodley DT, Peterson HD, Herzog SR, Sriklin GP, Bureson RE, Briggaman RA, et al. Burn wounds resurfaced by cultured epidermal autografts show abnormal reconstitution of anchoring fibrils. JAMA 1988;259:2566–671.

[34] De Luca M, Pellegrini G, Mavilio F. Gene therapy of inherited skin adhesion disorders: a critical overview. Br J Dermatol 2009;161(1):19–24.

[35] El Ghalbzouri A, Ponec M. Diffusible factors released by fibroblasts support epidermal morphogenesis and deposition of basement membrane components. Wound Rep Reg 2004;12(3):359–67.

[36] Boyce ST, Supp AP, Swope VB, Warden GD. Vitamin C regulates keratinocyte viability, epidermal barrier, and basement membrane in vitro, and reduces wound contraction after grafting of cultured skin substitutes. J Invest Dermatol 2002;118(4):565–72.

[37] Harriger MD, Warden GD, Greenhalgh DG, Kagan RJ, Boyce ST. Pigmentation and microanatomy of skin regenerated from composite grafts of cultured cells and biopolymers applied to full-thickness burn wounds. Transplantation 1995;59(5):702–7.

[38] Wojtowicz AM, Oliveira S, Carlson MW, Zawadzka A, Rousseau CF, Baksh D. The importance of both fibroblasts and keratinocytes in a bilayered living cellular construct used in wound healing. Wound Repair Regen 2014;22(2):246–55.

[39] Smiley AK, Klingenberg JM, Aronow BJ, Boyce ST, Kitzmiller WJ, Supp DM. Microarray analysis of gene expression in cultured skin substitutes compared with native human skin. J Invest Dermatol 2005;125(6):1286–301.
[40] Klingenberg JM, McFarland KL, Friedman AJ, Boyce ST, Aronow BJ, Supp DM. Engineered human skin substitutes undergo large-scale genomic reprogramming and normal skin-like maturation after transplantation to athymic mice. J Invest Dermatol 2010;130(2): 587–601.
[41] Boyce ST, Goretsky MJ, Greenhalgh DG, Kagan RJ, Rieman MT, Warden GD. Comparative assessment of cultured skin substitutes and native skin autograft for treatment of full-thickness burns. Ann Surg 1995;222(6):743–52.
[42] Supp DM, Karpinski AC, Boyce ST. Vascular endothelial growth factor overexpression increases vascularization by murine but not human endothelial cells in cultured skin substitutes grafted to athymic mice. J Burn Care Rehabil 2004;25(4):337–45.
[43] Kohn DB, Sadelain M, Dunbar C, Bodine D, Kiem HP, Candotti F, et al. American Society of Gene Therapy (ASGT) ad hoc subcommittee on retroviral-mediated gene transfer to hematopoietic stem cells. Mol Ther 2003;8(2):180–7.
[44] Supp DM, Wilson-Landy K, Boyce ST. Human dermal microvascular endothelial cells form vascular analogs in cultured skin substitutes after grafting to athymic mice. FASEB J 2002;16(8):797–804.
[45] Tremblay PL, Hudon V, Berthod F, Germain L, Auger FA. Inosculation of tissue-engineered capillaries with the host's vasculature in a reconstructed skin transplanted on mice. Am J Transplant 2005;5(5):1002–10.
[46] Gibot L, Galbraith T, Huot J, Auger FA. A preexisting microvascular network benefits in vivo revascularization of a microvascularized tissue-engineered skin substitute. Tissue Eng Part A 2010;16(10):3199–206.
[47] Zhang C, Chen Y, Fu X. Sweat gland regeneration after burn injury: is stem cell therapy a new hope? Cytotherapy 2014;14:10.
[48] Kurata R, Futaki S, Nakano I, Tanemura A, Murota H, Katayama I, et al. Isolation and characterization of sweat gland myoepithelial cells from human skin. Cell Struct Funct 2014;39(2):101–12.
[49] Lu CP, Polak L, Rocha AS, Pasolli HA, Chen SC, Sharma N, et al. Identification of stem cell populations in sweat glands and ducts reveals roles in homeostasis and wound repair. Cell 2012;150(1):136–50.
[50] Fradette J, Larouche D, Fugere C, Guignard R, Beauparlant A, Couture V, et al. Normal human Merkel cells are present in epidermal cell populations isolated and cultured from glabrous and hairy skin sites. J Invest Dermatol 2003;120(2):313–7.
[51] Harrison R, St-Pierre JP, Stevens M. Tissue engineering and regenerative medicine: a year in review. Tissue Eng Part B Rev 2014;20(1):1–16.
[52] Lei X, Liu B, Wu J, Lu Y, Yang Y. Matrigel-induced tubular morphogenesis of human eccrine sweat gland epithelial cells. Anat Rec (Hoboken) 2011;294(9):1525–31.
[53] Sheng Z, Fu X, Cai S, Lei Y, Sun T, Bai X, et al. Regeneration of functional sweat gland-like structures by transplanted differentiated bone marrow mesenchymal stem cells. Wound Repair Regen 2009;17(3):427–35.
[54] Rendl M, Fuchs E. Molecular dissection of mesenchymal-epithelial interactions in the hair follicle. PLoS Biol 2005;3(11):1910–24.
[55] Weinberg WC, Goodman LV, George C, Morgan DL, Ledbetter S, Yuspa SH, et al. Reconstitution of hair follicle development in vivo: determination of follicle formation, hair growth, and hair quality by dermal cells. J Invest Dermatol 1993;100(3):229–36.

[56] Sriwiriyanont P, Lynch KA, Maier EA, Hahn JM, Supp DM, Boyce ST. Morphogenesis of chimeric hair follicles in engineered skin substitutes with human keratinocytes and murine dermal papilla cells. Exp Dermatol 2012;21(10):783–5.

[57] Sriwiriyanont P, Lynch KA, McFarland KL, Supp DM, Boyce ST. Characterization of hair follicle development in engineered skin substitutes. PLoS One 2013;8(6):e65664.

[58] Wu X, Scott Jr L, Washenik K, Stenn K. Full-thickness skin with mature hair follicles generated from tissue culture expanded human cells. Tissue Eng Part A 2014;20(23–24):3314–21.

[59] Kocikova A, Kolesaric A, Koszik F, Stingl G, Elbe-Burger A. Murine Langerhans cells cultured under serum-free conditions mature into potent stimulators of primary immune responses in vitro and in vivo. J Immunol 1998;161(8):4033–41.

[60] Petersen MJ, Lessane B, Woodley DT. Characterization of cellular elements in healed cultured keratinocyte autografts used to cover burn wounds. Arch Dermatol 1990;126(2): 175–80.

[61] Maksimovic S, Nakatani M, Baba Y, Nelson AM, Marshall KL, Wellnitz SA, et al. Epidermal Merkel cells are mechanosensory cells that tune mammalian touch receptors. Nature 2014;509(7502):617–21.

[62] Anderson JR, Fear MW, Phillips JK, Dawson LF, Wallace H, Wood FM, et al. A preliminary investigation of the reinnervation and return of sensory function in burn patients treated with Integra(R). Burns 2011;37(7):1101–8.

[63] Swope VB, Supp AP, Boyce ST. Regulation of cutaneous pigmentation by titration of human melanocytes in cultured skin substitutes grafted to athymic mice. Wound Repair Regen 2002;10(6):378–86.

[64] Boyce ST, Zimmerman RL, Supp DM. Tumorigenicity testing in athymic mice of cultured human melanocytes for transplantation in engineered skin substitutes. Cell Transplant 2014;24(8):1423–9.

[65] Janson DG, Saintigny G, van AA, Mahe C, El GA. Different gene expression patterns in human papillary and reticular fibroblasts. J Invest Dermatol 2012;132(11):2565–72.

[66] Wang J, Dodd C, Shankowsky HA, Scott PG, Tredget EE. Deep dermal fibroblasts contribute to hypertrophic scarring. Lab Invest 2008;88(12):1278–90.

[67] Honardoust D, Ding J, Varkey M, Shankowsky HA, Tredget EE. Deep dermal fibroblasts refractory to migration and decorin-induced apoptosis contribute to hypertrophic scarring. J Burn Care Res 2012;33(5):668–77.

[68] Varkey M, Ding J, Tredget EE. Differential collagen-glycosaminoglycan matrix remodeling by superficial and deep dermal fibroblasts: potential therapeutic targets for hypertrophic scar. Biomaterials 2011;32(30):7581–91.

[69] Varkey M, Ding J, Tredget EE. Superficial dermal fibroblasts enhance basement membrane and epidermal barrier formation in tissue-engineered skin: implications for treatment of skin basement membrane disorders. Tissue Eng Part A 2014;20(3–4):540–52.

[70] Chang HY, Chi JT, Dudoit S, Bondre C, van de RM, Botstein D, et al. Diversity, topographic differentiation, and positional memory in human fibroblasts. Proc Natl Acad Sci USA 2002;99(20):12877–82.

[71] Rinn JL, Bondre C, Gladstone HB, Brown PO, Chang HY. Anatomic demarcation by positional variation in fibroblast gene expression programs. PLoS Genet 2006;2(7):e119.

[72] Rinn JL, Wang JK, Allen N, Brugmann SA, Mikels AJ, Liu H, et al. A dermal HOX transcriptional program regulates site-specific epidermal fate. Genes Dev 2008;22(3):303–7.

[73] Yamaguchi Y, Itami S, Tarutani M, Hosokawa K, Miura H, Yoshikawa K. Regulation of keratin 9 in nonpalmoplantar keratinocytes by palmoplantar fibroblasts through epithelial-mesenchymal interactions. J Invest Dermatol 1999;112(4):483–8.

[74] Compton CC, Nadire KB, Regauer S, Simon M, Warland G, O'Connor NE, et al. Cultured human sole-derived keratinocyte grafts re-express site-specific differentiation after transplantation. Differentiation 1998;64(1):45–53.

[75] Edalat F, Sheu I, Manoucheri S, Khademhosseini A. Material strategies for creating artificial cell-instructive niches. Curr Opin Biotechnol 2012;23(5):820–5.

[76] Powell HM, Supp DM, Boyce ST. Engineered skin substitutes fabricated with electrospun collagen inhibit wound contraction. Biomaterials 2007;28:1084–92.

[77] Powell HM, Supp DM, Boyce ST. Influence of electrospun collagen on wound contraction of engineered skin substitutes. Biomaterials 2008;29(7):834–43.

[78] Duranceau L, Genest H, Bortoluzzi P, Moulin V, Auger FA, Germain L. Successful grafting of a novel autologous tissue-engineered skin substitutes (dermis and epidermis) on twelve burn patients. J Burn Care Res 2014;35(3):S121.

[79] El-Ghalbzouri A, Commandeur S, Rietveld MH, Mulder AA, Willemze R. Replacement of animal-derived collagen matrix by human fibroblast-derived dermal matrix for human skin equivalent products. Biomaterials 2009;30(1):71–8.

[80] Webb A, Li A, Kaur P. Location and phenotype of human adult keratinocyte stem cells of the skin. Differentiation 2004;72(8):387–95.

[81] Clement AL, Moutinho Jr TJ, Pins GD. Micropatterned dermal-epidermal regeneration matrices create functional niches that enhance epidermal morphogenesis. Acta Biomater 2013;9(12):9474–84.

[82] Elsholz F, Harteneck C, Muller W, Friedland K. Calcium – a central regulator of keratinocyte differentiation in health and disease. Eur J Dermatol 2014;24(6):650–61.

[83] Wikramanayake TC, Stojadinovic O, Tomic-Canic M. Epidermal differentiation in barrier maintenance and wound healing. Adv Wound Care (New Rochelle) 2014;3(3):272–80.

[84] Feingold KR, Elias PM. Role of lipids in the formation and maintenance of the cutaneous permeability barrier. Biochim Biophys Acta 2014;1841(3):280–94.

[85] Haass NK, Smalley KS, Li L, Herlyn M. Adhesion, migration and communication in melanocytes and melanoma. Pigment Cell Res 2005;18(3):150–9.

[86] Gomperts BD, Kramer IM, Tatham PER. Signal transduction to and from adhesion molecules. In: Signal transduction. San Diego (CA): Elsevier; 2003. p. 315–43.

[87] Kippenberger S, Hofmann M, Zoller N, Thaci D, Muller J, Kaufmann R, et al. Ligation of β_4 integrins activates PKB/Akt and ERK1/2 by distinct pathways-relevance of the keratin filament. Biochim Biophys Acta 2010;1803(8):940–50.

[88] Wong VW, Rustad KC, Akaishi S, Sorkin M, Glotzbach JP, Januszyk M, et al. Focal adhesion kinase links mechanical force to skin fibrosis via inflammatory signaling. Nat Med 2011;18(1):148–52.

[89] Zhang Y, Tomann P, Andl T, Gallant NM, Huelsken J, Jerchow B, et al. Reciprocal requirements for Eda/Edar/NF-κB and Wnt/β-catenin signaling pathways in hair follicle induction. Dev Cell 2009;17(1):49–61.

[90] Ohyama M, Zheng Y, Paus R, Stenn KS. The mesenchymal component of hair follicle neogenesis: background, methods and molecular characterization. Exp Dermatol 2009;19(2): 89–99.

[91] Kiso M, Tanaka S, Saba R, Matsuda S, Shimizu A, Ohyama M, et al. The disruption of Sox21-mediated hair shaft cuticle differentiation causes cyclic alopecia in mice. Proc Natl Acad Sci USA 2009;106(23):9292–7.

[92] Ouji Y, Ishizaka S, Yoshikawa M. Dermal papilla cells serially cultured with Wnt-10b sustain their hair follicle induction activity after transplantation into nude mice. Cell Transplant 2012;21(10):2313–24.

[93] Hinde E, Haslam IS, Schneider MR, Langan EA, Kloepper JE, Schramm C, et al. A practical guide for the study of human and murine sebaceous glands in situ. Exp Dermatol 2013;22(10): 631–7.

[94] Niemann C, Horsley V. Development and homeostasis of the sebaceous gland. Semin Cell Dev Biol 2012;23(8):928–36.

[95] McNairn AJ, Brusadelli M, Guasch G. Signaling moderation: TGF-beta in exocrine gland development, maintenance, and regulation. BMC Dermatol 2013;13(2). http://dx.doi.org/10.1186/1471-5945-13-2.

[96] Lu C, Fuchs E. Sweat gland progenitors in development, homeostasis, and wound repair. Cold Spring Harb Perspect Med 2014;4(2):a015222.

[97] Holbrook KA. Structural abnormalities of the epidermally derived appendages in skin from patients with ectodermal dysplasia: insight into developmental errors. Birth Defects Orig Artic Ser 1988;24(2):15–44.

[98] Cui CY, Yin M, Sima J, Childress V, Michel M, Piao Y, et al. Involvement of Wnt, Eda and Shh at defined stages of sweat gland development. Development 2014;141(19):3752–60.

[99] Adameyko I, Lallemend F, Aquino JB, Pereira JA, Topilko P, Muller T, et al. Schwann cell precursors from nerve innervation are a cellular origin of melanocytes in skin. Cell 2009;139(2):366–79.

[100] Cichorek M, Wachulska M, Stasiewicz A, Tyminska A. Skin melanocytes: biology and development. Postepy Dermatol Alergol 2013;30(1):30–41.

[101] Aoki Y, Saint-Germain N, Gyda M, Magner-Fink E, Lee YH, Credidio C, et al. Sox10 regulates the development of neural crest-derived melanocytes in *Xenopus*. Dev Biol 2003;259(1): 19–33.

[102] Kaucka M, Adameyko I. Non-canonical functions of the peripheral nerve. Exp Cell Res 2014;321(1):17–24.

[103] Gagnon V, Larouche D, Parenteau-Bareil R, Gingras M, Germain L, Berthod F. Hair follicles guide nerve migration in vitro and in vivo in tissue-engineered skin. J Invest Dermatol 2011;131(6):1375–8.

[104] Biedermann T, Klar AS, Bottcher-Haberzeth S, Schiestl C, Reichmann E, Meuli M. Tissue-engineered dermo-epidermal skin analogs exhibit de novo formation of a near natural neurovascular link 10 weeks after transplantation. Pediatr Surg Int 2014;30(2):165–72.

[105] Paus R, Theoharides TC, Arck PC. Neuroimmunoendocrine circuitry of the 'brain-skin connection'. Trends Immunol 2006;27(1):32–9.

[106] Ito N, Ito T, Kromminga A, Bettermann A, Takigawa M, Kees F, et al. Human hair follicles display a functional equivalent of the hypothalamic-pituitary-adrenal axis and synthesize cortisol. FASEB J 2005;19(10):1332–4.

[107] Francois M, Koopman P, Beltrame M. *SoxF* genes: key players in the development of the cardio-vascular system. Int J Biochem Cell Biol 2010;42(3):445–8.

[108] Wat JJ, Wat MJ. Sox7 in vascular development: review, insights and potential mechanisms. Int J Dev Biol 2014;58(1):1–8.

[109] Bruderer M, Alini M, Stoddart MJ. Role of HOXA9 and VEZF1 in endothelial biology. J Vasc Res 2013;50(4):265–78.

[110] Caplan AI. All MSCs are pericytes? Cell Stem Cell 2008;3(3):229–30.

[111] Bourin P, Bunnell BA, Casteilla L, Dominici M, Katz AJ, March KL, et al. Stromal cells from the adipose tissue-derived stromal vascular fraction and culture expanded adipose tissue-derived stromal/stem cells: a joint statement of the International Federation for Adipose Therapeutics and Science (IFATS) and the International Society for Cellular Therapy (ISCT). Cytotherapy 2013;15(6):641–8.

[112] Kapur SK, Wang X, Shang H, Yun S, Li X, Feng G, et al. Human adipose stem cells maintain proliferative, synthetic and multipotential properties when suspension cultured as self-assembling spheroids. Biofabrication 2012;4(2):025004. http://dx.doi.org/10.1088/1758-5082/4/2/025004.

[113] Janani C, Ranjitha Kumari BD. PPAR gamma gene – a review. Diabetes Metab Syndr 2015;9(1):46–50.

[114] Kang S, Kong X, Rosen ED. Adipocyte-specific transgenic and knockout models. Methods Enzymol 2014;537:1–16. http://dx.doi.org/10.1016/B978-0-12-411619-1.00001-X.

[115] Harms M, Seale P. Brown and beige fat: development, function and therapeutic potential. Nat Med 2013;19(10):1252–63.

[116] University of Pennsylvania. Phase I trial to evaluate the safety of platelet derived growth factor B (PDGF-B) and a limb compression bandage in venous leg ulcers. National Institutes of Health; 2013. www.ClinicalTrials.gov.

[117] Tissue Repair Company. Growth factor gene therapy for wound healing. National Institutes of Health; 2007. www.ClinicalTrials.gov.

[118] Lane AT. A phase 1 single center trial of gene transfer for recessive dystrophic epidermolysis bullosa (RDEB) using the drug LZRSE-Col7A1 engineered autologous epidermal sheets (LEAES). National Institutes of Health; 2014. www.ClinicalTrials.gov.

[119] Supp DM, Supp AP, Bell SM, Boyce ST. Enhanced vascularization of cultured skin substitutes genetically modified to overexpress vascular endothelial growth factor. J Invest Dermatol 2000;114(1):5–13.

[120] Supp DM, Supp AP, Morgan JR, Boyce ST. Genetic modification of cultured skin substitutes by transduction of human keratinocytes and fibroblasts with platelet derived growth factor-A. Wound Repair Regen 2000;8:26–35.

[121] Comer AR, Rasmussen CA, Thomas-Virnig CL, Lokuta MA, Shaughnessy LM, Schlosser SJ, et al. Preclinical development and planned clinical evaluation of a human skin substitute engineered to secrete elevated levels of a host defense peptide. In: Proceedings of the innovations in wound healing conference, vol. 14. 2014.

[122] Nihongaki Y, Yamamoto S, Kawano F, Suzuki H, Sato M. CRISPR-Cas9-based photoactivatable transcription system. Chem Biol 2015;22(2):169–74.

[123] Osborn MJ, Gabriel R, Webber BR, DeFeo AP, McElroy AN, Jarjour J, et al. Fanconi anemia gene editing by the CRISPR/Cas9 system. Hum Gene Ther 2014;26(2):114–26.

[124] Abdul-Wahab A, Qasim W, McGrath JA. Gene therapies for inherited skin disorders. Semin Cutan Med Surg 2014;33(2):83–90.

[125] Cutlar L, Greiser U, Wang W. Gene therapy: pursuing restoration of dermal adhesion in recessive dystrophic epidermolysis bullosa. Exp Dermatol 2014;23(1):1–6.

[126] Umegaki-Arao N, Pasmooij AM, Itoh M, Cerise JE, Guo Z, Levy B, et al. Induced pluripotent stem cells from human revertant keratinocytes for the treatment of epidermolysis bullosa. Sci Transl Med 2014;6(264). 264ra164.

[127] Sebastiano V, Zhen HH, Derafshi BH, Bashkirova E, Melo SP, Wang P, et al. Human COL7A1-corrected induced pluripotent stem cells for the treatment of recessive dystrophic epidermolysis bullosa. Sci Transl Med 2014;6(264). 264ra163.

[128] US Food and Drug Administration. Guidance for industry: chronic cutaneous ulcer and burn wounds – developing products for treatment. Fed Regist 2006;65(125):39912.

[129] US Food and Drug Administration. Guidance for industry: guidance for human somatic cell therapy and gene therapy. Fed Regist 1998;63:36413.

[130] American Society for Testing and Materials. Standard guide for preservation of tissue engineered medical products. West Conshohocken, PA: American Society for Testing and Materials; 2007. Report No.: F 2386–04.

[131] American Society for Testing and Materials. Standard guide for classification of therapeutic skin substitutes. West Conshohocken, PA: ASTM International; 2009. Report No.: F 2311–08.

[132] US Food and Drug Administration. Guidance for industry: how to write a request for designation. 2011.

[133] Lohana P, Hassan S, Watson SB. Integra in burns reconstruction: our experience and report of an unusual immunological reaction. Ann Burns Fire Disasters 2014;27(1):17–21.

[134] Jansen LA, De CP, Guay NA, Lineaweaver WC, Shokrollahi K. The evidence base for the acellular dermal matrix AlloDerm: a systematic review. Ann Plast Surg 2013;70(5):587–94.

[135] Ellis CV, Kulber DA. Acellular dermal matrices in hand reconstruction. Plast Reconstr Surg 2012;130(5 Suppl. 2):256S–69S.

[136] Halim AS, Khoo TL, Mohd Yussof SJ. Biologic and synthetic skin substitutes: an overview. Indian J Plast Surg 2010;43(Suppl.):S23–8.

[137] DeVries HJC, Mekkes JR, Middelkoop E, Hinrichs WLJ, Wildevuur CHR. Dermal substitutes for full-thickness wounds in one-stage grafting model. Wound Rep Reg 1993;1:244–52.

[138] Gravante G, Delogu D, Giordan N, Morano G, Montone A, Esposito G. The use of Hyalomatrix PA in the treatment of deep partial-thickness burns. J Burn Care Res 2007;28(2):269–74.

[139] Myers SR, Partha VN, Soranzo C, Price RD, Navsaria HA. Hyalomatrix: a temporary epidermal barrier, hyaluronan delivery, and neodermis induction system for keratinocyte stem cell therapy. Tissue Eng 2007;13(11):2733–41.

[140] Kirsner RS, Warriner R, Michela M, Stasik L, Freeman K. Advanced biological therapies for diabetic foot ulcers. Arch Dermatol 2010;146(8):857–62.

[141] Carlson M, Faria K, Shamis Y, Leman J, Ronfard V, Garlick J. Epidermal stem cells are preserved during commercial-scale manufacture of a bilayered, living cellular construct (Apligraf(R)). Tissue Eng Part A 2011;17(3–4):487–93.

[142] Schurr MJ, Foster KN, Centanni JM, Comer AR, Wicks A, Gibson AL, et al. Phase I/II clinical evaluation of StrataGraft: a consistent, pathogen-free human skin substitute. J Trauma 2009;66(3):866–73.

[143] Schurr MJ, Foster KN, Lokuta MA, Rasmussen CA, Thomas-Virnig CL, Faucher LD, et al. Clinical evaluation of NIKS-based bioengineered skin substitute tissue in complex skin defects: phase I/IIa clinical trial results. Adv Wound Care (New Rochelle) 2012;1(2): 95–103.

[144] Frykberg RG, Marston WA, Cardinal M. The incidence of lower-extremity amputation and bone resection in diabetic foot ulcer patients treated with a human fibroblast-derived dermal substitute. Adv Skin Wound Care 2015;28(1):17–20.

[145] Harding K, Sumner M, Cardinal M. A prospective, multicentre, randomised controlled study of human fibroblast-derived dermal substitute (Dermagraft) in patients with venous leg ulcers. Int Wound J 2013;10(2):132–7.

[146] Gallico III GG, O'Connor NE, Compton CC, Kehinde O, Green H. Permanent coverage of large burn wounds with autologous cultured human epithelium. N Engl J Med 1984;311:448–51.

[147] Campanella SD, Rapley P, Ramelet AS. A randomised controlled pilot study comparing Mepitel(R) and SurfaSoft(R) on paediatric donor sites treated with Recell(R). Burns 2011;37(8):1334–42.

[148] Wood FM, Giles N, Stevenson A, Rea S, Fear M. Characterisation of the cell suspension harvested from the dermal epidermal junction using a ReCell(R) kit. Burns 2012; 38(1):44–51.

[149] Boyce ST, Simpson PS, Rieman MT, Warner P, Yakuboff KP, Bailey JK, et al. Autologous engineered skin substitutes reduce mortality and harvesting of skin graft donor sites for closure of extensive, full-thickness burns. J Burn Care Res 2015;36(3):S107.
[150] Morgan BA. The dermal papilla: an instructive niche for epithelial stem and progenitor cells in development and regeneration of the hair follicle. Cold Spring Harb Perspect Med 2014;4(7):a015180.
[151] Chueh SC, Lin SJ, Chen CC, Lei M, Wang LM, Widelitz R, et al. Therapeutic strategy for hair regeneration: hair cycle activation, niche environment modulation, wound-induced follicle neogenesis, and stem cell engineering. Expert Opin Biol Ther 2013;13(3):377–91.
[152] Clavel C, Grisanti L, Zemla R, Rezza A, Barros R, Sennett R, et al. *Sox2* in the dermal papilla niche controls hair growth by fine-tuning BMP signaling in differentiating hair shaft progenitors. Dev Cell 2012;23(5):981–94.

Chapter 12

Wound Healing: A Comprehensive Wound Assessment and Treatment Approach

William J. Ennis, Danielle Hill

Section of Wound Healing and Tissue Repair, University of Illinois Hospital and Health Sciences System, Chicago, IL, USA

Chapter Outline

WOUND HEALING, THEN AND NOW

Wound healing has been studied throughout the history of medicine, yet only over the past 50 years has there been a focused, research-based effort to expand the science, knowledge, and treatment options. Guido Majno wrote a comprehensive text on the history of man and wound care, which eloquently traces the evolution of the field [1]. Wound healing is inextricably linked

Skin Tissue Engineering and Regenerative Medicine. http://dx.doi.org/10.1016/B978-0-12-801654-1.00012-7

to military medicine. Throughout history, there have been major advances in products, techniques, and the overall science of wound healing that have emerged from military conflict [2]. Ambroise Pare and Alexis Carrell both advanced the field of wound care during wartime. Many of the techniques used today are simply modifications of their innovative practices [3,4]. Most wound care clinicians, however, cite the 1962 paper by George Winter as the sentinel event that led to advanced wound care [5]. This was the first time the concept of moist wound healing was proposed as an alternative to traditional wet-to-dry dressings and allowing wounds to remain open to the air. Despite the positive reviews of this paper, it took almost 20 years for the first advanced wound care products to emerge on the market. By the late 1980s, many clinicians were using hydrocolloid dressings, but physicians mostly deferred to the recommendations made by wound nurses. In the late 1990s and into the new millennium, advances in cell biology, tissue engineering, and genetics along with an explosion of outpatient wound markets, created a new market and field of medicine known as wound healing and tissue repair. Currently, wound care is an 18 billion dollar worldwide market with an expected compound growth rate of 7% [6]. Reasons for the explosion of the wound care market space include, but are not limited to, an aging society, the epidemic of obesity and diabetes, increases in chronic medical conditions, and an expansion of surgical procedures.

WOUND HEALING PROCESS

Wound healing is a complex process that requires the interaction of cells, cytokines, mechanical forces, and a functional scaffolding to allow for the interactions to take place. The functions of the skin, the largest organ in the body, are to protect us from the environment, allow for thermoregulation, fluid exchange and sensation, to act as a reservoir for vitamin D, and provide immunologic functions. The process of healing begins immediately after injury and is classically described as a process that proceeds through a number of interrelated sequential phases. These phases (hemostasis and coagulation, inflammation, proliferation, and remodeling) are described individually for teaching purposes, but overlap in vivo. The adult human heals a skin wound through the process of tissue repair. This process results in the formation of a scar that will only generate 70% of the tensile strength of the original tissue [7]. Alternatively, the process of regeneration results in the formation of tissue that is identical in form and function. There has been a tremendous interest in trying to achieve "scarless" healing, similar to that noted in the human fetus prior to 24 weeks gestation [8]. First described in the late 1970s, the physiology of fetal healing has been evaluated in an effort to design products and techniques that could recapitulate the regenerative healing process [9]. Factors that distinguish fetal from adult healing include a reduced inflammatory phase, increases in hyaluronic acid, molecular changes that include reduced transforming growth factor beta (TGFB) 1 and increases in TGFB3, and reduced collagen deposition [10]. Wound care products have been developed using amniotic membrane and hyaluronic acid in an effort to reproduce the fetal environment and decrease scar formation but have yet to completely reproduce fetal physiology in the adult wound healing process [11,12].

The first phase of wound healing involves hemostasis and coagulation. The formation of a platelet plug allows for a provisional matrix to be formed. This provisional scaffolding provides a temporary architecture for the recruitment of cells and soluble factors, while also initiating the inflammatory phase of healing. There is an initial vasoconstriction for 5–10 min post-injury, which assists in minimizing blood loss. This process is followed by a vasodilation, which allows for the ingress of cells and factors that assist in the healing cascade [13]. During the hemostasis phase, fibroblasts begin to transform into myofibroblasts, and the process of angiogenesis begins. This phase is clinically identified by an edematous wound bed and a local hyperemia around the perimeter of the wound, which is often clinically misdiagnosed as cellulitis.

The inflammatory phase is characterized by neutrophil recruitment in the early phase and macrophage migration and transformation in the late phase [14]. A chronic wound has been defined as one that does not progress through an orderly process in an expected time frame [15]. Most chronic wounds are thought to be stuck in the inflammatory phase of healing. This ultimately leads to delayed healing and excessive scar tissue with low tensile strength [16]. The chronic inflammatory microenvironment can also phenotypically alter macrophages from a pro-healing state into an inflammatory status [17].

During the proliferation phase there are many processes that are being coordinated at the same time. The angiogenesis process continues to mature, providing necessary nutrients to the area, and the wound edges begin the process of epithelialization. The restoration of the vascular system occurs through both angiogenesis and the de novo generation of blood vessels via the process of neovascularization. Endothelial cells from existing vessels detach from their basal lamina and migrate into the surrounding tissue in a process known as sprouting. These new sprouts ultimately form tubular canals that combine to form vascular loops [14]. The vessels mature into venules and arterioles and are further stabilized by surrounding pericytes. The vessels initially form an inner ring, directly at the wound edge, and are fed by radially located vessels in an outer location. This process leads to the above-described periwound hyperemia, which has been demonstrated in vitro via intravital fluorescence microscopy [18]. The vessels are initially immature and dilated, but with time, they contract and regress as the healing process progresses. New technology has allowed investigators to track this phenomenon in real time through the use of indocyanine green dye and a coupled camera system [19]. Epithelialization occurs from the edge of the wound and from epithelial stem cells that arise from hair follicles and/or sweat glands [20]. After eliminating hemidesmosome and desmosome connections that connect the epithelial cells to the basement membrane and each other, respectively, the epithelial cells migrate across the wound bed's provisional matrix along chemotactic gradients until they touch each other and stop migrating through the process of contact inhibition [21]. Granulation tissue formation occurs later in the proliferative phase and is characterized by beefy red tissue that is easily traumatized. The dominant cell types include fibroblasts, macrophages, and granulocytes. Myofibroblasts contribute to wound contraction and this process is ended via the process of apoptosis [22].

Lastly, wound healing proceeds through the process of remodeling. This can last upward of one year and results in a mature wound that is less vascular and cellular than during the other phases of healing. Collagen 3 is replaced over time by the stronger collagen 1, and the pattern of collagen deposition is altered to provide maximal tensile strength [23]. Ultimately, the healed wound is composed of scar tissue, absent skin appendages, and has less tensile strength than the surrounding native tissue. This is why the recidivism rate is so high for almost all wound etiologies. It also explains how chronic wound care adds a tremendous cost burden to the healthcare system. In the current payment model of fee-for-service, the clinicians and healthcare facilities are not economically incentivized to provide cost-effective, high-value wound care. As the system migrates to bundled payments and pay-for-performance, there likely will be a migration toward more effective therapies that provide longer-term solutions with low recurrence rates. These techniques and products may actually be more expensive in the short term but will provide a greater return on investment in the long run. Creating products and techniques that limit inflammation, decrease scar formation, and improve the structural integrity of the healed wound will be in great demand.

FACTORS IMPACTING WOUND HEALING

The above description of the normal healing process sets the stage to analyze potential problems that can arise and lead to the development of a nonhealing, chronic wound. Classically, the factors impeding healing are divided into three distinct categories (Figure 1). Local factors are those that impede the healing process at the actual wound bed level. These include, but are not limited to, the presence of bacteria, circulation defects, or a foreign body, moisture balance issues, nutrient supply, and oxygen levels. Systemic factors are those that affect the individual as a "whole" and therefore impact the wound indirectly. Comorbid illnesses, nutrition, obesity, and advancing age are cited as examples. The third category includes those that are iatrogenic, or clinician induced. The medications used, dressings applied, and modalities utilized all impact the healing process. It is the feeling of the authors that the cause of a nonhealing wound is frequently a combination of items from all three categories.

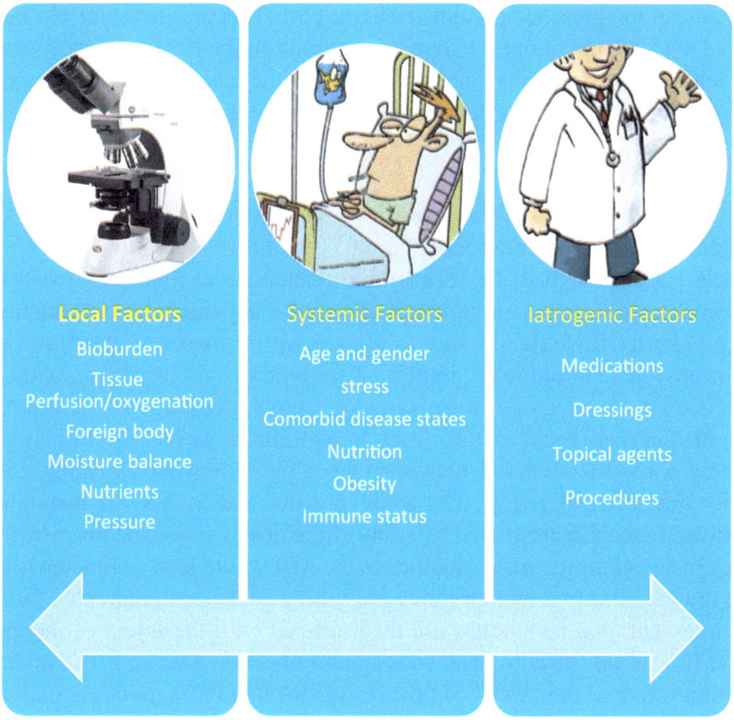

FIGURE 1 Factors impacting wound healing.

Local Factors

Bio-burden

Bacteria are present in all chronic wounds. There is a natural balance between the quantity of bacteria present (bio-burden) and the host's immune status. Any chronic wound will be colonized with bacteria, but if the inoculum of bacteria is increased (>10^6 organism/g tissue) or the host suffers a decrease in immunity, clinical infection may occur [24]. Many examples are cited in the literature describing the failure of skin grafts, delayed closures, and overall wound healing problems when the bacterial bio-burden exceeds 10^5 organism/g tissue [25]. The bacteria compete for nutrients and oxygen with host repair cells in the granulating bed, and when there is equilibrium between the host and the bacteria, a condition known as critical colonization may occur [26]. This can be diagnosed clinically when a previously healing wound (as determined by area reduction) fails to progress along the healing trajectory. Oftentimes, the use of a local therapy targeted at bio-burden reduction can reestablish a healing profile again [27]. Wound bio-burden and infection must be thought of as a continuum and not a point in time when a specific number of organisms are present. This concept is important to emphasize, because the presence of granulation tissue actually depends on the presence of some bacteria [28]. This discussion of bio-burden should not lead practitioners to blindly prescribe antibiotics for chronic wounds. The authors have previously published data on the potential negative impact of using antibiotics without a definitive diagnosis of wound infection [29].

Tissue Perfusion

The circulation can be divided into two distinctive vascular beds: the macrocirculation and the microcirculation. The macrocirculation refers to all vessels large enough to be seen by the unaided eye. Since the two vascular beds are connected in series, it should be apparent that a reduction in macrovascular flow should lead to a decrease in microvascular flow. What is not obvious, but clinically important for wound care providers, is that the presence of an adequate macrovascular supply does not always translate into adequate microcirculatory flow. In addition, a patient might be able to develop a robust collateral network that allows for wound healing despite a poor macrovascular flow. The ankle–brachial index (ABI) is the standard noninvasive examination used by many clinicians to assess the macrovascular status in an extremity. Abnormalities in the ABI would lead a clinician to evaluate the vascular tree by more invasive testing such as angiography. Surgical and interventional techniques are then utilized with the intent of increasing blood flow to an extremity and subsequently healing a wound. The microcirculation refers to the "web" of tiny vessels located throughout the body. The skin and chronic wounds are dependent on the microcirculation for oxygen, nutrients, and to eliminate metabolic wastes. Oxygen is a critical element that all healing tissues require. Decreased quantities of oxygen lead to decreased bacterial killing by leukocytes, decreased collagen production, and decreased epithelialization. There are, however, some transient beneficial effects of low oxygen, such as the stimulation of pro-angiogenic factors to be released from macrophages [30].

The concept of hypoxemia (decreased arterial oxygen) is different from hypoxia (decreased tissue oxygen content). Most healthcare providers are taught that the oxygen bound to hemoglobin is the most critical for wound healing. This concept is valid for metabolically active tissue like muscle but not completely appropriate for low oxygen consuming tissue like chronic wounds [31]. Oxygen extraction ratios in healthy volunteer wounds have been shown to be as low as 3% [32]. Therefore, the oxygen dissolved in plasma can be adequate to provide for wound healing, assuming tissue perfusion is adequate [33]. This concept is the basis for therapy utilizing hyperbaric oxygen [34]. Further support for this concept is found in the literature, which describes adequate healing in anemic patients [35]. The goal when treating a patient with chronic external stimuli such as hypothermia, stress, and pain, is to identify modifiable factors in order to decrease sympathetic tone resulting in increased tissue perfusion. Smoking, through the action of nicotine, can also result in decreased microcirculatory flow [36]. Several medications including beta-blockers can also adversely affect the microcirculation, while others (pentoxifylline) can be used to augment microcirculation and tissue perfusion [37]. These multiple factors help explain why local, systemic, and clinician-induced factors all combine to potentially affect the wound healing process at the cellular level.

Desiccation, Eschar, Foreign Body

Wound desiccation and eschar formation interfere with the healing process and should be prevented if possible. The impeding mechanism involves death of cells, inactivation of growth factors, a hostile local milieu for cellular functioning, and decreased epithelial migration under the eschar. There are several types of foreign matter that are frequently present in chronic wounds. Some items (mesh, sutures) are placed there by clinicians, and others (eschar, necrotic slough) appear naturally. The presence of necrotic tissue in a wound can significantly impair healing. Necrotic slough can provide a fertile environment for bacterial growth. Depending on the chemical composition of the debris, toxic metabolites can be released from the action of the bacteria on the slough material. The presence of necrotic debris in a wound bed slows the process of granulation tissue formation and wound contraction. If the debris is allowed to desiccate then the newly formed hardened eschar further impedes contraction by "stenting" the wound edges apart. Debridement should be performed to remove all necrotic slough, debris, and eschar. The debridement, however, must be followed by utilizing moist dressings that prevent desiccation from recurring and creating a cycle of repeated debridements.

The choice of sutures for primary or delayed secondary closure can impact wound healing. The amount of inflammation, incidence of infection, and cosmetic appearance of a wound are all, to a degree, determined by the choice of suture material used [38]. Specific surgical techniques including choice of closure, knot security, suture tension, tissue bite, and the placement of sutures outside the healing zone of a wound all impact healing [39].

Systemic Factors

Stress

The biochemical outcomes from various sources of stress are known to impede wound healing. This topic is included in the category of systemic factors but clearly impacts the wound healing process at a local level. The sympathetic nervous system and the adrenal axis play the major roles in the hormonal response to stress. Catecholamines, norepinephrine, and epinephrine are released in response to stress. The ultimate result of this hormone release is vasoconstriction and decreased tissue perfusion. In addition to decreased perfusion, evidence exists that elevated hormones act as fibroblast growth inhibitors, resulting in less granulation tissue formation [40]. There is also an immunoregulatory effect on lymphocyte cell lines with the hormonal effects of stress [41]. Laparoscopic cholecystectomy, for example, results in less stress as measured by cortisol and catecholamine levels than the standard open surgical procedure [42].

Psychological Stress

Psychological stress, much like surgical stress, can result in cellular dysfunctions [43]. This connection was proven in a report by Kiecolt-Glaser [44]. In this experiment, caregivers who were under stress underwent punch biopsies and received standard moist wound care. The caregivers demonstrated slower healing than matched controls who were not under stress. The stress of a chronic disease such as a nonhealing wound, therefore, may factor into a poor clinical response requiring the wound care provider's attention.

Obesity

The prevalence of obesity is increasing both in the United States and throughout the world. Obese patients have an increased risk of wound complications, including skin and wound infections, dehiscence, and seroma/hematoma formation [45]. Complications can arise in both bariatric and nonbariatric surgical procedures [46]. The relative hypoperfusion that occurs in subcutaneous tissue makes the tissue more vulnerable to pressure and infection and results in poor tissue penetration of antibiotics when they are used for treatment. Obesity can be linked to depression, stress, and anxiety that, taken together, negatively impact the immune system [47]. Recently, it has also been reported that adipose tissue can secrete a large amount of bioactive molecules including cytokines, chemokines, and hormones that are known to increase the overall inflammatory status of the patient [48].

Nutrition

In general, protein-calorie malnutrition is a more common problem in chronic wound care than an isolated nutrient or vitamin deficiency. A patient with a wound has a physiologic stress, with a subsequent neurohormonal response, not unlike that of the many other "stressors" already mentioned in this chapter. The hypermetabolic state that results from malnutrition is an adaptive response, which results in a significant increase in basal energy expenditure. Nutritional issues are as equally important when reviewing patients who are "at risk" for wounds as they are when planning therapy for a patient who already has a wound. Energy, carbohydrate, protein, fat, vitamin, and mineral metabolism all can affect the healing process [49]. There are many studies that show a relationship between the presence of ulcers and malnutrition, however, limited cause and effect, prospective data exist [50]. Research efforts continue in the field of nutritional supplementation and repletion therapeutics targeted specifically for wound care patients [51,52].

Comorbities

Comorbid illnesses can negatively impact the healing process in patients with chronic or acute wounds. It is of vital importance that the wound care clinician conduct a thorough history, physical examination, and appropriate laboratory testing prior to initiating wound care. This will ensure an accurate diagnosis for the wound etiology. Frequently, systemic therapy must be used in parallel with the local wound care, in order to achieve healing. Almost all medical conditions can impact some aspect of healing, and there are obviously too many to review in this chapter. A helpful approach to this seemingly overwhelming challenge is to utilize the LCD (least common denominator) model (Figure 2). Once you have established the diagnosis, evaluate the potential impact of each of the six components on the potential for healing. This approach helps you strategize your treatment without overlooking the impact of the various comorbid illnesses at the local level. Diabetes is a known risk factor for altered wound healing. Diabetes affects the macrovascular system and thus impacts the microcirculation as well, since these systems are connected in series. At the local level, this results in decreased tissue perfusion and oxygenation. Red blood cells become "sticky"

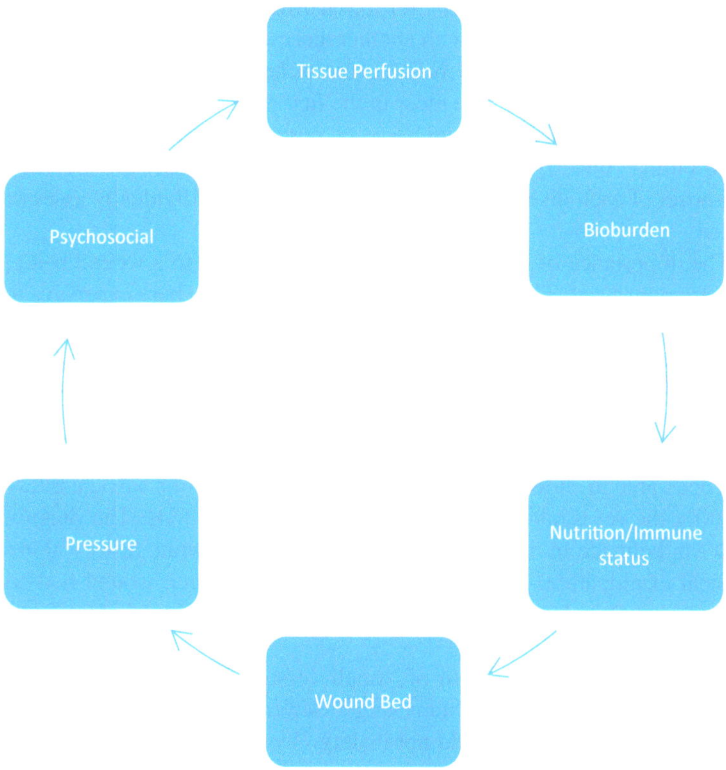

FIGURE 2 Least common denominator model.

and the subsequent sluggish flow adds to the decrease in perfusion through the microcirculation. Over time, the peripheral neuropathy associated with diabetes can lead to a loss of protective sensation on the plantar aspect of the feet. This sensory loss can result in callus formation and subsequent ulcerations, if careful follow-up is not carried out. Excess pressure applied to the plantar surface of the foot during ambulation is exacerbated by the structural and biomechanical abnormalities that occur as a result of a concomitant motor neuropathy. Taken together, the structural defect and sensory loss set the diabetic patient up for significant problems with foot ulcerations (pressure/neuropathy). At least 50% of the amputations performed each year in the United States are secondary to complications from diabetes [53]. Dressing selection, timing, and characteristics all need special attention when treating a diabetic patient.

Arterial Insufficiency

The presence of peripheral vascular disease (PVD) is a major comorbidity for wound healing. PVD can be present alone or in combination with any other disease state. It is imperative to assess the arterial circulation in all patients with chronic wounds, even though a "classic" appearance might be present for another cause (i.e., venous ulcer). It is often difficult to assign a wound etiology in cases where PVD coexists with another disease state. For example, a diabetic patient with a foot ulcer may have severe ischemia in addition to the neuropathic condition that led to the ulcer in the first place. In our clinic, we assign a primary etiology for each wound based on our clinical decision as to the major "contributing" underlying condition. All other comorbidities are recorded, and the impact of each disease on the LCD parameters is individually assessed and treated.

The importance of blood flow and oxygen delivery to a wound bed cannot be overemphasized. The authors have seen many cases where expensive topical medications and dressings have been used for a wound located on an ischemic limb. Clearly, without blood flow, there will be no healing. There has been a recent interest in describing the blood flow to the wound based on the angiosome theory. The goal of this approach is to provide specific blood supply to the anatomic region on the body where the wound is located. Some investigators have been able to show statistically higher healing and limb salvage rates when targeting the revascularization process in this manner [54]. The diagnosis of PVD can be made in the majority of cases with an accurate history alone. The physical examination, including the location of the ulcer, wound bed characteristics, and Doppler examination should confirm the suspicions raised in the history [55].

Logerfo dispelled the myth of "small vessel" disease in diabetics with a landmark paper in 1984 [56]. Many surgeons were previously under the impression that diabetic patients could not undergo bypass procedures because of the presence of atherosclerosis in the small vessels of the foot. There are now a myriad of open and endovascular options to treat patients with arterial disease as far distal as the foot. Teams are being created among podiatry, wound care, and vascular surgery in order to provide comprehensive efforts at limb salvage, with results that exceed those generated by individual practitioners [57].

Venous Insufficiency

Complications can arise on the venous side of the circulatory system as well, which also lead to potential leg ulcer formation. The pressure in the venous system should cycle up and down with normal ambulation and calf muscle pump function. When the pressure remains elevated (i.e., with venous valve incompetency, sedentary lifestyle, or venous obstruction/clot), hemodynamics are altered, and a cascade of events occur leading to decreased tissue oxygenation and ultimately skin ulceration. Theories describing the pathogenesis include fibrin cuffing, white blood cell trapping, growth factor trapping, and reperfusion injury [58–61].

An evaluation of the patient with a suspected venous ulcer should include a venous duplex scan, with air plethysmography (APG) added for functional assessments. An ABI should also be obtained to rule out underlying concomitant arterial disease, as this may significantly alter the treatment protocol (compression therapy used with caution with underlying arterial disease).

In summary, many diseases can impact wound healing. The clinician needs to be aware of all comorbid illnesses and analyze the impact each may have alone, or in combination, on the various phases of healing. Many diseases have a direct effect (e.g., vascular disease), while others manifest their effects through secondary means (e.g., liver and renal failure resulting in excess edema).

WORK-UP OF A PATIENT WITH A NONHEALING WOUND

The physician who deals with a recalcitrant wound must be skilled in communication, negotiation, and change management in order to coordinate an interdisciplinary team of specialists focused on a unified objective. The physician must possess an awareness and understanding of surgical concepts and treatment outcomes in plastic, orthopedic, vascular, and general surgery in order to make appropriate, timely referrals during the continuum of care. Regardless of the physician's primary background, the wound clinician must incorporate knowledge and skills from both surgical and medical fields. The physician must also command knowledge of dermatology, rheumatology, endocrinology, and general medicine in order to integrate the management of chronic systemic disease states into local wound care therapy. Communication with patients and family is paramount. An awareness of the social, psychological, spiritual, or existential concerns that are often involved with the possibility of limb loss and/or loss of independence and functionality is required to approach the wound patient in a holistic way. Many times a patient has already received conflicting therapeutic recommendations from other healthcare professionals, leading to frustration that can easily be misdirected toward the wound physician. Given the broad base of knowledge and skills required, coupled with a paucity of clinical experience most physicians have in wound care, the practice of wound management is as challenging, demanding, and complex as that of any other field in medicine. Clinical training programs need to be developed at the medical school and post-graduate levels, in order to address these deficits [62]. Fellowship programs are being developed currently, with the ultimate goal of creating a certifying process that allows the field to grow as an independent specialty [62,63].

Much information can be obtained by simply watching the patient walk into the clinic. Information concerning general appearance, body type, age, absence of a limb, ethnicity, and mobility can all be evaluated before the formal examination begins. The patient may deny pain verbally, but their body position and facial expressions may indicate otherwise. The formal process begins with a comprehensive history and review of systems. Wound care is not unlike other specialties; in that for a majority of cases, a working diagnosis can be achieved with a thorough history. Past medical history is important because many comorbid illnesses exist in a complex wound care patient. Most patients with rheumatoid arthritis (RA) and a leg ulcer on the medial side of the leg, for example, will be venous in origin, but the knowledge that the patient has RA will allow the clinician to consider a vasculitic wound in the differential diagnosis [64]. A history of coronary artery disease in a patient with a venous ulcer should alert the clinician to the possibility of PVD and a mixed etiology for the leg ulcer. This has implications for the work-up, expected time to healing, and the level of therapeutic compression to be used [65].

The social history will address confounding variables such as alcohol use, tobacco use, illicit drug use, occupational history, and level of activities. The patient must be questioned about the presence of any allergies, both to medications, environmental factors, and foods. A detailed list of all medications taken by the patient is also important. With an increasing number of patients using herbals, vitamins, homeopathics, and supplements, it is important to emphasize to the patient that all medications should be listed. It is well known that steroid use can negatively impact healing, but more recently, the commonly prescribed COX-2 inhibitors, ACE inhibitors, and calcium channel blockers have been implicated in slowing angiogenesis, a necessary component for normal healing [66,67]. A family history will help identify risk factors and might shed light on the patient's diagnosis; as oftentimes, the patient will not recognize the correlation between a disease in their family history with their own current problem.

A complete past surgical history is important. Patients with prior venous surgery are known to have wounds that are harder to heal [68]. Knowledge of this fact might influence the clinician to utilize more advanced technologies earlier in the treatment regimen to provide appropriate care beyond that considered standard of care for that wound type [69].

The patients' and their families' desired outcomes from the treatment program should be thoroughly understood and agreed upon from the onset of treatment.

For example, an elderly female with a highly exudative venous ulcer may be most concerned with controlling the drainage. Although the patient might like the wound to heal, their primary concern for exudate management might explain why she became more noncompliant as the treatment program progressed and "their" outcome was reached. In addition, there is a growing population of patients whose wounds, whether by choice, terminal illness, or chronic conditions, have reached a steady state and are not able to achieve healing. This group of patients might be better served with a palliative wound protocol that emphasizes the surrogate end points of pain control, exudates, and odor management, infection control, and quality of life outcomes rather than total healing [70,71].

Finally, a history of present illness (wound history) should be recorded. This part of the history needs to occur after all of the previously mentioned information has been gathered. The wound history should include the timing of ulcer occurrence, location, prior ulcer history, current and past treatments, size and appearance, complications, procedures performed, drainage, and pain.

After the history the clinician should begin a physical examination. Vital signs, a pain scale score, height, weight, and a comment on general appearance should be recorded. The blood pressure should be taken in both arms and legs at the initial visit, in order to calculate an ABI. A complete exam frequently uncovers findings that aid in the final diagnosis but also ensures appropriate documentation. After the complete history and physical exam is performed, the clinician should turn their attention to the wound itself. It is important to let the patient know that you intend to conduct a complete exam at the onset. Many patients assume they will simply have to remove their bandage and show the clinician their wound. They occasionally will become angry if much time is spent on what they perceive as "unnecessary" components to the office consultation. Frequently, they do not understand how important it is to get a complete medical picture before embarking on a wound treatment program.

The examination of the wound takes into account the important issues of wound location, drainage quality and quantity, surrounding skin condition, quality of wound bed tissue (e.g., granular, fibrinous, eschar, and so on), pain, odor, and condition of the wound perimeter (i.e., presence of undermining, and so on), as well as standard measurements (length, width, depth).

The wound work-up next focuses on what has been termed "the LCD model" (Figure 2) [29]. Composed of six subsections (tissue perfusion/oxygenation, infection, nutrition/immune status, psychosocial, pressure/neuropathy, and wound bed), this model ensures that the clinician consider all factors that impact healing for a wound of any etiology. Although nutrition and psychosocial aspects have been covered in the history and physical exam, these concepts are now revisited as they specifically relate to the wound and the potential for healing.

The most important aspect of the LCD model is the adequacy of tissue perfusion/oxygenation. It is imperative that tissue perfusion is analyzed within the paradigm of both the macro- and microvascular status. The initial evaluation focuses on the macrocirculation, with the palpation of peripheral pulses. Ankle pulses are, however, not sufficient to detect impaired arterial circulation, and additional testing is frequently required for the patient with leg ulcerations [72]. Performing an ABI is useful not only for wound healing prediction but also as an overall marker for cardiovascular health [73]. Only about 10% of patients presenting to an outpatient wound clinic will have isolated arterial disease as an etiology for their leg ulcer [74]. Arterial duplex scans, segmental pressures including toe pressures, pulse volume recordings, magnetic resonance angiography, and rapid sequence CT scans are other noninvasive macrocirculation studies that may be ordered. After a complete assessment of the macrocirculation, the clinician's attention must be turned to the status of the microcirculation. Adequacy of macrovascular flow does not ensure healing will occur.

Transcutaneous oximetry (TcPO_2) is a technique widely used to detect the skin oxygen tension. A dime-sized, Clarke-type, solid-state, polarographic electrode containing a platinum cathode with a reference electrode of silver chloride is housed in a probe tip along with a heater and thermistor. This procedure indirectly evaluates total microcirculation, without the ability to differentiate nutritive from the nonnutritive flow. Nutritive flow refers to the flow that relies on the perpendicular capillary flow toward the skin surface, not the parallel flow that is seen in the dermis and functions mainly as a shunt for thermoregulation. Some investigators feel TcPO_2 is more accurate than the ABI as a method to follow changes after a revascularization [75].

Laser Doppler perfusion imaging is a technique which utilizes a low-intensity laser (helium–neon) light [76]. The device measures the backscattering created by moving red blood cells over a specific rectangular area analyzing up to 4096 individual points. The wavelength of this monochromatic light is 670 nm, with a maximum accessible power of 1 mW. Two parameters of the returning laser light from the skin are analyzed. The first is the number of shifted photons, which relates to the concentration of moving red blood cells. The second is the mean Doppler shift of these photons, which relates to the blood cell velocity. These Doppler shift results are accumulated and translated into numeric values expressed in volts, and in an image colored map. Colors range from dark blue to burnt red and indicate minimum and maximum perfusion, respectively. The penetration of the laser beam reaches 500 μm when applied to intact skin; however, penetration can reach 2.5 times greater in other non-skin tissues like granulation tissue [77]. The shallow penetration of the laser beam allows for a closer approximation of nutritive flow and may allow for the quantification of the angiogenesis process in granulation tissue [78]. Using laser Doppler technology, the skin perfusion pressure can be measured. This technique is being used to assess pre- and post-revascularization success at the tissue level [79]. Other tests of the microcirculation include positron emission tomography, nuclear magnetic resonance spectroscopy, xenon washout, and

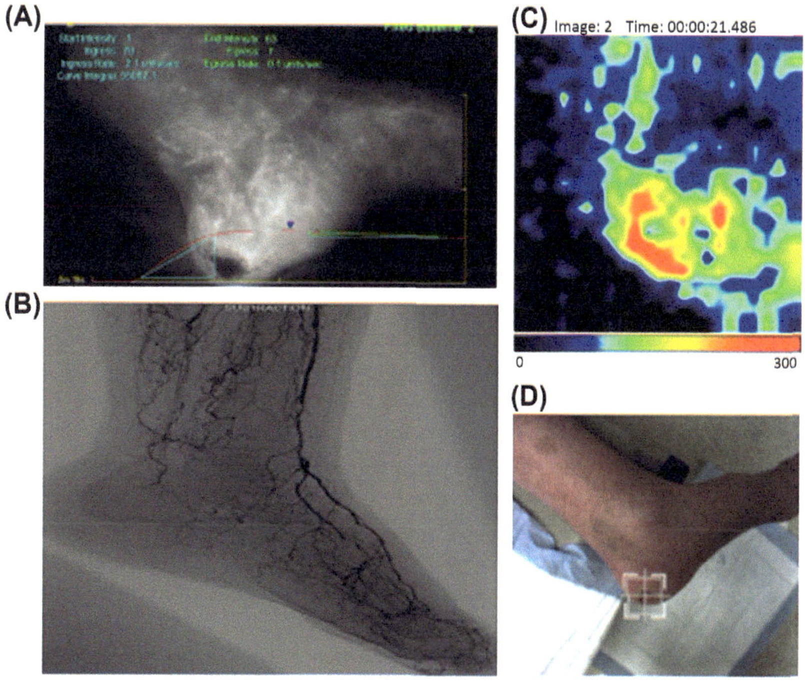

FIGURE 3 Heel ulcer. (A) Indocyanine green perfusion scan of microcirculation peri-wound. (B) Angio of the same patient demonstrating absence of macroflow. (C) Laser Doppler image showing high flow at wound margin. (D) Clinical picture showing wound bed eschar.

near infrared spectroscopy. Recently, there has been much attention paid to the use of indocyanine green dye in the outpatient setting, as a method to assess the microcirculation [80] (Figure 3).

If a patient has a leg ulcer and abnormal macrovascular flow studies but physiological studies (i.e., TcPO_2) indicate adequate values for healing, then a trial of aggressive local wound care is warranted. If, after a 4-week treatment course, there is no significant improvement in either wound dimensions or quality of tissue, then further invasive studies followed by revascularization might be necessary. Approaching the wound patient in this manner will avoid unnecessary high-risk procedures, and the limited treatment time will minimize potential harmful outcomes. If the microcirculatory studies are abnormal, then a trial of treatments aimed at enhancing the microcirculation (e.g., electrical stimulation, growth factors, bioengineered tissue, or therapeutic ultrasound) could be used, along with aggressive wound care for a short course. Systemic therapy and life style modifications should also be employed [81].

The second component of the LCD model refers to the burden placed on the host by the quantity and quality of bacteria located in the wound bed. This has been described previously in this chapter. A relatively recent area of focus in wound healing has been on the role of biofilms in nonhealing wounds. Biofilms have been described as a colony of bacteria sequestered in a glycocalyx that renders them impermeable to antibiotics [82]. They can be surgically removed through debridement but frequently penetrate deeper into the wound bed than previously thought and recur rapidly without additional therapy [83]. Cultures are notoriously not useful when trying to culture a biofilm, and many clinicians are beginning to utilize polymerase chain reaction (PCR)-based techniques to identify the entire bio-burden within a wound bed.

The third category within the LCD model focuses on the immunological state of the patient. Malnutrition is a major factor to consider here. Although the literature fails to provide us with statistically significant relationships between healing and nutritional status, it is obvious that a patient needs to be satisfied nutritionally in order to maximize their chances for healing. Patients are not considered for surgical wound closure until they have achieved nutritional support and lab testing confirms success (i.e., pre-albumen level, serum transferrin level, total lymphocyte count, albumen level, and so on). Every attempt is made to support patients with multivitamins, minerals, and enteral or parenteral support. Patients should be evaluated for other forms of immunosuppression such as the use of steroids, antimetabolic agents, overwhelming infection, and chronic disease states such as diabetes and HIV.

Psychosocial issues were discussed earlier in the chapter but should be reexamined at this point. Chronic depression has been shown to affect healing, and many wound patients suffer from psychiatric conditions if the clinician carefully probes during the history [84]. Pain can exacerbate underlying psychiatric issues and can delay healing in and of itself through the psycho-neuro-immunological connections [85]. As society ages and more patients live with chronic diseases, we will be faced with patients in whom healing options are limited and in whom maintenance of the wound bed, prevention of infection, exudates management, odor control, and wound pain issues will take priority over healing [86].

The LCD next addresses pressure on the wound and surrounding tissues. Pressure must be offloaded in order to maximize healing. This seems obvious when dealing with classic pressure ulcers located on the trunk but also applies to pressure from wheelchair leg rests, bed railings, oxygen tubing, and improperly fitting shoe-gear. There are many products available to offload the patient including mattresses, orthotics and prosthetics, and foam padding for wheelchairs and beds.

The wound bed is the last aspect of the LCD model. The peri-wound tissue should be assessed first. The skin can be painful to the touch, erythematous, macerated, dry and cracked, or edematous. A patient's continence status may play a pivotal role in wound healing and should be noted in the record. The wound bed must be assessed for the state of moisture balance, presence of necrotic tissue, and quality of granulation tissue. If the wound has been present for greater than 6 months to a year and/or has an abnormal appearance, then strong consideration should be given to performing a biopsy. Histology can help achieve a diagnosis, rule-out malignancy, or provide confirmation to a clinical suspicion. The threshold for biopsy should be low, as many wounds do not demonstrate classical features on presentation. After a "macroscopic" view of the wound bed, the clinician should consider microenvironmental issues. The microcirculation has already been reviewed and those results need to be documented in the patient's record as well. The biochemistry of the wound is becoming more important to wound healing, as new concepts in healing have been elucidated. A hostile wound environment with excess matrix metalloproteases can lead to the destruction of both endogenous and exogenous growth factors, leading to delayed healing. Point-of-service measurements are now becoming available to assess the matrix metalloprotease levels [87]. The bio-burden of the wound can lead to the presence of metabolic waste products in the wound bed. Inflammatory mediators and cytokines can create an environment that does not allow for proliferation to occur. Interestingly, this situation can develop with adequate macrovascular flow. Consider the patient with a venous ulcer, palpable pulses, but severe lipodermatosclerosis and relative dermal tissue hypoxia. This wound would be difficult to heal and might require modifying the biochemical composition of the wound bed to accelerate healing. Products are being developed to specifically target individual inflammatory molecules [88].

COMMON WOUND ETIOLOGIES/TREATMENT

Venous Leg Ulcers

The most common ulcer in the outpatient setting in the United States is a venous leg ulcer. There is a recently published consensus guideline describing the physiology, diagnosis, and treatment options [89]. The wounds are classically located on the medial ankle region but can be lateral. The patient has clinical signs including hemosiderin-stained skin, lipodermatosclerosis, evidence of varicosities, edema, irregularly shaped wound borders, and often a heavy exudate. The patients have highly sensitized peri-wound skin and are frequently allergic to multiple topical agents and dressings. The wounds are highly recurrent with increasing prevalence in the warm months. The etiology is based on venous insufficiency. This can be in either the superficial, deep, or perforator system or any combination thereof. In addition, recent evidence points to central vein obstruction and stenosis as playing a more prominent role in venous ulcer etiology than previously thought. Work-up follows the LCD model but is centered on the duplex exam, which can detect reflux in the various venous locations. It is of paramount importance to rule-out concomitant arterial disease (mixed ulcer) prior to beginning therapy for patients with venous ulcers. Treatment is focused on providing absorptive dressings and applying compression bandaging to combat the venous insufficiency. Debridement is often required early in the treatment course, and this often needs to be done in the operating room to have adequate pain management. Over the first 4 weeks of therapy, a 25% reduction in area is anticipated. If this is not achieved, additional advanced therapies can be considered including biological skin equivalents, autologous skin grafting and the use of oral pentoxyfylline therapy. Healing occurs in 70–80% of patient within 16–20 weeks, but there is up to a 70% recurrence rate. Recent data point to a potential lowering of wound recurrence with the use of saphenous vein ablation therapy [90].

Arterial Ulcers

These ulcers are usually distal on the foot and toes, although they can be anywhere on the leg. History is compatible with peripheral arterial disease, with claudication or rest pain being common. Wounds are usually small, punched out in appearance, and have little exudate. The pain associated with the wound is out of proportion to the size and appearance of the wound. Dangling the leg in a dependent position often relieves pain. The legs often take on a rubor-like color in dependency and completely blanch with elevation, a clinical sign known as a Beurger's sign. This is diagnostic of critical limb ischemia. Patients often have concomitant coronary disease or carotid disease and require medical clearance prior to initiating any interventional therapy. The work-up includes both noninvasive testing and angiography. Ultimately, the treatment requires a team approach with the patient at the center, in order to design a treatment plan that is optimal for the individual. Decisions concerning open vs. endovascular, angiosome directed vs. indirect revascularization, and the timing of debridements before or after the vascular procedure all must be decided ahead of time. After a successful revascularization, there is up to a 4-week period until the microcirculation stabilizes, and the wound should be carefully managed during this critical time period. Sometimes, the wound actually worsens after revascularization as a result of reperfusion injury. The concept between the revascularization process and the stabilization of the microcirculation has been referred to as the push–pull theory [91].

Pressure Ulcers

Pressure ulcers are caused by excessive external pressure being applied to the tissue over a prominent bone. Shear and friction forces also play a role in the etiology of pressure ulcers. The most common locations include the ishcial tuberosity, sacral region, greater trochanteric process, heel, elbow, and medial and lateral malleoli. Pressure ulcers have become a surrogate marker for quality of care, with prevalence and incidence rates being calculated at all hospitals and nursing homes in the United States. The formation of a pressure ulcer in an acute care hospitalized patient results in a "never" event, with Medicare not allowing any additional payments be attributed to the hospital for the treatment of that complication. Pressure ulcers require the same comprehensive work-up as described before in this chapter; however, there is even a greater emphasis on the nutritional status of the patient. Offloading must be achieved through the use of specialty support surfaces and pads. The wound bed must be debrided of all nonviable tissue, and dressings that provide a moist environment must be utilized. Over the past 10 years, there has been a tremendous use of negative pressure wound therapy to assist in the healing of pressure ulcers [92,93]. The goal standard treatment for pressure ulcers is to provide surgical coverage through the use of simple or advanced flap procedures. Despite excellent technique, there is a 70% recurrence rate for pressure ulcers with surgical closure, making them particularly difficult to heal long term. Spinal cord-injured patients are highly susceptible to these kinds of wounds, and prevention plays a very important role in this patient population.

Diabetic Foot Ulcers

With an ever-increasing number of diabetic and obese patients worldwide, there are a large number of diabetic foot ulcers each year. It is estimated that 15% of diabetic patients will have a wound sometime in their life, and that 85% of all amputations are preceded by a nonhealing diabetic foot wound [53]. The combination of sensory, motor, and autonomic neuropathy along with decreased circulation, creates a perfect environment for a wound to form. Often, due to neuropathy, the diabetic patient does not even know they have a wound, and it is only discovered on inspection or due to drainage in the patient's sock or shoe. Diabetic patients have a decreased ability to control local infection, and many patients have advanced infection on presentation to their clinician. In addition to glucose control, the diabetic patient with a foot ulcer needs advanced local wound care, offloading, debridement, an assessment of the vascular system, and frequent follow-up. Offloading can be provided with the use of custom-molded diabetic shoes, boots, or a total contact cast. Like pressure ulcers, diabetic foot ulcers have a high recurrence rate.

Atypical Wounds

Wounds that are not in classic anatomic locations or have atypical appearances might lead the clinician to categorize the wound as "atypical." These wounds can come from a variety of conditions such as sickle cell anemia, vasculitis, pyoderma gangrenosum, coagulopathy, drug-induced, self-inflicted, malignancy, and infectious diseases, to name just a few. The work-up is the same, but emphasis is placed on specific laboratory testing such as sedimentation rate, C-reactive protein, antinuclear antibody levels, sickle prep, standard chemistry profiles, and complete blood count. Biopsy of both the wound and the surrounding skin can often lead to an accurate diagnosis. Hematoxylin and eosin staining is adequate for the wound bed and the wound edge but intact peri-wound skin is needed for direct immunofluorescence studies to rule-in or rule-out vasculitis. Treatment is focused on local wound care for the ulcer, but more importantly, the systemic treatment of the causative medical condition is what ultimately heals the wound.

FUTURE DIRECTIONS

Much has been learned over the past 50 years in wound care. We have progressed from cover and protect to using products and medications that directly interact with the wound microenvironment. Energy-based wound modalities such as ultrasound, electrical stimulation, and ultraviolet light are useful adjuncts to wound care [94]. The components required to heal a wound include cells, cytokines, chemokines, a provisional matrix, electrical energy, and stretch and strain (mechano-transduction) [95]. Clinicians have at their disposal an almost overwhelming array of options to treat patients today. Without an accurate diagnosis, however, the treatment will not match up with the etiology, and a less than optimal outcome will be achieved. The complexity of the science behind wound healing physiology and treatment continues to grow, but the average medical student receives only 9.2 h of didactic education during their 4 year medical school curriculum [96]. The use of stem cells provides a potentially promising option for wound healing in the near future, with the goal of replicating regeneration instead of tissue repair [97,98]. Improvements in the field of tissue engineering will surely help the clinician obtain optimal outcomes for these complex patients in the future. Patient-centered outcomes such as pain, recidivism, functional status, and overall quality of life will play a central role in driving innovation and therapy in the future.

REFERENCES

[1] Majno G. The healing hand. Cambridge Harvard University Press; 1975.
[2] Gabriel R. Man and wound in the ancient world: a history of military medicine from Sumer to the fall of Constantinople. Washington DC: Potamac Books; 2012.
[3] Pare A. The apologie and treatise of ambroise pare. Chicago: University of Chicago Press; 1952.
[4] Carrel A. The treatment of infected wounds. New York: Paul Hoeber; 1917.
[5] Winter GD. Formation of the scab and the rate of epithelization of superficial wounds in the skin of the young domestic pig. Nature January 20, 1962;193:293–4.
[6] Crandall MA. Advanced wound care markets. 2014. Rockville, MD.
[7] Levenson SM, Geever EF, Crowley LV, Oates 3rd JF, Berard CW, Rosen H. The healing of rat skin wounds. Ann Surg February 1965;161:293–308.
[8] Colwell AS, Krummel TM, Longaker MT, Lorenz HP. An in vivo mouse excisional wound model of scarless healing. Plast Reconstr Surg June 2006;117(7):2292–6.
[9] Rowlatt U. Intrauterine wound healing in a 20 week human fetus. Virchows Arch A Pathol Anat Histol March 23, 1979;381(3):353–61.
[10] Walmsley GG, Maan ZN, Wong VW, Duscher D, Hu MS, Zielins ER, et al. Scarless wound healing: chasing the holy grail. Plast Reconstr Surg March 2015;135(3):907–17.

[11] Koob TJ, Rennert R, Zabek N, Massee M, Lim JJ, Temenoff JS, et al. Biological properties of dehydrated human amnion/chorion composite graft: implications for chronic wound healing. Int Wound J October 2014;10(5):493–500.

[12] Voigt J, Driver VR. Hyaluronic acid derivatives and their healing effect on burns, epithelial surgical wounds, and chronic wounds: a systematic review and meta-analysis of randomized controlled trials. Wound Repair Regen May–June 2013;20(3):317–31.

[13] Robson MC, Steed DL, Franz MG. Wound healing: biologic features and approaches to maximize healing trajectories. Curr Probl Surg February 2001;38(2):72–140.

[14] Reinke JM, Sorg H. Wound repair and regeneration. Eur Surg Res 2012;49(1):35–43.

[15] Lazarus GS, Cooper DM, Knighton DR, Percoraro RE, Rodeheaver G, Robson MC. Definitions and guidelines for assessment of wounds and evaluation of healing. Wound Repair Regen July 1994;2(3):165–70.

[16] Redd MJ, Cooper L, Wood W, Stramer B, Martin P. Wound healing and inflammation: embryos reveal the way to perfect repair. Philos Trans R Soc Lond B Biol Sci May 29, 2004;359(1445):777–84.

[17] Mirza RE, Fang MM, Weinheimer-Haus EM, Ennis WJ, Koh TJ. Sustained inflammasome activity in macrophages impairs wound healing in type 2 diabetic humans and mice. Diabetes March 2014;63(3):1103–14.

[18] Sorg H, Krueger C, Vollmar B. Intravital insights in skin wound healing using the mouse dorsal skin fold chamber. J Anat December 2007;211(6):810–8.

[19] Braun JD, Trinidad-Hernandez M, Perry D, Armstrong DG, Mills Sr JL. Early quantitative evaluation of indocyanine green angiography in patients with critical limb ischemia. J Vasc Surg May 2014;57(5):1213–8.

[20] Lau K, Paus R, Tiede S, Day P, Bayat A. Exploring the role of stem cells in cutaneous wound healing. Experimental dermatology November 2009;18(11):921–33.

[21] Jacinto A, Martinez-Arias A, Martin P. Mechanisms of epithelial fusion and repair. Nat Cell Biol May 2001;3(5):E117–23.

[22] Hinz B. Formation and function of the myofibroblast during tissue repair. J Invest Dermatol March 2007;127(3):526–37.

[23] Gurtner GC, Evans GR. Advances in head and neck reconstruction. Plast Reconstr Surg September 2000;106(3):672–82. quiz 83.

[24] Robson MC, Payne WG, Ko F, Mentis M, Donati G, Shafii SM, et al. Hypochlorous acid as a potential wound care agent: Part II. Stabilized hypochlorous acid: Its role in decreasing tissue bacterial bioburden and overcoming the inhibition of infection on wound healing. J Burns Wounds 2007;6:e6.

[25] Robson MC. Wound infection. A failure of wound healing caused by an imbalance of bacteria. Surg Clin North Am June 1997;77(3):637–50.

[26] White RJ, Cutting KF. Critical colonization–the concept under scrutiny. Ostomy Wound Manage November 2006;52(11):50–6.

[27] Robson MC, Mannari RJ, Smith PD, Payne WG. Maintenance of wound bacterial balance. Am J Surg November 1999;178(5):399–402.

[28] Burke JF. Effects of inflammation on wound repair. J Dent Res March–April 1971;50(2):296–303.

[29] Ennis WJ, Meneses P. Clinical evaluation: outcomes, benchmarking, introspection, and quality improvement. Ostomy Wound Manage November–December 1996;42(10A Suppl.):40S–7S.

[30] Knighton DR, Hunt TK, Scheuenstuhl H, Halliday BJ, Werb Z, Banda MJ. Oxygen tension regulates the expression of angiogenesis factor by macrophages. Science September 23, 1983;221(4617):1283–5.

[31] Hunt TK, Hopf HW. Wound healing and wound infection. What surgeons and anesthesiologists can do. Surg Clin North Am June 1997;77(3):587–606.
[32] Gottrup F, Firmin R, Rabkin J, Halliday BJ, Hunt TK. Directly measured tissue oxygen tension and arterial oxygen tension assess tissue perfusion. Crit Care Med November 1987;15(11):1030–6.
[33] Chang N, Goodson 3rd WH, Gottrup F, Hunt TK. Direct measurement of wound and tissue oxygen tension in postoperative patients. Ann Surg April 1983;197(4):470–8.
[34] Thom SR. Hyperbaric oxygen: its mechanisms and efficacy. Plast Reconstr Surg January 2012;127(Suppl. 1):131S–41S.
[35] Jonsson K, Jensen JA, Goodson 3rd WH, Scheuenstuhl H, West J, Hopf HW, et al. Tissue oxygenation, anemia, and perfusion in relation to wound healing in surgical patients. Ann Surg November 1991;214(5):605–13.
[36] Silverstein P. Smoking and wound healing. Am J Med July 15, 1992;93(1A):22S–4S.
[37] Rendell M, Bamisedun O. Skin blood flow and current perception in pentoxifylline-treated diabetic neuropathy. Angiology October 1992;43(10):843–51.
[38] Spotnitz WD, Falstrom JK, Rodeheaver GT. The role of sutures and fibrin sealant in wound healing. Surg Clin North Am June 1997;77(3):651–69.
[39] Carlson MA. Acute wound failure. Surg Clin North Am June 1997;77(3):607–36.
[40] Saito T, Tazawa K, Yokoyama Y, Saito M. Surgical stress inhibits the growth of fibroblasts through the elevation of plasma catecholamine and cortisol concentrations. Surg Today 1997;27(7):627–31.
[41] Landmann RM, Muller FB, Perini C, Wesp M, Erne P, Buhler FR. Changes of immunoregulatory cells induced by psychological and physical stress: relationship to plasma catecholamines. Clin Exp Immunol October 1984;58(1):127–35.
[42] Karayiannakis AJ, Makri GG, Mantzioka A, Karousos D, Karatzas G. Systemic stress response after laparoscopic or open cholecystectomy: a randomized trial. Br J Surg April 1997;84(4):467–71.
[43] Esterling BA, Kiecolt-Glaser JK, Bodnar JC, Glaser R. Chronic stress, social support, and persistent alterations in the natural killer cell response to cytokines in older adults. Health Psychol July 1994;13(4):291–8.
[44] Kiecolt-Glaser JK, Marucha PT, Malarkey WB, Mercado AM, Glaser R. Slowing of wound healing by psychological stress. Lancet November 4, 1995;346(8984):1194–6.
[45] Guo S, Dipietro LA. Factors affecting wound healing. J Dent Res March 2010;89(3):219–29.
[46] Greco 3rd JA, Castaldo ET, Nanney LB, Wendel JJ, Summitt JB, Kelly KJ, et al. The effect of weight loss surgery and body mass index on wound complications after abdominal contouring operations. Ann Plast Surg September 2008;61(3):235–42.
[47] Wilson JA, Clark JJ. Obesity: impediment to postsurgical wound healing. Adv Skin Wound Care October 2004;17(8):426–35.
[48] Calabro P, Yeh ET. Obesity, inflammation, and vascular disease: the role of the adipose tissue as an endocrine organ. Subcell Biochem 2007;42:63–91.
[49] Arnold M, Barbul A. Nutrition and wound healing. Plast Reconstr Surg June 2006;117 (7 Suppl.):42S–58S.
[50] Pinchcofsky-Devin GD, Kaminski Jr MV. Correlation of pressure sores and nutritional status. J Am Geriatr Soc June 1986;34(6):435–40.
[51] Schols JM, Heyman H, Meijer EP. Nutritional support in the treatment and prevention of pressure ulcers: an overview of studies with an arginine enriched oral nutritional supplement. J Tissue Viability August 2009;18(3):72–9.

[52] Heyman H, Van De Looverbosch DE, Meijer EP, Schols JM. Benefits of an oral nutritional supplement on pressure ulcer healing in long-term care residents. J Wound Care November 2008;17(11):476–8. 80.
[53] Braun L, Kim PJ, Margolis D, Peters EJ, Lavery LA. What's new in the literature: an update of new research since the original WHS diabetic foot ulcer guidelines in 2006. Wound Repair Regen September 2014;22(5):594–604.
[54] Neville RF, Attinger CE, Bulan EJ, Ducic I, Thomassen M, Sidawy AN. Revascularization of a specific angiosome for limb salvage: does the target artery matter? Ann Vasc Surg May–June 2009;23(3):367–73.
[55] Ennis WJ, Meneses P. Leg ulcers: a practical approach to the leg ulcer patient. Ostomy Wound Manage August 1995;41(7A Suppl.):52S–62S. discussion 3S.
[56] LoGerfo FW, Coffman JD. Current concepts. Vascular and microvascular disease of the foot in diabetes. Implications for foot care. N Engl J Med December 20, 1984;311(25):1615–9.
[57] Sumpio BE, Armstrong DG, Lavery LA, Andros G. The role of interdisciplinary team approach in the management of the diabetic foot: a joint statement from the Society for Vascular Surgery and the American Podiatric Medical Association. J Vasc Surg June 2010;51(6):1504–6.
[58] Burnand KG, Whimster I, Naidoo A, Browse NL. Pericapillary fibrin in the ulcer-bearing skin of the leg: the cause of lipodermatosclerosis and venous ulceration. Br Med J (Clin Res Ed) October 16, 1982;285(6348):1071–2.
[59] Browse NL, Burnand KG. The cause of venous ulceration. Lancet July 31, 1982;2(8292):243–5.
[60] Falanga V, Eaglstein WH. The "trap" hypothesis of venous ulceration. Lancet April 17, 1993;341(8851):1006–8.
[61] Coleridge Smith PD, Thomas P, Scurr JH, Dormandy JA. Causes of venous ulceration: a new hypothesis. Br Med J (Clin Res Ed) June 18, 1988;296(6638):1726–7.
[62] Ennis WJ. Wound care specialization: the current status and future plans to move wound care into the medical community. Adv Wound Care (New Rochelle) October 2012;1(5): 184–8.
[63] Ennis WJ. The status of wound care education. Adv Wound Care (New Rochelle) October 2012;1(5):183.
[64] Magro CM, Crowson AN. The spectrum of cutaneous lesions in rheumatoid arthritis: a clinical and pathological study of 43 patients. J Cutan Pathol January 2003;30(1):1–10.
[65] Vowden K, Vowden P. Mixed aetiology ulcers. J Wound Care January 2001;10(1):520.
[66] Masferrer J. Approach to angiogenesis inhibition based on cyclooxygenase-2. Cancer J November–December 2001;7(Suppl. 3):S144–50.
[67] Qiu JG, Factor S, Chang TH, Knighton D, Nadel H, Levenson SM. Wound healing: captopril, an angiogenesis inhibitor, and *Staphylococcus aureus* peptidoglycan. J Surg Res August 2000;92(2):177–85.
[68] Margolis DJ, Berlin JA, Strom BL. Risk factors associated with the failure of a venous leg ulcer to heal. Arch Dermatol August 1999;135(8):920–6.
[69] Ennis WJ, Meneses P. Standard, appropriate and advanced care and medicolegal considerations; Part 2 venous ulcerations. Wounds 2003;15(4):107–22.
[70] Ennis WJ, Li WW. Patient centered outcomes: a report of summit proceedings. American College of Wound Healing and Tissue Repair.
[71] Corbett LQ, Ennis WJ. What do patients want? patient preference in wound care. Adv Wound Care (New Rochelle) August 1, 2014;3(8):537–43.
[72] Moffatt C, O'Hare L. Ankle pulses are not sufficient to detect impaired arterial circulation in patients with leg ulcers. J Wound Care March 1995;4(3):134–8.

[73] Newman AB, Shemanski L, Manolio TA, Cushman M, Mittelmark M, Polak JF, et al. Ankle-arm index as a predictor of cardiovascular disease and mortality in the cardiovascular health study. The cardiovascular health study group. Arterioscler Thromb Vasc Biol March 1999;19(3):538–45.

[74] Hafner J, Schaad I, Schneider E, Seifert B, Burg G, Cassina PC. Leg ulcers in peripheral arterial disease (arterial leg ulcers): impaired wound healing above the threshold of chronic critical limb ischemia. J Am Acad Dermatol December 2000;43(6):1001–8.

[75] Pardo M, Alcaraz M, Bernal FL, Felices JM, Achel GD, Canteras M. Transcutaneous oxygen tension measurements following peripheral transluminal angioplasty procedure has more specificity and sensitivity than ankle brachial index. Br J Radiol February 2015;88(1046):20140571.

[76] Kernick DP, Shore AC. Characteristics of laser Doppler perfusion imaging in vitro and in vivo. Physiol Meas May 2000;21(2):333–40.

[77] Kolarova H, Ditrichova D, Wagner J. Penetration of the laser light into the skin in vitro. Lasers Surg Med 1999;24(3):231–5.

[78] Christ F, Bauer A, Brugger D. Different optical methods for clinical monitoring of the microcirculation. Eur Surg Res January–April 2002;34(1–2):145–51.

[79] Okamoto S, Iida O, Nakamura M, Yamauchi Y, Fukunaga M, Yokoi Y, et al. Postprocedural skin perfusion pressure correlates with clinical outcomes 1 year after endovascular therapy for patients with critical limb ischemia. Angiology February 4, 2015;66(9):862–6.

[80] Benitez E, Sumpio BJ, Chin J, Sumpio BE. Contemporary assessment of foot perfusion in patients with critical limb ischemia. Semin Vasc Surg March 2014;27(1):3–15.

[81] Conte MS, Pomposelli FB, Clair DG, Geraghty PJ, McKinsey JF, Mills JL, et al. Society for vascular surgery practice guidelines for atherosclerotic occlusive disease of the lower extremities: management of asymptomatic disease and claudication. J Vasc Surg March 2015;61 (3 Suppl.):2S–41S.

[82] Black CE, Costerton JW. Current concepts regarding the effect of wound microbial ecology and biofilms on wound healing. Surg Clin North Am December 2010;90(6):1147–60.

[83] Attinger C, Wolcott R. Clinically addressing biofilm in chronic wounds. Adv Wound Care (New Rochelle) June 2012;1(3):127–32.

[84] Cole-King A, Harding KG. Psychological factors and delayed healing in chronic wounds. Psychosom Med March–April 2001;63(2):216–20.

[85] Tournier JN, Hellmann AQ. Neuro-immune connections: evidence for a neuro-immunological synapse. Trends Immunol March 2003;24(3):114–5.

[86] Ennis WJ. Healing: can we? Must we? Should we?. Ostomy Wound Manage September 2001. 47(9):6, 8.

[87] Snyder RJ, Driver V, Fife CE, Lantis J, Peirce B, Serena T, et al. Using a diagnostic tool to identify elevated protease activity levels in chronic and stalled wounds: a consensus panel discussion. Ostomy Wound Manage December 2011;57(12):36–46.

[88] Veves A, Sheehan P, Pham HT. A randomized, controlled trial of promogran (a collagen/ oxidized regenerated cellulose dressing) vs standard treatment in the management of diabetic foot ulcers. Arch Surg July 2002;137(7):822–7.

[89] O'Donnell TFPM, Marston WA, Ennis WJ, Dalsing M, Kistner RL. Management of venous leg ulcers: clinical practice guidelines of the Society for Vascular Surgery and the American Venous Forum. J Vasc Surg 2014;60(2 Suppl.):3s–59s.

[90] Barwell JR, Davies CE, Deacon J, Harvey K, Minor J, Sassano A, et al. Comparison of surgery and compression with compression alone in chronic venous ulceration (ESCHAR study): randomised controlled trial. Lancet June 5, 2004;363(9424):1854–9.

[91] Ennis WJ, Meneses P, Borhani M. Push-pull theory: using mechanotransduction to achieve tissue perfusion and wound healing in complex cases. Gynecol Oncol November 2008;111 (2 Suppl.):S81–6.

[92] Wlaschek M, Scharffetter-Kochanek K. Oxidative stress in chronic venous leg ulcers. Wound Repair Regen September–October 2005;13(5):452–61.

[93] Fette A. Treatment of pressure ulcers with topical negative pressure versus traditional wound management methods: a research sampler. Plast Surg Nurs October–December 2005;25(4): 176–80.

[94] Ennis WJ, Lee C, Plummer M, Meneses P. Current status of the use of modalities in wound care: electrical stimulation and ultrasound therapy. Plast Reconstr Surg January 2011;127(Suppl. 1): 93S–102S.

[95] Gurtner GC, Werner S, Barrandon Y, Longaker MT. Wound repair and regeneration. Nature May 15, 2008;453(7193):314–21.

[96] Patel NP, Granick MS. Wound education: American medical students are inadequately trained in wound care. Ann Plast Surg July 2007;59(1):53–5; discussion 5.

[97] Ennis WJ, Sui A, Bartholomew A. Stem cells and Healing: impact on inflammation. Adv Wound Care (New Rochelle) September 2014;2(7):369–78.

[98] Lee S, Szilagyi E, Chen L, Premanand K, DiPietro LA, Ennis W, et al. Activated mesenchymal stem cells increase wound tensile strength in aged mouse model via macrophages. J Surg Res May 1, 2012;181(1):20–4.

Chapter 13

Current Innovations for the Treatment of Chronic Wounds

B. Lynn Allen-Hoffmann[1,2,3], Peggy J. Rooney[1]

[1]*Stratatech Corporation, Madison, WI, USA;* [2]*Department of Pathology and Laboratory Medicine, School of Medicine and Public Health, University of Wisconsin, Madison, WI, USA;* [3]*Department of Surgery, School of Medicine and Public Health, University of Wisconsin, Madison, WI, USA*

Chapter Outline

BURDEN OF DIABETIC FOOT ULCERS

Chronic nonhealing ulcers are a significant medical problem, and the incidence of these wounds is expected to increase as the population ages. It is projected that about 1,400,000 diabetics, in the United States alone, will suffer from chronic, non-healing diabetic foot ulcers (DFU) in 2015 [1]. Moreover, a DFU on the heel is associated with significantly longer healing times [2]. In 2010, 73,000 amputations were performed in adults over 20 years of age with the rates increasing about 10% annually [3,4]. A recent study has shown that about 28% of patients with infected DFU require amputations [5]. The sense of well-being for individuals suffering from chronic wounds is significantly affected by the pain, infection, odor, diminished self-image, and increased social isolation that these wounds cause [6,7].

Unlike other chronic wounds, the development of a DFU is often associated with, and complicated by, the neuropathy and peripheral artery disease associated with diabetes. DFU can be classified as neuropathic, ischemic, or

Skin Tissue Engineering and Regenerative Medicine. http://dx.doi.org/10.1016/B978-0-12-801654-1.00013-9

neuroischemic. Neuropathic wounds result from a loss of protective sensation, leading to undetected or repetitive injuries from external causes such as poorly fitting footwear, physical, chemical, or thermal trauma, or anatomic changes leading to the application of abnormal pressure on protuberances. Ischemic wounds occur from poor arterial blood supply to the diabetic foot. DFU most commonly result from multiple risk factors, and those with noted perfusion issues often present as neuroischemic wounds of mixed etiology. Complications from microvascular dysfunction and/or peripheral artery disease contribute to poor healing and increased risk of infection from the inability to meet increased demand for blood at sites of injury. These complicated wounds present practitioners with management challenges that require a new and potentially diverse set of treatment modalities.

WOUND HEALING

Human skin is one of the body's renewable tissues. The dermal compartment of skin is composed of a collagen-rich dermal matrix of glycoproteins that provides resiliency and elasticity and is synthesized, secreted, and processed by dermal cells, primarily fibroblasts. Overlying the dermis are layers of epidermal keratinocytes possessing a continuum of replicative potential and differentiation characteristics that result in a stratified squamous epithelium (Figure 1). Keratinocytes found in the innermost epidermal layers have replicative potential, and those keratinocytes juxtaposed to the dermis actively contribute to the synthesis of basement membrane proteins that reside between and anchor the dermal and epidermal compartments of skin. Keratinocytes in the outermost layers are terminally differentiated, nonviable squames that become part of a thick, oily, lipid-rich layer that prevents water loss and exposure to foreign invaders. Skin also contains specialized epidermal appendages such as hair follicles, sebaceous glands, and sweat glands. Disruption of the epithelial barrier triggers a healing response involving complex interactions between epidermal and dermal cells, immune cells, the extracellular matrix, plasma-derived proteins, and controlled angiogenesis coordinated by an array of cytokines and growth factors. This dynamic process is classically divided into three overlapping phases: inflammation, proliferation, and remodeling of the extracellular matrix [8–12]. Upon injury, cytokines and growth factors provide chemotactic cues to recruit circulating inflammatory cells to the wound. Proteases released during the acute inflammatory phase assist with the removal of damaged and denatured cellular and extracellular matrix components. As the epidermal cells at the wound margin or within epidermal appendages proliferate and migrate to reepithelialize the denuded wound surface, the inflammatory phase declines, and fibroblasts produce new extracellular matrix in the form of contractile granulation tissue that draws the wound margins together. Once the wound surface has been covered by a monolayer of keratinocytes, epidermal migration ceases, the underlying

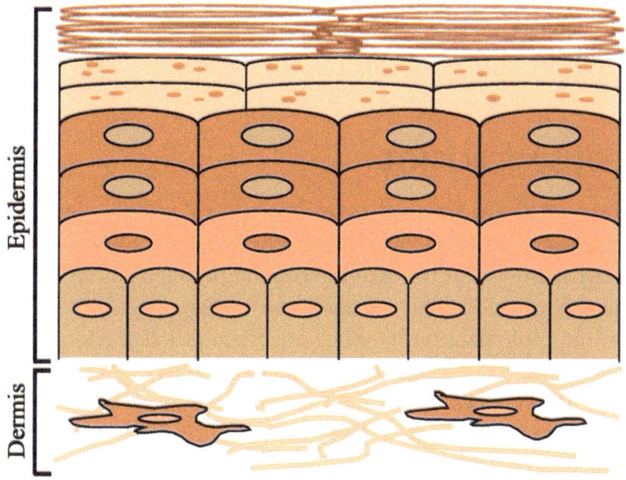

FIGURE 1 **Schematic of the architecture of normal intact skin.** The dermis is the deepest layer of skin and is comprised of a matrix of collagen and other extracellular matrix glycoproteins secreted and organized by dermal fibroblasts to provide resiliency and elasticity. The epidermis is composed of keratinocytes and is attached to the dermis via a basement membrane derived from basal keratinocytes and the dermal fibroblasts. Basal keratinocytes (depicted as columnar cells at the dermal junction) both divide for renewal and differentiate to form the stratified layers of the epidermis. The barrier function of skin is dependent on this process of differentiation, during which keratinocytes organize and secrete specialized lipids into the intercellular spaces of the stratum corneum. As the keratinocytes are moved upward, they lose replicative potential, undergo terminal differentiation, and eventually enucleate to form mature squames (flattened cells on the surface), which are ultimately lost due to friction.

basement membrane is reestablished, and a new stratified epidermis is established through terminal differentiation of the epidermal keratinocytes. Wound closure not only provides a barrier to environmental stresses, such as infectious agents, but provides protection from wound reoccurrence [13]. Collagen production within the dermal compartment continues for several weeks after wound closure, and subsequent extracellular matrix remodeling may continue for 2 years or more.

In a chronic wound environment, these complex cellular and molecular processes are disordered, resulting in a failure to form granulation tissue thereby impeding subsequent reepithelialization. Diabetic patients, a population prone to the development of chronic cutaneous ulcers of the feet, are likely to have an altered immune response that may contribute to poor healing and infection of wounds. Features common to chronic wounds include sustained levels of pro-inflammatory cytokines, diminished growth factor activity, increased proteolytic activity, and the dysregulation of the cells' response to molecular regulators [14–17]. Factors known to contribute to this condition include repeated ischemia–reperfusion injury, local tissue hypoxia, cellular and systemic changes that occur with aging, and polymicrobial colonization that perpetuates the pro-inflammatory host response [18]. The disruption of skin function and impaired healing associated with chronic wounds often lead to complications such as infection, and devastating consequences such as amputation.

CUTANEOUS MICROBIOTA

The surface of the skin houses a diverse microbial population that is both an asset and a reservoir for potential pathogens upon disruption of the epithelial barrier. Many members of the skin microbiota produce antibacterial peptides, bacteriocins, bacteriocin-like compounds, or antifungal products that are thought to limit colonization by other microbes [19,20]. Along with nutrient competition, the active prevention of colonization by other, potentially more pathogenic microbes has led to the theory that some members of the cutaneous microbiota are mutualists benefitting the host. While members of the normal skin microbiota exhibit traits allowing persistence, it is changes in the host immune response or epithelial barrier function, allowing introduction due to breaks, on implanted devices, or on foreign bodies, that lead to the switch from resident flora to opportunistic pathogen upon access to subepidermal tissue [12,20].

Intact skin prevents local infection of the dermis or other underlying tissues through several means. Pathogens attempting to invade through the skin are captured in and among epidermal squames that are continuously removed by friction, cleaning, and other minor trauma. The stratum corneum is also cooler, slightly acidic, and drier than mucosal surfaces, creating an inhospitable environment for many pathogenic microbes [20]. In addition to these passive control mechanisms, as keratinocytes differentiate they produce host defense peptides (HDP). HDP serve as a critical component of innate defenses against wound infection, acting locally within the skin to protect against a broad range of bacteria, fungi, and viruses, including skin flora that may act as pathogens after disruption of the epithelial barrier [12]. These defenses against microbial invasion are compromised within chronic wounds, creating an environment at high risk for clinical infection [21]. Diabetic wounds in particular have shown increased susceptibility to infection [22].

STANDARD OF CARE

Current first-line management of DFU includes initial wound assessment, restoration of blood flow as needed, identification and treatment of infection, cleansing and debridement of devitalized tissue, dressing selection, pressure-relieving strategies, and diabetes control. Topical antibacterial products are often used in conjunction with dressings, as bacterial colonization of DFU is common and can impair wound healing [10,23]. Metronidazole, neomycin, gentamycin, mupirocin, and silver-impregnated dressings are examples of topical antibiotic products used to prevent or control infection in the treatment of DFU [10].

These standard therapies, however, cure only about 30% of DFU after 20 weeks [24–26]. In addition, the prolonged use of traditional antibiotics may enhance the development of drug-resistant pathogenic bacteria. As commonly encountered bacterial strains continue to develop antibiotic resistance, clinicians are often limited to more potent antibiotics that can have deleterious effects on the viability and migration of keratinocytes [27]. As a result, the need to use these stronger antibiotics actually prolongs the wound healing process and the time to durable wound closure [28]. Together, these underscore the need for innovative treatment approaches that shorten the time to wound closure and supplement or reduce our dependence on the antibiotic treatment regimens used in open wound therapy.

ADVANCED BIOENGINEERED TREATMENT OPTIONS

The advent of advanced therapies in the treatment of non-healing DFU has shown significant advantages beyond traditional treatment alone [29]. In addition to systemic therapies such as the use of hyperbaric oxygen or device-based options like negative-pressure wound therapy, several biological skin substitute products with select skin tissue features have been developed for the treatment of chronic wounds (Table 1). These products provide wound coverage, acting as a barrier to prevent bacterial contamination or additional trauma and reducing the pain of exposed nerve endings. The majority of skin substitutes approved in the United States offer only a dermal equivalent. The dermis of healthy skin is primarily composed of fibroblasts and an extracellular matrix of collagen, elastin, fibronectin, hyaluronic acid, and proteoglycans that afford strength and flexibility to the tissue. Dermal substitutes are often constructed from purified, animal-derived collagen or decellularized human or animal tissues, which give the wound a framework for the invasion of the patient's own cells and the growth of capillaries during healing.

Examples of acellular dermal analogs for the treatment of chronic wounds include CellerateRx®, Integra®, PriMatrix®, AlloDerm®, Graftjacket®, and EpiFix®. CellerateRx (Wound Management Technologies, Addison, TX) is comprised of hydrolyzed bovine collagen alone, whereas Integra (Integra LifeSciences, Plainsboro, NJ) is a mixture of bovine collagen and shark cartilage glycosaminoglycan, and includes a silicone membrane covering, mimicking the fluid-retaining property normally provided by an intact epidermis. PriMatrix (TEI Biosciences, Waltham, MA) is an acellular matrix created from fetal bovine dermis. PriMatrix Ag is a derivative of PriMatrix that incorporates silver ions to reduce bacterial colonization of the dermal substitute. This is the only marketed product enhanced to address bacterial colonization, a known impediment to the healing of chronic wounds. With each of these products and others containing bovine-derived materials, there is regulatory concern and industrial vigilance regarding the transmission of bovine spongiform encephalopathy (BSE).

TABLE 1 Bioengineered Therapeutics for Chronic Cutaneous Wounds

Name	Description	Regulation[a]	Indications[b]	References[c]
ACell® UBM Technology	Porcine collagen dressing	510(k)	D, V, P, Vasc, B	http://acell.com/
AlloDerm® Regenerative Tissue Matrix/Cymetra® Micronized AlloDerm Tissue	Processed acellular dermal matrix from donated human skin. Elastin and collagen. Available as tissue or micronized injectable form	HCT/P	Damaged integumental tissue	http://www.lifecell.com/health-care-professionals/lifecell-products/
AlloSkin™/AlloSkin™ RT	Meshed processed human allograft from cadaveric epidermis and dermis	HCT/P	Acute and chronic wounds, B	http://www.allosource.org/products/
Apligraf®	Bilayered skin substitute derived from bovine type 1 collagen, allogeneic human fibroblasts, and keratinocytes from neonatal foreskins.	PMA	D, V	http://www.apligraf.com/
Atlas Wound Matrix	Decellularized processed porcine collagen dermal material	510(k)	D, V, P, Vasc, B	http://www.ahrq.gov/research/findings/ta/skinsubs/HCPR0610_skinsubst-final.pdf
Avagen Wound Dressing	Cross-linked bovine tendon collagen and glycosaminoglycan	510(k)	D, V, P, Vasc, B	http://www.ahrq.gov/research/findings/ta/skinsubs/HCPR0610_skinsubst-final.pdf
CellerateRx	Activated fragments of type I bovine collagen in gel formulation	510(k)	D, V, P, Vasc, B	http://www.celleraterx.com/
CollaCare™ Powder/Film; Film marked in United States as Collieva®	Type I renatured collagen from equine or bovine Achilles tendon	510(k)	D, V, P, B	http://www.innocollinc.com/collacare-powder.html

CollaRX® collagen Sponge (Innocoll)/Cogenzia gentamicin-containing collagen sponge	Porous collagen matrix sponge	510(k)	D, V, P	http://innocollinc.com/
CollaSorb™	Bovine collagen and calcium alginate	510(k)	D, V, P, Vasc, B	http://us.hartmann.info/collasorb.php
CollaWound™	Particulate porcine collagen dressing	510(k)	D, V, P, Vasc, B	http://collamatrix.com/CWH-E.htm
CoreLeader Colla-Pad	Porous, sponge matrix of cross-linked bovine collagen	510(k)	D, V, P, Vasc, B	http://www.coreleaderbio.com/product_view.php?cid=797&id=46
DermADAPT™ Biomatrix Wound Dressing	Decelluarized collagen matrix	510(k)	D, V, P, Vasc, B	http://www.nursingcenter.com/journalarticle?Article_ID=701680; http://www.accessdata.fda.gov/scripts/cdrh/cfdocs/cfPMN/pmn.cfm?ID=K061494; http://www.medscape.com/viewarticle/546388
Dermagraft®	Cryopreserved human fibroblast-derived dermal substitute of fibroblasts, extracellular matrix, and a bioabsorbable scaffold	PMA	D	http://www.dermagraft.com/home/
Endoform Dermal Template™	Basement membrane and propria–submucosa from ovine forestomach	510(k)	D, V, P, Vasc, B	http://www.mesynthes.com/index.php?option=com_content&view=category&layout=blog&id=15&Itemid=17

Continued

TABLE 1 Bioengineered Therapeutics for Chronic Cutaneous Wounds—cont'd

Name	Description	Regulation[a]	Indications[b]	References[c]
EpiFix®	Dehydrated human amnion/chorion membrane allograft	HCT/P	D, V, P, B	http://mimedx.com/products?qt-product_tabs=2#qt-product_tabs=&quicktabs-product_tabs=2
Excellagen®	Pharmaceutically formulated fibrillar type I bovine collagen gel	510(k)	D, V, P, Vasc, B	http://excellagen.com/
ExpressGraft-C9T1 skin tissue	Bilayered skin substitute containing allogeneic human fibroblasts and NIKS keratinocytes genetically modified for enhanced cathelicidin production	IND	D	http://stratatechcorp.com/products/expressgraft.php
EZ Derm®	Porcine xenograft	510(k)	D, P, B	http://www.molnlycke.com/advanced-wound-care-systems/skin-grafting-systems/ez-derm/#confirm
FortaDerm™ Wound Dressing (formerly called Graftpatch™)	Fenestrated sheet of porcine intestinal collagen	510(k)	D, V, P, Vasc, B	http://www.ahrq.gov/research/findings/ta/skinsubs/HCPR0610_skinsubst-final.pdf
GammaGraft®	Irradiated cadaveric human skin allograft	HCT/P	D, full-thickness ulcers, B	http://www.gammagraft.com/
Grafix®	Cryopreserved human placental membrane	HCT/P	D, V, P, B	http://www.osiris.com/grafix/healthcare-professionals/

Graftjacket® Regenerative Tissue Matrix	Processed, freeze-dried, acellular donated human dermal allograft	HCT/P	D, V, P	http://graftjacketbykci.com/graftjacket.html
HA Absorbent Wound Dressing	Absorbent fibrous fleece of hyaluronic acid benzyl ester HYAFF 11p75™	510(k)	D, P, leg ulcers, B	http://www.ahrq.gov/research/findings/ta/skinsubs/HCPR0610_skinsubst-final.pdf
Helicoll	Semi-occlusive, adherent sheet of ovine or bovine type 1 collagen	510(k)	D, V, P, Vasc, B	http://www.encoll.com/images/Helicoll.pdf
Hyalogran®	Absorbent dressing of fibrous benzyl ester of hyaluronic acid (HYAFF®) and sodium alginate	510(k)	D, P, leg ulcers	http://www.anikatherapeutics.com/products/dermal/hyalogran/
Hyalomatrix® (Laserskin®)	Bi-layered, nonwoven pad of fibrous benzyl esters of hyaluronic acid (HYAFF®) with outer semipermeable silicone membrane	510(k)	D, V, P, B	http://www.anikatherapeutics.com/products/dermal/hyalomatrix/
Integra™ Bilayer Wound Matrix	Dermal matrix of cross-linked bovine tendon collagen and glycosaminoglycan with a semipermeable silicone membrane covering	510(k)	D, V, P, B	http://www.ilstraining.com/bmwd/bmwd/bmwd_it_00.html
Jaloskin®	Semipermeable film dressing of HYAFF® 11, a benzyl ester of hyaluronic acid	510(k)	D, V, P, Vasc, B	http://www.ahrq.gov/research/findings/ta/skinsubs/HCPR0610_skinsubst-final.pdf
MatriStem® wound Matrix/Multilayer Wound Matrix	Porcine urinary bladder-derived lyophilized extracellular matrix sheet	510(k)	D, V, P, Vasc, B	http://acell.com/matristem-devices/current-indications/wound-care/
MatriStem MicroMatrix®	Porcine urinary bladder-derived lyophilized extracellular matrix in particle form	510(k)	D, V, P, Vasc, B	http://acell.com/matristem-devices/current-indications/wound-care/

Continued

TABLE 1 Bioengineered Therapeutics for Chronic Cutaneous Wounds—cont'd

Name	Description	Regulation[a]	Indications[b]	References[c]
Matrix Collagen Wound Dressing	Absorbent animal-derived type I collagen membrane matrix for topical use, available as particles, sponge or film	510(k)	D, V, P, B	http://www.collagenmatrix.com/products-woundcare.htm
MediHoney®	Wound care dressing with *Leptospermum* honey and hydrogel	510(k)	D, P, leg ulcers, B	http://www.dermasciences.com/medihoney
MemoDerm™	Acellular dermal matrix derived from human allograft skin tissue	HCT/P	D, B	http://mmi.ourassets.net/memoderm/index.html
OASIS® Wound Matrix (formerly SIS Wound Dressing II)/OASIS® Ultra TriLayer Matrix	Porcine small intestinal submucosa matrix	510(k)	D, V, P, Vasc, B	http://oasiswoundmatrix.com/
PriMatrix® Dermal Repair Scaffold (formerly DressSkin™)	Acellular, type III collagen-rich dermal matrix derived from fetal bovine dermis	510(k)	D, V, P, B	http://www.primatrix.com/
SS Matrix™	Porcine collagen matrix	510(k)	D, V, P, Vasc, B	http://www.ahrq.gov/research/findings/ta/skinsubs/HCPR0610_skinsubst-final.pdf
Stimulen™ Collagen Woundcare	Soluble modified bovine collagen dressing, available as gel or powder	510(k)	D, V, P, B	http://www.elastogel.com/product-catalog/wound-care/stimulen-collagen-wound-care

StrataGraft®	Bilayered skin substitute containing nonbovine collagen, allogeneic human fibroblasts, and NIKS keratinocytes	IND	B	http://stratatechcorp.com/products/stratagraft.php
Suprathel®	Synthetic tri-polymer, bioresorbable dermal covering	510(k)	D, V, P, B	http://www.silon.com/products/advanced-woundcare/suprathel/; http://www.accessdata.fda.gov/cdrh_docs/pdf9/K090160.pdf
Talymed™	Matrix of shortened fibers of poly-*N*-acetylglucosamine from microalgae	510(k)	D, V, P, Vasc, B	http://talymed.com/
Theraform™	Absorbable porcine-derived collagen membrane	510(k)	D, V, P, chronic ulcers, B	http://www.rmsbio.net/html/RegenGraft/TheraForm.asp
TheraSkin®	Meshed, cryopreserved bilayered human skin donor allograft	HCT/P	D, V, P, B	http://www.solublesystems.com/Products/TheraSkin
Unite® Biomatrix	Collagen-based dressing of decellularized equine pericardium	510(k)	D, V, P, B	https://www.clinicaltrials.gov/ct2/show/NCT00958711; http://www.reuters.com/article/2009/07/21/idUS195460+21-Jul-2009+BW20090721

[a]*Mechanism of regulation by the U.S. Food and Drug Administration. Animal-derived and synthetic products are regulated under the 510(k) process. Human cells, tissue, and cellular and tissue-based products (HCT/P) are regulated as human tissue for transplantation if consistent with definitions cited in Title 21 of the Code of Federal Regulations, Parts 1270 and 1271. Human and human/animal-derived devices are regulated through a premarket approval (PMA) process. Biologic material is regulated under an Investigational New Drug (IND) application.*
[b]*D, diabetic ulcer; V, venous ulcer; P, pressure ulcer; Vasc, vascular ulcer; B, burn. This does not represent an exhaustive list of indications, and does not highlight specific parameters such as duration of the wound or burn wound classification.*
[c]*Additional information for several products was obtained from http://www.woundsource.com and http://www.ahrq.gov/research/findings/ta/skinsubs/HCPR0610_skinsubst-final.pdf.*

AlloDerm and Graftjacket (Acelity, San Antonio, TX) are made from decellularized human cadaveric skin tissues. These products bear risk of disease transmission in addition to variable product performance, since each product lot is derived from a different tissue donor. As a result of harvesting and limited testing, this type of allogeneic product has been associated with specific safety concerns including disease transmission and manufacture recalls due to suspect sterility results and donor-screening procedures [30–32].

EpiFix Amniotic Membrane Allograft (MiMedx, Marietta, GA) is a composite tissue of human amniotic membrane and underlying chorion that has been processed to retain an intact extracellular matrix and is then dehydrated and sterilized [33]. This product does not contain viable cells. In a multicenter trial evaluating the treatment of DFU, healing promoted by EpiFix was more rapid than either standard of care or a bilayered skin substitute (Apligraf®) [34]. EpiFix is considered a minimally manipulated human cellular and tissue-based product, and therefore it has not been held to the Food and Drug Administration (FDA)'s rigorous premarket requirements for devices or biologics. The use of EpiFix is contraindicated on infected wounds.

In contrast to products based on acellular or dehydrated dermal analogs, select skin substitutes contain living human cells that, in addition to structural integrity and a barrier to infection, may actively produce growth factors to stimulate cellular recruitment and promote healing within the wound bed. Two examples of living skin substitutes providing select skin features are Dermagraft® (Organogenesis Inc., Canton, MA) and Grafix® (Osiris Therapeutics Inc., Columbia, MD). Dermagraft contains human foreskin fibroblasts within a polyglactin mesh scaffold. Dermagraft is approved for use as a second-line therapy in chronic DFU and contraindicated for use in infected wounds. Grafix is derived from human placental membranes. It contains viable fibroblasts, epithelial cells, and mesenchymal stem cells, and showed promising clinical results in patients with DFU [25]. Though formerly believed to fit the classification of a human cells and tissue product (HCT/P), recent FDA review determined that it is dependent on the metabolic activity of living cells for its primary function and is not intended for autologous use [35]. As a result, Grafix will ultimately require a valid biologic license.

Bilayered skin substitutes attempt to recapitulate the very nature of intact skin, including the provision of mechanical integrity, the secretion of growth factors, and a barrier to fluid loss, toxin exposure, and microbial invasion. The vapor barrier function of skin is dependent on the differentiation of keratinocytes, which organize and secrete specialized lipids into the intercellular spaces of the stratum corneum. Bioengineered bilayered skin substitutes composed of stratified human keratinocytes grown on dermal analogs containing living human fibroblasts reproduce many of the structural and biological features of intact human skin. As a living tissue, they are anticipated to provide continuous production of bioactive molecules that may promote healing. First-generation cultured skin substitutes have shown efficacy as second-line therapies in the treatment of noninfected chronic wounds, and their use for these indications has gained widespread acceptance [36,37]. Apligraf (Organogenesis Inc., Canton, MA), indicated as a second-line treatment for chronic skin wounds, is the only currently available skin substitute comprised of both dermal and epidermal components that is approved for use in the United States. In contrast to results of a recent head-to-head trial, a comparative retrospective analysis of Apligraf and EpiFix treatment of DFU performed using the WoundExpert electronic medical record database indicated that Apligraf-treated wounds closed in less time than those treated with EpiFix [34,38]. Production of Apligraf requires periodic sourcing of cells from human tissue, which introduces variability in product performance and necessitates costly adventitious agent testing for each new cell bank. Furthermore, infection was a major adverse event in clinical trials on venous stasis ulcers; 29.6% of patients receiving Apligraf had a suspected wound infection versus 14.0% in the control [39]. Because the shelf life of Apligraf is only 10 days, by necessity, it is released for clinical use prior to the completion of sterility tests on the final product.

At best, advanced modality therapies achieve ~60% healing at 3 months [24–26,40]. Some of the marketed products for the treatment of chronic wounds have in fact been associated with higher rates of infection, and all are currently contraindicated for the treatment of infected wounds [39,41,42]. Although some represent an improvement over traditional treatment modalities, there is considerable room for improvements that enhance the efficient and durable closure of these recalcitrant wounds and reduce complications.

NEXT-GENERATION SKIN SUBSTITUTES

A recently developed bilayered skin substitute provides the springboard for the next generation of therapeutic skin substitutes promoting a wound healing microenvironment and reducing time to closure, designed for the medical management of chronic cutaneous ulcers. In StrataGraft skin tissue (Stratatech Corp., Madison, WI), a human keratinocyte epidermis is grown on a dermal layer comprised of human fibroblasts in a matrix of collagen obtained from a closed, nonbovine animal source, resulting in a thick tissue that can be meshed prior to placement and sutured in place. A clinical trial evaluating the safety and tolerability of escalating doses of StrataGraft skin tissue used as a temporary cover for full-thickness, complex, cutaneous injuries was performed [43]. In this study, each subject served as their own control. A single area of full-thickness skin loss was temporarily covered with StrataGraft on one half, while the other half received cadaveric allograft. After 7 days, both tissues were removed, and the wound beds were treated with autograft. The safety of StrataGraft was assessed by autograft acceptance 2 weeks after placement. Exposure to StrataGraft was well tolerated, and autograft take was similar between the areas temporarily covered by StrataGraft skin tissue and by cadaver skin [43,44]. The appearance of StrataGraft tissue (pink and adherent) was significantly better than cadaver skin during the 7 days of wound coverage [43]. These data supported a recently completed study evaluating prolonged exposure to StrataGraft skin tissue as an alternative to autograft treatment for the management of complex burns, showing a similarly positive safety profile as well as effectiveness at promoting wound closure (manuscript in preparation).

NIKS keratinocytes are the core of StrataGraft skin tissue technology. NIKS are long-lived human keratinocytes that demonstrate a normal pattern of epidermal differentiation and stratification, including the development of a cornified envelope with lipid production thereby producing squames and providing barrier function [45]. Importantly, the clonal nature and progenitor characteristics render NIKS keratinocytes amenable to stable genetic manipulation by nonviral means [46–48]. This combined with their ability to form the epidermal compartment of a bilayered skin substitute promoting wound healing make them the ideal starting material for the creation of genetically modified skin substitutes expressing wound healing factors. The creation of designer tissues has the potential to revolutionize wound healing and to customize treatment options to address the three major challenges in the medical management of DFU: (1) the reduction of microbial infection, both directly and through enhancement of a productive immune response; (2) the restoration of a constructive wound healing microenvironment; and (3) the induction of sufficient revascularization.

TAPPING INNATE ANTIMICROBIALS

The use of multifunctional HDP has emerged as a new therapeutic approach to enhance innate host defense mechanisms against invading microorganisms [49–51]. As part of the first line of defense against microbial invasion, the innate immune system mounts a comprehensive response to prevent wound infection, including the production of HDP by both keratinocytes and leukocytes in response to epithelial compromise [49,52,53]. HDP are generally 6–50 amino acids in length and possess immunomodulatory functions, while also killing or inhibiting the growth of microorganisms. Most contain two to nine positive lysine and/or arginine residues, are composed of up to 50% hydrophobic amino acids, and have these hydrophobic and charged regions spatially separated [49]. In mammals, HDP are primarily stored in the granules of neutrophils, but are also found in monocytes, B cells, natural killer cells, mast cells, sweat and salivary glands, testes, human keratinocytes, and various squamous epithelial tissues [49,54,55]. The two major families of small cationic HDP in mammalian skin are β-defensins and cathelicidins. To enhance expression of these antimicrobial compounds within wounds and the context of a skin tissue promoting wound closure, NIKS keratinocytes were separately modified to enhance the expression of human beta defensin-3 (hBD-3) and human cathelicidin for the generation of antimicrobial skin substitute tissues [46,47].

The hBD-3 antimicrobial peptide is up-regulated in infected tissue, and has been shown to promote wound healing and the reduction of bacterial growth when delivered by viral vector to a porcine model of infected diabetic wounds [56,57]. To assess the effectiveness of the nonviral delivery of hBD-3 from a bioengineered skin substitute, NIKS were modified to express hBD-3 ($NIKS^{hBD\text{-}3}$) and used to generate the epidermal portion of full-thickness skin tissue. These tissues exhibited normal tissue growth and differentiation, matching those generated with unmodified NIKS, but contained enhanced levels of hBD-3 [47]. Conditioned medium from $NIKS^{hBD\text{-}3}$ tissues retarded the in vitro growth of an antimicrobial peptide-sensitive strain of *Staphylococcus aureus* by twofold over that of medium from unmodified NIKS tissue [47]. Further, using a mouse model in which third-degree burns were inoculated with *S. aureus*, debrided, and transplanted with $NIKS^{hBD\text{-}3}$ or control NIKS tissue, mice receiving $NIKS^{hBD\text{-}3}$ tissue showed a 90% decrease in bacterial burden 3 days post-graft placement [47]. These data demonstrated the in vivo antimicrobial activity of the genetically modified tissue, and the potential of antimicrobial tissue for the healing of infected wounds.

CATHELICIDIN, A MULTIFUNCTIONAL HDP

In addition to several beta defensins, humans express a single cathelicidin antimicrobial peptide. This protein is natively produced as an 18 kDa holoprotein (hCAP-18) in healthy human skin and other squamous epithelia, where it is stored in its inactive pro-form until injury or infection triggers its secretion. It is found in neutrophils, mast cells, salivary glands, lung, seminal fluid, squamous epithelia, keratinocytes, and urogenital and gastrointestinal tracts [58]. Proteases secreted from infiltrating neutrophils and resident keratinocytes in response to injury, infection, or inflammation cleave extracellular hCAP-18 to generate the active, α-helical, 4 kDa antimicrobial peptide consisting of 37 residues starting with a pair of leucines (LL-37) [55,58–60].

The LL-37 antimicrobial peptide provides a chemical barrier and a first-line defense against invading pathogens, exhibiting antimicrobial activity against a broad range of bacteria, fungi, and viruses [61–64]. LL-37 also impacts bacterial biofilms such as those found in chronic skin ulcers. These nonmotile communities of bacteria form an outer shell of exopolysaccharide and have a distinct protein secretion profile, which together result in resistance to traditional antimicrobial agents and host immune cells [59,65,66]. Studies by several groups have demonstrated that LL-37 inhibits biofilm formation by clinically relevant strains such as *Staphylococcus epidermidis*, *S. aureus*, *E. coli*, and *Pseudomonas aeruginosa* in a dose-dependent manner, as well as disrupts established biofilms [66–68].

LL-37 also acts as a critical component of innate defenses against wound infection by enhancing leukocyte recruitment and activation [62,69–73], and has been reported to promote vascularization and keratinocyte migration in model systems [74–77]. Association of LL-37 with the FPRL1 receptor on endothelial cells induces both angiogenesis (sprouting of small capillaries) and arteriogenesis (growth of preexisting vessels), which were shown to be enhanced by LL-37 injection in a rabbit wound model [74]. Synthetic LL-37 similarly increased the proliferation, migration, and formation of capillary-like structures by endothelial cells in vitro, and accelerated healing and new vessel formation after topical application to murine wounds with impaired healing [78]. Using an ex vivo wounded skin model, LL-37 was also shown to enhance wound reepithelialization, establishing it as a critical component of the overall wound healing process [21]. In a diabetic mouse model, electroporation of a cathelicidin-encoding plasmid accelerated epithelialization and improved wound closure [77]. Human LL-37 in a gel formulation (by Pergamum AB, Sweden) is being evaluated as a topical treatment for venous leg ulcers. In a double-blind multicenter phase I/II trial, a doses up to 1.6mg/mL were well tolerated, exhibited no safety concerns, and displayed a statistically significant improved healing rate (3–6 times higher) as compared to placebo [79]. The multifunctional cathelicidin is an ideal candidate for expression by a skin substitute tissue intended to promote the closure of chronic wounds.

A PROMISING SKIN TISSUE WITH ENHANCED CATHELICIDIN EXPRESSION

The level of cathelicidin was shown to be reduced in chronic cutaneous ulcers [21]. Restoration of cathelicidin to the chronic wound environment from a bioengineered skin tissue providing immediate wound coverage has the potential to address the underlying pathophysiology of diabetic ulcers, namely infection, poor circulation, and impaired wound healing. To enhance the expression of human cathelicidin, a DNA expression fragment encoding the full-length precursor protein hCAP-18 under control of the human involucrin promoter was stably introduced into NIKS keratinocytes [46]. The involucrin promoter targets gene expression to nonproliferating, suprabasal keratinocytes, limiting cathelicidin expression to the differentiated upper layers of developed skin tissues by design. Like NIKS^{hBD-3} tissues, NIKS$^{hCAP-18}$ skin tissues appeared similar to unmodified NIKS-based tissues on histologic examination [46]. NIKS$^{hCAP-18}$ skin tissues expressed and secreted elevated levels of full-length hCAP-18, and processing to the active LL-37 peptide form was detected in tissue lysates. Conditioned medium from NIKS$^{hCAP-18}$ skin tissues reduced the outgrowth of a nonpathogenic *Staphylococcus* species in vitro in comparison to conditioned medium from control NIKS skin tissues, demonstrating the increased antimicrobial activity of the cathelicidin overexpressing tissues [46]. The in vivo efficacy of NIKS$^{hCAP-18}$ skin tissues was tested against a multidrug-resistant clinical isolate of *Acinetobacter baumannii*, a significant nosocomial pathogen of skin and soft tissue that forms biofilms on clinically relevant surfaces and shows increased development of antibiotic resistance [80–82]. Using a burned mouse model, wounds were infected with *A. baumannii* and grafted with NIKS or NIKS$^{hCAP-18}$ tissues. The NIKS$^{hCAP-18}$ tissue showed a two-log reduction in wound bed bacterial burden in comparison to unmodified NIKS tissue [46]. These data demonstrate that NIKS$^{hCAP-18}$ skin tissue produced physiologically relevant levels of active cathelicidin in vivo.

NIKS$^{hCAP-18}$ skin tissue, like StrataGraft skin tissue, is an allogeneic, fully differentiated, stratified skin substitute containing both epidermal and dermal layers, and is predicted to possess barrier function comparable to that of intact human skin. This modified skin substitute meets the physical and functional characteristics of an optimal skin substitute secreting many native wound healing factors, while also expressing increased levels of human cathelicidin. A skin substitute tissue with enhanced expression of cathelicidin not only provides immediate wound coverage, but may positively impact wound healing beyond the effects of the base tissue alone. Published preclinical studies support clinical investigation of the safety and efficacy of NIKS$^{hCAP-18}$ skin tissue in the healing of chronic cutaneous wounds, including DFU.

SAFETY FEATURES OF GENETICALLY MODIFIED SKIN TISSUE

Proceeding into the realm of genetically modified therapeutics must be done with caution, as the introduction of foreign DNA for gene therapy and the use of viral vectors for its delivery may result in severe consequences, including insertional mutagenesis and deleterious immune responses [83,84]. NIKS keratinocytes offer the advantage of providing stable gene modification by nonviral means. Clinical safety assessment of unmodified StrataGraft skin tissue alone uncovered no significant, targeted host response to the tissue. Unlike autologous transplanted tissue, allogeneic skin substitutes offer the unique advantage of a limited lifespan on the recipient. The allogeneic cells of other skin substitutes were shown to be gradually replaced by the recipients own cells. Specifically, the cells comprising Apligraf were not detectable 2 months after application to a human wound bed [85–87]. Similarly, molecular analysis showed StrataGraft DNA was not present after 3 months (manuscript in preparation). This provides an added level of patient protection, restricting exposure to elevated levels of the transgene product to the acute stages of wound healing immediately following graft placement, when the activity of an antimicrobial peptide is most needed. Cathelicidin is also native to humans, limiting the potential for development of host responses against this protein.

DESIGNER TISSUES FOR SPECIFIC NEEDS

Clinical trials of genetically modified bioengineered skin tissue are eagerly anticipated to provide proof-of-concept for this novel type of therapeutic for the management of recalcitrant wounds. If deemed safe and efficacious, alternative skin substitute tissues could be designed to combat other wound healing deficiencies, such as poor vascularization. Venous leg ulcers comprise up to 70% of all leg ulcers, affect as much as 1% of the adult population, and are characterized by avascularization and lack of angiogenesis [88,89]. A skin substitute tissue modified to overexpress vascular endothelial growth factor (VEGF), a central mediator of angiogenesis in hypoxic environments, resulted in an increase in the number of dermal blood vessels and a decrease in time to graft vascularization in an animal model of engraftment [90]. The targeted, transient overexpression of VEGF from a proven, safe skin substitute tissue may hold tremendous therapeutic potential for promoting the healing of poorly vascularized chronic wounds, by stimulating the restoration of vascularization and deposition of granulation tissue necessary to produce definitive wound closure.

CONCLUSIONS

Advances in biomedical sciences have led to the development of numerous therapeutics for the vast chronic wound market, yet no advanced therapeutic has been uniformly adopted in replacement of the current standard of care. Only recently have head-to-head comparisons of advanced biologic therapeutics for chronic wound care been featured in clinical trials; however, the range of wound characteristics including ulcer size, depth, location, infection status, and underlying medical conditions of the patient complicate such studies. It is likely that specific wounds may require select, matched interventions to achieve rapid, definitive wound closure. The next generation of skin substitute tissues offering the expression of bioactive molecules preventing or combatting complications such as infection, or targeting known deficits in the environment of these wounds like poor vascularization, may hold the key to the next leap forward in the care of non-healing cutaneous ulcers, as well as skin loss resulting from burns, disease states, and trauma.

ACKNOWLEDGMENTS

The authors thank Mary A. Lokuta, Allen R. Comer, Cathy A. Rasmussen, Christina L. Thomas-Virnig, Kelly Barbeau, and Stuart Mohoney for contributing to this chapter.

REFERENCES

[1] SmartTRAK BioMedGPS database. BioMedGPS, LLC; 2012.
[2] Pickwell KM, Siersma VD, Kars M, Holstein PE, Schaper NC. Diabetic foot disease: impact of ulcer location on ulcer healing. Diabetes Metab Res Rev 2013;29(5):377–83.
[3] Allie DE, Patlola RR, Herbert CJ, Walker CM. Critical limb ischemia and diabetes: creative limb salvage revascularization strategies. J Vasc Ultrasound 2008;32(1):27–33.
[4] NIDDK. National diabetes statistics report. 2014.
[5] Pickwell K, Siersma V, Kars M, et al. Predictors of lower-extremity amputation in patients with an infected diabetic foot ulcer. Diabetes Care 2015;38(5):852–7.
[6] Vileikyte L. Diabetic foot ulcers: a quality of life issue. Diabetes Metab Res Rev 2001;17(4):246–9.
[7] Herber OR, Schnepp W, Reiger MA. A systematic review on the impact of leg ulceration on patients' quality of life. Health Qual Life Outcomes 2007;5:44.
[8] Martin P. Wound healing–aiming for perfect skin regeneration. Science 1997;276(5309):75–81.

[9] Diegelmann RF, Evans MC. Wound healing: an overview of acute, fibrotic and delayed healing. Front Biosci 2004;9:283–9.
[10] Falanga V. Wound healing and its impairment in the diabetic foot. Lancet 2005;366(9498):1736–43.
[11] Broughton 2nd G, Janis JE, Attinger CE. The basic science of wound healing. Plast Reconstr Surg 2006;117(Suppl. 7):12S–34S.
[12] MacLeod A, Mansbridge JN. The innate immune system in acute and chronic wounds. Adv Wound Care 2015:1–14.
[13] Pastar I, Stojadinovic O, Yin NC, et al. Epithelialization in wound healing: a comprehensive review. Adv Wound Care (New Rochelle) 2014;3(7):445–64.
[14] Mendez MV, Raffetto JD, Phillips T, Menzoian JO, Park HY. The proliferative capacity of neonatal skin fibroblasts is reduced after exposure to venous ulcer wound fluid: a potential mechanism for senescence in venous ulcers. J Vasc Surg 1999;30(4):734–43.
[15] Mendez MV, Stanley A, Phillips T, Murphy M, Menzoian JO, Park HY. Fibroblasts cultured from distal lower extremities in patients with venous reflux display cellular characteristics of senescence. J Vasc Surg 1998;28(6):1040–50.
[16] Agren MS, Steenfos HH, Dabelsteen S, Hansen JB, Dabelsteen E. Proliferation and mitogenic response to PDGF-BB of fibroblasts isolated from chronic venous leg ulcers is ulcer-age dependent. J Invest Dermatol 1999;112(4):463–9.
[17] Eming SA, Martin P, Tomic-Canic M. Wound repair and regeneration: mechanisms, signaling, and translation. Sci Transl Med 2014;6(265):265sr6.
[18] Mustoe TA, O'Shaughnessy K, Kloeters O. Chronic wound pathogenesis and current treatment strategies: a unifying hypothesis. Plast Reconstr Surg 2006;117(Suppl. 7):35S–41S.
[19] Cogen AL, Nizet V, Gallo RL. Skin microbiota: a source of disease or defence? Br J Dermatol 2008;158(3):442–55.
[20] Christensen GJ, Bruggemann H. Bacterial skin commensals and their role as host guardians. Benef Microbes 2014;5(2):201–15.
[21] Heilborn JD, Nilsson MF, Kratz G, et al. The cathelicidin anti-microbial peptide LL-37 is involved in re-epithelialization of human skin wounds and is lacking in chronic ulcer epithelium. J Invest Dermatol 2003;120(3):379–89.
[22] Hirsch T, Spielmann M, Zuhaili B, et al. Enhanced susceptibility to infections in a diabetic wound healing model. BMC Surg 2008;8:5.
[23] Kavitha KV, Tiwari S, Purandare VB, Khedkar S, Bhosale SS, Unnikrishnan AG. Choice of wound care in diabetic foot ulcer: a practical approach. World J Diabetes 2014;5(4):546–56.
[24] Margolis DJ, Kantor J, Berlin JA. Healing of diabetic neuropathic foot ulcers receiving standard treatment. A meta-analysis. Diabetes Care 1999;22(5):692–5.
[25] Lavery LA, Fulmer J, Shebetka KA, et al. The efficacy and safety of Grafix((R)) for the treatment of chronic diabetic foot ulcers: results of a multi-centre, controlled, randomised, blinded, clinical trial. Int Wound J 2014;11(5):554–60.
[26] Shah S. Clinical and economic benefits of healing diabetic foot ulcers with a rigid total contact cast. Wounds 2012;24(6):152–9.
[27] Gibson AL, Schurr MJ, Schlosser SJ, Comer AR, Allen-Hoffmann BL. Comparison of therapeutic antibiotic treatments on tissue-engineered human skin substitutes. Tissue Eng Part A 2008;14(5):629–38.
[28] Boyce ST, Supp AP, Swope VB, Warden GD. Topical sulfamylon reduces engraftment of cultured skin substitutes on athymic mice. J Burn Care Rehabil 1999;20(1 Pt 1):33–6.
[29] Greer N, Foman N, et al. Advanced wound care therapies for non-healing diabetic, venous, and arterial ulcers: a systematic review. Washington (DC): Department of Veterans Affairs; 2012.

[30] FDA. Recall of human tissue products-LifeCell Corp. September 30, 2005. 2005. http://www.fda.gov/BiologicsBloodVaccines/SafetyAvailability/Recalls/ucm053666.htm.

[31] FDA. Enforcement report for April 2, 2008. 2008. http://www.fda.gov/Safety/Recalls/EnforcementReports/2008/ucm120500.htm.

[32] FDA. Enforcement report for April 16, 2008. 2008. http://www.fda.gov/Safety/Recalls/EnforcementReports/2008/ucm120502.htm.

[33] Koob TJ, Rennert R, Zabek N, et al. Biological properties of dehydrated human amnion/chorion composite graft: implications for chronic wound healing. Int Wound J 2013;10(5):493–500.

[34] Zelen CM, Gould L, Serena TE, Carter MJ, Keller J, Li WW. A prospective, randomised, controlled, multi-centre comparative effectiveness study of healing using dehydrated human amnion/chorion membrane allograft, bioengineered skin substitute or standard of care for treatment of chronic lower extremity diabetic ulcers. Int Wound J 2014.

[35] FDA. Osiris Therapeutics, Inc. – untitled letter. 2013. http://www.fda.gov/BiologicsBlood-Vaccines/GuidanceComplianceRegulatoryInformation/ComplianceActivities/Enforcement/UntitledLetters/ucm371540.htm.

[36] Langer A, Rogowski W. Systematic review of economic evaluations of human cell-derived wound care products for the treatment of venous leg and diabetic foot ulcers. BMC Health Serv Res 2009;9:115.

[37] Kamel RA, Ong JF, Eriksson E, Junker JP, Caterson EJ. Tissue engineering of skin. J Am Coll Surg 2013;217(3):533–55.

[38] Kirsner RS, Sabolinski ML, Parsons NB, Skornicki M, Marston WA. Comparative effectiveness of a bioengineered living cellular construct vs. a dehydrated human amniotic membrane allograft for the treatment of diabetic foot ulcers in a real world setting. Wound Repair Regen 2015;23(5):737–44.

[39] FDA okays artificial skin for treatment of venous leg ulcers. Biotechnol Law Rep 1998;17(4):519–20.

[40] Blozik E, Scherer M. Skin replacement therapies for diabetic foot ulcers: systematic review and meta-analysis. Diabetes Care 2008;31(4):693–4.

[41] Organogenesis. Apligraf safety information, http://www.apligraf.com/patient/what_is_apligraf/safety_information.html.

[42] WoundSource. EpiFix® amniotic membrane allograft. 2015. http://www.woundsource.com/print/product/epifix-amniotic-membrane-allograft.

[43] Schurr MJ, Foster KN, Centanni JM, et al. Phase I/II clinical evaluation of StrataGraft: a consistent, pathogen-free human skin substitute. J Trauma 2009;66(3):866–73.

[44] Centanni JM, Straseski JA, Wicks A, et al. StrataGraft skin substitute is well-tolerated and is not acutely immunogenic in patients with traumatic wounds. Ann Surg 2011;253(4):672–83.

[45] Allen-Hoffmann BL, Schlosser SJ, Ivarie CA, Sattler CA, Meisner LF, O'Connor SL. Normal growth and differentiation in a spontaneously immortalized near-diploid human keratinocyte cell line, NIKS. J Invest Dermatol 2000;114(3):444–55.

[46] Thomas-Virnig CL, Centanni JM, Johnston CE, et al. Inhibition of multidrug-resistant *Acinetobacter baumannii* by nonviral expression of hCAP-18 in a bioengineered human skin tissue. Mol Ther 2009;17(3):562–9.

[47] Gibson AL, Thomas-Virnig CL, Centanni JM, et al. Nonviral human beta defensin-3 expression in a bioengineered human skin tissue: a therapeutic alternative for infected wounds. Wound Repair Regen 2012;20(3):414–24.

[48] De Abrew KN, Thomas-Virnig CL, Rasmussen CA, Bolterstein EA, Schlosser SJ, Allen-Hoffmann BL. TCDD induces dermal accumulation of keratinocyte-derived matrix metalloproteinase-10 in an organotypic model of human skin. Toxicol Appl Pharmacol 2014;276(3):171–8.

[49] Hirsch T, Jacobsen F, Steinau HU, Steinstraesser L. Host defense peptides and the new line of defence against multiresistant infections. Protein Pept Lett 2008;15(3):238–43.

[50] Kaus A, Jacobsen F, Sorkin M, et al. Host defence peptides in human burns. Burns 2008;34(1):32–40.

[51] Steinstraesser L, Koehler T, Jacobsen F, et al. Host defense peptides in wound healing. Mol Med 2008;14(7–8):528–37.

[52] Kim JE, Kim BJ, Jeong MS, et al. Expression and modulation of LL-37 in normal human keratinocytes, HaCaT cells, and inflammatory skin diseases. J Korean Med Sci 2005;20(4): 649–54.

[53] Braff MH, Di Nardo A, Gallo RL. Keratinocytes store the antimicrobial peptide cathelicidin in lamellar bodies. J Invest Dermatol 2005;124(2):394–400.

[54] Zanetti M. Cathelicidins, multifunctional peptides of the innate immunity. J Leukoc Biol 2004;75(1):39–48.

[55] Gallo RL, Hooper LV. Epithelial antimicrobial defence of the skin and intestine. Nat Rev Immunol 2012;12(7):503–16.

[56] Hirsch T, Spielmann M, Zuhaili B, et al. Human beta-defensin-3 promotes wound healing in infected diabetic wounds. J Gene Med 2009;11(3):220–8.

[57] Brogden NK, Mehalick L, Fischer CL, Wertz PW, Brogden KA. The emerging role of peptides and lipids as antimicrobial epidermal barriers and modulators of local inflammation. Skin Pharmacol Physiol 2012;25(4):167–81.

[58] Sorensen OE, Follin P, Johnsen AH, et al. Human cathelicidin, hCAP-18, is processed to the antimicrobial peptide LL-37 by extracellular cleavage with proteinase 3. Blood 2001;97(12):3951–9.

[59] Vandamme D, Landuyt B, Luyten W, Schoofs L. A comprehensive summary of LL-37, the factotum human cathelicidin peptide. Cell Immunol 2012;280(1):22–35.

[60] Yamasaki K, Schauber J, Coda A, et al. Kallikrein-mediated proteolysis regulates the antimicrobial effects of cathelicidins in skin. Faseb J 2006;20(12):2068–80.

[61] Nijnik A, Hancock RE. The roles of cathelicidin LL-37 in immune defences and novel clinical applications. Curr Opin Hematol 2009;16(1):41–7.

[62] Agerberth B, Charo J, Werr J, et al. The human antimicrobial and chemotactic peptides LL-37 and alpha-defensins are expressed by specific lymphocyte and monocyte populations. Blood 2000;96(9):3086–93.

[63] Gwyer Findlay E, Currie SM, Davidson DJ. Cationic host defence peptides: potential as antiviral therapeutics. BioDrugs 2013;27(5):479–93.

[64] Wang G, Mishra B, Epand RF, Epand RM. High-quality 3D structures shine light on antibacterial, anti-biofilm and antiviral activities of human cathelicidin LL-37 and its fragments. Biochim Biophys Acta 2014;1838(9):2160–72.

[65] Hoiby N, Bjarnsholt T, Givskov M, Molin S, Ciofu O. Antibiotic resistance of bacterial biofilms. Int J Antimicrob Agents 2010;35(4):322–32.

[66] Duplantier A, van Hoek ML. The human cathelicidin antimicrobial peptide LL-37 as a potential treatment for polymicrobial infected wounds. Front Immunol 2013;4(143):1–14.

[67] Dean SN, Bishop BM, van Hoek ML. Susceptibility of *Pseudomonas aeruginosa* biofilm to alpha-helical peptides: D-enantiomer of LL-37. Front Microbiol 2011;2:128.

[68] Jacobsen AS, Jenssen H. Human cathelicidin LL-37 prevents bacterial biofilm formation. Future Med Chem 2012;4(12):1587–99.

[69] Yang D, Chen Q, Schmidt AP, et al. LL-37, the neutrophil granule- and epithelial cell-derived cathelicidin, utilizes formyl peptide receptor-like 1 (FPRL1) as a receptor to chemoattract human peripheral blood neutrophils, monocytes, and T cells. J Exp Med 2000;192(7): 1069–74.

[70] Niyonsaba F, Iwabuchi K, Someya A, et al. A cathelicidin family of human antibacterial peptide LL-37 induces mast cell chemotaxis. Immunology 2002;106(1):20–6.

[71] Davidson DJ, Currie AJ, Reid GS, et al. The cationic antimicrobial peptide LL-37 modulates dendritic cell differentiation and dendritic cell-induced T cell polarization. J Immunol 2004;172(2):1146–56.

[72] Elssner A, Duncan M, Gavrilin M, Wewers MD. A novel P2X7 receptor activator, the human cathelicidin-derived peptide LL37, induces IL-1 beta processing and release. J Immunol 2004;172(8):4987–94.

[73] Tjabringa GS, Ninaber DK, Drijfhout JW, Rabe KF, Hiemstra PS. Human cathelicidin LL-37 is a chemoattractant for eosinophils and neutrophils that acts via formyl-peptide receptors. Int Arch Allergy Immunol 2006;140(2):103–12.

[74] Koczulla R, von Degenfeld G, Kupatt C, et al. An angiogenic role for the human peptide antibiotic LL-37/hCAP-18. J Clin Invest 2003;111(11):1665–72.

[75] Shaykhiev R, Beisswenger C, Kandler K, et al. Human endogenous antibiotic LL-37 stimulates airway epithelial cell proliferation and wound closure. Am J Physiol Lung Cell Mol Physiol 2005;289(5):L842–8.

[76] Steinstraesser L, Ring A, Bals R, Steinau HU, Langer S. The human host defense peptide LL37/hCAP accelerates angiogenesis in PEGT/PBT biopolymers. Ann Plastic Surg 2006;56(1):93–8.

[77] Steinstraesser L, Lam MC, Jacobsen F, et al. Skin electroporation of a plasmid encoding hCAP-18/LL-37 host defense peptide promotes wound healing. Mol Ther 2014;22(4): 734–42.

[78] Ramos R, Silva JP, Rodrigues AC, et al. Wound healing activity of the human antimicrobial peptide LL37. Peptides 2011;32(7):1469–76.

[79] Gronberg A, Mahlapuu M, Stahle M, Whately-Smith C, Rollman O. Treatment with LL-37 is safe and effective in enhancing healing of hard-to-heal venous leg ulcers: a randomized, placebo-controlled clinical trial. Wound Repair Regen 2014;22(5):613–21.

[80] Gaddy JA, Actis LA. Regulation of *Acinetobacter baumannii* biofilm formation. Future Microbiol 2009;4(3):273–8.

[81] Gales AC, Jones RN, Forward KR, Linares J, Sader HS, Verhoef J. Emerging importance of multidrug-resistant *Acinetobacter* species and *Stenotrophomonas maltophilia* as pathogens in seriously ill patients: geographic patterns, epidemiological features, and trends in the SENTRY Antimicrobial Surveillance Program (1997–1999). Clin Infect Dis 2001;32(Suppl. 2):S104–13.

[82] Rice LB. Challenges in identifying new antimicrobial agents effective for treating infections with *Acinetobacter baumannii* and *Pseudomonas aeruginosa*. Clin Infect Dis 2006;43(Suppl. 2): S100–5.

[83] Louis Jeune V, Joergensen JA, Hajjar RJ, Weber T. Pre-existing anti-adeno-associated virus antibodies as a challenge in AAV gene therapy. Human Gene Ther Methods 2013;24(2): 59–67.

[84] Deakin CT, Alexander IE, Kerridge I. Accepting risk in clinical research: is the gene therapy field becoming too risk-averse? Mol Ther 2009;17(11):1842–8.

[85] Phillips TJ, Manzoor J, Rojas A, et al. The longevity of a bilayered skin substitute after application to venous ulcers. Arch Dermatol 2002;138(8):1079–81.

[86] Griffiths M, Ojeh N, Livingstone R, Price R, Navsaria H. Survival of Apligraf in acute human wounds. Tissue Eng 2004;10(7–8):1180–95.

[87] Hu S, Kirsner RS, Falanga V, Phillips T, Eaglstein WH. Evaluation of Apligraf persistence and basement membrane restoration in donor site wounds: a pilot study. Wound Repair Regen 2006;14(4):427–33.

[88] Markova A, Mostow EN. US skin disease assessment: ulcer and wound care. Dermatol Clin 2012;30(1):107–11. ix.

[89] Herouy Y, Kreis S, Mueller T, et al. Inhibition of angiogenesis in lipodermatosclerosis: implication for venous ulcer formation. Int J Mol Med 2009;24(5):645–51.

[90] Supp DM, Supp AP, Bell SM, Boyce ST. Enhanced vascularization of cultured skin substitutes genetically modified to overexpress vascular endothelial growth factor. J Invest Dermatol 2000;114(1):5–13.

Chapter 14

The Surgical Management of Burn Wounds

Jeffrey E. Carter, James H. Holmes IV
Department of Surgery, Wake Forest University School of Medicine, Winston-Salem, NC, USA

Chapter Outline

EPIDEMIOLOGY OF BURN INJURIES

In the United States, over 486,000 patients with burn injuries receive medical treatment each year [1]. In general, patients are relatively young with an average age of 32 years and male (69%) [2]. Patients less than 5 years of age make up 19% of the burn admissions, while patients greater than age of 60 years comprise 13% of admissions. Injuries occur at home in 73% of patients and commonly are deemed "accidental."

Flame burns remain the most common mechanism of injury at 43%, followed by scalds (34%), contact (9%), electrical (4%), and chemical (3%). Infants commonly present with scald injuries, while young children have similar mechanisms to adults.

Nearly 75% of all burn admissions have an injury ≤10% total body surface area (TBSA) and an associated mortality of 0.6, which is significantly lower than 3.3–5.9% mortality for larger burn injuries. The length of stay for a burn injury is variable and ranges from 1.62 days/%TBSA to 4.78 days/%TBSA. This has produced a length of stay for all survivors of roughly 1 day/%TBSA. Inhalation injury occurs in ~10% of all burn patients and dramatically increases mortality, with a 16-fold increase in mortality when associated with a ≤20% TBSA burn.

Skin Tissue Engineering and Regenerative Medicine. http://dx.doi.org/10.1016/B978-0-12-801654-1.00014-0

The most frequent complication in burns patients is pneumonia, which occurs in 5.8% of all flame burns.

PRESENTATION OF BURN INJURIES

In the United States, severe burn injuries, or those with complicating factors, are typically transferred to one of over 120 burn centers for definitive care by surgeons, nurses, therapists, and many others with specialty training in burn care and resources aligned with the severity of the injuries. In an effort to ensure that patients with burn injuries were appropriately referred to a burn center, the American Burn Association created referral criteria in 1990 [3]. Referral criteria included assessment of the wound size and location for partial- and full-thickness injuries, mechanism of injury, preexisting illnesses, concomitant trauma, and resources of the referring facilities (Table 1).

Once an injury has occurred, a patient should be evaluated by a health-care provider, ensuring hemodynamic and airway stability and careful removal of burned or contaminated clothing. Wounds with blisters or areas appearing whitish, translucent, or charred are included in the calculation of wound size, noting location. A TBSA calculation includes the percentage of involved areas using a Lund–Browder diagram or other methods of size calculation. The American Burn Association created the Advanced Burn Life Support provider course to train health-care providers in the management of patients with burn injuries for the first 24h and is available through the American Burn Association. Following initial stabilization, there are three phases of care: resuscitation, repair, and rehabilitation.

TABLE 1 Burn Center Referral Criteria

1. Partial-thickness burns greater than 10% total body surface area (TBSA).
2. Burns that involve the face, hands, feet, genitalia, perineum, or major joints.
3. Third-degree burns in any age group.
4. Electrical burns, including lightning injury.
5. Chemical burns.
6. Inhalation injury.
7. Burn injury in patients with preexisting medical disorders that could complicate management, prolong recovery, or affect mortality.
8. Any patient with burns and concomitant trauma (such as fractures) in which the burn injury poses the greatest risk of morbidity or mortality. In such cases, if the trauma poses the greater immediate risk, the patient may be initially stabilized in a trauma center before being transferred to a burn unit. Physician judgment will be necessary in such situations and should be in concert with the regional medical control plan and triage protocols.
9. Burned children in hospitals without qualified personnel or equipment for the care of children.
10. Burn injury in patients who will require special social, emotional, or rehabilitative intervention.

RESUSCITATION

While many burn centers vary in their subtle management of complicated burn resuscitations based upon endemic factors or biases, the American Burn Association recommends that patients with burn injuries with ≥20%TBSA undergo fluid resuscitation using crystalloid solution to start [4]. The goal of resuscitation is to replace volume deficits lost as a result of the burn injury causing increased capillary permeability that can lead to organ dysfunction and death. Inadequate or excessive resuscitation can have negative consequences, such as increased tissue edema or poor tissue perfusion.

The rate of resuscitation varies between 2 and 4cc/kg of body weight/% TBSA burn during the first 24h and typically incorporates lactated Ringer's solution, which is slightly hypotonic relative to plasma [6]. By this method, larger patients or patients with larger injuries will receive larger volume replacement relative to smaller patients or smaller injuries. Resuscitation is not a fixed process, however. Close monitoring of heart rate, blood pressure, and especially urine output is required hourly during the first 24–48h. Resuscitation is commonly increased or decreased if the goal urine output, defined as 0.5–1.0mL/kg/h in adults and 1.0–1.5mL/kg/h in children, is not met. Frequently, patients with inhalation injuries, delays in presentation, or severe full-thickness injuries will require increased volume requirements (Figure 1). Of note, children require additional maintenance fluid beyond the 2–4cc/kg of body weight/% TBSA burn, due to variation in glycogen storage and limited physiological reserve [5].

Adaptations to resuscitations include administration of colloid-containing fluid such as plasma or albumin, high-dose ascorbic acid administration, and plasma exchange therapy. These options are used by burn providers as adjuncts

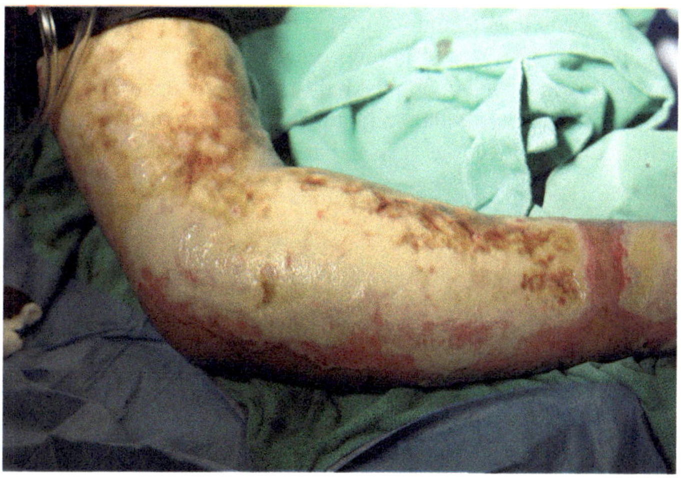

FIGURE 1 Full-thickness burn injury to the arm.

to augment end-organ perfusion or reduce further crystalloid administration and improve organ dysfunction. While promising, many are still undergoing continued investigation, and widespread use remains variable.

Complications in resuscitation can include "fluid creep" secondary to over-resuscitation, acute respiratory distress syndrome, or hypotension secondary to myocardial ischemia in those with coronary artery disease or reduced cardiac output during early burn shock [7,8]. Successful resuscitation reduces organ dysfunction and allows for the next stage of care—repair.

Wound Care

Burn wounds present the patient with an inherent risk for infection and complications and require consummate care. Wound care is focused on reducing nonviable tissue from the wound bed, providing patient comfort and protection especially considering wound drainage and painful stimuli, and applying dressing materials with antibacterial properties resulting in a moist, warm wound bed ready to facilitate re-epithelialization.

Wound care with topical antimicrobial agents is focused on prophylactic coverage with minimal toxicity. Silver sulfadiazine is a sodium sulfadiazine and silver nitrate water-based compound that is applied as a cream once or twice daily to the wound, with in vitro activity against common cutaneous organisms. Complications include transient leukopenia and skin sensitivity in 5–15% of patients [9,10]. Mafenide acetate serves as an alternative antibacterial topical agent, with good eschar penetration as an 11.1% cream base or excellent coverage as a 5% solution. Unfortunately, mafenide can function as a carbonic anhydrase inhibitor in some patients, resulting in a metabolic acidosis that can be dose and therapy duration related [11]. Alternatives include polymyxin B and silver-based dressings. Silver-based dressings include gels or woven fiber where silver is temporarily bonded until activated by wound exudate and released into the wound bed. To date, relatively few complications have been reported using silver-based dressings, and continued industry development leaves great promise for a dressing that providers and patients will find both comfortable and effective at reducing infectious risks in burn wounds.

REPAIR

Following successful resuscitation, patients and providers are faced with the challenge of repairing damaged skin and soft tissue. Surgical intervention typically involves excision of the nonviable tissue with either immediate wound closure using full- or split-thickness autografts (STAGs) or staged surgery in which the nonviable tissue is excised and wounds are then covered with skin substitutes. Early excision and grafting has repeatedly been demonstrated to improve mortality and reduce the inpatient length of stay for adults and children [12–14]. Following excision, clinical efforts are directed to healing the

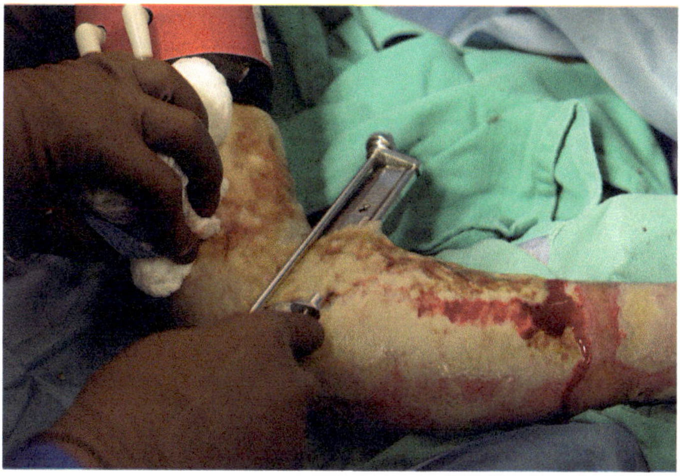

FIGURE 2 Excision with tourniquet and Watson knife.

wound with minimal collateral damage. Small wounds may be amenable to primary closure with adjacent skin flaps, while larger wounds require a transfer of skin from another area of the body in the form of a skin graft.

Excision

Burn excision removes nonviable tissue that would delay wound healing, due to the burn eschar's strong inflammatory response and increased risk of bacterial colonization and infection. Excision is continued until a plane of well-vascularized tissue, with punctate bleeding, is reached and necrotic material or debris is removed. Tangential excision typically occurs with hydrosurgical or sharp excision and helps maintain body contour/form but is prone to blood loss unless performed under tourniquet. Fascial excision results in less blood loss, given the continuous use of cautery, and allows quick removal of deep burns that would otherwise require serial attempts using tangential excision.

Excision can be performed using a sharp blade, electrocautery, or hydrosurgical instruments. Sharp excision has traditionally been performed using a hand-held dermatome with a disposable blade and adjustable guard that aids in determining the depth of excision and preventing deep excision potentially damaging vital subcutaneous neurovascular structures. The Weck or Goulian knife has disposable blades that are covered with an exchangeable guard of various preset depths. A larger version, with a built-in adjustable guard, is the Braithwaite, Cobbett, or Watson knife (Figure 2). Electrocautery excision is performed using monopolar instruments to dissect and cauterize where deep burns have compromised the entire dermis. Hydrosurgical excision combines irrigation and mechanical forces to remove tissue using a thin, high-velocity, saline jet to excise necrotic tissue and create an adjacent vacuum to remove

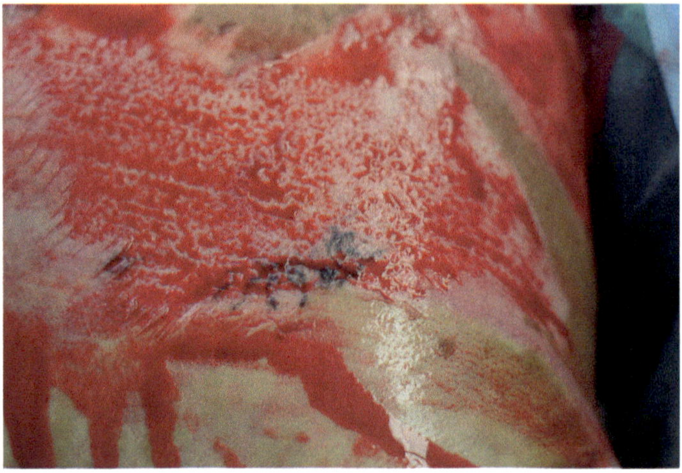

FIGURE 3 Wound bed following sharp excision.

debris, while minimizing collateral damage of healthy tissue. Areas of excisional research include ultrasonic dissection using high-frequency vibration to create local hydrodynamic forces and topical drugs that can aid in the removal of necrotic tissue.

Blood loss is frequently controlled with topical vasoactive drugs such as epinephrine in 1:100,000, tourniquet use in extremities, tumescence with vasoactive drugs in truncal injuries, thrombin sealants, and goal-directed resuscitation.

Following excision, wound closure becomes the primary goal. Many wounds are immediately amenable to autografting. Patients undergoing fascial excision or patients with large wounds and severe comorbid conditions prohibiting single-stage operations can be excised and treated using "skin-substitutes," with the goal being to either optimize the wound bed into a confluent vascularized foundation more amenable to autografting or, at minimum, to function as a barrier preventing insensible heat and fluid loss (Figure 3).

Autografts

Autograft skin closure techniques date back to 1871, with radical advancements since then including improved harvesting techniques, donor site recovery, and skin expansion techniques [15]. Autografts allow patients to serve as their own donors for full-thickness autografts or STAGs. Full-thickness autografts are typically used for small areas, such as the face or hand, given that the donor site will require surgical closure following the harvest. STAGs are used on larger areas and harvested using an air- or electric-powered dermatome, which oscillates a blade in a covered device. They are usually taken at 0.008–0.012 inch from areas easily covered by clothing, such as thigh or back. STAGs have the advantage of leaving a donor site that will heal in 2–3 weeks on average and allow reharvesting, if necessary. Nonetheless, STAG donor sites invariably scar and are exquisitely painful.

Once harvested, STAGs can be used as a sheet graft with no perforations or meshed to allow greater expansion and egress of wound exudate that might otherwise result in a hematoma or seroma with graft loss. Meshing ratios range from 0.25:1 to 6:1, allowing STAGs to cover significantly more area than just the size of the native donor site and aid in closing large wounds. The larger the meshing ratio, the larger the interstices are in the autograft, resulting in longer wound healing time as the epidermal cells migrate to cover the interstices.

Techniques including cultured epithelial autograft (CEA) and autologous cell-suspension devices offer decreased donor site morbidity. CEAs are commercially available, patient-specific, epidermal autografts grown to order from a skin biopsy. While promising, CEAs are noted to increase the costs of burn care and require several weeks of cultivation prior to graft placement. Autologous cell suspension-harvesting devices, like ReCell® (Avita Medical, Royston, UK), allow broad expansion through enzymatic "meshing," so that 80 cm^2 of wound is covered for each 1 cm^2 of donor skin, and it is performed in the operating room at the time of excision [16,17].

Allografts

Allografts are harvested from cadaveric sources either locally or cryopreserved commercially and maintained in skin banks. Allografts rapidly adhere to the wound bed and reduce heat and fluid loss, prevent wound desiccation, and improve pain control, while marginally decreasing the hypermetabolic demands of the open wounds and potentially providing a dermal template for regeneration. Indications for allograft include: placement following excision of full-thickness burn injuries, coverage of widely meshed autografts, coverage for exfoliative disease processes, or testing the viability of the wound bed before proceeding with autografts (Figure 4). Allograft use as a coverage for widely meshed STAGs was first described in 1981, and this technique allows the allograft to serve as a protective layer to prevent desiccation and disruption of fragile autografts [18].

Xenografts

Porcine-derived xenograft is the primary xenograft used today. Split-thickness porcine xenografts are harvested and either cryopreserved or glycerol preserved to offer readily available wound coverage. Xenograft does not vascularize to the degree of allograft, but is effective at reducing pain in partial-thickness injuries such as scalds [19]. Xenograft has also been used for exfoliative skin disorders, like toxic epidermal necrolysis [20]. While the role of xenografts and allografts

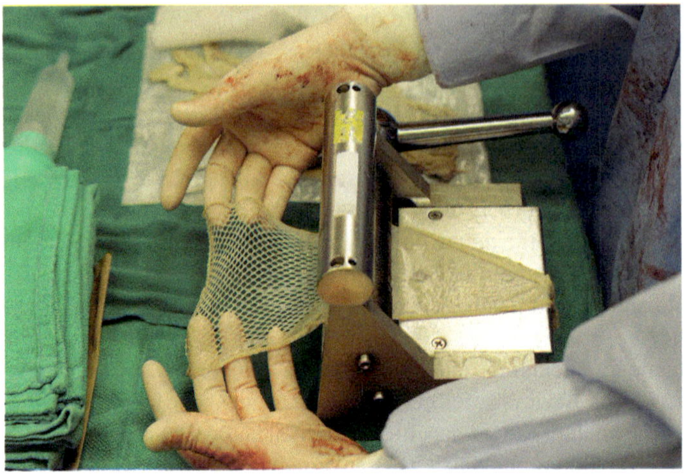

FIGURE 4 2:1 mesh autograft.

overlap, the costs of each do not. Prior research demonstrates that commercially available xenograft is approximately one-tenth the price of commercially available allograft [21].

Dermal Substitutes

Dermal substitutes for use in burn surgery have been commercially available since 1996, when Integra® (Integra LifeSciences, Plainsboro, NJ) was initially approved by the US FDA. Today, the number of dermal substitutes with indications for burns numbers 5–10 worldwide. Despite their rapid development and expansion into the burn market over the last 2 decades, dermal substitutes are still not widely used on a regular basis either in the United States or worldwide. Potential explanations for the apparent failure of their widespread adoption in burns are as follows: (1) There are no high-quality data to support their efficacy or demonstrating reduced mortality/morbidity; (2) They are very expensive; (3) There are critical, surgeon-specific, human factors associated with their successful use resulting in a learning curve and the potential for wide interuser variability in results.

Nonetheless, all available dermal substitutes carry the same promise and hope—to reduce the morbidity of autografting and possibly reduce mortality. Dermal substitutes are derived from bovine, porcine, or human sources. They all enable the application of an ultra-thin STAG (0.006–0.008 inches), called an epidermal autograft, to be applied to an appropriately vascularized dermal substitute. Most dermal substitutes require 1–3 weeks for vascularization following application to an excised burn wound bed. In theory, utilization of epidermal autografts with dermal substitutes allows more rapid donor site healing, with reharvesting if needed, quicker total wound closure, and possibly decreased lengths of stay in the hospital. All available dermal substitutes rely on the same treatment paradigm, with subtle nuances existing between them with respect to their successful use. A burn surgeon typically chooses to use the dermal substitute that works best for him/her.

There have clearly been significant scientific and technological advancements in the past few years, leading to a wide range of dermal substitutes on the market. However, clinical science has not kept pace, and there is virtually no data to effectively support and justify the use of dermal substitutes in burns. Perhaps the great advancement of this decade will be large, prospective, randomized multicenter clinical trials assessing the efficacy of, and possible superiority for, a currently available dermal substitute.

REFERENCES

[1] National Hospital Ambulatory Medical Care Survey. Emergency department summary tables. http://www.cdc.gov/nchs/ahcd/web_tables.html#2011; 2011[accessed on January 22.01.15].

[2] National Burn Registry. NBR annual report (American burn association web site). 2014. Available from: http://www.ameriburn.org/2014NBRAnnualReport.pdf. Internet [accessed 24.04.15].

[3] American College of Surgeons. American college of surgeons committee on trauma: guidelines for the operation of burn units: resources for optimal care of the injured patient. Chicago: American College of Surgeons; 1990.

[4] Pham TN, Cancio LC, Gibran NS. American burn association practice guidelines burn shock resuscitation. J Burn Care Rehabil 2008;29:257–66.

[5] Aynsley-Green A, McGann A, Deshpande S. Control of intermediary metabolism in childhood with special reference to hypoglycaemia and growth hormone. Acta Paediatr 1991;377:43–52.

[6] Burn Resuscitation: the results of the ISBI/ABA survey. Burn 2010;36(2):176–82.

[7] Saffle JR. The phenomenon of "fluid creep" in acute burn resuscitation. J Burn Care Res 2007;28:555–65.

[8] Horton JW, White J, Maass D. Protein kinase C inhibition improves ventricular function after thermal trauma. J Trauma 1998;44:254–64.

[9] Choban PS, Marshall WL. Leukopenia secondary to silver sulfadiazine: frequency, characteristics, and clinical consequences. Am Surg 1987;53:515–7.

[10] Gamelli RL, Paxton TP, O'Reilly M. Bone marrow toxicity by silver sulfadiazine. Surg Gynecol Obstet 1993;177:115–20.

[11] Asch MJ, White MG, Pruitt Jr BA. Acid-base changes associated with topical sulfamylon therapy: retrospective study of 100 burn patients. Ann Surg 1970;172:946–50.

[12] Muller MJ, Herndon DN. The challenge of burns. Lancet 1994;343:216–20.

[13] Burke JF, Bandoc CC, Quinby WE. Primary burn excision and immediate grafting: a method for shortening illness. J Trauma 1974;14:389–95.

[14] Herndon DN, Barrow RE, Rutan RL, et al. A comparison of conservative versus early excision therapies in severely burned patients. Ann Surg 1989;209:547–53.

[15] Reverdin JL. Sur la greffe epidermique. CR Acad Science 1871;73:1280.

[16] Gravante F, Di Fede MC, Araco A, Grimaldi M, De Angelis B, Arpino A, Cervelli V, Montone A. A randomized trial comparing ReCell® system of epidermal cells delivery versus classic skin grafts for the treatment of deep partial thickness burns. Burns 2007;33(8):966–72.

[17] Wood FM, Giles N, Stevenson A, Rea S, Fear M. Characterisation of the cell suspension harvested from the dermal epidermal junction using a ReCell® Kit. Burns 2012;38(1):44–51.
[18] Alexander JW, MacMillan BG, Law E, et al. Treatment of severe burns with widely meshed skin autograft and widely meshed skin allograft overlay. J Trauma 1981;21:433–8.
[19] Bukovcan P, Koller J. Treatment of partial-thickness scalds by skin xenografts- a retrospective study of 109 cases in a three-year period. Acta Chir Plast 2010;52:7–12.
[20] Bradley T, Brown RE, Kucan JO, smoot EC, Hussmann J. Toxic epidermal necrolysis: a review and report of the successful use of biobrane for early wound coverage. Ann Plast Surg 1995;35:124–32.
[21] Yovino SJ, Dominic WJ. Reduction in cost for temporary skin coverings. Suppl J Burn Care Res 2014;35(3):S69.

Chapter 15

Advances in Isolation and Expansion of Human Cells for Clinical Applications

Fiona Wood
University of Western Australia, Crawley, WA, Australia

Chapter Outline

INTRODUCTION

Skin integrity is essential for normal physiological function. The complex three-dimensional structure is a result of the interplay of cells from all defined embryological layers and the supporting extracellular matrix (ECM) [1]. The skin has a capacity for regeneration and is responsive to the external environment and internal signaling. All too frequently, the skin is damaged, such that the capacity to regenerate is overwhelmed, and the integrity of the cutaneous surface is restored by scarring [2]. Cutaneous scarring is distinct from skin in terms of the structure of the architectural framework, the chemistry of the ECM proteins, and the phenotypic expression of the cells within the scar construct. The limited capacity to heal by regeneration and the sequelae of scarring, both functionally and aesthetically, have driven the exploration of techniques for expedient wound treatment to restore the cutaneous barrier in a regenerative pattern [3].

Skin grafting is the well-established method to achieve healing post skin injury; however, as the area of injury increases, the available noninjured skin on the body for use as donor sites is clearly reduced [4]. There are techniques to expand the area of coverage achieved from a given donor site, such as meshing [5], micro-grafting [6], and the use of buried tissue expanders [7]. Although widely practiced and effective, the limitations of these techniques

Skin Tissue Engineering and Regenerative Medicine. http://dx.doi.org/10.1016/B978-0-12-801654-1.00015-2

and the desire to improve the therapeutic opportunities for healing and scar reduction led to the exploration of tissue-engineering solutions. The use of laboratory tissue expansion, based on tissue culture techniques, has been used clinically to repair cutaneous wounds since the early 1980s [8].

Initially, in situations of severe donor site restriction in major burn injury, the cells harvested from the dermo-epidermal junction (DEJ), grown in a keratinocyte selective media into confluent cultured epidermal autograft (CEA) cell sheets, greatly expanded the surface area of wound coverage from a restricted donor site [9]. However, there are issues with CEA that have limited its clinical use and include:

- the time taken to achieve confluence [10].
- the risks and expense associated with a prolonged laboratory culture period [11].
- the resulting skin repair is solely epidermal (i.e., lacking dermis) and so fragility is a problem [12].

The subsequent work to address these issues has taken essentially two paths:

- the development of a complex tissue-engineered solution aiming to provide a layered, cell-seeded skin construct prior to grafting [13].
- the harvesting of cells for rapid clinical use delivered to the wound via a suspension or on the surface of a carrier dressing material with tissue generation at the wound site [14].

The cell populations harvested from the skin that form the base in either situation remain a complex mix of keratinocytes, fibroblasts, melanocytes, endothelial cells, Langerhans cells, neural cells, and glandular cells [15].

The following sections consider the isolation, expansion, and clinical application of the key cells of the skin for use as a single-cell population, or in combination, with a focus on the clinical use.

SKIN HARVESTING

The source of the cells for current clinical use is either from allograft donor skin or autograft skin. The cells are harvested from the skin construct as a mixed population, including skin progenitor cells and those that are fully differentiated [16]. There is no universal stem cell in current clinical use though there is potential to explore the use of mesenchymal stem cells from allograft source [17], and harvesting the patient's own cells that retain the capacity for proliferation from a range of tissues such as fat [18]. However, the discussion of the use of a selected stem cell approach is beyond the scope of this chapter.

The key advantage of an allograft source of cells is to have the cells readily available, without the inherent delay associated with the cell culture process [19]. Harvested skin may be from live donors, such as in the development of Transcyte [20]: a construct of silicone, collagen, and neonatal fibroblasts sourced from neonatal circumcision specimens. Skin from surgical excisions may be used as a source, or skin may be harvested from organ donors. In all cases, compliance with the regulation framework is essential to ensure that the risks associated with infection transmission are addressed [21].

The harvesting of the skin must be done under aseptic conditions, with antimicrobial skin cleaning, sterile drapes, and instrumentation. The conditions favorable for skin cell viability and proliferation are also favorable for microbial contaminates. Therefore, maintaining a sterile field is essential. Once harvested, the skin needs to be maintained in moist conditions in a sterile container and kept between 4 and 10 °C for transport or refrigerated until cell harvesting. The viability of the cells within the skin will decrease over time, and the rate of cell damage can be reduced by cooling, also the transport in gauze moistened by cell culture media is advocated to reduce the rate of reduction of cell viability [22]. Some advocate the addition of antibiotics to the solution for transfer. The use of antimicrobials is also used by some in the culture process; however, the aim should be to not have contamination and therefore to maintain aseptic conditions along the continuum of the process from harvest to clinical delivery.

Macroscopically, the skin over different body sites differs related directly to the functions of the skin in that area [23]. Donor skin for donor cells needs to consider site matching whenever possible for optimum skin restoration. It should be noted that younger skin will yield greater numbers of viable cells than skin from an older donor [24]. The functionality of the cells harvested will be influenced by the pathophysiology of the host, for example, poor nutrition, presence of malignancy, drugs, and alcohol, as well as age [25].

The donor skin must be harvested, stored, and transported to the processing area, with meticulous attention to detail at every stage [26].

The practical technique of harvesting the skin can impact on the process of cell isolation. A full-thickness piece of skin has the advantage that the donor site may be able to be sutured and donor site healed by primary intention. However, the skin will require a prolonged immersion in the enzyme solution to facilitate cell dissociation [27].

The alternative is a split-thickness skin graft harvesting the epidermis with papillary dermis and a varying amount of reticular dermis depending on the thickness of the sample. The thinner the split-thickness skin is, the less time required in the enzyme solution for dissociation. The cell yield will change as the depth increases specifically to include the region of the hair follicle bulb [28]. The donor site wound will heal by secondary intention and can therefore be harvested again at a later date if required.

The cells around the hair follicle can be specifically harvested by plucking the hair [29]. To harvest sufficient cell numbers requires multiple hairs, and the time in culture is prolonged to achieve useful cell numbers.

Once the skin is harvested, clearly the less time taken to commence the cell isolation process, the better the preservation of cell functionality. The whole process needs to be planned in the context of the clinical problem, specifically the body site needing repair, the total surface area, depth of skin damage, and so on. Understanding the clinical context is essential in terms of the optimal timing of the potential cell-based therapy options.

CELL ISOLATION

The cells are the engine room of the skin, producing the surrounding specialized tissue construct ECM driving the morphology of the skin. Their intimate relationship within the tissue can be disrupted enzymatically, chemically, and physically in order to isolate the cell populations [30].

It is well established that fragments of skin can be maintained in vitro and subsequently returned to a donor [31]. The conditions of the incubation of the skin can be manipulated to facilitate cell proliferation, the premise of healing by secondary intention [32]. Initial work used skin fragments with the harvest of the cells from the edge of the explant [33]. Agitation was used as a method to release the cells from the advancing edge. The media for incubation developed in complexity from glucose-enriched electrolyte solutions to those introducing growth factors to stimulate cell activity [34]. However, the explant technique has limited utility in terms of the differential growth of fibroblasts over keratinocytes and the time taken to generate large numbers of cells for clinical use.

The key to harvesting cells is to do so retaining their viability. The use of trypsin as a method of enzymatic harvest of keratinocytes was reported in the 1950s, opening the way to explore single-cell suspension in culture conditions [35]. Trypsin is a serine protease produced by the pancreas of mammals, there are now synthetically produced analogs; by cleaving the peptide bonds it disrupts the adhesion between cells and the cells and the ECM at the level of the suprabasal desmesomes. It is frequently used in combination with EDTA as a chelating agent of calcium, which would otherwise bind to the trypsin and limit its activity.

Collagenases to isolate fibroblasts from dermal samples may be used in isolation and in combination with trypsin, such that the concentration of the enzymes in the solution can be kept to a minimum to reduce the impact on cell viability [36].

Dispase is a bacterial neutral protease that has been used to selectively dissociate the skin at the basement membrane level of the DEJ by selective action on the hemidesmesomes [37]. Although secondary cell isolation is needed, this allows separation of the epidermis from the dermis facilitating the isolation of either the keratinocyte or fibroblast without cross-contamination. Dispase is also used for releasing cells from the surface of the tissue culture system post expansion [38].

Thermolysin, a neutral metalloprotease of bacterial origin, has been explored, and with action similar to dispase, it facilitates the separation of the dermis from the epidermis. The combination of enzymatic dissociation and physical forces are reported to improve the cell yield and stem cell functionality in culture [39]. Again, the secondary trypsinization for less time at a lower concentration allows isolation of keratinocyte cell populations with less fibroblast contamination.

The focus is on the retention of multiple keratinocyte phenotypes in a state of proliferation for expansion [16]. The keratinocyte population will include the basal cells, capable of ongoing proliferation, all the way through to those that are terminally differentiated. As the cells are cultured and passaged, the characteristics of the cell phenotype within the keratinocyte population will change as those with colony-forming capability are retained.

The conditions used for cell expansion are integral to the development of a specific cell population; for example, the fibroblast feeder layers are known to inhibit fibroblast proliferation and facilitate keratinocyte proliferation [40].

Melanocytes are also a cell population of interest for both research and clinical purposes. The melanocytes are present in the cells harvested from the DEJ, and they can be selectively stimulated by modification of the culture conditions [41]. The melanocytes can be used clinically in isolation in the treatment of conditions such as vitiligo [42] and in conjunction with other skin cell populations applied directly onto the wound for repair or seeded into a three-dimensional scaffold for composite skin transfer.

The cells within the skin construct can be identified by morphological characteristics and cytoplasmic and cell membrane features related to their function assisting in their isolation:

- Keratinocytes are a highly specialized responsive cells intimately linked with the dendritic cells, the ECM, and proteins of the basement membrane of the DEJ. They form the first line of defense of the innate immune system and differentiate to produce a multilayered waterproof structure, the epidermis. They express a range of proteins on the cell surface (e.g., integrins and defensins), linked with the functional demands on the cell at the time [43]. They also express a range of cytokeratins that are specific to the body site (e.g., cytokeratin 9 related to the epidermis of glaberous skin) [44]. They have a capacity to migrate across a breach in the surface to heal a wound and reestablish epidermal continuity. Within the keratinocyte linage are specialized cells related to skin adnexal structure such as hair, sebacous glands, and nails. The understanding of the characteristics of the skin at a given donor site is linked to the clinical utility of the cells and is a key issue to consider.
- Fibroblasts are the main cell of the dermis, providing the structural integrity to the skin tissue. They produce the protein of the ECM, predominately collagen and elastin [45]. They are motile and have a contractile capacity. Again their function replies in an intimate relationship with the ECM which needs to be disrupted in order to isolate the cells.
- Melanocytes are cells with a dendritic morphology arising from the neural crest. They are located at the DEJ and hair follicle region and predominately produce melanin in response to sunlight stimulation. The melanin is transferred to the keratinocyte requiring a close relationship for the effective transport mechanism. The cells can be identified by the presence of a range of markers (e.g., DOPA and S100) [46].

The use of enzymatic dissociation, predominately with trypsin-based proteases, remains the mainstay of cellular harvesting for clinical use. The choice of donor material and cell isolation procedure can vary depending on the clinical situation. Allograft has the potential to supply a bank of cells continually ready for harvest. The laboratory-based procedures for clinical use require a fully accredited GMP facility, compliant with the regulatory framework [47].

The time to cell availability for clinical use remains a key issue in the use of autograft cells. The isolation and culture to a confluent sheet takes about 3 weeks to prepare [38]. CEA has been widely used in a range of areas, specifically major burn injury. The laboratory time can be reduced by harvesting the cells in culture when they are subconfluent [14]. They then need to be delivered to the wound using a carrier or via an aerosol [48].

Seeded skin composites take up to 6 weeks to prepare for clinical use [49]. The inkjet-printed composites require the time to expand the cell populations prior to the printing [50]. In a planned reconstructive surgical repair of the skin such as scar revision, there is time to produce cells via laboratory expansion methods with small donor sites.

For immediate clinical use, the isolated activated cells from the DEJ may be used clinically with no prior cultured phase, as harvested using the ReCell device [51]. This allows modest expansion, up to 1 to 80, as would be achieved in the first passage of the culture process, but achieved on the wound surface rather than in the laboratory. A mixed-cell population is harvested from the DEJ with no efforts to segregate the cells into their types, but returned to the wound immediately to facilitate wound healing.

The focus of cell harvesting for clinical use to date has been on the restoration of:

- epidermis using keratinocytes
- dermis using fibroblasts
- pigment using melanocytes

The cells can be used in isolation or in combination, related to the timing of intervention, based on the needs of the clinical problem.

The endothelial cells of the vasculature within the skin may be harvested and used to establish a vascularized construct for skin tissue transfer not yet in the clinical arena [52]. The cells of the sebaceous units, sebocytes, can be isolated and cultured; currently, they are not used clinically but have been used as a model in drug testing [53]. The hair follicle is an area of much research, with a view to hair regeneration but limited success since the initial reports [54]. The cells around the hair follicle can be used as a source of cells with minimal donor site morbidity [29]. The cells of the inflammatory and immune system are responsive and invade rapidly from the host in the wound healing response and are not explored further [55].

The use of pleuripotential cells are the subject of much research, with the aim of identifying a universal cell that can be driven down the lineage of the cell phenotypes of the skin. Currently, no such cell is in clinical use for skin repair, and is beyond the scope of this chapter.

Within the isolated cell populations are cells with a phenotype for replication that retain the capacity to proliferate in appropriate culture conditions facilitating expansion of cell numbers for clinical use.

CELL EXPANSION

The drive for rapid skin healing to save morbidity and mortality underpins a body of research focusing on tissue expansion. In vivo skin expansion is widely practiced culturally and therapeutically. With the development of inert balloon materials, tissue expanders can be placed subcutaneously and slowly enlarged [7]. The resulting skin is as a result of recruitment from the surrounding area, stretch, and also tissue growth with a measurable increase in dry weight DNA. However, the cells within the skin tissue construct are restrained and the time taken to develop new tissue is in the order of months. Cell isolation into discrete populations was the first step in exploring the potential of cell-based tissue expansion. The clinical need to augment skin healing rapidly in situations such as major burns continues to be a key driver of cell-based in vitro tissue expansion research [56].

The first step in the restoration of the skin using tissue-engineering principles is a source of cells capable of differentiation into the specific layers of skin, along with a scaffold capable of maintaining the cell viability [57]. In the absence of a single universal cell, it was clear in the early stages of the research that each cell type could be optimized in culture by the conditions of the media and the tissue culture environment. The successful expansion of a cell population requires consideration of the environment (aseptic, temperature controlled and optimally at 5% level of carbon dioxide), the makeup of the cell culture media and the substrate for cell attachment.

Currently, there are commercially available media specific to a broad range of cell lines, most are well characterized and aimed at the cell culture in the research and development and testing environments [58]. When considering cell expansion for clinical use, it is vital to look at all source materials, many are animal origin from highly regulated herds (e.g., porcine trypsin and fetal bovine serum). In all cases, a clear risk–benefit analysis must be taken into consideration when clinically implementing the technology.

The early attempts at culturing cells from skin fragments maintained moisture with an isotonic electrolyte solution to which glucose was added to provide an energy source [59]. The basic media, Dulbecco's minimal essential media (DMEM), was enriched by the use of biological fluids such as ascites or serum [60]. With the development of techniques of large molecule separation and identification, cytokines with the capacity to specifically target cell populations and stimulate cell proliferation were introduced into the media. As the media become more complex and specific for the target cell, they could be manipulated to reduce proliferation of the nontarget cells (e.g., manipulating the calcium level will drive keratinocytes to differentiate rather than proliferate [61] or the use of tetradecanoyl phorbol acetate (TPA) in melanocyte culture which is unfavorable to keratinocytes) [62].

Currently, media are characterized as natural containing biological fluids or artificial defined media. Though natural media is widely used and effective, the use of biological source material such as fetal bovine serum and bovine pituitary extract are not 100% reproducible and have the issue of potential transmission of disease [63]. The knowledge of source material is essential when considering the clinical context. Bovine spongiform encephalopathy was a clear demonstration of why the regulatory framework is so stringent in this regard [64]. The desire to control the constituents of the media and its activity has led to the development of artificial defined media, with clarity around the constituents in terms of concentration and activity. Using serum-free media requires the secondary introduction of stimulating cytokines to establish an equivalent rate of cell proliferation. It is reproducible and again, the source materials of the additives need consideration [65].

Media manipulation in isolation was not successful in achieving large-scale expansion beyond the explant concept. Understanding the need for cell attachment to a surface for proliferation was also a key element in enhancing cell culture for successful cell expansion [66].

A major step forward was made by Rhienwald and Green in 1975 with the development of a system for keratinocyte culture, considering both the media and attachment substrate and including DMEM with Hams F12 media with EGF and cAMP-elevating agent, cholera toxin, on an irradiated 3T3 layer of mouse fibroblasts [38]. The control of the environment within the culture system is essential for optimal expansion. The use of pH, buffers with color indicating change in pH, guides the changing of the media and is intended to keep the cell in a proliferative state. This remains one of the most efficient methods of keratinocyte culture. All components of the system have been rigorously analyzed over the decades since the initial clinical use in the early 1980s [67]. However, synthetic source materials continue to be explored to reduce the risks of biological origin materials and remain an area of significant activity, many of which are now commercially available.

In addition, the surface layer for cell attachment has developed considerably from the irradiated 3T3 murine fibroblasts, to human fibroblasts, and on to ECM proteins such as collagen, lamanin, hyaluronic acid, fibrin, and fibronectin; all suitable to sustain cell growth and provide a mechanism for cell transfer [68]. Cells have also been cultured on microsurfaces as beads manufactured from both synthetic and biological materials, to facilitate cell transfer [69].

The laboratory processes involved in cell culture require a specialized workforce. The upscaling to large cell numbers may be done using cell factory systems. Large-scale media changes can be done in a sealed system, which reduces the risk of contamination [70]. Large numbers of cells are needed for seeding into constructs to develop skin composite materials. Also, the system of inkjet printing has a significant dead space within the printing device and large-scale automated cell culture is well suited to link with the printing mechanism [71].

The environment of optimal cell expansion may arguably be the prepared wound surface. The laboratory-based cell expansion techniques are achieved by replicating key aspects of the wound conditions in a controlled environment. The secondary harvest of the cell populations is directed by the clinical utility and the clinical need.

The concerns with respect to contamination can be avoided by using the isolated cells immediately. Initial expansion can be achieved by seeding the cells harvested from the DEJ directly onto the prepared wound bed as an activated mixed-cell population harvested at the time of the procedure to augment skin healing.

Further, expansion is achieved in the culture environment by selectively culturing cells for a specific purpose. The keratinocyte-selective systems are used and can be passaged to continuously expand the available cell population. Again, the time for secondary harvest can be planned to provide a cell suspension, or for the cells to be seeded onto a specialized surface that acts as a temporary attachment substrate, a carrier system, and a primary dressing [72].

The culture environment can be physically manipulated to drive differentiation. By lifting a confluent sheet of cells to an air–liquid interface, the keratinocytes will develop a keratin layer and be more robust for transfer [73].

Carrier surfaces, such as allograft dermis or single proteins most frequently collagen, can be seeded for expansion and subsequent delivery.

Bringing the cell groups back together can also facilitate expansion; for example, melanocytes, which have become senescent in culture, have improved viability when keratinocytes are reintroduced into the system [74].

As the cells are serially seeded into a composite structure, they can be maintained in a proliferative state to continue the cell expansion process. The differential cell turnover rate dictates the timing of the introduction of the cell type, allowing the establishment of the keratinocyte layer before introduction of the fibroblasts and adjusting the media prior to clinical transfer [75].

CLINICAL APPLICATION

The therapeutic use of human skin cells has been made possible by a body of research that began over a century ago [76]. The understanding of the possibility of cell culture running parallel with increasing knowledge of cutaneous wound healing and scarring has fostered a wide range of therapeutic options.

Integral to the clinical implementation of cell-based therapies is an understanding of the impact of time on the therapeutic opportunities and clinical outcome:

- The time taken to harvest the cells from donor tissue, expand the cell population, develop them into a tissue construct, and deliver it to the wound.
- The time pressure for optimal healing from a lifesaving procedure to treat a major burn, balanced with a secondary scar replacement undertaken in a planned reconstructive procedure.

Clearly, if a purely allograft solution was possible, then cells and potentially tissue constructs could be maintained and stored for off-the-shelf use. The problem with such an approach is the stimulation of rejection as the cells trigger immune system responses. There have been significant advances in therapeutic immune system modulation facilitating allograft persistence in more complex cases, as in face and hand transplants. More routinely, in current clinical practice, the use of allograft may provide temporary assistance or act as a biologic dressing to stimulate healing, used with the knowledge of the risk–benefit analysis that also needs to consider costs.

The therapeutic application of autologous cells is achieved using a wide range of techniques:

- immediate cell delivery as an aerosol [77]
- the use of cell carrier dressings seeded at the preconfluent stage of culture [78]
- CEA sheets [38]
- cell-seeded scaffolds with increasing complexity in terms of time cultures and cell combinations [79]
- inkjet printing of scaffold cell combinations [80]

Understanding the design features required to repair skin with the dynamics of acute and chronic wound healing is essential to the appropriate clinical utilization of the technology available. A complete history of the mechanism of the skin damage, with systemic and local examination, and a focus on factors affecting healing forms the basis of the clinical decision. Assessing the wound in terms of body site, the surface area involved, and the depth of skin loss is fundamental and necessary information in planning the appropriate cell-based therapy solution. Understanding the patient's needs will drive selection of therapy; for example, keratinocytes are capable of rapidly reestablishing their barrier properties but does not address the repair of the dermal elements. For successful use of a skin construct, the wound needs to be an environment that facilitates integration into surrounding tissue, specifically the vascular, nervous, and immune systems. Therefore, prior to delivery of any advanced wound care option, meticulous attention to detail in wound bed preparation will optimize the system [23].

With autologous cells, they can be used directly after enzymatic and physical dissociation to cover an area larger than the donor site. The cell suspension is a mixed population harvested from the DEJ, including cells with the capacity for self-renewal as well as terminally differentiated cells, such that the cells are active on the wound surface. On analysis of the suspension, the majority of the cells are keratinocyte phenotypes; despite melanocyte being only 4% of the total cell yield, they are active within the wound as indicated by repigmentation [81]. The resulting cell suspensions can be delivered using a range of techniques. The same delivery techniques are used also for cell suspensions that are precultured in a laboratory environment and harvested prior to confluence. The phenotype of the cells delivered will be directed by the environment of culture chosen and by the clinical indication.

The cells in suspension can be introduced under occlusive dressings, dripped onto cell carrier dressing systems and applied to the wound, such as Suprathel, Polymedics Germany [82], or Biobrane, Smith & Nephew USA [83], or introduced into scaffolds in a tissue-guided regenerative approach, as well as via an aerosol. The key to cell delivery is viability and attachment. The system of delivery must focus on maintaining cell viability as high as possible when delivering cells through the system. Cell damage needs to be minimized.

The aerosol generated is as a result of the pressure through the system against the vortex and aperture of the exit nozzle. The power to generate the flow may be mechanical, as when clipping the ReCell (AvitaMedical Australia) nozzle onto a syringe of cells. Using a 5 mL syringe the cell viability is maintained as opposed to the cell damage seen using a 2 mL syringe with higher pressure generation. Air flow can be used linking an air compressor to the syringe system. This technique is also used to mix the cells with a second agent such as fibrin [84]. The cells are delivered to the prepared wound surface, such that the surface is favorable for cell attachment and proliferation.

Cell carrier systems can be used to facilitate laboratory-based expansion and delivery to the wound. The optimal design of a cell carrier is one that provides an environment for cell proliferation and allows transfer of the cells preferentially to the wound bed. Layered dressing systems combining synthetic material with a biological material as a coating have been successfully used [85]. The key issues are the release of the cells from the carrier surface and effective migration onto the wound surface to achieve healing. Carrier systems have been designed to be biodegradable, as in the use of fibrin microspheres coated with fibroblasts, or synthetic and shed from the surface as the epithelial barrier is restored [86]. The development of cell-compatible biomaterials has been exponential and is well covered elsewhere in this book.

Biomaterials have been developed to support cell proliferation [87], transfer, and tissue generation. Integra, a bovine, bilayered, collagen and glycosaminoglycan construct was first used in the 1980s as a dermal-regeneration template [88]. There are currently a number of dermal scaffolds on the market that can also be used to develop composite skin constructs, with introduction into the wound bed after a period of laboratory-based cell culture. The time taken to generate such composite skin has improved with upscaling cell expansion, but their routine clinical use remains to be defined. The composite skin constructs have the advantage of transferring an established layered tissue for grafting, which addresses both dermal and epidermal elements and more recently pigment match and vascularity [89].

The concept of inkjet printing of tissue structures has opened up possibilities of placing cells and ECM more accurately, in attempts to replicate the skin structures. This clearly requires an in depth knowledge of the skin histology and access to the base materials of the ECM and cells of the appropriate phenotype [90]. As technology advances, it may be possible to identify subsets of cells and place them in the appropriate environmental niche for optimum functionality [91].

As the technology advances in laboratory-based tissue generation, the vascularity of the construct is pivotal. The ultimate aim would be to establish a vascular bed within the tissue to facilitate microsurgical vascularization of the tissue-engineered construct when incorporation into the wound bed as a standard grafting procedure is exceeded [92].

When planning the use of a cell-based therapy, consideration needs to be given to the conditions for transport of the cells to the bedside if produced in a remote facility. Maintaining consistent conditions, such as cooling, and reducing the time of transport are essential in optimizing the utility of all methods. There are well-established systems in place with appropriate conditions maintained, but it is clear that when using biological systems attention to the planning for expedient use is essential.

CONCLUSION

The flexibility of the systems of cell isolation, proliferation, and delivery now provide a range of options, such that the use of cell-based therapies can be incorporated into clinical practice. The clinical outcomes of the technologies are now widely reported in the literature, with improved understanding of their impact on cutaneous healing [93]. Their use in any specific situation needs to consider the patient's needs, the availability of the technology, and the likely outcome. Understanding both the risk–benefit analysis and the cost–benefit analysis underpins the planning.

Despite widespread use, there remain many challenges such as the need for large cell numbers rapidly. The role of alternative pluripotent cell populations, such as adipocytes, may solve the problem of cell availability in large numbers for rapid upscaling [94]. The reproduction of the detailed structural environment may be solved by improved understanding of self-assembly of scaffold and the drivers to self-organization of tissue generation [95].

Work in understanding the cellular basis for life began in previous centuries, with increasing knowledge we are able to harness the capacity of cells to repair skin in ever more complex situations [96]. Tissue engineering of skin has progressed remarkably in the recent decades, with the future of healing by regeneration, a challenging but realistic goal.

REFERENCES

[1] Cormack DH. The integument system. Essential histology. Lippincott-Raven Publishers; 1997. p. 255–68.

[2] Diegelmann RF, Evans MC. Wound healing an overview of acute, fibrotic and delayed healing. Front Biosci 2004;218:13–29.

[3] Bello YM, Phillips TJ. Recent advances in wounds healing. JAMA 2000;283:716–9.

[4] Freshwater MF, Krizek TJ. George David Pollock and the development of skin grafting. Ann Plast Surg 1978;1:96–102.

[5] Alexander JW, MacMillan BG, Law E, Kittur DS. Treatment of severe burns with widely meshed skin autograft and meshed allograft overlay. J Trauma 1981;21:433–8.

[6] Cairns BA, de Serres S, Peterson HD, Meyer AA. Skin replacements. The biotechnological quest for optimal wound closure. Arch Surg 1993;128:1246–52.

[7] Cherry GW, Austad E, Pasyk K, McClatchey K, Rohrich RJ. Increased survival and vascularity of random-pattern skin flaps elevated in controlled, expanded skin. Plast Reconstr Surg 1983;72(5):680–7.

[8] O'Connor NE, Mulliken JB, Banks-Schlegel S, Kehinde O, Green H. Grafting of burns with cultured epithelium prepared from autologous epidermal cells. Lancet 1981;1(8211):75–8.

[9] Wood FM, Kolybaba ML, Allen P. The use of cultured epithelial autograft in the treatment of major burn injuries: a critical review of the literature. Burns 2006;32(4):395–401.

[10] Arons JA, Wainwright DJ, Jordon RE. The surgical applications and implications of cultured human epidermis: a comprehensive review. Surgery 1992;111:4–11.

[11] Wood FM, Kolybaba ML, Allen P. The use of cultured epithelial autograft in the treatment of major burn wounds: eleven years of clinical experience. Burns 2006;32(5):538–44.

[12] Desai MH, Mlakar JM, McCauley RL. Lack of long-term durability of cultured keratinocyte burn wound coverage. J Burn Care Rehabil 1991;12:540–5.

[13] Boyce ST, Goretsky MJ, Greenhalgh DG, Kagan RJ, Reiman MT, Warden GD. Comparative assessment of cultured skin substitutes and native skin autograft for treatment of full-thickness burns. Ann Surg 1995;222:743–52.

[14] Worst PK, MacKenzie IC, Fusenig NE. Reformation of organized epidermal structure by transplantation of suspensions and cultures of epidermal and dermal cells. Cell Tissue Res 1982;225:65–77.

[15] Cruickshank CN, Cooper JR, Hooper C. The cultivation of cells from adult epidermis. J Invest Dermatol 1960;34:339–42.

[16] Michel M, Torok N, Godbout MJ, Lussier M, Gaaudreau P, Royal A, et al. Keratin 19 as a biological marker of skin stem cells in vivo and vitro: keratin 19 expressing cells are differentially localized in function of anatomical sites and their number varies with donor age and culture stage. J Cell Sci 1996;109:1017–28.

[17] Nakagawa H, Akita S, Fukui M, Fujii T, Akiro L. Human mesenchymal stem cells successfully improve skin substitute wound healing. Br J Dermatol 2005;153(1):29–36.

[18] Foubert P, Barillas S, Gonzalez AD, Alfonso Z, Zhao S, Hakim I, et al. Uncultured adipose derived regenerative cells seeded in collagen scaffold improves dermal regeneration, enhancing early vascularization and structural organization following thermal burns. Burns 2015. http://dx.doi.org/10.1016/j.burns.2015.05.004.

[19] Fratianne R, Papay F, Housini I, Lang C, Schafer IA. Keratinocyte allografts accelerate healing of split-thickness donor sites: applications for improved treatment of burns. J Burn Care Rehabil 1993;14:148–54.

[20] Purdue GF. Dermagraft TC pivotal safety and efficacy study. J Burn Care Rehabil 1996;18:S13–4.
[21] Hellman KB. Biomedical applications if tissue engineering technology: regulatory issues. Tissue Eng 1997;1(2):203–16.
[22] Wilkins LM, Watson SR, Prosky SJ, Meunier SE, Patenteau NL. Development of a bilayered skin construct for clinical applications. Biotechnol Bioeng 1994;43:747–56.
[23] Wood FM. Tissue engineering of skin. Clin Plast Surg 2012;39(1):21–32.
[24] Slott Jensen H. Skin viability studies in vitro. Scand J Plast Rec Surg 1984;18(1):55–9.
[25] Stoner ML, Wood FM. Systemic factors influencing growth of cultured epithelial autograft. Burns 1996;22(3):197–9.
[26] Ben-Bassat H, Chaouat M, Segal N, Zumai E, Wexler MR, Eldad A. How long can cryopreserved skin be stored to maintain adequate graft performance. Burns 2001;27(5):425–31.
[27] Hybbinette S, Bostrom M, Lindbery K. Enzymatic dissociation of keratinocytes from human skin biopsies for in vitro cell propagation. Exp Dermatol 1999;8:30–8.
[28] Aasen T, Belmonte JCI. Isolation and cultivation of human keratinocytes from skin or plucked hair for the generation of induced pluripotent stem cells. Nat Protoc 2010;5(2):371–82.
[29] Limat A, French LE, Blal L, Saurat JH, Hunziker T, Salomon D. Organotypic cultures of autologous hair follicle keratinocytes for the treatment of recurrent leg ulcers. J Am Acad Dermatol 2003;48:207–14.
[30] Medawart PB. The fate and behaviour of skin homografts in rabbits. J Anat 1944;78:176–84.
[31] Flaxman BA, Lutzner MA, Van Scott EJ. Cell maturation and tissue organization in epithelial outgrowths from skin and buccal mucosa in vitro. J Invest Dermatol 1967;49:322–5.
[32] Johnson SR, McMinn RMH. The cytology of wound healing of the body surface in mammals. Biol Rev 1960;35:364–71.
[33] Parshley MS, Simm HS. Cultivation of adult skin epithelial cells in vitro. Am J Anat 1950;86:163–9.
[34] Waymouth C. Construction and use of synthetic media. Cells and tissues in culture. London Academic Press; 1965. p. 99–110.
[35] Moscana A. Inhibition by trypsin inhibition of dissociation of embryonic tissue by trypsin. Nature 1963;199:379–80.
[36] Layman DL, McGoodwin EB, Martin GR. The nature of the collagen synthesized by cultured human fibroblasts. Proc Natl Acad Sci USA 1971;68(2):454–8.
[37] Kaviani M, Geramizadeh B, Rahsaz M, Marzban S. Considerations in the improvement of human epidermal keratinocyte culture in vitro. Exp Clin Transplant 2015;13(1):366–70.
[38] Rheinwald J, Green H. Serial cultivation of strains of human epidermal keratinocytes: the formation of keratinizing colonies from single-cells. Cell 1975;6:331–44.
[39] Park HS, Kang HJ, Kim CH, Kim CH, Han ES, Han K, et al. Application of physical force is essential to enrich for epidermal stem cells in primary human keratinocyte isolation. Tissue Eng 2004;10(3–4):343–51.
[40] Fusenig NE. Epithelial-mesenchymal interactions regulate keratinocyte growth and differentiation in vitro. The keratinocyte handbook. Cambridge University Press; 1994. p. 77–94.
[41] Abdel Naser MB. Differential effects on melanocyte growth and mealanization of low vs. high calcium keratinocyte conditioned media. Br J Dermatol 1999;140:50–5.
[42] Mulekar SV, Ghwish B, Al Eisa A. Treatment of vitiligo lesions by ReCell vs. conventional melanocyte keratinocyte transplantation: a pilot study. Br J Dermatol 2008;158:45–9.
[43] Niyonsaba F, Ushio H, Nagaoka I, Okumura K, Ogawa H. The human beta-defensins (-1, -2, -3, -4) and cathelicidin LL-37 induce IL-18 secretion through p38 and ERK MAPK activation in primary human keratinocytes. J Immunol 2005;175(3):1776–84.

[44] Stoner ML, Wood FM. Cultured epithelial autograft "take" confirmed by the pressure of cytokeratin 9. J Invest Dermatol 1999;112(3):391–2.

[45] Driskell RR, Lichtenberger BM, Hoste E. Distinct fibroblast lineages determine dermal architecture in skin development and repair. Nature 2013;504:277–81.

[46] Gilchrest BA, Vrabel MA, Flynn E, Szabo G. Selective cultivation of human melanocytes from newborn and adult epidermis. J Invest Dermatol 1984;83:370–6.

[47] Wood FM. Skin regeneration: the complexities of translation into clinical practise. Int J Biochem Cell Biol 2014;56:133–40.

[48] Navarro FA, Stoner ML, Park CS, Huertas JCLH, Wood FM, Orgill DP. Sprayed keratinocyte suspension accelerate epidermal coverage in a porcine microwound model. J Burn Care Rehabil 2000;21:513–8.

[49] Sander EA, Lynch KA, Boyce ST. Development of the mechanical properties of engineered skin substitutes after grafting to full-thickness wounds. J Biomech Eng 2014;136(5):81–7.

[50] Lee V, Singh G, Trasatti JP, Bjornsson C, Xu X, Tran TN, et al. Design and fabrication of human skin by three-dimensional bioprinting. Tissue Eng Part C 2014;20(6):473–84.

[51] Wood FM. Clinical potential of cellular autologous epithelial suspension. Wounds 2002; 15:16–22.

[52] Sahota PS, Burn JL, Heaton M, Freedlander E, Suvarna SK, Brown NJ, et al. Development of a reconstructed human skin model for angiogenesis. Wound Repair Regen 2003;11:275–84.

[53] Xia L, Zouboulis CH, Detmar M, Mayer de Silva A, Stadler R, Orfanos CE. Isolation of human sebaceous glands and cultivation of sebaceous gland derived cells as an in vitro model. J invest Dermatol 1989;93:315–21.

[54] Jahoda CA, Horne KA, Oliver RF. Induction of hair growth by implantation of cultured dermal papilla cells. Nature 1984;311:560–2.

[55] Cross KJ, Mustoe TA. Growth factors in wound healing. Surg Clin North Am 2003;13:4–14.

[56] Brusselaers N, Pirayesh A, Hoeksema H, Richters CD, Verbelen J, Beele H, et al. Skin replacement in burn wounds. J Trauma 2010;68(2):490–5.

[57] MacNeil S. Progress and opportunities for tissue-engineered skin. Nature 2007;445(7130):874–80.

[58] Stark HJ, Baur M, Breitkreutz D, Mirancea N, Fusenig NE. Organotypic keratinocyte coculture in defined medium with regular epidermal morphogenesis and differentiation. J Invest Dermatol 1999;112:681–91.

[59] Bettger WJ, Ham RG. The nutrient requirements of mammalian cells grown in culture. Adv Nutr Res 1982;4:249–86.

[60] Supp DM, Boyce ST. Engineered skin substitutes: practices and potentials. Clin Dermatol 2005;23(4):403–12.

[61] Boyce ST, Ham RG. Calcium regulated differentiation of normal human keratinocytes in chemically defined clonal culture and serum-free culture. J Invest Dermatol 1983;82:33–40.

[62] Boyce ST. Fabrication, quality assurance, and assessment of cultured skin substitutes for treatment of skin wounds. Biochem Eng J 2004;20(2):107–12.

[63] Durfor CN. Biotechnology biomaterials: a global regulatory perspective for tissue engineered products, summary report and future directions. Tissue Eng 1997;3(1):115–20.

[64] Fink DW. FDA regulation of stem-cell-based products. Science 2009;324:1662–3.

[65] Metcalfe AD, Ferguson MW. Tissue engineering of replacement skin: the crossroads of biomaterials, wound healing, embryonic development, stem cells and regeneration. J R Soc Interface 2007;4(14):413–37.

[66] Bradshaw M, Ho D, Fear MW, Gelain F, Wood FM, Iyer KS. Designer self-assembling hydrogel scaffolds can impact skin cell proliferation and migration. Sci Rep 2014;11(4):6903. http://dx.doi.org/10.1038/srep06903.

[67] Atiyeh BS, Costagliola M. Cultured epithelial autograft (CEA) in burn treatment: three decades later. Burns 2007;33(4):405–13.
[68] Watt FM. Role of integrins in regulating epidermal adhesion, growth and differentiation. EMBO J 2002;21:3919–26.
[69] Gustafson CJ, Birgisson A, Junker J, Huss F, Salemark JH, Kratz G. Employing human keratinocytes cultured on macroporous gelatin spheres to treat full-thickness-wounds: an in vivo study on athymic rats. Burns 2007;33(6):726–35.
[70] Ehrlich HP. Understanding experimental biology of skin equivalent: from laboratory to clinical use in patients with burns and chronic wounds. Am J Surg 2004;187(5A):29S–33S.
[71] Koch L, Deiwick A, Schlie S, Michael S, Gruene M, Coger V, et al. Skin tissue generation by laser cell printing. Biotechnol Bioeng 2012;109(7):1855–63.
[72] Berthod F, Germain L, Li H, Xu W, Damour O, Auger FA. Collagen fibril network and elastic system remodeling in a reconstructed skin transplanted on nude mice. Matrix Biol 2001;20(7):463–73.
[73] van Dorp AG, Verhoeven MC, Nat-Van Der Meij TH, Koerten HK, Ponec M. A modified culture system for epidermal cells for grafting purposes: an in vitro and in vivo study. Wound Repair Regen 1999;7(4):214–25.
[74] Swope BV, Supp AP, Boyce ST. Regulation of cutaneous pigmentation by titration of human melanocytes in cultured skin substitutes grafted to athymic mice. Wound Rep Regen 2002;10(6):378–86.
[75] Boyce ST, Kagan RJ, Meyer NA, Yakuboff KP, Warden GD. Cultured skin substitutes combined with Integra artificial skin to replace native skin autograft and allograft for the closure of excised full-thickness burns. J Burn Care Rehabil 1999;20(6):453–61.
[76] Andreassi L. History of keratinocyte cultivation. Burns 1992;18:S2–4.
[77] Fredriksson C, Kratz G, Huss F. Transplantation of cultured human keratinocytes in single-cell suspension: a comparative in vitro study of different application techniques. Burns 2008;34(2):212–9.
[78] CastroV P, Moreira LF, Kashima S, Covas DT. Can pleuripotential stem cells be used in cell-based therapy. Cell Reprogram 2014;16(2):98–107.
[79] Kremer M, Lang E, Berger AC. Evaluation of dermal-epidermal skin equivalents ('composite-skin') of human keratinocytes in a collagen-glycosaminoglycan matrix(Integra artificial skin). Br J Plast Surg 2000;53(6):459–65.
[80] Michael S, Sorg H, Peck CT, Koch L, Deiwick A, Chichkov B, et al. Tissue engineered skin substitutes created by laser-assisted bioprinting form skin-like structures in the dorsal skin fold chamber in mice. PLoS One 2013;8(3):e57741. http://dx.doi.org/10.1371/journal.pone.0057741.
[81] Wood FM, Giles N, Stevenson A, Rea S, Fear M. Characterisation of the cell suspension harvested from the dermal epidermal junction using a ReCell® kit. Burns 2012;38(1):44–51.
[82] Schwarze H, Küntscher M, Uhlig C, Hierlemann H, Prantl L, Ottomann C, Hartmann B. Suprathel, a new skin substitute, in the management of partial-thickness burn wounds: results of a clinical study. Ann Plast Surg 2008;60(2):181–5.
[83] Wood F, Martin L, Lewis D, Rawlins J, McWilliams T, Burrows S, et al. A prospective randomised clinical pilot study to compare the effectiveness of Biobrane® synthetic wound dressing, with or without autologous cell suspension, to the local standard treatment regimen in paediatric scald injuries. Burns 2012;38(6):830–9.
[84] Stark GB, Kaiser HW, Horch RE, Kopp J, Spilker G. Cultured autologous kerstinocyres suspended in fibrin glue with allogenic overgraft for definitive burn wound coverage. Eur J Plast Surg 1995;18:267–71.

[85] Rennekampff HO, Hansbrough JF, Kiessig V, Abiezzi S, Woods V. Wound closure with human keratinocytes cultured on a polyurethane dressing overlaid on a cultured human dermal replacement. Surgery 1996;120(1):16–22.

[86] Myers SR, Partha VN, Soranzo C, Price RD, Navsaria HA. Hyalomatrix: a temporary epidermal barrier, hyaluronan delivery, and neodermis induction system for keratinocyte stem cell therapy. Tissue Eng 2007;13(11):2733–41.

[87] Zeng Q, Macri LK, Prasad A, Clark RAF, Zeugolis DI, Hanley C, et al. Skin tissue engineering, comprehensive biomaterials. Philadelphia (PA): Elsevier; 2011. p. 467–99.

[88] Yannis IV, Burke JF, Orgill DP, Skrabut EM. Wound tissue can utilize a polymeric template to synthesize a functional extension of skin. Science 1982;215:174–6.

[89] Boyce ST. Design principles for composition and performance of cultured skin substitutes. Burns 2001;27(5):523–33.

[90] Wang Y, Xu R, He W, Yao Z, Li H, Zhou J, et al. Three dimensional histological structures of the human dermis. Tissue Eng Part C 2015:1–13.

[91] Toma JG, McKenzie IA, Bagli D, Miller FD. Isolation and characterisation of multipotent skin derived precursors from human skin. Stem Cells 2005;23:727–37.

[92] Dew L, MacNeil S, Chong CK. Vascularization strategies for tissue engineers. Regen Med 2015;10(2):211–24.

[93] Ebersole G, Anderson P, Powell H. Epidermal differentiation governs engineered skin biomechanics. J Biomech 2010;43(16):3183–90.

[94] Macri L, Clark RA. Tissue engineering for cutaneous wounds: selecting the proper time and space for growth factors, cells and the extracellular matrix. Skin Pharmacol Physiol 2009;22(2):83–93.

[95] Sun T, Mai S, Norton D, Haycock JW, Ryan AJ, MacNeil S. Self-organization of skin cells in three-dimensional electrospun polystyrene scaffolds. Tissue Eng 2005;11(7–8):1023–33.

[96] Rosenblatt B. Meeting regulatory challenges for cell-based therapies. Bioprocess Int 2012;10:5–8.

Chapter 16

Cutaneous Applications of Stem Cells for Skin Tissue Engineering

Sean V. Murphy[1], Mohammad Z. Albanna[2,3]

[1]*Wake Forest Institute for Regenerative Medicine (WFIRM), Winston-Salem, NC, USA;* [2]*Wake Forest Baptist Medical Center (WFBMC), Department of General Surgery, Winston-Salem, NC, USA;* [3]*Pinnacle Transplant Technologies, Research & Development, Phoenix, AZ, USA*

Chapter Outline

INTRODUCTION: STEM CELLS

The international Society for Cellular Therapy (ISCT) recommends the use of the term "multipotent mesenchymal stromal cells" (MSCs) instead of mesenchymal stem cells. MSCs are characterized as multipotent cells capable of self-renewing and differentiating into at least osteocytes, chondrocytes, and adipocytes, adhering to culture plates, acquiring a mature phenotype that is different from their tissue of origin, and expressing and lacking surface antigens. While several surface antigens have been identified and are commonly used to identify stem cells based on the tissue of origin, the lack of specific markers renders the characterization of MSCs difficult and often ambiguous. Also, these markers can change for fresh versus cultured or even propagated MSCs.

Nonetheless, MSCs have been isolated and expanded from different tissues and organs including bone marrow, fat, cord blood, cord tissue, placental tissue, amniotic membrane, amniotic fluid, and skin. Despite some similarities between the different stem cell sources, several distinctions have been revealed

Skin Tissue Engineering and Regenerative Medicine. http://dx.doi.org/10.1016/B978-0-12-801654-1.00016-4

concerning at least the proliferative capacity, secretory profile, and immunomodulatory properties of each stem cell source. These discrepancies warranted extensive investigation of the safety and efficacy of each stem cell source.

Each type of stem cell contributes to one or more phases of wound healing by (1) recruiting inflammatory cell precursors from bone marrow or skin, (2) reepithelialization, (3) transition from mesenchymal to epithelial phenotypes, and (4) revascularization by resident and circulating endothelial cell precursors. The chapter addresses the underlying mechanism of how stem cells contribute to each phase of wound healing based on the research efforts to date.

SKIN-DERIVED STEM CELLS

Skin is in constant regeneration throughout life. To date, more than 20 different cell types have been identified residing within the skin, each performing a number of critical functions necessary for human survival. Several locations within the skin are believed to contain stem cells, including the hair follicle and interfollicular epidermis, with varying levels of contribution to wound regeneration [1]. Some reports described up to eight different types of stem cells found in the hair follicle and surrounding skin, including epidermis stem cells, follicle multipotent stem cells, melanocyte stem cells, mesenchymal stem cell-like cells, dermal sheath stem cells, neural crest stem cells, hematopoietic stem cells (HSCs), and endothelial stem cells [2].

Epidermal stem cells are characterized by a prolonged self-renewal and differentiation capacity into multiple lineages derived from the tissue of origin [3]. The bulge of the hair follicle serves as the main repository of skin stem cells, is located at the base of the follicle, and does not degenerate during the hair cycle. Skin stem cells are more quiescent than other skin cells, including keratinocytes, have a high proliferative potential, and can terminally differentiate into the three differentiation lineages [4]. Dermal papilla (DP) is the dermal part of the hair follicle within the skin. It consists of a small, densely packed cluster of mesenchymal cells. The interfollicular epidermis contains progenitor cells that ensure continuous renewal in the absence of an injury. The hair follicle contains multipotent stem cells that are activated at the start of the hair cycle and mobilized upon injury to provide the necessary cells for the regeneration and repair of the epidermis. The bulge stem cells respond rapidly to epidermal wounding during the acute wound repair phase and acquire an epidermal phenotype but are eliminated from the epidermis over several weeks after injury, which could indicate that these stem cells contribute to the wound repair but not to the homeostasis of the epidermis [5]. Potential markers to identify hair follicle stem cells include β1-integrin, keratin 19, α6-integrin, CD71, p63, and CD34 [6]. However, in vitro markers might not be identical to those in vivo, as hair follicle stem cells might have already been activated after they left the stem cell niche. Precise laser capture microdissection and microarray techniques have helped to identify novel cell surface markers [7]. Bulge cells maintain their stem cell characteristics after propagation in vitro and regenerate hair follicles, epidermis, and sebaceous glands when combined with neonatal dermal cells in vitro [8].

Isolation and functional assessment of epidermal stem cells have been described in numerous research reports [9,10]. Epidermal stem cells can be isolated by rapid adherence to culture plates coated with collagen or other matrix proteins, or other techniques such as gravity-assisted cell sorting [11]. Cells were shown to have high levels of α6-integrin and low levels of CD71. Epidermal stem cells have been used for the construction of skin equivalents [12]. These skin-derived stem cells have been shown to differentiate into all hair follicle lineages, including sebaceous gland, neuron, glial cells, Schwann cells, smooth muscle cells, dermal fibroblasts, melanocytes, and other cell types [13].

Additionally, multipotent stem cells with mesodermal progeny, termed skin-derived precursors (SKPs), have been isolated and characterized from human and rodent dermis, taking about 2 weeks to isolate and expand [14,15]. SKPs were differentiated into skeletogenic cell types [16], Schwann cells [15], neuron-like cells [17], islet-like insulin-producing cell clusters [18,19], and hepatocytes [20]. Most recently, some preliminary animal studies showed that these cells promote angiogenesis in ischemic regions after cerebral infraction [21]. This could potentially contribute to healing in chronic wounds with vascular insufficiency. The implantation of SKPs that differentiated into Schwann cells in a rat model was shown to improve functional recovery after incomplete cervical spinal cord injury [22]. Additionally, skin-derived stem cells have been shown to inhibit tumor growth and extend animal survival [23]. While these cells seem to be neural progenitors in nature, they may find application in cutaneous wound applications. SKPs are still in the investigational research phase, with inconclusive results about their promise before transitioning to clinic.

While these tissue-specific stem cells hold great promise to contribute to all phases of wound healing, very limited and inconclusive data related to the role and effectiveness of epidermal stem cells in wound healing are available and warrant further investigation to ensure safety and effectiveness in wound healing therapies.

ADULT STEM CELLS

Bone Marrow-Derived Stem Cells

Human bone marrow is a heterogeneous reservoir of stromal cell populations that are different in cell morphologies, growth rates, differentiation abilities, migration mechanisms, paracrine effects, and immunomodulations [24]. Isolation and expansion of these stromal/stem cells through different protocols revealed cells with different multipotential properties, committed progenitors, and terminal differentiations [25].

HSCs and nonhematopoietic mesenchymal stem/progenitor cells (MSCs) originate from precursor cells in the bone marrow stroma [26]. MSCs in the bone marrow represent only 0.001–0.01% of the nucleated cells, 10-fold less abundant than HSCs. Despite this low percentage, the bone marrow supplies MSCs at 30% higher rate than cord blood [27].

MSCs can differentiate into several mesenchymal lineage cells, such as osteocytes, chondrocytes, adipocytes, and other cell lineages including tendocytes, myotubes, astrocytes, and hematopoiesis-supporting stroma [28]. The properties of these cells have been characterized extensively in the literature [29].

In cutaneous wound healing, MSCs play an important role in the host defense and inflammatory processes, matrix deposition principally collagen type I and III, angiogenesis, and recruitment of endogenous cells for dermal and epidermal reconstruction. While wound reepithelialization and dermal cell repopulation have long been thought to be driven by uninjured cells in the periphery of the wound, early infiltration of inflammatory cells and fibroblasts originate from the bone marrow [30]. Others showed that the major source of deposited collagen is coming from the resident cells in the skin, and bone marrow-derived stem cells are contributing partially to collagen production during fibrogenesis [31,32].

A strong and evident body of work demonstrated that MSCs have a profound therapeutic potential for skin defects, such as chronic wounds and burns [33]. MSCs contribute to wound healing by (1) accelerating wound closure, (2) differentiating into keratinocytes evident by expression of keratin 14 in wounds, (3) promoting angiogenesis by releasing vascular endothelial growth factor (VEGF) and recruiting endothelial cells and endothelial progenitor cells (EPCs) to the wound, (4) promoting regeneration of appendages by releasing cytokines such as epidermal growth factor (EGF) and keratinocytes growth factor (KGF), and lastly (5) influencing the inflammatory cells by releasing cytokines such as macrophage inflammatory protein-1 (MIP-1) and monocytes chemoattractant protein (MCP).

While there is a general consensus that bone marrow-derived stem cells contribute to accelerated cutaneous wound healing through enhanced reepithelialization, granulation tissue formation, and neovascularization, the mechanism is still debatable. One theory is that stem cells help to recruit endogenous cells and secrete growth factors and cytokines, then migrate from the wound after the reepithelialization is complete rather than direct participation and incorporation in the tissue or transdifferentiation [34,35]. A second theory is that stem cells contribute to the wound healing through direct differentiation [36–38]. The reason backing up the second theory could be related to fibrocytes.

Fibrocytes are inactive mesenchymal cells derived from bone marrow progenitors and constitute 0.1–0.5% of peripheral blood cells. Fibrocytes circulate in the peripheral blood vessels and are capable of secreting matrix proteins, including collagen types I and III, with the potential for further differentiation into mature mesenchymal cells, such as fibroblasts, myofibroblasts, and adipocytes [39,40]. They are capable of producing inflammatory cytokines, growth factors, and chemokines and contribute to tissue fibrosis [41]. Myofibroblasts may play an important role in exerting a contractile force to close cutaneous wounds [42]. However, the modulation of fibrocyte migration, proliferation, and differentiation through intercellular signals is only partially defined. Until several fundamental questions on the mechanism controlling fibrocytes and migration and differentiation of other bone-derived stromal cells are addressed, the mechanism of how stem cells contribute to wound healing and the use of these cells in a safe and effective clinical application will remain debatable. Other mechanisms may involve the effect of physical cues such as mechanical stress [43], or biological cues such as the presence of other primary cells [44] or growth factors [45], which affect proliferation and differentiation of stem cells and subsequently the wound healing process.

Autologous bone marrow-derived cultured stem cells accelerated wound healing in patients with chronic and acute wounds when delivered in a fibrin hydrogel [46]. Bone marrow-derived stem cells have been shown to improve dermal rebuilding and lead to closure of nonhealing chronic wounds, whether applied directly [47] or injected into the wound periphery [48]. Similar observations were seen in patients with critical limb ischemia who received bone marrow stem cells injected intramuscularly or systemically [49]. Other studies showed that intradermal injection yielded a faster wound healing in full-thickness wounds compared to systemic injection of stem cells [50]. The optimal route of delivery of bone marrow stem cells and whether in vitro expansion of bone marrow stem cells to purify the heterogeneous cell population and yield reliable growth factors require further studies in large patient populations to ensure maximum benefit of the stem cells and enhanced the clinical response.

Some studies showed that wound healing in whole bone marrow was superior to stem cells derived from bone marrow [51]. Autologous bone marrow mononuclear cells have been shown to increase blood perfusion in patients with advanced critical limb ischemia due to peripheral arterial diseases [52], and could be utilized as a safe and effective treatment for critical limb ischemia [53,54]. Therefore, careful consideration to the selection of what cell population (mononuclear cells, MSCs, endothelial progenitor cells, or fibrocytes) within the bone marrow aspirate could be used for wound healing is required. In addition to their role in wound healing, bone marrow-derived MSCs may play an important role in enhancing the fibrotic behavior of deep dermal fibroblasts and have a possible involvement in the pathogenesis of hypertrophic scarring [55,56]. For example, MSCs from bone marrow secrete high levels of VEGF, which is linked to excessive keloid scarring in patients who are susceptible to keloid scarring [57]. Several clinical trials have been conducted with different bone marrow-derived cell populations for chronic wounds and showed promising results [58,59]; however, head-to-head studies are still needed.

Adipose-Derived Stem Cells

Several nomenclatures have been used to describe plastic-adherent, multipotent, and self-renewal stem cells isolated from fat tissue including adipose-derived mesenchymal stem/stromal cells, lipoblasts, pericytes, preadipocytes, and many other names. While these names led to much confusion in the literature, the International Fat Applied Technology Society adopted the term "adipose-derived stem cells" (ADSCs) to identify stem cells isolated from fat tissue. Several protocols have been developed to isolate and characterize ADSCs as surface-positive antigens for CD9, CD34, CD49, CD54, CD166, CD44, CD10, CD90, CD146, CD55, CD59, HLA-ABC, CD 105, CD29, CD44, CD73, CD90, and CD166 and surface-negative antigens for CD11b, CD18, CD50, CD56, CD62, CD104, CD16, CD14, CD31, CD45, and HLA-DR, with variable differentiation potentials to adipocyte, cardiomyocyte, chondrocyte, endothelial, myocyte, neuronal-like, and osteoblast among many other cell types [60]. Expansion of ADSC populations in culture yield 100–1000 times more cells per cm^3 than bone marrow, which make them an attractive allogeneic or autologous source [61].

In addition to the fact that ADSCs can be easily isolated, possess immunomodulatory characteristics and great potential of differentiation, they also have other attractive properties including protection, survival, and differentiation of variety of endogenous cells and tissues [62]. These cells have broad secretory profiles of various angiogenic, hematopoietic, and pro-inflammatory growth factors and cytokines [63]. Fibroblast proliferation can be enhanced if co-cultured with ADSCs due to the paracrine effect of ADSCs [64]. The secretory effect of ADSCs is still pronounced in the conditioned media used to culture fat tissue. It has been shown that the conditioned media of adipose tissue enhanced ADSC proliferation and induced cell sprouting of endothelial cells could be used as a safe topical treatment of skin wounds without the need to use cells [65]. The effect of conditioned media of ADSCs increases proliferation and migration of vascular endothelial cells, fibroblasts, and keratinocytes, up-regulates transcription of type I procollagen in fibroblasts and increases the contraction of fibroblasts populated in a collagen lattice [66–68]. It has also been demonstrated that the conditioned media of ADSCs has potent antioxidant activity and can protect dermal fibroblasts from oxidative stress injury by decreasing cell apoptosis [64].

Adipose-derived stem cells have been shown to enhance the survival of fat grafted onto the face [69] or full-thickness [70] wounds. Additionally, they enhance wound healing of diabetic ulcers, pressure ulcers, and full-thickness wounds in rats through angiogenesis, epithelialization, ability to differentiate into endothelial cells in situ, and secretion of growth factors such as VEGF and transforming growth factor-β3 (TGF-β3) [70–72]. It has also been demonstrated that ADSCs have the ability to differentiate in vivo to epithelial and endothelial cells, indicating that ADSCs not only contribute to the cutaneous wound regeneration, but also to neovascularization [73] and could be detected in the wound 2 weeks postimplantation [74]. There have been some scientific reports on the ability of ADSCs to stay viable and at the site of injury postinjection. One study reported that viable ADSCs could be detected 1 year postinjection [75].

In addition to their use as a cell suspension or seeded on a matrix, ADSC sheets have been manufactured and studied for full-thickness wound healing and showed promising results as off-the-shelf, three-dimensional constructs [76,77].

However, the prolonged expansion time, cost, and fragility might hinder the transition of such technique to the clinic.

Despite the abundant source of stem cells from fat tissue, ease of isolation and high proliferation capacity, and secretion of pro-inflammatory and pro-angiogenic growth factors, the use of ADSCs as an autologous cell therapy option might not be attainable in some cases due to various reasons, such as patient age, current health, and comorbid diseases such as diabetes. The health status of ADSCs, whether they are harvested from a healthy or a diabetic subject, can affect their proliferation capacity, apoptosis rate, and secretory properties and could result in impaired vascular stabilization [78]. Diabetic ADSCs have lower expression of stem cell markers, lower amounts of secreted growth factors playing an important role in skin repair, and reduced capability to promote fibroblast and keratinocyte proliferation and migration [79]. The lack of local secretion of donor growth factors, not necessarily related to skin repair such as hepatocyte growth factors (HGFs), can impair the function and potency of ADSCs [80].

ADSCs can differentiate into pericytes, which have protective and regenerative potentials for wound healing [81]. Pericytes have gained much attention lately due to their ability to differentiate into other mesenchymal cell types including smooth muscle cells and fibroblasts, interact with other cell types such as endothelial cells, and contribute to the angiogenesis process during wound healing [82].

When compared to fibroblasts, ADSCs seeded in collagen gel were less effective in wound closure of a full-thickness wound than fibroblasts seeded in collagen gel [83]. However, ADSCs respond vigorously to factors present in the wounds. These factors influence their proliferation and increase their secretion of angiogenesis and granulation tissue formation factors [84]. It has been demonstrated that the use of bone marrow- and adipose-derived stem cells can minimize wound contraction, which could be due to the lower expression of α-smooth muscle actin (α-SMA) [85]. When compared to other neonatal-derived stem cells such as amnion or adult stem cells such as bone marrow-derived MSCs, ADSCs were superior at promoting human dermal fibroblasts migration, up-regulation of VEGF, bFGF, KGF, and TGF-β1, and protein level of collagen type 1 [86]. It has also been demonstrated that better blood recovery of ADSCs compared to BMSCs can be achieved in ischemic limbs [87]. Interestingly, similar findings were seen in the treatment of spinal cord injury [88] and regeneration of the nervous system [89]. Such findings warrant investigation of the cellular mechanisms of stem cells, route of injection, expansion protocols, purity, and identity for therapeutic applications.

PERINATAL TISSUES AND STEM CELLS

Amniotic Membrane

Gestational tissue offers a considerable advantage as a source of therapeutic biological material, as this tissue is often discarded following birth and is readily available without an invasive biopsy or the manipulation of embryos or somatic cells [90–92]. This means that there are minimal ethical and legal considerations associated with collection and use for clinical therapies.

The amnion is the innermost of the placental membranes lining the amniotic cavity and consists of five layers. A cell monolayer of amnion epithelial cells is in contact with the fetus during the first stages of development, and is known to have important roles in fetal development, immune regulation, and defense against infection [93]. The epithelial monolayer rests on a basement membrane (BM) attached to the connective tissue layers, comprised of a compact layer, a fibroblast layer, and a spongy layer. The connective tissue layers are in direct contact with the cellular layer of the chorion.

Laboratory investigations have revealed that the BM of amnion shares major BM components with human skin, and the BM zone morphologically resembles human skin [94]. The epithelial side of denuded amnion has been shown to support the proliferation, spreading, and differentiation of corneal, mucosal, and bronchial epithelial cells [95–97]. Furthermore, the stroma of amnion can serve as a dermal matrix in which fibroblasts show good adherence and proliferation [98,99].

The placenta and placental membranes have been used for medicinal applications for over 100 years, with the earliest modern use of placental membranes for wound healing applications described in 1910 by William Thornton and Staige Davis, at Johns Hopkins Hospital [100]. Since this initial report, clinical trials and animal experiments have documented that amnion is superior to allograft and xenograft skin as a therapeutic adjunct to accelerate healing of burns, wounds, and ulcers [101,102]. Amnion has also been found to be an effective biologic dressing for burns, as it diminishes the loss of plasma, fluid, protein, and heat [95,103]. The use of amniotic membranes has also been shown to cause fewer complications than synthetic skin grafts [104,105]. The following section describes some of these clinical reports and experiences using amnion for the treatment of burns and wounds.

In 1980, Fauk et al. reported the treatment of 15 patients with chronic leg ulcers using cultured amnion dressings [106]. Treatment with amnion resulted in the formation of profuse granulation tissue, decreased connective tissue fibers, and increased capillary density compared to ulcers receiving standard treatments. Sawhney et al. evaluated fresh amnion application on 45 patients with partial-thickness burns in three subgroups of 15 patients each: superficial, intermediate, and deep dermal burns [95]. This study found that the mean healing time was significantly faster for all amnion-treated burns and resulted in less scarring in some patients. Additionally, dressing changes were less painful in patients with amnion dressings, and wound exudate was reduced. Mohammadi et al. studied 50 patients who had two similar donor sites on both lower extremities, treated either with standard procedures or with fresh amnion [107]. The mean healing time was about 6 days in amnion-treated wounds and 13 days in the control group, and pain was significantly lower in amnion-treated groups. These studies also concluded that a single application of amnion significantly reduces patient discomfort and costs by eliminating the need for multiple dressing changes. In another study, this group evaluated the effect of amnion as an adjunct in split-thickness skin graft on itching and hypertrophic scar formation in burn patients. This double-blind, randomized, controlled, clinical trial showed 59.25% decreased itching and 64.81% less hypertrophic scar formation in the skin grafted area covered with amnion in comparison to areas that were managed by conventional skin grafting methods.

A recent publication has compared preparation methods for amnion and evaluated its effectiveness as a dressing for partial-thickness burns [108]. In a study comprising 50 patients, half of the wound area was covered with glycerol-preserved or irradiated amnion. Application of irradiated amnion was much easier and more comfortable, compared with application of glycerol-preserved membranes. Healing times were comparable between treatments, ranging from 10 to 14 days in all wound samples. Exudate fluid developed under the membranes in four patients with irradiated amnion and five patients with glycerol-preserved membranes. None of the patients suffered from restrictions in joint movement; however, three patients with irradiated amnion and four patients with glycerol-preserved membranes developed hypertrophic scarring. Gajiwala and Gajiwala investigated in two studies the coverage of 24 and 35 patients with second- and third-degree burns with freeze-dried or irradiated amnion dressings, respectively [109]. In this case-control study, the use of amnion resulted in reduced exudation, reduced induration, more comfort, less pain, and improved reepithelialization.

Branski et al. compared decellularized amnion dressings with standard topical treatment with antibiotic cream for the management of pediatric second-degree burns in a prospective randomized trial [110]. Fifty-three patients with burns on the face, head, and neck region received amnion with topical antibiotic cream, and 49 patients received topical antibiotic cream alone. Patients receiving amnion treatment had significantly faster total healing time (62 days vs 82 days) and significantly fewer dressing changes than control patients. However, no differences were measured for scar formation, infection rate, and length of hospital stay between treatment groups.

The collective evidence of the clinical trials using amnion have favored its use in burns for promotion of wound healing and for increasing patient comfort and reduced requirement for dressing changes [111]. Additionally, many of these studies demonstrated the antimicrobial effects, pain relief, reduction of fluid, and reduced scar formation with amnion treatments.

Placental MSCs

It was first postulated in 1980, by Faulk et al. that the therapeutic properties of the amnion may be due to factors secreted by the amnion-derived cells influencing angiogenesis or other healing mechanisms [106]. With this potential therapeutic mechanism in mind, the isolated cells of the placenta and amnion have recently been evaluated for the treatment of burns and wounds. There are several cell types that have been investigated in preclinical and clinical trials for treating burns and wounds, and these are summarized below.

Populations of cells matching the identification criteria of MSCs have also been isolated from the amnion [112] and chorion [113], as well as from the placental tissues [114]. One method to isolate an MSC-like population from the placenta involves mechanically separating the fetal placental specimen from the maternal decidua by blunt dissection, then the placental tissue is minced, digested with a collagenase/dispase mixture, filtered, and plated on coverslips [114]. The resulting placental MSCs share with adult bone marrow-derived MSCs a spindle-shaped fibroblast appearance, the ability to adhere to plastic, and the ability to expand ex vivo. However, it has been reported that placental MSCs expand faster in vitro than adult MSCs and appear to be both less immunogenic and more immunosuppressive than their adult counterparts [115,116]. These properties make placental MSCs attractive as a potential clinical therapy for the treatment of burns and chronic nonhealing wounds.

Placenta-derived MSCs were evaluated for the treatment of wounds in diabetic Goto-Kakizaki (GK) rats by Kong et al. [117]. In this study, a full-thickness circular excisional wound received 1 million MSCs injected intradermally around the wound. After complete wound healing, full-thickness skin samples were taken from the wound sites for histological evaluation of the volume and density of vessels. The study found that transplantation of MSCs significantly increased the wound healing rate, and histological evaluation demonstrated an improved condition of the remaining scar. Interestingly, the authors found a higher microvessel density in wound bed biopsy sites and that transplanted MSCs appeared to be localized to the wound tissue and were incorporated into recipient vasculature.

Amnion Epithelial Cells

A highly multipotent epithelial cell population can be isolated from the amnion [118,119]. Human amnion epithelial cells (hAECs) have now been widely applied for many therapeutic applications [91], with reports that epithelial cells derived from human term amnion possess multipotent differentiation ability [120], low immunogenicity [113], and anti-inflammatory functions [115]. To isolate hAEC, term human amnion is manually separated from the chorion and washed in a saline solution to remove blood. Epithelial cells are isolated from the amnion by enzymatic digestion in TrypZean (an animal product-free recombinant trypsin). About 120 million viable epithelial cells can be isolated from a single amnion, and these cells can be maintained in serum-free culture conditions. While hAECs lack the proliferative capacity of placental MSCs, they are highly multipotent, differentiating into cells representing all three germ layers, specifically osteocytes, adipocytes, neurons, lung epithelial cells, cardiomyocytes, myocytes, hepatocytes, and pancreatic cells [118,120]. The immunosuppressive properties of hAECs have been well characterized both in vitro and in vivo [121,122]. It has been proposed that soluble factors produced by hAECs have anti-inflammatory effects and act to inhibit both the innate and adaptive immune systems. Thus, as first proposed by Faulk et al., hAECs may have potent regenerative properties when used to treat burns and wounds.

Xing et al. evaluated the amnion epithelial cells to accelerate and improve laparotomy wound healing in rat abdominal walls [123]. A 5-cm full-thickness laparotomy incision is made through the linea alba at the musculo-tendinous layer of the abdominal wall and repaired with a running suture. Cell-treated animals received 1 million amnion epithelial cells injected along 5 cm of the linea alba in 200 μL of saline carrier. In this study, wound mechanical properties and incidence and severity of laparotomy wound failure was measured over 28 days. GFP-labeled amnion epithelial cells were identified in the wound area over the 28 day study, although the numbers of cells appeared to decrease over time. No GFP expressing cellular elements were found in the liver, lymph nodes, spleen, or testis, suggesting migration from the wound area did not occur. Tensiometric analysis found that wounds treated with amnion epithelial cells developed increased breaking strength in the early postoperative fascial incision. Additionally, cell-treated wounds showed evidence of higher vascularization, more granulation tissue, and organized fibro-proliferation. The authors also suggested that the intensity of the wound inflammatory response was modulated and less severe.

Amniotic Fluid Stem Cells

Human amniotic fluid can be obtained during amniocentesis and is mainly performed in pregnancies in which the fetus has a congenital abnormality or to determine characteristics such as sex [124]. The amniotic fluid facilitates fetal growth, as it contains nutrients and growth factors, functions to provide mechanical cushioning and antimicrobial properties that protect the fetus, and allows assessment of fetal maturity and disease [125]. Amniotic fluid also contains a mixture of different cell types, and various origins have been suggested for these cells [126], with identification of cells derived from the developing fetus, sloughed from the fetal amnion and skin, as well as the alimentary, respiratory, and urogenital tracts.

Recently, researchers have isolated and characterized highly multipotent cells from the amniotic fluid, called amniotic fluid–derived stem cells (AFSCs) [127]. Multiple studies have now demonstrated that they have the potential to differentiate into various cell lineages, including hematopoietic, adipogenic, osteogenic, myogenic, endothelial, hepatogenic, chondrocytic, pulmonary, cardiac, and neurogenic [127–129]. Additionally, AFSCs have been shown to be anti-inflammatory [130], suggesting potential clinical applications of these cells to treat diseases such as bone defects, lung disease, neurological disorders, kidney disease, and heart disease.

There have been several reports of the therapeutic application of AFSCs for the treatment of skin wounds. Using bioprinting technology, Skardal et al. delivered 5 million AFSCs to treat full-thickness 2×2 cm mouse skin wounds [131]. The bioprinter was used to deposit two layers of a fibrin–collagen gel by depositing a layer of thrombin, a layer of fibrinogen/collagen, a second layer of thrombin, a second layer of fibrinogen/collagen, and a final layer of thrombin. This technique achieved 100% hydrogel coverage containing a uniform distribution of cells. Over the course of the 2-week study, wound closure was determined using weekly image analysis. At each time point wound closure and reepithelialization were significantly greater in AFSC-treated wounds compared to control groups. Histological examination showed increased microvessel density and capillary diameters, suggesting that AFSCs were promoting wound healing and angiogenesis. Interestingly, tracking of fluorescently labeled AFSCs revealed that the cells did not permanently integrate into the tissue, and the authors suggested that the wound healing properties of the cells were more likely due to production of secreted paracrine factors, rather than direct cell contribution to new tissue formation.

This hypothesis is supported by research from Yoon et al. who examined the paracrine factors released by human AFSCs and their ability to accelerate wound healing [132]. Examination of the soluble factors secreted by AFSCs into the (serum-free) culture medium revealed that AFSCs produce growth factors, cytokines, and chemokines such as TNF-α, VEGF, TGF-β, leptin, IL8, and IL-6. This AFSC-conditioned medium was capable of significantly increasing the migratory capacity of dermal fibroblasts, which also produced greater amounts of extracellular matrix molecules, including collagen type III, vitronectin, fibronectin, MMP1, syndecan 2, syndecan 4, SPP 1, and elastin. To examine whether AFSC-conditioned medium could enhance wound healing in vivo, the authors of this study injected and topically applied 100 μL of AFSC-conditioned medium to 2-mm full-thickness excisional skin wounds on the dorsal surface of ICR mice. Wound analysis revealed that treatment with AFSC-conditioned medium significantly accelerated wound closure and reepithelialization compared to treatment with vehicle control medium.

SUMMARY AND FUTURE APPLICATIONS OF STEM CELLS IN WOUND HEALING

Several types of stem cells have been isolated from various perinatal and adult human tissues and organs and show promising results as a cellular therapy for skin regeneration and repair. These cells differ in their proliferation capacity, immunomodulatory properties, secretory profile, tissue regeneration capacity, degree of reliability, ease of isolation, yield, expansion cost, clinical practicality, mode of delivery, and so on. However, the underlying molecular and biological mechanisms of each stem cell type are still not fully defined. Despite the tremendous body of work available in the literature on the use of stem cells for skin tissue engineering and wound healing, there is no general consensus on the efficacy of each stem cell type in wound healing, and substantial discrepancies between studies can be seen. These discrepancies are attributed to many factors including the strain and species of the animal model used, the immunologic aspects of the donor and recipient cells, methods of cell isolation, propagation, storage/freezing, timing of cell delivery, route of delivery, and many other factors. Each of these factors has tremendous potential to significantly impact the findings and conclusion of a given study. Lack of head-to-head studies comparing different sources of stem cells in a reliable preclinical model hinders the clinical translation of these therapies to patients. However, clinical trials using stem cells have already started for diseases that are particularly difficult to treat. Nonetheless, caution should be taken given that inadequate preclinical trials are all that is available for many applications of stem cells. Another important aspect to keep in mind during the development of a cell-based therapy is the regulatory pathway consisting of guidelines and manufacturing considerations. Early thoughts about the feasibility and cost of Good Manufacturing Practice will save substantial time and effort and expedite the transition into the clinic. Also, the engagement of end-users at an early stage during the development of a product has proven to enhance proper and suitable utilization of the developed product on patients.

REFERENCES

[1] Vagnozzi AN, Reiter JF, Wong SY. Hair follicle and interfollicular epidermal stem cells make varying contributions to wound regeneration. Cell Cycle 2015;23:0.

[2] Shi C, Zhu Y, Su Y, Cheng T. Stem cells and their applications in skin-cell therapy. Trends Biotechnol 2006;24(1):48–52.

[3] Blanpain C, Fuchs E. Epidermal stem cells of the skin. Annu Rev Cell Dev Biol 2006;22: 339–73.

[4] Tumbar T, Guasch G, Greco V, Blanpain C, Lowry WE, et al. Defining the epithelial stem cell niche in skin. Science 2004;303(5656):359–63.

[5] Ito M, Liu Y, Yang Z, Nguyen J, Liang F, et al. Stem cells in the hair follicle bulge contribute to wound repair but not to homeostasis of the epidermis. Nat Med 2005;11(12):1351–4.

[6] Ma DR, Yang EN, Lee ST. A review: the location, molecular characterisation and multipotency of hair follicle epidermal stem cells. Ann Acad Med Singapore 2004;33(6):784–8.

[7] Ohyama M. Hair follicle bulge: a fascinating reservoir of epithelial stem cells. J Dermatol Sci 2007;46(2):81–9.

[8] Cotsarelis G. Epithelial stem cells: a folliculocentric view. J Invest Dermatol 2006;126(7): 1459–68.

[9] Doucet YS, Owens DM. Isolation and functional assessment of cutaneous stem cells. Methods Mol Biol 2015;1235:147–64. http://dx.doi.org/10.1007/978-1-4939-1785-3_13.

[10] Reiisi S, Esmaeili F, Shirazi A. Isolation, culture and identification of epidermal stem cells from newborn mouse skin. In Vitro Cell Dev Biol Anim 2010;46(1):54–9.

[11] Fujimori Y, Izumi K, Feinberg SE, Marcelo CL. Isolation of small-sized human epidermal progenitor/stem cells by Gravity Assisted Cell Sorting (GACS). J Dermatol Sci 2009;56(3):181–7.

[12] Kim DS, Cho HJ, Choi HR, Kwon SB, Park KC. Isolation of human epidermal stem cells by adherence and the reconstruction of skin equivalents. Cell Mol Life Sci 2004;61(21):2774–81.
[13] Lau K, Paus R, Tiede S, Day P, Bayat A. Exploring the role of stem cells in cutaneous wound healing. Exp Dermatol 2009;18(11):921–33.
[14] Toma JG, McKenzie IA, Bagli D, Miller FD. Isolation and characterization of multipotent skin-derived precursors from human skin. Stem Cells 2005;23(6):727–37.
[15] Biernaskie JA, McKenzie IA, Toma JG, Miller FD. Isolation of skin-derived precursors (SKPs) and differentiation and enrichment of their Schwann cell progeny. Nat Protoc 2006;1(6):2803–12.
[16] Lavoie JF, Biernaskie JA, Chen Y, Bagli D, Alman B, et al. Skin-derived precursors differentiate into skeletogenic cell types and contribute to bone repair. Stem Cells Dev 2009;18(6):893–906.
[17] Fernandes KJ, Kobayashi NR, Gallagher CJ, Barnabe-Heider F, Aumont A, et al. Analysis of the neurogenic potential of multipotent skin-derived precursors. Exp Neurol 2006;201(1):32–48.
[18] Mehrabi M, Mansouri K, Hosseinkhani S, Yarani R, Yari K, et al. Differentiation of human skin-derived precursor cells into functional islet-like insulin-producing cell clusters. In Vitro Cell Dev Biol Anim 2015;51(6):595–603.
[19] Guo W, Miao C, Liu S, Qiu Z, Li J, et al. Efficient differentiation of insulin-producing cells from skin-derived stem cells. Cell Prolif 2009;42(1):49–62.
[20] Rodrigues RM, De Kock J, Branson S, Vinken M, Meganathan K, et al. Human skin-derived stem cells as a novel cell source for in vitro hepatotoxicity screening of pharmaceuticals. Stem Cells Dev 2014;23(1):44–55.
[21] Mao D, Yao X, Feng G, Yang X, Mao L, et al. Skin-derived precursor cells promote angiogenesis and stimulate proliferation of endogenous neural stem cells after cerebral infarction. Biomed Res Int March 26, 2015;2015:10. Article ID: 945846.
[22] Vasudeva VS, Abd-El-Barr MM, Chi JH. Implantation of neonatal skin-derived precursor schwann cells improves outcomes after incomplete cervical spinal cord injury in rats. Neurosurgery August 2015;77(2):N15–7. http://dx.doi.org/10.1227/01.neu.0000467294.10763.69.
[23] Pisati F, Belicchi M, Acerbi F, Marchesi C, Giussani C, et al. Effect of human skin-derived stem cells on vessel architecture, tumor growth, and tumor invasion in brain tumor animal models. Cancer Res 2007;67(7):3054–63.
[24] Bianco P, Riminucci M, Gronthos S, Robey PG. Bone marrow stromal stem cells: nature, biology, and potential applications. Stem Cells 2001;19(3):180–92.
[25] Squillaro T, Peluso G, Galderisi U. Clinical trials with mesenchymal stem cells: an update. Cell Transplant 2015;29:29.
[26] Krebsbach PH, Kuznetsov SA, Bianco P, Robey PG. Bone marrow stromal cells: characterization and clinical application. Crit Rev Oral Biol Med 1999;10(2):165–81.
[27] Burd A, Ahmed K, Lam S, Ayyappan T, Huang L. Stem cell strategies in burns care. Burns 2007;33(3):282–91.
[28] Minguell JJ, Conget P, Erices A. Biology and clinical utilization of mesenchymal progenitor cells. Braz J Med Biol Res 2000;33(8):881–7.
[29] Boxall SA, Jones E. Markers for characterization of bone marrow multipotential stromal cells. Stem Cells Int 2012:12. Article ID: 975871.
[30] Rea S, Giles NL, Webb S, Adcroft KF, Evill LM, et al. Bone marrow-derived cells in the healing burn wound–more than just inflammation. Burns 2009;35(3):356–64.
[31] Higashiyama R, Nakao S, Shibusawa Y, Ishikawa O, Moro T, et al. Differential contribution of dermal resident and bone marrow-derived cells to collagen production during wound healing and fibrogenesis in mice. J Invest Dermatol 2011;131(2):529–36.

[32] Higashiyama R, Moro T, Nakao S, Mikami K, Fukumitsu H, et al. Negligible contribution of bone marrow-derived cells to collagen production during hepatic fibrogenesis in mice. Gastroenterology 2009;137(4):1459–66.
[33] Wu Y, Zhao RC, Tredget EE. Concise review: bone marrow-derived stem/progenitor cells in cutaneous repair and regeneration. Stem Cells 2010;28(5):905–15.
[34] Fathke C, Wilson L, Hutter J, Kapoor V, Smith A, et al. Contribution of bone marrow-derived cells to skin: collagen deposition and wound repair. Stem Cells 2004;22(5):812–22.
[35] Javazon EH, Keswani SG, Badillo AT, Crombleholme TM, Zoltick PW, et al. Enhanced epithelial gap closure and increased angiogenesis in wounds of diabetic mice treated with adult murine bone marrow stromal progenitor cells. Wound Repair Regen 2007; 15(3):350–9.
[36] Wu Y, Chen L, Scott PG, Tredget EE. Mesenchymal stem cells enhance wound healing through differentiation and angiogenesis. Stem Cells 2007;25(10):2648–59.
[37] Wu Y, Wang J, Scott PG, Tredget EE. Bone marrow-derived stem cells in wound healing: a review. Wound Repair Regen 2007;15(1).
[38] Sasaki M, Abe R, Fujita Y, Ando S, Inokuma D, et al. Mesenchymal stem cells are recruited into wounded skin and contribute to wound repair by transdifferentiation into multiple skin cell type. J Immunol 2008;180(4):2581–7.
[39] Bellini A, Mattoli S. The role of the fibrocyte, a bone marrow-derived mesenchymal progenitor, in reactive and reparative fibroses. Lab Invest 2007;87(9):858–70.
[40] Abe R, Donnelly SC, Peng T, Bucala R, Metz CN. Peripheral blood fibrocytes: differentiation pathway and migration to wound sites. J Immunol 2001;166(12):7556–62.
[41] Quan TE, Cowper SE, Bucala R. The role of circulating fibrocytes in fibrosis. Curr Rheumatol Rep 2006;8(2):145–50.
[42] Fan X, Liang HP. Circulating fibrocytes: a potent cell population in antigen-presenting and wound healing. Chin J Traumatol 2010;13(2):111–6.
[43] Kobayashi M, Spector M. In vitro response of the bone marrow-derived mesenchymal stem cells seeded in a type-I collagen-glycosaminoglycan scaffold for skin wound repair under the mechanical loading condition. Mol Cell Biomech 2009;6(4):217–27.
[44] Zhao J, Zeiai S, Ekblad A, Nordenskjold A, Hilborn J, et al. Transdifferentiation of autologous bone marrow cells on a collagen-poly(ε-caprolactone) scaffold for tissue engineering in complete lack of native urothelium. J R Soc Interface 2014;11(96):6.
[45] Shokrgozar MA, Fattahi M, Bonakdar S, Ragerdi Kashani I, Majidi M, et al. Healing potential of mesenchymal stem cells cultured on a collagen-based scaffold for skin regeneration. Iran Biomed J 2012;16(2):68–76.
[46] Falanga V, Iwamoto S, Chartier M, Yufit T, Butmarc J, et al. Autologous bone marrow-derived cultured mesenchymal stem cells delivered in a fibrin spray accelerate healing in murine and human cutaneous wounds. Tissue Eng 2007;13(6):1299–312.
[47] Badiavas EV, Falanga V. Treatment of chronic wounds with bone marrow-derived cells. Arch Dermatol 2003;139(4):510–6.
[48] Rogers LC, Bevilacqua NJ, Armstrong DG. The use of marrow-derived stem cells to accelerate healing in chronic wounds. Int Wound J 2008;5(1):20–5.
[49] Klepanec A, Mistrik M, Altaner C, Valachovicova M, Olejarova I, et al. No difference in intra-arterial and intramuscular delivery of autologous bone marrow cells in patients with advanced critical limb ischemia. Cell Transplant 2012;21(9):1909–18.
[50] El Sadik AO, El Ghamrawy TA, Abd El-Galil TI. The effect of mesenchymal stem cells and chitosan gel on full thickness skin wound healing in albino rats: histological, immunohistochemical and fluorescent study. PLoS One 2015;10(9).

[51] Rodriguez-Menocal L, Shareef S, Salgado M, Shabbir A, Van Badiavas E. Role of whole bone marrow, whole bone marrow cultured cells, and mesenchymal stem cells in chronic wound healing. Stem Cell Res Ther 2015;6(24). http://dx.doi.org/10.1186/s13287-015-0001-9.
[52] Amann B, Luedemann C, Ratei R, Schmidt-Lucke JA. Autologous bone marrow cell transplantation increases leg perfusion and reduces amputations in patients with advanced critical limb ischemia due to peripheral artery disease. Cell Transplant 2009;18(3):371–80.
[53] Li M, Zhou H, Jin X, Wang M, Zhang S, et al. Autologous bone marrow mononuclear cells transplant in patients with critical leg ischemia: preliminary clinical results. Exp Clin Transplant 2013;11(5):435–9.
[54] Gabr H, Hedayet A, Imam U, Nasser M. Limb salvage using intramuscular injection of unfractionated autologous bone marrow mononuclear cells in critical limb ischemia: a prospective pilot clinical trial. Exp Clin Transplant 2011;9(3):197–202.
[55] Ding J, Ma Z, Shankowsky HA, Medina A, Tredget EE. Deep dermal fibroblast profibrotic characteristics are enhanced by bone marrow-derived mesenchymal stem cells. Wound Repair Regen 2013;21(3):448–55.
[56] Wang J, Dodd C, Shankowsky HA, Scott PG, Tredget EE. Deep dermal fibroblasts contribute to hypertrophic scarring. Lab Invest 2008;88(12):1278–90.
[57] Wilgus TA, Ferreira AM, Oberyszyn TM, Bergdall VK, Dipietro LA. Regulation of scar formation by vascular endothelial growth factor. Lab Invest 2008;88(6):579–90.
[58] Yang M, Sheng L, Zhang TR, Li Q. Stem cell therapy for lower extremity diabetic ulcers: where do we stand? Biomed Res Int 2013:8. Article ID: 462179.
[59] Kirby GT, Mills SJ, Cowin AJ, Smith LE. Stem cells for cutaneous wound healing. Biomed Res Int 2015:11. Article ID: 285869.
[60] Gimble JM, Katz AJ, Bunnell BA. Adipose-derived stem cells for regenerative medicine. Circ Res 2007;100(9):1249–60.
[61] Butler KL, Goverman J, Ma H, Fischman A, Yu YM, et al. Stem cells and burns: review and therapeutic implications. J Burn Care Res 2010;31(6):874–81.
[62] Salgado AJ, Reis RL, Sousa NJ, Gimble JM. Adipose tissue derived stem cells secretome: soluble factors and their roles in regenerative medicine. Curr Stem Cell Res Ther 2010;5(2):103–10.
[63] Kilroy GE, Foster SJ, Wu X, Ruiz J, Sherwood S, et al. Cytokine profile of human adipose-derived stem cells: expression of angiogenic, hematopoietic, and pro-inflammatory factors. J Cell Physiol 2007;212(3):702–9.
[64] Kim WS, Park BS, Sung JH, Yang JM, Park SB, et al. Wound healing effect of adipose-derived stem cells: a critical role of secretory factors on human dermal fibroblasts. J Dermatol Sci 2007;48(1):15–24.
[65] Kober J, Gugerell A, Schmid M, Zeyda M, Buchberger E, et al. Wound healing effect of conditioned media obtained from adipose tissue on human skin cells: a comparative in vitro study. Ann Plast Surg 2014;30:30.
[66] Lee SH, Jin SY, Song JS, Seo KK, Cho KH. Paracrine effects of adipose-derived stem cells on keratinocytes and dermal fibroblasts. Ann Dermatol 2012;24(2):136–43.
[67] Zhao J, Hu L, Liu J, Gong N, Chen L. The effects of cytokines in adipose stem cell-conditioned medium on the migration and proliferation of skin fibroblasts in vitro. Biomed Res Int 2013:11. Article ID: 578479.
[68] Hu L, Zhao J, Liu J, Gong N, Chen L. Effects of adipose stem cell-conditioned medium on the migration of vascular endothelial cells, fibroblasts and keratinocytes. Exp Ther Med 2013;5(3):701–6.

[69] Koh KS, Oh TS, Kim H, Chung IW, Lee KW, et al. Clinical application of human adipose tissue-derived mesenchymal stem cells in progressive hemifacial atrophy (Parry-Romberg disease) with microfat grafting techniques using 3-dimensional computed tomography and 3-dimensional camera. Ann Plast Surg 2012;69(3):331–7.

[70] Zografou A, Tsigris C, Papadopoulos O, Kavantzas N, Patsouris E, et al. Improvement of skin-graft survival after autologous transplantation of adipose-derived stem cells in rats. J Plast Reconstr Aesthet Surg 2011;64(12):1647–56.

[71] Zografou A, Papadopoulos O, Tsigris C, Kavantzas N, Michalopoulos E, et al. Autologous transplantation of adipose-derived stem cells enhances skin graft survival and wound healing in diabetic rats. Ann Plast Surg 2013;71(2):225–32.

[72] Strong AL, Bowles AC, MacCrimmon CP, Frazier TP, Lee SJ, et al. Adipose stromal cells repair pressure ulcers in both young and elderly mice: potential role of adipogenesis in skin repair. Stem Cells Transl Med 2015;4(6):632–42.

[73] Nie C, Yang D, Xu J, Si Z, Jin X, et al. Locally administered adipose-derived stem cells accelerate wound healing through differentiation and vasculogenesis. Cell Transplant 2011;20(2):205–16.

[74] Huang SP, Hsu CC, Chang SC, Wang CH, Deng SC, et al. Adipose-derived stem cells seeded on acellular dermal matrix grafts enhance wound healing in a murine model of a full-thickness defect. Ann Plast Surg 2012;69(6):656–62.

[75] Koellensperger E, Lampe K, Beierfuss A, Gramley F, Germann G, et al. Intracutaneously injected human adipose tissue-derived stem cells in a mouse model stay at the site of injection. J Plast Reconstr Aesthet Surg 2014;67(6):844–50.

[76] Lin YC, Grahovac T, Oh SJ, Ieraci M, Rubin JP, et al. Evaluation of a multi-layer adipose-derived stem cell sheet in a full-thickness wound healing model. Acta Biomater 2013;9(2):5243–50.

[77] McLaughlin MM, Marra KG. The use of adipose-derived stem cells as sheets for wound healing. Organogenesis 2013;9(2):79–81.

[78] Cronk SM, Kelly-Goss MR, Ray HC, Mendel TA, Hoehn KL, et al. Adipose-derived stem cells from diabetic mice show impaired vascular stabilization in a murine model of diabetic retinopathy. Stem Cells Transl Med 2015;4(5):459–67.

[79] Cianfarani F, Toietta G, Di Rocco G, Cesareo E, Zambruno G, et al. Diabetes impairs adipose tissue-derived stem cell function and efficiency in promoting wound healing. Wound Repair Regen 2013;21(4):545–53.

[80] Cai L, Johnstone BH, Cook TG, Liang Z, Traktuev D, et al. Suppression of hepatocyte growth factor production impairs the ability of adipose-derived stem cells to promote ischemic tissue revascularization. Stem Cells 2007;25(12):3234–43.

[81] Mendel TA, Clabough EB, Kao DS, Demidova-Rice TN, Durham JT, et al. Pericytes derived from adipose-derived stem cells protect against retinal vasculopathy. PLoS One 2013;8(5).

[82] Gerhardt H, Betsholtz C. Endothelial-pericyte interactions in angiogenesis. Cell Tissue Res 2003;314(1):15–23.

[83] Lee SH, Lee JH, Cho KH. Effects of human adipose-derived stem cells on cutaneous wound healing in nude mice. Ann Dermatol 2011;23(2):150–5.

[84] van den Broek LJ, Kroeze KL, Waaijman T, Breetveld M, Sampat-Sardjoepersad SC, et al. Differential response of human adipose tissue-derived mesenchymal stem cells, dermal fibroblasts, and keratinocytes to burn wound exudates: potential role of skin-specific chemokine CCL27. Tissue Eng Part A 2014;20(1–2):197–209.

[85] Uysal CA, Tobita M, Hyakusoku H, Mizuno H. The effect of bone-marrow-derived stem cells and adipose-derived stem cells on wound contraction and epithelization. Adv Wound Care 2014;3(6):405–13.
[86] Liu X, Wang Z, Wang R, Zhao F, Shi P, et al. Direct comparison of the potency of human mesenchymal stem cells derived from amnion tissue, bone marrow and adipose tissue at inducing dermal fibroblast responses to cutaneous wounds. Int J Mol Med 2013;31(2): 407–15.
[87] Kim Y, Kim H, Cho H, Bae Y, Suh K, et al. Direct comparison of human mesenchymal stem cells derived from adipose tissues and bone marrow in mediating neovascularization in response to vascular ischemia. Cell Physiol Biochem 2007;20(6):867–76.
[88] Zhou Z, Chen Y, Zhang H, Min S, Yu B, et al. Comparison of mesenchymal stromal cells from human bone marrow and adipose tissue for the treatment of spinal cord injury. Cytotherapy 2013;15(4):434–48.
[89] Han C, Zhang L, Song L, Liu Y, Zou W, et al. Human adipose-derived mesenchymal stem cells: a better cell source for nervous system regeneration. Chin Med J 2014;127(2):329–37.
[90] Serikov V, Hounshell C, Larkin S, Green W, Ikeda H, et al. Human term placenta as a source of hematopoietic cells. Exp Biol Med (Maywood) 2009;234(7):813–23.
[91] Murphy S, Wallace E, Jenkin G. In: Appasani K, Appasani RK, editors. Placental-derived stem cells: potential clinical applications – stem cells & regenerative medicine. Humana Press; 2011. p. 243–63.
[92] Troyer DL, Weiss ML. Wharton's jelly-derived cells are a primitive stromal cell population. Stem Cells 2008;26(3):591–9.
[93] Bowen JM, Chamley L, Keelan JA, Mitchell MD. Cytokines of the placenta and extra-placental membranes: roles and regulation during human pregnancy and parturition. Placenta 2002;23(4):257–73.
[94] Tauzin H, Rolin G, Viennet C, Saas P, Humbert P, et al. A skin substitute based on human amniotic membrane. Cell Tissue Bank 2014;15(2):257–65.
[95] Sawhney CP. Amniotic membrane as a biological dressing in the management of burns. Burns 1989;15(5):339–42.
[96] Haberal M, Oner Z, Bayraktar U, Bilgin N. The use of silver nitrate-incorporated amniotic membrane as a temporary dressing. Burns Incl Therm Inj 1987;13(2):159–63.
[97] Morykwas MJ, Thornton JW, Bartlett RH. Zeta potential of synthetic and biological skin substitutes: effects on initial adherence. Plast Reconstr Surg 1987;79(5):732–9.
[98] Yang L, Shirakata Y, Tokumaru S, Xiuju D, Tohyama M, et al. Living skin equivalents constructed using human amnions as a matrix. J Dermatol Sci 2009;56(3):188–95.
[99] Gajiwala K, Gajiwala AL. Evaluation of lyophilised, gamma-irradiated amnion as a biological dressing. Cell Tissue Bank 2004;5(2):73–80.
[100] Davis JS. Skin transplantation, vol. 15. Johns Hopkins Hospital Reports; 1910. p. 307–96.
[101] Bose B. Burn wound dressing with human amniotic membrane. Ann R Coll Surg Engl 1979;61(6):444.
[102] Bennett J, Matthews R, Faulk WP. Treatment of chronic ulceration of the legs with human amnion. Lancet 1980;315(8179):1153–6.
[103] Atanasov W, Mazalova J, TodorodR SK, Trencheva W. Use of amniotic membrane as biological dressings in contemporary treatment of burns. Ann Medi Burns Club 1994;7(4).
[104] Fisher JC. Amniotic membranes as a temporary wound dressing. Plast Reconstr Surg 1973;52(5):601.
[105] Ward D, Bennett J. The long-term results of the use of human amnion in the treatment of leg ulcers. Br J Plast Surg 1984;37(2):191–3.

[106] Faulk WP, Stevens P, Burgos H, Matthews R, Bennett J, et al. Human amnion as an adjunct in wound healing. The Lancet 1980;315(8179):1156–8.

[107] Mohammadi AA, Mohammadi MK. How does human amniotic membrane help major burn patients who need skin grafting: new experiences. INTECH Open Access Publisher; 2011. p. 265–76.

[108] Singh R, Purohit S, Chacharkar M, Bhandari P, Bath A. Microbiological safety and clinical efficacy of radiation sterilized amniotic membranes for treatment of second-degree burns. Burns 2007;33(4):505–10.

[109] Gajiwala K, Gajiwala AL. Use of banked tissue in plastic surgery. Cell Tissue Bank 2003;4(2–4):141–6.

[110] Branski LK, Herndon DN, Celis MM, Norbury WB, Masters OE, et al. Amnion in the treatment of pediatric partial-thickness facial burns. Burns 2008;34(3):393–9.

[111] Kesting MR, Wolff K-D, Hohlweg-Majert B, Steinstraesser L. The role of allogenic amniotic membrane in burn treatment. J Burn Care Res 2008;29(6):907–16.

[112] Alviano F, Fossati V, Marchionni C, Arpinati M, Bonsi L, et al. Term amniotic membrane is a high throughput source for multipotent mesenchymal stem cells with the ability to differentiate into endothelial cells in vitro. BMC Dev Biol 2007;7(1):11.

[113] Bailo M, Soncini M, Vertua E, Signoroni PB, Sanzone S, et al. Engraftment potential of human amnion and chorion cells derived from term placenta. Transplantation 2004;78(10):1439–48.

[114] Steigman SA, Fauza DO. Isolation of mesenchymal stem cells from amniotic fluid and placenta. Curr Protoc Stem Cell Biol 2007: 1E. 2.1–1E. 2.12.

[115] Chang Dong L, Zhang WY, He Lian L, Jiang XX, Zhang Y, et al. Mesenchymal stem cells derived from human placenta suppress allogeneic umbilical cord blood lymphocyte proliferation. Cell Res 2005;15(7):539–47.

[116] Kern S, Eichler H, Stoeve J, Klüter H, Bieback K. Comparative analysis of mesenchymal stem cells from bone marrow, umbilical cord blood, or adipose tissue. Stem Cells 2006;24(5):1294–301.

[117] Kong P, Xie X, Li F, Liu Y, Lu Y. Placenta mesenchymal stem cell accelerates wound healing by enhancing angiogenesis in diabetic Goto-Kakizaki (GK) rats. Biochem Biophys Res Commun 2013;438(2):410–9.

[118] Murphy S, Rosli S, Acharya R, Mathias L, Lim R, et al. Amnion epithelial cell isolation and characterization for clinical use. Curr Protoc Stem Cell Biol 2010. [Chapter 1] (Unit 1E 6).

[119] Miki T, Marongiu F, Dorko K, Ellis E, Strom SC. Isolation of amniotic epithelial stem cells. Curr Protoc Stem Cell Biol 2010: 1E. 3.1–1E. 3.10.

[120] Ilancheran S, Michalska A, Peh G, Wallace EM, Pera M, et al. Stem cells derived from human fetal membranes display multilineage differentiation potential. Biol Reprod 2007;77(3): 577–88.

[121] Li H, Niederkorn JY, Neelam S, Mayhew E, Word RA, et al. Immunosuppressive factors secreted by human amniotic epithelial cells. Invest Ophthalmol Vis Sci 2005;46(3):900–7.

[122] Hori J, Wang M, Kamiya K, Takahashi H, Sakuragawa N. Immunological characteristics of amniotic epithelium. Cornea 2006;25:S53–8.

[123] Xing L, Franz MG, Marcelo CL, Smith CA, Marshall VS, et al. Amnion-derived multipotent progenitor cells increase gain of incisional breaking strength and decrease incidence and severity of acute wound failure. J Burns Wounds 2007;7.

[124] Hoehn H, Bryant EM, Karp LE, Martin GM. Cultivated cells from diagnostic amniocentesis in second trimester pregnancies. Clin Genet 1975;7(1):29–36.

[125] Underwood MA, Gilbert WM, Sherman MP. Amniotic fluid: not just fetal urine anymore. J Perinatol 2005;25(5):341–8.

[126] Medina-Gomez P, Del Valle M. The culture of amniotic fluid cells. An analysis of the colonies, metaphase and mitotic index for the purpose of ruling out maternal cell contamination. Ginecol Obstet Mex 1988;56:122–6.

[127] De Coppi P, Bartsch G, Siddiqui MM, Xu T, Santos CC, et al. Isolation of amniotic stem cell lines with potential for therapy. Nat Biotechnol 2007;25(1):100–6.

[128] Scherjon SA, Kleijburg-van der Keur C, Noort WA, Claas FH, Willemze R, et al. Amniotic fluid as a novel source of mesenchymal stem cells for therapeutic transplantation. Blood 2003;102(4):1548–9.

[129] Tsai M-S, Hwang S-M, Tsai Y-L, Cheng F-C, Lee J-L, et al. Clonal amniotic fluid-derived stem cells express characteristics of both mesenchymal and neural stem cells. Biol Reprod 2006;74(3):545–51.

[130] Moorefield EC, McKee EE, Solchaga L, Orlando G, Yoo JJ, et al. Cloned, CD117 selected human amniotic fluid stem cells are capable of modulating the immune response. PLoS One 2011;6(10):e26535.

[131] Skardal A, Mack D, Kapetanovic E, Atala A, Jackson JD, et al. Bioprinted amniotic fluid-derived stem cells accelerate healing of large skin wounds. Stem Cells Transl Med 2012;1(11):792–802.

[132] Yoon BS, Moon J-H, Jun EK, Kim J, Maeng I, et al. Secretory profiles and wound healing effects of human amniotic fluid–derived mesenchymal stem cells. Stem Cells Dev 2009;19(6):887–902.

Chapter 17

Advances in Biopharmaceutical Agents and Growth Factors for Wound Healing and Scarring

Ferdinand V. Lali, Yella H. Martin, Anthony D. Metcalfe
Blond McIndoe Research Foundation, Queen Victoria Hospital, West Sussex, UK; The Brighton Centre for Regenerative Medicine, The University of Brighton, Brighton, UK

Chapter Outline

INTRODUCTION

Both tissue repair and regeneration are characterized by a constantly changing environment in which cells are exposed to a complex pattern of molecular cues and signals, which impart positional information necessary for tissue reconstruction [1]. These cell signals result in a series of events that control a complex interplay between cell proliferation, differentiation, and cell death, such that a specific tissue can be delineated and its edges specifically defined [1]. Since these molecules are often major components of early developmental pathways that result in the formation of a fully functioning organism, they represent potential candidates for therapeutic manipulation when the repair process is not as perfect as it might be.

Skin Tissue Engineering and Regenerative Medicine. http://dx.doi.org/10.1016/B978-0-12-801654-1.00017-6

There are obviously many types of wounds, and most undergo similar overlapping phases of repair that are mediated by cytokines, growth factors, and other chemotactic factors within the tissue. The duration of each stage of healing varies with the wound type but the resolution can still take months to years to resolve. One of the common outcomes of this resolution is the development of fibrous tissue or scars. These scars can form as a consequence of trauma, elective surgery, burn, or other injury. This often results in adverse aesthetics, loss of function, restriction of tissue movement, and adverse psychological effects [1,2]. Additional to the cosmetic appearance of a scar, pain, redness, pruritus, paresthesia, and functional impairment are all related symptoms that can have a negative impact on the patient's quality of life [3]. Current treatments are often unreliable, and there are no prescription drugs for the prevention or treatment of scarring in the skin.

Growth factors and cytokines are signaling molecules that help orchestrate the activities of cells through intercellular communication. Numerous families of diverse molecules are categorized as cytokines and/or growth factors. Members of these families bind to specific receptors on the surface of their target cells and induce intracellular signaling pathways, in turn regulating diverse biological processes such as cell activation, proliferation, differentiation, and migration. In recent years, there have been a number of biopharmaceutical agents and growth factors used to try to improve wound healing outcomes and to reduce scars. In this chapter, we consider some examples of the approaches attempted using these agents and factors.

THE TRANSFORMING GROWTH FACTOR-β FAMILY

The transforming growth factor-β (TGF-β) superfamily consists of structurally and functionally related cytokines that signal through two transmembrane serine–threonine kinase receptors known as type I and type II receptors (TβRI and TβRII), which activate intracellular Smad transcription factors to mediate a diversity of downstream signaling events (reviewed extensively in Finnson et al. [4]).

In the animal kingdom, there are five distinct forms of TGF-β but only TGF-β1, -β2, and -β3 isoforms are expressed in mammalian tissues [5]. TGF-β family members are expressed in all cells and have functions that control diverse cellular behaviors such as cell proliferation, differentiation, apoptosis, migration, epithelial-to-mesenchymal transition (EMT), extracellular matrix (ECM) remodeling, immune function, and roles in cancer [6–8]. These different roles have a number of important effects on both embryonic and adult tissues. Some of these effects can be important for correct development, and others are involved in the pathogenesis of diseases such as fibrosis, cardiovascular illness, and cancer. Here we consider the intricate balance and interplay that the various TGF-β isoforms have during wound healing and how that has been used for therapeutic intervention for scar reduction therapies.

Wound healing is orchestrated by many growth factors and cytokines, prominent among which are TGF-β1 and -β2. After activation, the TGF-β receptor (comprised of two transmembrane serine/threonine kinases—TβRI and TβRII) phosphorylates and activates Smad2 and Smad3 proteins that complex with co-Smad (Smad4), translocate to the nucleus, and transcribe genes whose products regulate scarring (reviewed by Finnson et al. [4]). All three major isoforms appear to be present during wound healing, though with very subtle but important differences in expression between embryos and adults. Embryonic wounds express very high levels of TGF-β3 and very low levels of TGF-β1 and TGF-β2. By contrast, adult wounds contain predominantly TGF-β1 (and TGF-β2), which is derived initially from degranulating platelets and subsequently from inflammatory cells such as monocytes and macrophages (reviewed in Metcalfe and Ferguson [1]). Although TGF-β isoforms (β1, β2, and β3) signal through the same cell surface receptors, they display distinct functions during wound healing in vivo through mechanisms that have not been fully elucidated [4]. It is, however, the elucidation of the complex relationship between TGF-β1 and TGF-β3 that led to TGF-β3 being developed as a potential scar reducing therapeutic following surgery [2,9–12]. From these early seminal discoveries over 20 years ago, careful and extensive preclinical studies and clinical trials have been undertaken. These studies showed that a clinical form of recombinant human TGF-β3, avotermin (Juvista, Shire), had the potential to reduce scarring [9,11,13]. The most ambitious and wide-ranging number of Phase I and Phase II clinical trials were undertaken by the company, Renovo, investigating a range of surgical interventions. These clinical trials demonstrated that intradermal injection of Juvista™ to skin wounds created either by surgical incision or excision, at or shortly after the time of surgery, resulted in improved subsequent scar appearance. These results finally led to the company embarking on a Phase III clinical trial using Juvista™ in scar revision surgery. This trial had a total of 350 patients in about 55 centers, in 10 countries across the European Union. On unblinding the results of the European Phase III clinical trial, disappointingly, it did not meet its primary or secondary efficacy end-points. As Agren and Danielsen state [14], Renovo should be acknowledged for their contribution to the wound-healing field. The investment of huge resources into developing and validating wound-healing models, as well as assessment tools, is not lost—it will become an invaluable resource for the wound healing community and can be used in future studies [14].

OTHER TGF-β-BASED APPROACHES: DECORIN AND MANNOSE 6 PHOSPHATE

Other approaches to scar reduction have focused on blocking the TGF-β signaling cascade, with a focus on inhibiting the pro-fibrotic members, TGF-β1 and/or -β2. One such approach involved using blocking antibodies to TGF-β1 and -β2 [9]. Natural inhibitors of TGF-β1 include the molecules decorin and mannose-6-phosphate (M6P). These natural inhibitors have structural similarity to beta-glycan (TGF-β type-III receptor) and may competitively bind TGF-β (Figures 1 and 2).

Decorin is a small, leucine-rich, chondroitin–dermatan sulfate proteoglycan expressed by neurons and astrocytes in the central nervous system [15]. It is both antifibrotic and antiinflammatory and attenuates the formation and partial dissolution of established and chronic scars [15]. Decorin's name is a derivative of the fact that it "decorates" collagen.

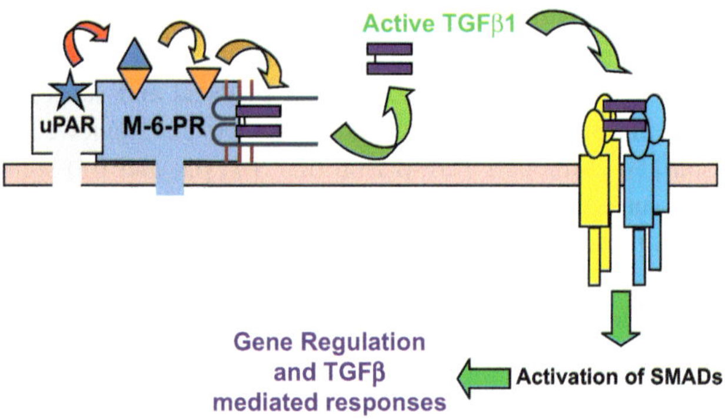

FIGURE 1 Binding of inactive TGF-β1 at the M6P receptor site permits the release of active TGF-β1, which can bind to TGF-β receptors at the cell membrane, activating SMADs, and permitting further gene regulation and TGF-β1-mediated responses.

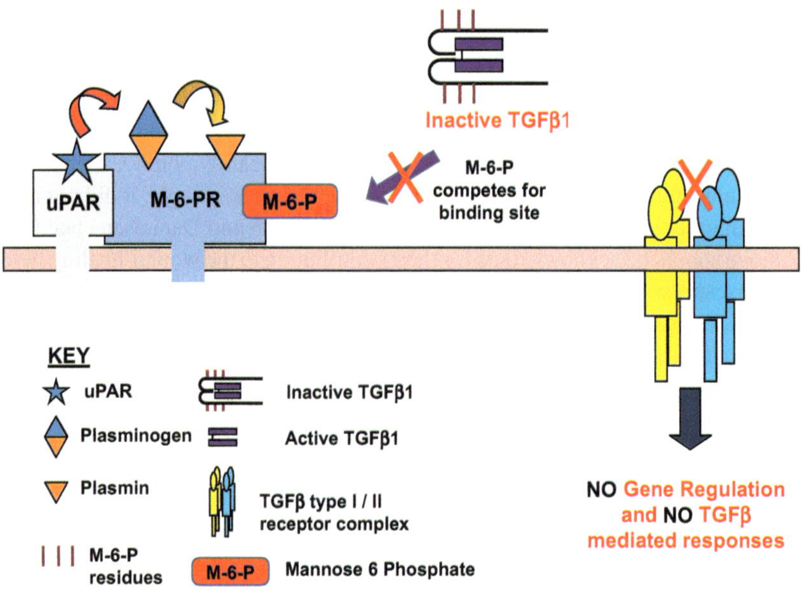

FIGURE 2 Prevention of the binding of inactive TGF-β1 at the M6P receptor by M6P reduces levels of active TGF-β1 and subsequent TGF-β1-mediated responses.

M6P is a molecule bound by lectin in the immune system. M6P is converted to fructose 6-phosphate by mannose phosphate isomerase and has been demonstrated to reduce active TGF-β1 expression on cultured tendon fibroblasts and improve the range of movement in a rabbit flexor tendon injury model [16]. M6P also facilitates early peripheral nerve regeneration [17,18]. Studies of M6P, in relation to skin scarring, have demonstrated improvement in scar cosmesis and accelerated return of normal dermal architecture [19]. However, the mechanisms by which M6P reduce adhesion formation is still unclear, and it is questionable whether its mode of action is solely via the inhibition of the TGF-β1 pathway [20].

MODULATION OF SMAD3/SMAD7 SIGNALING

Following activation of the TGF-β receptors, intracellular signal transduction is mediated by a variety of Smad proteins. These proteins play a crucial role in TGF-β signaling and in terminating the TGF-β signal by a negative feedback loop through Smad6 and -7 [21]. Like other elements of the wound healing cascade, it is unclear exactly what role the downstream TGF-β signaling molecules have in the pathogenesis of scarring. Smad3 acts as an activator of signal transduction, whereas Smad7 has an inhibitory effect. The inhibition of the Smad2/3 pathway is a critical step involved in the down-regulation of type I collagen synthesis, thus preventing scar formation in tissue [22] and in keloid fibroblasts [23]. Inhibitors of Smad3 represent promising therapies to treat hypertrophic scars and keloids [24]. Some of the inhibitors of Smad3 include halfuginone, quercitin, P144, and paclitaxel [25–27].

Recent reports have suggested that Smad7 might act as a critical regulator of TGF-β activity in human disease, where aberrant expression of Smad7 is implicated in inflammatory bowel disease and scleroderma. This has led to the postulation that modulation of Smad7 expression could provide a novel therapeutic basis for TGF-β-family modulation and associated disorders (reviewed in Nakao et al. [28]) [29]. Interferon-γ (IFN-γ) is known to induce endogenous Smad7 and, as a consequence, antagonizes TGF-β signaling [30]. A number of years ago, a few small clinical trials suggested that IFN-γ could be used as a therapeutic agent in treating abnormal dermal fibroses [31,32].

EPIDERMAL GROWTH FACTOR FAMILY

The EGF family comprises a number of members. In wound healing, EGF, TGF-α, and heparin-binding EGF (HB-EGF) bind to the EGF receptor (EGFR), which activates a tyrosine kinase signaling cascade. EGFR is found throughout the epidermis, particularly in the basal layer. During wound healing, EGF ligand-mediated EGFR receptor signaling promotes reepithelialization by stimulating keratinocyte proliferation and migration. EGF ligands are also involved in paracrine signaling between keratinocytes, fibroblasts, and inflammatory cells.

In chronic wounds, EGFR is down-regulated or mislocalized to the cytoplasm at the edge of nonhealing wounds. This results in impairment of reepithelialization, dysfunctional granulation tissue formation, and failed angiogenic response (reviewed in [33–35]).

Topical application of EGF in clinical studies has been shown to increase reepithelialization and shorten healing time in venous leg ulcers and diabetic foot ulcers [33,36–43]. Early works also showed similar effects in skin graft donor sites [44].

FIBROBLAST GROWTH FACTOR FAMILY

Within the fibroblast growth factor (FGF) family, FGF-2 (basic (b)FGF), FGF-7 (keratinocyte (K) GF), and FGF-10 have been identified as key players in wound healing. They are produced by keratinocytes, fibroblasts, endothelial cells, smooth muscle cells, and mast cells. Signaling is achieved through binding to one of the four FGF receptors (FGFR), which initiates tyrosine kinase signaling. FGF ligands have important roles in reepithelialization through promotion of keratinocyte proliferation and migration, granulation tissue formation, angiogenesis, and tissue remodeling (reviewed in [33,35,45]).

bFGF is known to be present at reduced levels in chronic wounds [46]. A pilot study testing the topical application of bFGF in diabetic ulcers failed to show improved wound healing [47], although a more recent study has reported accelerated wound healing compared with placebo [48]. However, the use of bFGF in the treatment of pressure ulcers has been reported in two studies to result in improved healing [49,50]. The topical application of recombinant human KGF (repifermin) has also been reported to accelerate wound healing in venous leg ulcers [51]. In the treatment of superficial and partial-thickness burn injuries, treatment with recombinant bovine bFGF was reported to improve reepithelialization and granulation tissue formation compared with placebo treatment [52].

PLATELET-DERIVED GROWTH FACTOR FAMILY

Platelet-derived growth factors (PDGFs) are produced by keratinocytes, fibroblasts, platelets, macrophages, and vascular endothelium and play different roles during different stages of wound healing. PDGFs act through binding to alpha and beta PDGF receptors (PDGFR), which initiates tyrosine kinase signaling. PDGF mediates cellular infiltration during the inflammatory stage of wound healing through chemotaxis. It also stimulates the release of TGF-β from macrophages and mediates tissue debridement. It is further implicated in the processes of granulation tissue formation, angiogenesis, and reepithelialization. In addition, it is known to mediate fibroblast and myofibroblast proliferation, wound contraction, and up-regulation of matrix metalloproteinase MMP production (reviewed in [33,35,53]). PDGF levels are decreased in chronic wounds.

Becaplermin (Regranex, Smith and Nephew Inc.) is a recombinant PDGF approved by the FDA for the treatment of diabetic ulcers that extend into the subcutaneous tissue and have adequate blood supply [35]. It has been shown to improve wound closure in diabetic ulcers [54–57], as well as pressure ulcers [49,58–61]. A safety warning by the FDA has been in place since 2008 advising that patients who have received three or more tubes of Regranex have an increased risk of cancer [62].

Novel approaches to improve the efficacy of PDGF treatment include the use of gene therapy to target PDGF expression to venous leg ulcers [63].

In the treatment of burn injuries, becaplermin application to split-thickness skin graft donor sites in red Duroc pigs showed increased reepithelialization over untreated wounds [64]. Gene therapy with PDGF was also shown to improve reepithelialization in large white pigs with burn wounds [65].

GRANULOCYTE MACROPHAGE-COLONY STIMULATING FACTOR

Granulocyte macrophage-colony stimulating factor (GM-CSF) is a cytokine involved in the different stages of wound healing. In the early phase, it is upregulated in keratinocytes and is known to stimulate neutrophils during inflammation. It is a potent mitogen for keratinocytes and upregulates proliferation and migration in keratinocytes, fibroblasts, endothelial cells, monocytes, macrophages, and dendritic cells, which is thought to be mediated by stimulating release of other cytokines and growth factors (reviewed in [33,35,66]). In mice that overexpress GM-CSF, faster reepithelialization and increased angiogenesis have been observed [66].

Sagramostin and its analogue molgramostim (Leucomax, Schering-Plough) are commercially available, injectable recombinant human GM-CSF formulations originally developed for use in combination with cancer chemotherapy [35]. Improved healing has been reported in clinical studies in patients with venous leg ulcers, pressure ulcers, diabetic foot ulcers, and diabetic foot infections [67–73], which is thought to be mediated through increased angiogenesis. In deep, partial-thickness burn wounds, topical application of rh-GM-CSF has also been reported to improve wound healing in two clinical studies [74,75].

CONNECTIVE TISSUE GROWTH FACTOR

Connective tissue growth factor (CTGF) is a heparin-binding protein and is part of the CCN protein family. A specific CTGF receptor has not been identified. Instead, CTGF acts via directly binding to integrins and ECM proteins. TGF-β is a potent inducer of CTGF and promotes CTGF expression in fibroblasts, although CTGF can act through TGF-β-independent mechanisms. CTGF is known to be involved in granulation tissue formation and reepithelialization, as well as matrix formation and remodeling (reviewed in [33,76]).

CTGF, in conjunction with TGF-β, is implicated in scar formation [76]. Fibroblasts from hypertrophic and keloid scars have elevated levels of CTGF, and a reduction in CTGF protein levels by application of RNA interference (RNAi) in pig fibroblasts has been shown to result in reduced levels of pro-fibrotic ECM components and other molecules [77]. Reduction in CTGF levels with antisense oligonucleotide technology showed reduced hypertrophic scarring in a preclinical rabbit model [78].

A self-delivering RNAi (ds-rxRNA) compound is in development by RXi Pharmaceuticals to reduce CTGF in patients. Two phase I studies showed that this compound was safe and well tolerated and reduced both CTGF protein and mRNA levels in intradermal injections in abdominal sites, as well as adjacent to incisional wounds made in abdominal sites [79].

A separate approach to reducing CTGF level is being taken by Pfizer. Through the acquisition of Excaliard Pharmaceuticals, which developed an antisense technology to reduce CTGF levels in conjunction with Isis Pharmaceuticals, Pfizer has to date completed five Phase II studies [80,81].

HEPATOCYTE GROWTH FACTOR

Hepatocyte growth factor (HGF) regulates cell growth, cell motility, and morphogenesis by activating a tyrosine kinase signaling cascade, after binding to the proto-oncogenic c-Met receptor. Its ability to stimulate cell growth and motility, and matrix invasion gives it a central role in diverse processes such as angiogenesis, tumorigenesis, and tissue regeneration. Additionally HGF is thought to play an important role in scar formation. HGF can improve tissue fibrosis and reverse the imbalance of collagen metabolism. An in vivo study in a rabbit ear model of hypertrophic scarring has used HGF to decrease scar formation [82].

HGF has been implicated in peripheral wound healing by regulating TGF-β signaling, which is thought to be a potent inducer of glial scar formation in the central nervous system [83]. These authors also found that HGF completely blocked secretion of TGF-β1 and -β2 from activated astrocytes in culture [83].

INTERLEUKIN 10

Interleukin 10 (IL-10) was originally described as cytokine synthesis inhibitory factor [84,85]. It is a homodimeric cytokine that signals via the signal transducer and activator of transcription 3 (STAT-3) to induce the expression of suppressor of cytokine signaling (SOCS) proteins, which dock on the intracellular domains of pro-inflammatory cytokine receptors, blocking downstream signaling (Figure 3). This mechanism makes IL-10 a powerful, broad-spectrum, anti-inflammatory cytokine helping initiate the post-inflammatory phases of wound healing. IL-10 is one of a family of antiinflammatory cytokines expressed in the skin. The importance of IL-10 in limiting inflammatory damage has been shown in studies of knock-out mice, which develop colitis and arthritis of the

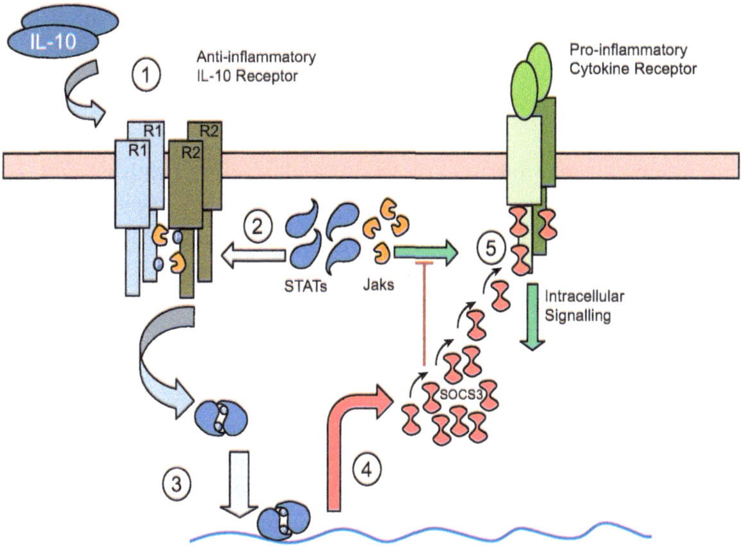

1. IL-10 binds to its receptor and activates receptor phosphorylation by associated Jaks.
2. STAT monomers are recruited to consensus domains on receptor and are themselves phosphorylated.
3. Phosphorylated STATs dimerise and translocate to the nucleus.
4. IL-10 target genes including SOCS3 are induced.
5. SOCS3 binds pro-inflammatory cytokine receptor domains uncoupling the activation of STATs and downstream signalling

FIGURE 3 IL-10 signaling via binding to its receptor complex.

joints [86]. Treatment with viral IL-10 showed improved healing in full-thickness murine wound models [87].

In the embryo IL-10 is important for scarless wound repair. Preclinical studies in rat incisional wounds revealed that IL-10 treatment resulted in healing with decreased inflammation, better scar histology, and better macroscopic scar appearance [88]. These authors also describe a phase II randomized controlled clinical trial where four incisions were made on each arm of 175 healthy volunteers. These four incisions received four different concentrations of recombinant human IL-10 (rhIL-10), which were matched with four control incisions that received either standard care or placebo. IL-10-treated human incisions at low concentrations healed with better macroscopic scar appearance and less red scars. These studies were carried out in predominantly Caucasian volunteers [88].

In a further study by Kieran et al., the effect of IL-10 on cutaneous scarring was investigated in human volunteers of African ancestral origin in an exploratory, single-center, within-subject, double-blind randomized controlled trial [89]. A range of doses of IL-10 were administered. Subjects showed a trend toward favoring treatment with 5 ng/100 μL/linear cm IL-10 between 5 and 9 months post-excision [89]. As Agren and Danielsen observed the discrepancy in response to IL-10 in relation to skin type is striking and indicates some fundamental biological differences in wound healing mechanisms that are currently unknown [14].

CONNEXINS

Connexins are junctional proteins expressed at high levels in stratified epithelia and avascular tissue (Figure 4). They form hemi-channels on the cell membrane and are aligned between adjacent cells to form intercellular channels, through which small molecules of less than 1 kDa can pass. This permits the exchange of a vast array of molecules important in metabolism and signal transduction. A number of connexins (Cx) are expressed in the skin, and these include Cx26, Cx30, and Cx43, named according to their molecular weights. These junctional proteins are expressed in both the epidermis and the dermis, and they are acutely regulated following injury to the skin. It has been shown in animal studies that the normal regulation of connexin expression is lost in nonhealing wounds, such as diabetic foot ulcers in rodents [90] and burn wounds in mice [91]. Studies targeting these proteins have shown promise in modulating the process of wound healing in terms of wound closure and inflammation.

Two approaches have been used to target Cx43. One is the down-regulation of protein expression, and the other is functional antagonism of the assembled Cx43 hemichannels in the gap junction. Down-regulation of protein expression was explored via in vivo knock-out of protein expression by genetic manipulation [92] and knock-down by topical antisense application in gel form [93].

The second approach is to functionally antagonize Cx43 activity by blocking critical domains for association with other proteins. Gap27 is a Cx43 mimetic peptide that binds to the extracellular domain of hemichannels, thus destabilizing gap junctional structures. Another functional antagonist is a peptide mimetic

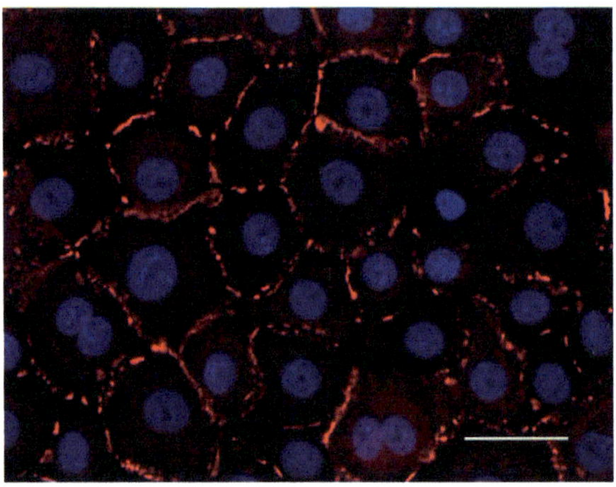

FIGURE 4 Connexin 43 staining in a confluent monolayer of human primary keratinocytes grown in defined medium. Scale bar represents 30 μm.

of the intracellular COOH terminus, tethered to an intracellular targeting peptide domain. This ACT1 fusion peptide destabilizes the intracytoplasmic bonding between Cx43 and ZO-1 proteins, thereby dissociating the functional units. In a prospective, multicenter clinical trial, adults with chronic venous leg ulcers (VLUs) were randomized to treatment with an ACT1 gel formulation plus conventional standard-of-care [94]. A significantly greater reduction in mean percent ulcer area from baseline to 12 weeks was associated with the incorporation of ACT1 therapy when compared to compression bandage therapy alone [94].

After wounding, the increased levels of Cx43 in keratinocytes are lost within hours. This typically lasted up to 48 h before the levels recovered. During this period of Cx43 down-regulation, keratinocytes migrated and proliferated to reepithelialize the wound. The transient loss of Cx43 is therefore a hallmark of normal wound healing. These changes are not seen in the diabetic foot ulcer, and Cx43 targeting in diabetic wounds has shown promise [95,96]. It is noteworthy that fibroblasts derived from diabetic ulcers are not responsive to Cx43 targeting Gap27 until after several passages [97]. This is reminiscent of the imprinted disease phenotype, which reverts following extensive passaging.

The role of the other connexins (Cx26 and Cx30) in wound healing has also been investigated. While these connexins are expressed at a lower level in comparison to Cx43, their levels initially drop following injury but later they are further elevated in the periphery of the wound margin during the proliferative phase. Targeting Cx26 and Cx30 may therefore impact the proliferation and differentiation of keratinocytes [98].

Because the inflammatory cells recruited to the wound area have been shown to express Cx43, this connexin may have an important role in communication between cells in the wound area. The levels of neutrophils and macrophages are lower in Cx43 targeted wounds, suggesting a possible role for Cx43 in their recruitment and retention. PU.1 null mice, which lack both neutrophils and macrophages, repair cutaneous wounds with better scarring outcomes [99]. The use of topical antineutrophil antiserum also promotes reepithelialization [100]. Targeting of connexins may therefore lead to the reduction of these inflammatory cells in the wound area and improve on healing outcomes.

OTHER APPROACHES UNDER INVESTIGATION FOR SCAR REDUCTION

There are a number of other approaches that are being investigated as scar reduction therapies. DS-SILY is a decorin mimetic and synthetic collagen-binding peptidoglycan that has been developed and inhibits MMP-1- and -13-mediated collagen degradation [101]. The peptidoglycan treatment of rat incisional wounds resulted in a significant reduction in scar tissue at 21 days, as measured by histology and visual analysis [101].

Increasing evidence also suggests that Wnt signaling, through its effector β-catenin, is the key regulator of fibroproliferation. As a key mediator of canonical Wnt signaling, β-catenin levels have been shown to regulate the number of mesenchymal cells that contribute to wound repair and scar size [102]. From this work, SCX-001 was identified as a possible scar reducing agent and works by decreasing the level of β-catenin produced during wound healing, thereby reducing the amount of postsurgical scarring. SCX-001 is based on Nefopam (Acupan, Silentan, Nefadol and Ajan, Riker Laboratories) which has been used as an oral and intravenous analgesic in Europe for more than 30 years. According to the company Scar X Therapeutics, SCX-001 cream has demonstrated antiscarring effects in two in vivo models. They also state that the safety of SCX-001 cream as a topical agent has also been confirmed in nonclinical in vitro and in vivo studies.

DISCUSSION AND FUTURE DIRECTIONS

Evolution has led to a rapid wound closure process, and while it is efficient, it is not functionally ideal. Epidermal appendages such as hair follicles, sweat and sebaceous glands, as well as melanocytes do not regenerate, and the uninjured dermal architecture is replaced by dense parallel arranged bundles of collagen that lead to the formation of scar tissue. Scars can sometimes display dyspigmentation, and have poor vasculature and innervation. Over the past two to three decades, a great deal of time and resource has been expended in trying to understand the intricacies of the wound healing process. From these efforts, a greater insight has developed into some of the key growth factors and cytokines involved. However, there are still many undiscovered processes that have to be elucidated. Some of this may come from understanding individuals and their responses to scar reducing therapeutics. Single nucleotide polymorphisms (SNPs), pharmacogenomics, and next-generation sequencing are likely to reveal interesting biological insights. These techniques are likely to improve our understanding of the mechanisms of scarless repair, and will likely lead to the development of improved in vitro, ex vivo, and preclinical models with a greater ability to define targets for scar prevention. Key to this will be to understand both the similarities and the differences in the wound repair process across different model systems, skin types, and genetic backgrounds, followed by carefully designed, high-quality clinical trials. The complex interactions that occur between growth factors during the phases of wound healing need to be better characterized alongside determining the subtle differences between repair and regeneration. This understanding may hold the key to the development of better therapeutics for scar-free healing.

LIST OF ABBREVIATIONS

CTGF connective tissue growth factor
Cx connexin
ECM extracellular matrix

EGF epidermal growth factor
EMT epithelial-to-mesenchymal transition
FGF fibroblast growth factor
GM-CSF granulocyte macrophage-colony stimulating factor
HGF hepatocyte growth factor
IL-10 interleukin 10
M6P mannose 6 phosphate
MMP matrix metalloproteinases
PDGF platelet-derived growth factor
SNP single nucleotide polymorphism
SOCS3 suppressor of cytokine signaling 3
STAT-3 signal transducer and activator of transcription 3
TGF-β transforming growth factor-β

REFERENCES

[1] Metcalfe AD, Ferguson MW. Tissue engineering of replacement skin: the crossroads of biomaterials, wound healing, embryonic development, stem cells and regeneration. J R Soc Interface 2007;4(14):413–37.
[2] Ferguson MW, O'Kane S. Scar-free healing: from embryonic mechanisms to adult therapeutic intervention. Philos Trans R Soc Lond B Biol Sci 2004;359(1445):839–50.
[3] Bock O, Schmid-Ott G, Malewski P, Mrowietz U. Quality of life of patients with keloid and hypertrophic scarring. Arch Dermatol Res 2006;297(10):433–8.
[4] Finnson KW, McLean S, Di Guglielmo GM, Philip A. Dynamics of transforming growth factor beta signaling in wound healing and scarring. Adv Wound Care (New Rochelle) 2013;2(5):195–214.
[5] Arno AI, Gauglitz GG, Barret JP, Jeschke MG. New molecular medicine-based scar management strategies. Burns 2014;40(4):539–51.
[6] Leask A, Abraham DJ. TGF-beta signaling and the fibrotic response. FASEB J 2004;18(7):816–27.
[7] Laverty HG, Wakefield LM, Occleston NL, O'Kane S, Ferguson MW. TGF-beta3 and cancer: a review. Cytokine Growth Factor Rev 2009;20(4):305–17.
[8] Xu J, Lamouille S, Derynck R. TGF-beta-induced epithelial to mesenchymal transition. Cell Res 2009;19(2):156–72.
[9] Shah M, Foreman DM, Ferguson MW. Neutralising antibody to TGF-beta 1,2 reduces cutaneous scarring in adult rodents. J Cell Sci 1994;107(Pt 5):1137–57.
[10] Shah M, Foreman DM, Ferguson MW. Control of scarring in adult wounds by neutralising antibody to transforming growth factor beta. Lancet 1992;339(8787):213–4.
[11] Ferguson MW, Duncan J, Bond J, Bush J, Durani P, So K, et al. Prophylactic administration of avotermin for improvement of skin scarring: three double-blind, placebo-controlled, phase I/II studies. Lancet 2009;373(9671):1264–74.
[12] Occleston NL, O'Kane S, Laverty HG, Cooper M, Fairlamb D, Mason T, et al. Discovery and development of avotermin (recombinant human transforming growth factor beta 3): a new class of prophylactic therapeutic for the improvement of scarring. Wound Repair Regen 2011;19(Suppl. 1):s38–48.
[13] Occleston NL, Metcalfe AD, Boanas A, Burgoyne NJ, Nield K, O'Kane S, et al. Therapeutic improvement of scarring: mechanisms of scarless and scar-forming healing and approaches to the discovery of new treatments. Dermatol Res Pract 2010;2010.

[14] Agren MS, Danielsen PL. Antiscarring pharmaceuticals: lost in translation? Wound Repair Regen 2014;22(3):293–4.

[15] Esmaeili M, Berry M, Logan A, Ahmed Z. Decorin treatment of spinal cord injury. Neural Regen Res 2014;9(18):1653–6.

[16] Bates SJ, Morrow E, Zhang AY, Pham H, Longaker MT, Chang J. Mannose-6-phosphate, an inhibitor of transforming growth factor-beta, improves range of motion after flexor tendon repair. J Bone Joint Surg Am 2006;88(11):2465–72.

[17] Harding AJ, Christmas CR, Ferguson MW, Loescher AR, Robinson PP, Boissonade FM. Mannose-6-phosphate facilitates early peripheral nerve regeneration in thy-1-YFP-H mice. Neuroscience 2014;279:23–32.

[18] Ngeow WC, Atkins S, Morgan CR, Metcalfe AD, Boissonade FM, Loescher AR, et al. The effect of Mannose-6-phosphate on recovery after sciatic nerve repair. Brain Res 2011;1394:40–8.

[19] Renovo A. single-site, Placebo-controlled, double-blind trial to investigate the safety, tolerability, systemic exposure and scar- improvement efficacy of intradermal applications of three doses of Juvidex (5.64 mg/100µl, 2.82 mg/100µl and 11.28 mg/100µl), given once only or three times within and between male and female subjects aged. 2008. p. 18–45.

[20] Wong JK, Metcalfe AD, Wong R, Bush J, Platt C, Garcon A, et al. Reduction of tendon adhesions following administration of adaprev, a hypertonic solution of mannose-6-phosphate: mechanism of action studies. PLoS One 2014;9(11):e112672.

[21] Yu H, Bock O, Bayat A, Ferguson MW, Mrowietz U. Decreased expression of inhibitory SMAD6 and SMAD7 in keloid scarring. J Plast Reconstr Aesthet Surg 2006;59(3):221–9.

[22] Cho JW, Il KJ, Lee KS. Downregulation of type I collagen expression in silibinin-treated human skin fibroblasts by blocking the activation of Smad2/3-dependent signaling pathways: potential therapeutic use in the chemoprevention of keloids. Int J Mol Med 2013;31(5):1148–52.

[23] Wang Z, Gao Z, Shi Y, Sun Y, Lin Z, Jiang H, et al. Inhibition of Smad3 expression decreases collagen synthesis in keloid disease fibroblasts. J Plast Reconstr Aesthet Surg 2007;60(11):1193–9.

[24] Roberts AB, Russo A, Felici A, Flanders KC. Smad3: a key player in pathogenetic mechanisms dependent on TGF-beta. Ann N Y Acad Sci 2003;995:1–10.

[25] McGaha TL, Phelps RG, Spiera H, Bona C. Halofuginone, an inhibitor of type-I collagen synthesis and skin sclerosis, blocks transforming-growth-factor-beta-mediated Smad3 activation in fibroblasts. J Invest Dermatol 2002;118(3):461–70.

[26] Liu X, Zhu S, Wang T, Hummers L, Wigley FM, Goldschmidt-Clermont PJ, et al. Paclitaxel modulates TGFbeta signaling in scleroderma skin grafts in immunodeficient mice. PLoS Med 2005;2(12):e354.

[27] Maciver AH, McCall MD, Edgar RL, Thiesen AL, Bigam DL, Churchill TA, et al. Sirolimus drug-eluting, hydrogel-impregnated polypropylene mesh reduces intra-abdominal adhesion formation in a mouse model. Surgery 2011;150(5):907–15.

[28] Nakao A, Miike S, Hatano M, Okumura K, Tokuhisa T, Ra C, et al. Blockade of transforming growth factor beta/Smad signaling in T cells by overexpression of Smad7 enhances antigen-induced airway inflammation and airway reactivity. J Exp Med 2000;192(2):151–8.

[29] Tang B, Zhu B, Liang Y, Bi L, Hu Z, Chen B, et al. Asiaticoside suppresses collagen expression and TGF-beta/Smad signaling through inducing Smad7 and inhibiting TGF-betaRI and TGF-betaRII in keloid fibroblasts. Arch Dermatol Res 2011;303(8):563–72.

[30] Dooley S, Said HM, Gressner AM, Floege J, En-Nia A, Mertens PR. Y-box protein-1 is the crucial mediator of antifibrotic interferon-gamma effects. J Biol Chem 2006;281(3):1784–95.

[31] Granstein RD, Rook A, Flotte TJ, Haas A, Gallo RL, Jaffe HS, et al. A controlled trial of intralesional recombinant interferon-gamma in the treatment of keloidal scarring. Clinical and histologic findings. Arch Dermatol 1990;126(10):1295–302.

[32] Larrabee Jr WF, East CA, Jaffe HS, Stephenson C, Peterson KE. Intralesional interferon gamma treatment for keloids and hypertrophic scars. Arch Otolaryngol Head Neck Surg 1990;116(10):1159–62.

[33] Barrientos S, Stojadinovic O, Golinko MS, Brem H, Tomic-Canic M. Growth factors and cytokines in wound healing. Wound Repair Regen 2008;16(5):585–601.

[34] Pastore S, Mascia F, Mariani V, Girolomoni G. The epidermal growth factor receptor system in skin repair and inflammation. J Invest Dermatol 2008;128(6):1365–74.

[35] Barrientos S, Brem H, Stojadinovic O, Tomic-Canic M. Clinical application of growth factors and cytokines in wound healing. Wound Repair Regen 2014;22(5):569–78.

[36] Brown GL, Curtsinger L, Jurkiewicz MJ, Nahai F, Schultz G. Stimulation of healing of chronic wounds by epidermal growth factor. Plast Reconstr Surg 1991;88(2):189–94.

[37] Falanga V, Eaglstein WH, Bucalo B, Katz MH, Harris B, Carson P. Topical use of human recombinant epidermal growth factor (h-EGF) in venous ulcers. J Dermatol Surg Oncol 1992;18(7):604–6.

[38] Tsang MW, Wong WK, Hung CS, Lai KM, Tang W, Cheung EY, et al. Human epidermal growth factor enhances healing of diabetic foot ulcers. Diabetes Care 2003;26(6): 1856–61.

[39] Hong JP, Jung HD, Kim YW. Recombinant human epidermal growth factor (EGF) to enhance healing for diabetic foot ulcers. Ann Plast Surg 2006;56(4):394–8.

[40] Fernandez-Montequin JI, Infante-Cristia E, Valenzuela-Silva C, Franco-Perez N, Savigne-Gutierrez W, Artaza-Sanz H, et al. Intralesional injections of Citoprot-P (recombinant human epidermal growth factor) in advanced diabetic foot ulcers with risk of amputation. Int Wound J 2007;4(4):333–43.

[41] Mohan VK. Recombinant human epidermal growth factor (REGEN-D 150): effect on healing of diabetic foot ulcers. Diabetes Res Clin Pract 2007;78(3):405–11.

[42] Fernandez-Montequin JI, Betancourt BY, Leyva-Gonzalez G, Mola EL, Galan-Naranjo K, Ramirez-Navas M, et al. Intralesional administration of epidermal growth factor-based formulation (Heberprot-P) in chronic diabetic foot ulcer: treatment up to complete wound closure. Int Wound J 2009;6(1):67–72.

[43] Berlanga-Acosta J, Gavilondo-Cowley J, Lopez-Saura P, Gonzalez-Lopez T, Castro-Santana MD, Lopez-Mola E, et al. Epidermal growth factor in clinical practice - a review of its biological actions, clinical indications and safety implications. Int Wound J 2009;6(5):331–46.

[44] Brown GL, Nanney LB, Griffen J, Cramer AB, Yancey JM, Curtsinger III LJ, et al. Enhancement of wound healing by topical treatment with epidermal growth factor. N Engl J Med 1989;321(2):76–9.

[45] Carter EP, Fearon AE, Grose RP. Careless talk costs lives: fibroblast growth factor receptor signalling and the consequences of pathway malfunction. Trends Cell Biol 2014;25.

[46] Robson MC. The role of growth factors in the healing of chronic wounds. Wound Repair Regen 1997;5(1):12–7.

[47] Richard JL, Parer-Richard C, Daures JP, Clouet S, Vannereau D, Bringer J, et al. Effect of topical basic fibroblast growth factor on the healing of chronic diabetic neuropathic ulcer of the foot. A pilot, randomized, double-blind, placebo-controlled study. Diabetes Care 1995;18(1):64–9.

[48] Uchi H, Igarashi A, Urabe K, Koga T, Nakayama J, Kawamori R, et al. Clinical efficacy of basic fibroblast growth factor (bFGF) for diabetic ulcer. Eur J Dermatol 2009;19(5):461–8.

[49] Robson MC, Phillips LG, Lawrence WT, Bishop JB, Youngerman JS, Hayward PG, et al. The safety and effect of topically applied recombinant basic fibroblast growth factor on the healing of chronic pressure sores. Ann Surg 1992;216(4):401–6.

[50] Ohura T, Nakajo T, Moriguchi T, Oka H, Tachi M, Ohura Jr N, et al. Clinical efficacy of basic fibroblast growth factor on pressure ulcers: case-control pairing study using a new evaluation method. Wound Repair Regen 2011;19(5):542–51.

[51] Robson MC, Phillips TJ, Falanga V, Odenheimer DJ, Parish LC, Jensen JL, et al. Randomized trial of topically applied repifermin (recombinant human keratinocyte growth factor-2) to accelerate wound healing in venous ulcers. Wound Repair Regen 2001;9(5):347–52.

[52] Fu X, Shen Z, Chen Y, Xie J, Guo Z, Zhang M, et al. Randomised placebo-controlled trial of use of topical recombinant bovine basic fibroblast growth factor for second-degree burns. Lancet 1998;352(9141):1661–4.

[53] Andrae J, Gallini R, Betsholtz C. Role of platelet-derived growth factors in physiology and medicine. Genes Dev 2008;22(10):1276–312.

[54] Steed DL. Clinical evaluation of recombinant human platelet-derived growth factor for the treatment of lower extremity diabetic ulcers. Diabetic Ulcer Study Group. J Vasc Surg 1995;21(1):71–8.

[55] Wieman TJ, Smiell JM, Su Y. Efficacy and safety of a topical gel formulation of recombinant human platelet-derived growth factor-BB (becaplermin) in patients with chronic neuropathic diabetic ulcers. A phase III randomized placebo-controlled double-blind study. Diabetes Care 1998;21(5):822–7.

[56] Smiell JM, Wieman TJ, Steed DL, Perry BH, Sampson AR, Schwab BH. Efficacy and safety of becaplermin (recombinant human platelet-derived growth factor-BB) in patients with non-healing, lower extremity diabetic ulcers: a combined analysis of four randomized studies. Wound Repair Regen 1999;7(5):335–46.

[57] Margolis DJ, Bartus C, Hoffstad O, Malay S, Berlin JA. Effectiveness of recombinant human platelet-derived growth factor for the treatment of diabetic neuropathic foot ulcers. Wound. Repair Regen 2005;13(6):531–6.

[58] Robson MC, Phillips LG, Thomason A, Altrock BW, Pence PC, Heggers JP, et al. Recombinant human platelet-derived growth factor-BB for the treatment of chronic pressure ulcers. Ann Plast Surg 1992;29(3):193–201.

[59] Mustoe TA, Cutler NR, Allman RM, Goode PS, Deuel TF, Prause JA, et al. A phase II study to evaluate recombinant platelet-derived growth factor-BB in the treatment of stage 3 and 4 pressure ulcers. Arch Surg 1994;129(2):213–9.

[60] Pierce GF, Tarpley JE, Allman RM, Goode PS, Serdar CM, Morris B, et al. Tissue repair processes in healing chronic pressure ulcers treated with recombinant platelet-derived growth factor BB. Am J Pathol 1994;145(6):1399–410.

[61] Rees RS, Robson MC, Smiell JM, Perry BH. Becaplermin gel in the treatment of pressure ulcers: a phase II randomized, double-blind, placebo-controlled study. Wound Repair Regen 1999;7(3):141–7.

[62] FDA. Regranex. 2008. http://www.fda.gov/ForConsumers/ConsumerUpdates/ucm048471.htm.

[63] Margolis DJ, Morris LM, Papadopoulos M, Weinberg L, Filip JC, Lang SA, et al. Phase I study of H5.020CMV.PDGF-beta to treat venous leg ulcer disease. Mol Ther 2009;17(10):1822–9.

[64] Travis TE, Mauskar NA, Mino MJ, Prindeze N, Moffatt LT, Fidler PE, et al. Commercially available topical platelet-derived growth factor as a novel agent to accelerate burn-related wound healing. J Burn Care Res 2014;35(5):e321–9.

[65] Branski LK, Masters OE, Herndon DN, Mittermayr R, Redl H, Traber DL, et al. Pre-clinical evaluation of liposomal gene transfer to improve dermal and epidermal regeneration. Gene Ther 2010;17(6):770–8.

[66] Mann A, Breuhahn K, Schirmacher P, Blessing M. Keratinocyte-derived granulocyte-macrophage colony stimulating factor accelerates wound healing: stimulation of keratinocyte proliferation, granulation tissue formation, and vascularization. J Invest Dermatol 2001;117(6):1382–90.

[67] El Saghir NS, Bizri AR, Shabb NS, Husami TW, Salem Z, Shamseddine AI. Pressure ulcer accelerated healing with local injections of granulocyte macrophage-colony stimulating factor. J Infect 1997;35(2):179–82.

[68] Gough A, Clapperton M, Rolando N, Foster AV, Philpott-Howard J, Edmonds ME. Randomised placebo-controlled trial of granulocyte-colony stimulating factor in diabetic foot infection. Lancet 1997;350(9081):855–9.

[69] Marques da Costa R, Jesus FM, Aniceto C, Mendes M. Double-blind randomized placebo-controlled trial of the use of granulocyte-macrophage colony-stimulating factor in chronic leg ulcers. Am J Surg 1997;173(3):165–8.

[70] Jaschke E, Zabernigg A, Gattringer C. Recombinant human granulocyte-macrophage colony-stimulating factor applied locally in low doses enhances healing and prevents recurrence of chronic venous ulcers. Int J Dermatol 1999;38(5):380–6.

[71] Da Costa R, Jesus FM, Aniceto C, Mendes M. Randomized, double-blind, placebo-controlled, dose ranging study of granulocyte-macrophage colony-stimulating factor in patients with chronic venous leg ulcers. Wound Repair Regen 1999;7(1):17–25.

[72] Bianchi L, Ginebri A, Hagman JH, Francesconi F, Carboni I, Chimenti S. Local treatment of chronic cutaneous leg ulcers with recombinant human granulocyte-macrophage colony-stimulating factor. J Eur Acad Dermatol Venereol 2002;16(6):595–8.

[73] Cianfarani F, Tommasi R, Failla CM, Viviano MT, Annessi G, Papi M, et al. Granulocyte/macrophage colony-stimulating factor treatment of human chronic ulcers promotes angiogenesis associated with de novo vascular endothelial growth factor transcription in the ulcer bed. Br J Dermatol 2006;154(1):34–41.

[74] Zhang L, Chen J, Han C. A multicenter clinical trial of recombinant human GM-CSF hydrogel for the treatment of deep second-degree burns. Wound Repair Regen 2009;17(5):685–9.

[75] Yan H, Chen J, Peng X. Recombinant human granulocyte-macrophage colony-stimulating factor hydrogel promotes healing of deep partial thickness burn wounds. Burns 2012;38(6):877–81.

[76] Shi-Wen X, Leask A, Abraham D. Regulation and function of connective tissue growth factor/CCN2 in tissue repair, scarring and fibrosis. Cytokine Growth Factor Rev 2008;19(2):133–44.

[77] Wang JF, Olson ME, Ma L, Brigstock DR, Hart DA. Connective tissue growth factor siRNA modulates mRNA levels for a subset of molecules in normal and TGF-beta 1-stimulated porcine skin fibroblasts. Wound Repair Regen 2004;12(2):205–16.

[78] Sisco M, Kryger ZB, O'Shaughnessy KD, Kim PS, Schultz GS, Ding XZ, et al. Antisense inhibition of connective tissue growth factor (CTGF/CCN2) mRNA limits hypertrophic scarring without affecting wound healing in vivo. Wound Repair Regen 2008;16(5):661–73.

[79] RXi Pharmaceuticals. RXI-109 trial. 2014. http://www.rxipharma.com/clinical-trials/#phaseone.

[80] Isis Pharmaceuticals. EXC 001 trials. 2014. http://www.isispharm.com/Pipeline/Therapeutic-Areas/Inflammatory-Disease.htm#EXC001.

[81] ClinicalTrialsgov. EXC 001. 2014. http://clinicaltrials.gov/ct2/results?term=exc+001&Search=Search.

[82] Xiao Z, Xi C. Hepatocyte growth factor reduces hypertrophy of skin scar: in vivo study. Adv Skin Wound Care 2013;26(6):266–70.

[83] Jeong SR, Kwon MJ, Lee HG, Joe EH, Lee JH, Kim SS, et al. Hepatocyte growth factor reduces astrocytic scar formation and promotes axonal growth beyond glial scars after spinal cord injury. Exp Neurol 2012;233(1):312–22.

[84] Ding L, Linsley PS, Huang LY, Germain RN, Shevach EM. IL-10 inhibits macrophage costimulatory activity by selectively inhibiting the up-regulation of B7 expression. J Immunol 1993;151(3):1224–34.

[85] Moore KW, de Waal MR, Coffman RL, O'Garra A. Interleukin-10 and the interleukin-10 receptor. Annu Rev Immunol 2001;19:683–765.

[86] Kuhn R, Lohler J, Rennick D, Rajewsky K, Muller W. Interleukin-10-deficient mice develop chronic enterocolitis. Cell 1993;75(2):263–74.

[87] Wise LM, Stuart GS, Real NC, Fleming SB, Mercer AA. Orf virus IL-10 accelerates wound healing while limiting inflammation and scarring. Wound Repair Regen 2014;22(3):356–67.

[88] Kieran I, Knock A, Bush J, So K, Metcalfe A, Hobson R, et al. Interleukin-10 reduces scar formation in both animal and human cutaneous wounds: results of two preclinical and phase II randomized control studies. Wound Repair Regen 2013;21(3):428–36.

[89] Kieran I, Taylor C, Bush J, Rance M, So K, Boanas A, et al. Effects of interleukin-10 on cutaneous wounds and scars in humans of African continental ancestral origin. Wound Repair Regen 2014;22(3):326–33.

[90] Romanelli M, Vuerstaek JD, Rogers LC, Armstrong DG, Apelqvist J. Economic burden of hard-to-heal wounds. European Wound Management Association (EWMA). Position document: hard-to-heal wounds: a holistic approach. London MEP, Ltc; 2008. http://www.woundsinternational.com/pdf/content_45.pdf.

[91] Coutinho P, Qiu C, Frank S, Wang CM, Brown T, Green CR, et al. Limiting burn extension by transient inhibition of Connexin43 expression at the site of injury. Br.J Plast Surg 2005;58(5):658–67.

[92] Kretz M, Euwens C, Hombach S, Eckardt D, Teubner B, Traub O, et al. Altered connexin expression and wound healing in the epidermis of connexin-deficient mice. J Cell Sci 2003;116(Pt 16):3443–52.

[93] Qiu C, Coutinho P, Frank S, Franke S, Law LY, Martin P, et al. Targeting connexin43 expression accelerates the rate of wound repair. Curr Biol 2003;13(19):1697–703.

[94] Ghatnekar GS, Grek CL, Armstrong DG, Desai SC, Gourdie RG. The effect of a connexin43-based peptide on the healing of chronic venous leg ulcers: a multicenter, randomized trial. J Invest Dermatol 2014;135.

[95] Brandner JM, Houdek P, Husing B, Kaiser C, Moll I. Connexins 26, 30, and 43: differences among spontaneous, chronic, and accelerated human wound healing. J Invest Dermatol 2004;122(5):1310–20.

[96] Wang CM, Lincoln J, Cook JE, Becker DL. Abnormal connexin expression underlies delayed wound healing in diabetic skin. Diabetes 2007;56(11):2809–17.

[97] Evans WH, Boitano S. Connexin mimetic peptides: specific inhibitors of gap-junctional intercellular communication. Biochem Soc Trans 2001;29(Pt 4):606–12.

[98] Labarthe MP, Bosco D, Saurat JH, Meda P, Salomon D. Upregulation of connexin 26 between keratinocytes of psoriatic lesions. J Invest Dermatol 1998;111(1):72–6.

[99] Martin P, D'Souza D, Martin J, Grose R, Cooper L, Maki R, et al. Wound healing in the PU.1 null mouse–tissue repair is not dependent on inflammatory cells. Curr Biol 2003;13(13):1122–8.

[100] Dovi JV, He LK, Dipietro LA. Accelerated wound closure in neutrophil-depleted mice. J Leukoc Biol 2003;73(4):448–55.

[101] Stuart K, Paderi J, Snyder PW, Freeman L, Panitch A. Collagen-binding peptidoglycans inhibit MMP mediated collagen degradation and reduce dermal scarring. PLoS One 2011;6(7):e22139.

[102] Poon R, Hong H, Wei X, Pan J, Alman BA. A high throughput screen identifies Nefopam as targeting cell proliferation in beta-catenin driven neoplastic and reactive fibroproliferative disorders. PLoS One 2012;7(5):e37940.

Chapter 18

Skin Models for Drug Development and Biopharmaceutical Industry

Heinz Ruffner[1], Ursula Graf-Hausner[2], Stephanie Mathes[2]
[1]*Novartis Institutes for BioMedical Research, Basel, Switzerland;* [2]*Zurich University of Applied Sciences, Waedenswil, Switzerland*

Chapter Outline

OUTLINE OF THE CHAPTER

Human organotypic skin models that show typical characteristics of a disease state of interest are necessary to develop treatment strategies and to increase expectations of being successful in in vivo application. However, to reach this lofty goal, several aspects have to be considered to ensure biological relevance,

Skin Tissue Engineering and Regenerative Medicine. http://dx.doi.org/10.1016/B978-0-12-801654-1.00018-8

predictivity, and feasibility. In this chapter these aspects are highlighted and discussed to illustrate the strengths and limitations of in vitro grown skin models for drug development, as well as pharmacokinetic (PK) and pharmacodynamic (PD) studies.

INTRODUCTION

In vitro created human skin equivalents already have a relatively long history and have gone through multiple rounds of optimization and adaptation. Driven by the demands of regulatory authorities and customers, the cosmetics industry has particularly supported the establishment of human skin models to serve as predictive tools for safety assessment studies. The results of toxicological screening tests performed with in vitro skin equivalents were correlated with in vivo data to enable a robust predictive model. Today, reconstructed human epidermal models are routinely used to assess the risk of skin irritancy and corrosion. Beginning with the cultivation of keratinocytes on feeder layers to epidermal equivalents and co-culture systems, a broad range of research and development has been undertaken to improve human skin model creation, as well as relevant end-point definitions. These in vitro created skin equivalents are not only applicable for safety tests, but also serve as functional models that allow deeper insight into, and understanding of, cellular interactions. Organotypic constructs intended for the evaluation of tissue homeostasis and cell interactions of different cell types (e.g., T_H17 cells, Langerhans cells, melanocytes, tumor cells, and so on) need to exhibit a higher level of complexity to replicate in vivo conditions. A great deal of knowledge has also been gathered by the use of ex vivo explant cultures, which feature all important cells, organelles, and matrix components, but they are generally not feasible for standard tests, due to donor variability and limited availability. As the title of the chapter already indicates, the use of skin models for the biopharmaceutical industry is presented and discussed.

The question that instantly arises is why might the biopharmaceutical industry be interested in this type of tissue model? However, it can be seen that the establishment and integration of human skin models is important for the pharmaceutical industry for several reasons:

1. There is still a reasonable potential for dermatologically acting drugs on the market. Skin diseases like psoriasis, atopic dermatitis, decubitus ulcer, eczema, acne, vitiligo, skin cancer, and abnormal scar formation after trauma can only be treated to a certain degree, and treatment methods can be improved by a holistic understanding of the disorder.
2. There are no satisfactory standardized models of skin disorders available. Xenografts are different from human skin (absence of hair, composition of the stratum corneum, SC), and the majority of human skin disorders have no counterpart in animals. They can be introduced into an animal by transplantation, but in such approaches, a negative control is lacking.

3. As stated above, the composition of the SC is species specific, and even pigskin, which is used as a substitute for human skin, differs significantly in terms of lipid content [1]. This fact has to be considered whenever such models are used for investigating transdermal drug penetration. Nevertheless, it should be pointed out that human skin models established so far also show significant shortcomings with respect to penetration behavior due to an insufficient barrier function.

This wide field of developmental process in biopharmaceutical drug development using human skin models can, in essence, be broken down into three main issues: (1) composition/definition of the test system, (2) data acquisition for drug evaluation, and (3) translation to in vivo systems (Figure 1).

If we agree that it is worthwhile and important for the biopharmaceutical industry to use human skin models for drug development in order to facilitate new therapeutic approaches, the question of what is applicable as a predictive

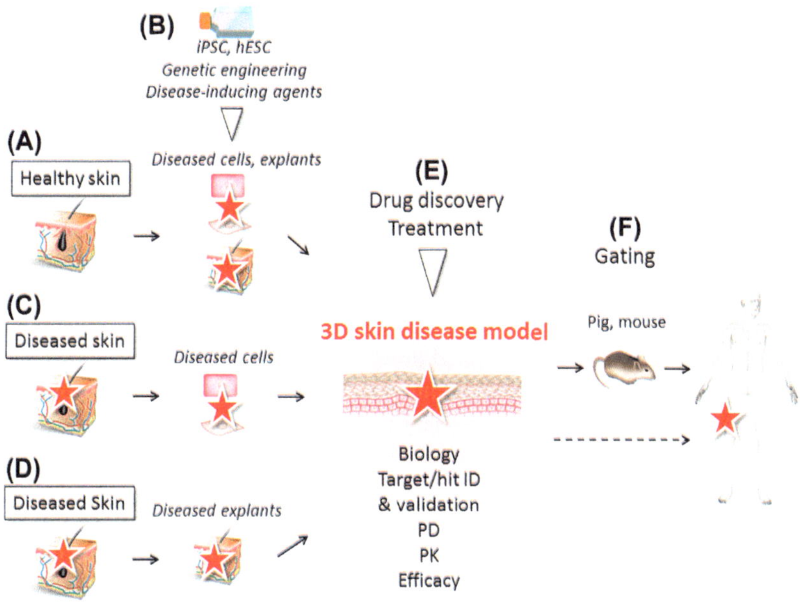

FIGURE 1 Realization of human disease skin models. 3D skin disease models can be generated using multiple strategies: individual cell types or explant cultures can be generated from healthy skin biopsies (A) or by using iPSC (from healthy or diseased patient samples) and hESC technologies (B) and treated in vitro to induce disease phenotypes by genetic-engineering approaches as well as by addition of disease-inducing agents. Alternatively, patient-derived diseased skin can be either processed into individual cell types (C) or propagated as explant cultures in vitro (D). The resulting 3D skin disease models serve as platform to address multiple drug discovery demands (E). The data obtained facilitate gating into preclinical in vivo animal models and potentially even human clinical studies (F). The red stars in denote the disease state of skin, either manifested on patient skin or experimentally induced in vitro/ex vivo.

model comes up. First of all, as for all other test systems, in vitro created tissue models need to fulfill the fundamental parameters for a validation including accuracy, precision, specificity, sensitivity, reproducibility, and stability. Secondly, the model should be thoroughly predefined: in essence, test conditions, threshold values for relevant end-points, and exclusion criteria need to be set. In recent years, a good deal of knowledge has been gained clinically and by basic research concerning specific triggers and markers for different disease states of human skin. These data need to be correlated to findings of approaches using 3D skin tissues. As a third requirement, the applicability and subsequent analysis of test substances has to be ensured in order to assess pharmacodynamics, as well as PK effects.

As previously stated, the degree of tissue model complexity correlates directly to enhanced comparability with the in vivo situation. The choice of cells incorporated within the tissue model is particularly significant: not only the interaction between fibroblasts and keratinocytes but also between melanocytes [2] and even endothelial cells [3] may trigger a biological response to stimuli. Moreover, the origin of the cells can predetermine the fate of the developmental process and outcome of the 3D tissue. Depending upon the rationale of the test system, stem cell-like cells, cell lines, primary cells from one or multiple donors (pooled), or primary cells from diseased skin biopsies may be used. But the more complex the system, the more the technical hurdles arise and batch-to-batch variations may occur. Nevertheless, compared to ex vivo skin, data reproducibility is, in most cases, higher where in vitro produced tissue equivalents are used.

Organotypic skin models are mainly interesting for the biopharmaceutical industry to enable transdermal drug delivery for systemic action or to support development of new drugs aimed at certain disorders. These disorders may arise for different reasons, for example, due to an external effect (e.g., trauma, chemical action of a substance, and infections), endogenous effect (imbalance of molecular events that govern communication between cells, immunological and nonimmunological responses to foreign agents and autoimmune events, and venous insufficiency), or a combination of both (pressure or shear forces on skin combined with venous insufficiency).

Taking into account the multiple factors and circumstances that can lead to a need for therapeutic interventions, the translation of in vitro human skin models simulating the different states of homeostatic imbalance for drug development is challenging.

Basic Considerations

Cellular model systems represent approximations of healthy and diseased tissues or organs present in animals or humans. By virtue of the culture conditions in artificial tissue cultures over a period of time, these cellular systems suffer from a number of shortcomings in comparison to tissues in organisms, such as a lack of blood supply, lymphatic clearance, innervation, and hormonal regulation. Therefore, cell models can only capture a number of characteristics of tissue physiology. The better the models are in representing a particular tissue state, the more the key parameters can be represented in the models. It is, therefore, of crucial importance to understand the complexity of tissue and organ biology, in order to define the conditions that need to be established in culture. The appropriate cell types and their relative and spatial orientation belong to these conditions, as well as extracellular matrix (ECM) components, medium composition, and additives. A similar challenge exists for the establishment of disease models. Not only do we need to understand disease mechanisms, we also need to recognize what cellular and molecular consequences arise on normal tissue when the disease is present. Replication of such complex relationships is a very demanding task.

There are several ways to assess healthy and diseased tissue states in a culture. One comprises the cultivation of healthy or diseased tissues directly isolated from healthy individuals or animals as well as from patients or disease animal models in culture for a certain period of time. Such cultures will subsequently be referred to as "normal" or "disease" explant cultures. One of the advantages of explant cultures is that the composition of cells in their natural environment fully reflects the in vivo situation, at least at the start of the cultivation. It is important for such cultures that the cultivation conditions allow the perpetuation of the healthy or disease state for as long as possible over the required duration of an experiment. Another approach is co-cultivating cells in a defined manner, resembling normal tissue architecture (e.g., human 3D dermo-epidermal skin equivalents). Here, a simplified dermal compartment is represented by normal human primary fibroblasts embedded in a collagen matrix, whereas a simplified epidermal compartment is substituted by normal human primary keratinocytes, which are allowed to stratify in an air–liquid interphase (ALI). To convert such reconstituted "healthy" cultured tissue into a disease model, disease stimuli need to be introduced in a disease-specific manner, as further addressed below. Yet another approach involves a combination of reconstructed models consisting of cells from healthy individuals and from diseased tissues. In the simplest version, cells from healthy individuals and from patients are compared with each other for a certain biological function upon cultivation. For models with higher complexity, cells from healthy individuals are used to generate 3D skin models, in which patient-derived cells harvested from diseased tissue are subsequently introduced to study the effects of diseased cells on tissue function, thus constituting in vitro disease models (Figure 1).

Target and Hit Identification and Validation

Advances in cell culture techniques over the past decades and increased understanding of the molecular signaling pathways have enabled the generation of tissue culture models to study normal and pathophysiological processes in vitro. A suite of technologies is available, allowing the characterization and functional analysis of complex in vitro skin cultures [4]. The feasibility of genetically altering target cells allows for rapid adaptation of cellular assay systems toward target validation, without the necessity to create transgenic mouse models.

Primary human cells, including keratinocytes, are generally difficult to either transfect or transduce by conventional transfection reagents. Several procedures have been described, however, that allow for sufficient transfection or transduction rates following careful optimization of conditions for each cellular system. It is of critical importance that the effects caused by the transferred nucleic acid, for example, knockdown levels of mRNA species targeted by transfected small interfering RNAs (siRNAs) or by transduced small hairpin RNAs (shRNAs), are maintained as long as necessary, in some cases for more than 1 week, so that the genetic manipulation results in a cellular disease phenotype. In this respect, knockdown of the membrane-type serine protease-1, matriptase-1, using a siRNA-based approach has caused a phenotype reminiscent of the epidermal phenotype observed in the corresponding knockout mice [5]. The high knockdown efficiency observed over at least 1 week in this organotypic model underscores the feasibility of loss-of-function (LOF) studies in the epidermal compartments of full-thickness (FT) skin models. An extension of such an LOF approach enables reproduction of disease-associated manifestations in the epidermis by knocking down the messenger RNA (mRNA) encoding filaggrin (FLG), known to associate with the respective disease if mutated [6]. In addition to investigating disease phenotypes in vitro to identify small molecules modulating the phenotypes, such models also allow identification of novel links between targets and cellular functions, such as FLG-induced photoprotection of the epidermis. Another study using an shRNA-mediated knockdown approach in a human 3D reconstructed epidermis model revealed that FLG-2 is important for proper cornification [7]. In addition to serving as experimental research tools, genetic reagents might also be applied as therapeutic agents, given the easy accessibility of skin for topical treatment. A reconstructed human pigmented skin model encompassing melanocytes has demonstrated the feasibility of siRNA-mediated depigmentation, showing the potential of such complex in vitro models as validation and possibly even screening tools [8].

HUMAN SKIN MODELS TO ASSESS PHARMACOLOGICALLY RELEVANT DATA

Table 1 summarizes the approaches, with respect to tissue-engineered skin models, that play an important role in biopharmaceutical development processes. The authors are conscious of the incompleteness of the list, but have focused on models that showed the most promising data supporting their ability to serve as predictive models.

With respect to pharmacokinetics and pharmacodynamics, different requirements need to be defined for appropriate test systems. It should be mentioned that the term "pharmacokinetics" is interpreted here as a snapshot of actions,

TABLE 1 Summary of Skin Models for Assessment of Drug Permeation, Metabolism and Efficacy

For PK Studies (Drug Delivery and Metabolism)			
Requirements for Skin Models	**End-Points to Be Investigated**	**Correlation to**	**Examples**
Effective barrier function	Flux, lag time, and permeation coefficient (Franz cells) Skin adsorption/penetration (Raman)	Human skin (epidermis sheets) Pigskin	EpiDerm™ EpiSkin™, skin Ethic™ [78–80], Phenion® FT model [81] (Henkel, Duesseldorf, Germany), (NativeSkin® [9])
Xenobiotic-metabolizing enzymes (cytochrome P450 and non-CYP-xenobiotic-metabolizing enzymes)	Enzyme activities in nmol/min/mg protein or nmol/h/cm^2	Human skin	EpiDerm™ EpiSkin RHE™, EpiSkin™ FTM, skin Ethic™ RHEPhenion® FT,StrataTest® (Stratatech Corporation, Madison, USA) [16]
For Efficacy Studies			
Replication Symptoms of	**End-Points to Be Investigated**	**Correlation to**	**Examples**
Infected wounds	Antimicrobial efficacy, cytotoxicity, microbial colonization, cytokine release (IL-8, IL-6), gene expression TLR2 β-defensin-2, -3, and RNAse7	Murine and porcine burn wounds	Haisma [33,82] (FT models with thermal wounds, infection with methicillin-resistant or multi-resistant strains of *S. aureus* and eradication by antibiotic mupirocin) Burned FT models colonized with *P. aeruginosa* or *S. aureus* to evaluate new polymers [32]

Continued

TABLE 1 Summary of Skin Models for Assessment of Drug Permeation, Metabolism and Efficacy—cont'd

Replication Symptoms of	End-Points to Be Investigated	Correlation to	Examples
Cutanous candidosis	Morphological alterations Increased expression of IL1α, exodus-2, TNF- α, and PSL, decrease of TGF- β transcripts	Noninfected RHE models	Infection of RHE models with *C. albicans* [34]
Skin melanoma	S100, Ki-67, caspase-3/apoptosis, drug efflux transporter ABCB5, histone H3 demethylase JARID1B, size of tumors, and number of tumor nests	Monocultures, skin models with or without cancer cells	Spheroids embedded in FT model [42] MatTek 3-DSR model human epidermal keratinocytes, and A375 melanoma [83]
Squamous cell carcinoma	Immunohistochemical staining of Ki67, cytokeratins 5, 6, 14, 10, 16, 17, involucrin, and basement membrane proteins	Human skin biopsies from SSC	SSC-25 cells integrated into FT model [44] Integration of SCC-12B2, and SCC-13 cell lines in FT models or induction by EGF supplementation [84,85]
Reepithelialization after wounding	Reepithelialization rate (assessed by optical coherence tomography or multiphoton microscopy), enhanced mRNA expression of MMP3, IL1α, KRTAP4–12, and decreased Expression of S100A7, immunohistological analysis of keratins (KRT1/10, KRT 6/16), proliferation markers (Ki67), basement membrane, and hemidesmosomal proteins	Mouse, rat, or pigskin, epiboly	Van Kilsdonk [86] (excisional 2–3 mm wound) Laplante [23] (excisional 6 mm wound) Marquart [87] (mechanical, thermal, and irradiation-induced lesions and subsequent topical treatment with various dexpanthenol-containing ointments) Impact of fibroblasts for reepithelialization [25] Influence of morphine and solid lipid nanoparticles on CO_2-laser irritated of EpiDermFT™ models [29]

For PD Studies			
Psoriatic skin	Secretion and expression of IL-8, IL-23, (VEGF)-A, RANTES, S100A7, SPRR2A (small proline-rich proteins), chemokines (CCL20, CXCL8), TNF-a	Human psoriatic skin	Stimulation RHE by IL-17 ixekizumab [65] Stimulation of RHE by IL-17, Il-22, and TNFα [88] Stimulation of EpiDerm™ IL-22 and calcitriol treatment [89] Stimulation of Phenion™ IL-22 and evaluation of calcipotriol treatment [90] RHE stimulation by TNFα, IL-1, IL-6, and IL-22 effect of retinoic acid [35]
Atopic dermatitis	Expression of FLG (protein and mRNA), TSLP production/release, chemokine (CXCL10, CCL5) production/release, TLR3/RLRs signaling: NFkB, IRF3 translocation/ phosphorylation, IL-13RA2 expression, caspase-14, eotaxin-3	Human biopsies of AD lesions	Influence of *S. aureus* population on marker expression in RHE models [51] AD induced by application of IL-31 in RHE models [91] Knockdown of FLG in N/TERT FT models [19] Knockdown of FLG in primary keratinocytes [6]
Netherton syndrome	Appearance of parakeratotic stratum corneum, lack of corneodesmosomes, mRNA expression level of SPINK5, occurrence of pro-filaggrin	Wild type	FT models generated with SPINK5 knockdown keratinocytes (siRNA transfected) [92]
Harlequin ichthyosis (HI)	Absence of ABCA12, expression of proteases kallikrein 5, and cathepsin D as well as late differentiation markers	Biopsy from patient suffering from HI	Organotypic skin models using shRNA silencing system to suppress expression of ABCA12 [17]
Autosomal-recessive congenital ichthyosis (ARCI)	Penetration of lucifer yellow FLG expression, basement membrane integrity	Wild type	Transfected normal human keratinocytes with small interfering RNA against TGM1, ALOX12B, ALOXE3, NIPAL4, CYP4F22, and ABCA12 [18]

rather than kinetics, resulting in drug accumulation in the dermal and epidermal layers, and enzymatic activities of the cells.

PK Evaluation on Human Reconstructed Skin Models

PK investigations aim to evaluate the actions of a tissue to an administered drug. As healthy skin represents a barrier to the environment, permeation of drugs is influenced in accordance with chemical characteristics such as polarity, molar mass, and lipophilicity. Penetration of a substance can be modulated by formulation of the drug and is discussed later in this chapter. Since penetration behavior of molecules is also an important end-point for cosmetic actives, extensive studies were performed using validated reconstructed epidermal models and have led to comprehensive understanding of the permeation characteristics of multiple chemicals. Finally, it should be stated that the reconstructed skin models available today are useful for estimating the rank order of percutaneous absorption of a series of compounds with different physicochemical properties [8]. It is a fact that reconstructed models have a relatively weak barrier function, mainly due to differences in lipid lamellae packaging and triacylglycerol metabolism. This limits their applicability for dermal penetration studies. However, whether reconstruction of a full barrier in vitro is feasible or if ex vivo models such as NativeSkin® (Genoskin, Toulouse, France) can indeed serve as the "missing link" [9] still needs to be investigated.

More crucial for drug discovery is the xenobiotic metabolism of a drug and the presence of participating enzymes in reconstructed models. If we assume epidermal application of an active substance to the SC and penetration into viable layers and compartments, the xenobiotic substrate may be altered by enzymes expressed in keratinocytes and/or fibroblast before it enters systemic circulation. Compounds that show no genotoxic effects per se may be converted into genotoxic ones such as 4-nitroquinoline-*N*-oxide (4NQO), cyclophosphamide (CP), dimethylbenzanthracene (DMBA), dimethylnitrosamine (DMN), dibenzanthracene (DBA), or benzo(a)pyrene (BaP). These conversions were imitated in EpiDerm™ (MatTek Corporation, Ashland, USA) models by the human reconstructed skin micronucleus assay [10]. Interestingly, the above-mentioned substances present substrates of xenobiotic-metabolizing CYP enzymes (mainly CYP1A1 and CYP1B1), which are constitutively expressed in very low levels in reconstructed skin models. Higher levels were obtained by application of 3-methylcholanthrene. This molecule was able to induce EROD (7-ethoxyresorufin O-deethylase, CYP 1 family) activity in five of six batches of EpiDerm™, two of five batches EpiSkin™ (EpiSkin, Lyon, France), and in two of six batches SkinEthic™ (EpiSkin, Lyon, France) [11]. CPYs of the families 1–3 play a major role in phase I detoxifying processes in the liver, but their contribution to metabolite conversion in the skin is much less clear. It is the action of non-CYP oxidoreductases, such as FMOs (flavin-dependent monooxygenases) and cyclooxigenases (COX) that is of higher relevance in respect to xenobiotic oxidative reactions in the skin. FMO expression profiles in reconstructed skin models are quite different from those detected in human skin, and COX activity determined for EpiDerm™ turned out to be lower, but was about twofold inducible [12]. Especially COX may become of major importance for oxidative xenobiotic metabolism, as COX is able to act on many lipophilic substrates such as amines [13]. Other oxidoreductases that may be involved in the xenobiotic metabolism are alcohol dehydrogenase and NADH/NADPH quinone reductase (NQR). In the case of alcohol dehydrogenase, no comparison between reconstructed models and human skin is feasible, but NQR activities seem to be represented in the in vitro produced skin models. Especially, EpiDerm™ and EpiSkin™ show close similarities with respect to NQR activities in human skin [14].

A profound study focusing on esterase activities of different reconstructed skin models revealed the usefulness of reconstructed epidermis, as well as FT models, to assess xenobiotic metabolism. Occurrence of the main metabolite prednisolone from prednicarbate and fluorescein from fluorescein diacetate was used to test for esterase activity, which was found to be increased when compared to native human skin [15].

Detoxifying phase II enzymes are well represented in human skin, and it is of special importance to evaluate the activities of conjugating enzymes such as glutathione S-transferase, UTG-glucoronosyl-transferase, and *N*-acetyltransferase. For all these enzymes, some level of activity was demonstrated in reconstructed skin models [16].

Oesch et al. comprehensively summarized and compared xenobiotic-metabolizing enzymes, their activities focusing on reconstructed skin models and skin cell cultures, as well as animal skin models with human skin [16]. The authors found specific gaps that still impede the predictability of drug metabolism in human-native skin. For human reconstructed skin models, the amount of data is still low, making it difficult to draw firm conclusions.

Within the context of pharmacokinetics, it should be kept in mind that drug metabolism may in turn affect penetration behavior. Especially esterase action on compounds leads to the conversion of lipophilic esters to more hydrophilic alcohols and carboxylic acids, and also reduces the molecular size of a molecule. As stated above, lipophilicity and molecular size influence the permeation behavior of a given compound.

REALIZATION OF DISEASE MODELS

Models Composed of Normal Cells in the Presence of Disease Stimuli

Genetically Engineered Disease Models

Skin disorders, which are evoked by specific mutations in the genome, are particularly amenable and well suited to being modeled in vitro. These mutations mostly lead to a deficiency in the normal function of the affected protein.

The direct application of patient-derived diseased cells represents one viable option for skin model generation, as is discussed in one of the following sections. Another approach makes use of the opportunity to interfere with normal gene expression by specifically targeting mRNA molecules by RNAi or by targeting extracellular protein domains by inhibitory antibodies. These approaches are generically applicable, since normal human cell cultures, which are often commercially available, can be used, eliminating the need to establish access to clinical specimens (Figure 1(A) and (B)).

Genetic approaches, such as RNAi, to establish 3D skin models of a disease are particularly useful for keratinocyte disorders. A 3D organotypic skin model for the study of morphological and biochemical alterations in Harlequin ichthyosis (HI), a severe form of recessive congenital ichthyosis, was created by retrovirus-mediated shRNA transfer targeting the ABC transporter protein ABCA12, for which mutations are known to be associated with HI [17]. Normal and diseased clinical skin specimens were analyzed in parallel, whereby the normal samples served as references. The model captured several features of the human disease, such as an epidermis with increased thickness, abnormal lipid content, and protein expression, indicative of premature terminal differentiation. The authors were able to draw conclusions on the role of ABCA12 in the epidermis, the regulation of keratinocyte differentiation and events associated with desquamation, as well as the respective consequences of the loss of ABCA12.

A recent study has described the generation of a variety of 3D human FT skin disease models for autosomal-recessive congenital ichthyosis (ARCI) in FT human skin models by siRNA-mediated knockdown of TGM1, ALOX12B, ALOXE3, NIPAL4, CYP4F22, and ABCA12 in normal human keratinocytes [18]. These models captured several features observed in the histopathology of ARCI, and the altered expression of differentiation markers was congruent with impaired barrier function, as determined experimentally. Therefore, these RNAi-based skin models represent a versatile and robust platform for the experimental interrogation of ARCI in vitro.

Ichthyosis vulgaris and atopic dermatitis represent two additional keratinocyte disorders that are manifested by impaired skin barrier function. Both disorders are associated with mutations in the differentiation-linked FLG gene. An FLG siRNA-based approach in normal human epidermal keratinocytes was used to create an organotypic dermo-epidermal skin model in vitro [6]. While keratinocyte differentiation appeared largely normal, loss of keratohyalin granules, impaired lamellar body formation, decreased urocanic acid production, and abnormal diffusion barrier function were evident. Importantly, FLG deficiency caused increased UV sensitivity. Nevertheless, knockdown of FLG in another 3D skin model did not affect epidermal morphogenesis or lipid organization and composition in the SC [19]. Hence, absence of FLG seems not to be the only reason for an impaired barrier function. Functional testing of siRNA-mediated TGM1 (transglutaminase) knockdown in a similar FT human skin model revealed impaired barrier function by increased permeation of testosterone and caffeine [20]. In addition, the study showed enhanced penetration of the disease model by the test substance nile red incorporated into nanoparticles, demonstrating a major effect on the formulation of substance availability in the diseased tissue. The importance of formulation and drug delivery toward therapeutic approaches is further discussed below. Clearly, these 3D skin disease models are not limited to cells of human origin. An organotypic culture model comprised of Tgm1-shRNA-transfected rat epidermal keratinocytes revealed several features of human diseases, such as hyperkeratosis [21]. Gene expression analysis in the Tgm1-shRNA-expressing cultures identified IL1A as the most highly up-regulated gene in comparison to control transfected cells, and these findings were validated by demonstrating increased expression of IL1A in skin samples from lamellar ichthyosis patients. Treatment of normal keratinocytes and organotypic cultures with recombinant IL1A protein consistently led to a hyperkeratotic phenotype, and treatment of Tgm1-shRNA-expression skin models with IL1RA, a soluble competitive inhibitor of the IL-1 receptor, reversed hyperkeratosis. These findings demonstrate that molecular profiling of 3D disease skin models can enable the identification of key pathway nodes involved in disease, and that these models may directly serve as useful test systems for therapeutic approaches targeting identified pathways.

Models Generated by Application of Disease-Inducing Agents

A multitude of stimulus-induced skin models are available that replicate features of some human skin disorders. We highlight a selection of these models in the section below.

Cutaneous Wounds

Nonhealing, so-called "chronic," wounds constitute major forms of skin disorders. Non-healing wounds are on the rise, since we are now experiencing an aging population, which has a higher incidence of chronic wounds, such as diabetic foot ulcers, venous stasis ulcers, and pressure sores. To study the biology of and the effects of therapeutic agents on wounds, skin models must be generated that contain artificially introduced wounds. There are two ways to create such models: (1) a wound is either introduced into an organotypic model following maturation or (2) a model is prepared in such a manner that a cavity is generated during model maturation, for example, by an inserted stopper, which is removed at the end of maturation process. Concerning the former approach, physically damaged FT skin models were achieved by using a skin mesher. Following wounding, the model was placed on a dermal compartment in order to follow reepithelialization. It was shown that these wounded models were reepithelialization competent, with a keratinocyte shift from proliferation to migration, capable of eliciting an early pro-inflammatory cytokine and a late growth factor response, thus mimicking normal acute healing [22]. Another method was developed using a biopsy punch tool. Punched skin equivalents were added to the top of a dermal compartment, and careful evaluation allowed an in-depth biological understanding of the reepithelialization process to be gained [23]. Effects of topically applied exogenous substances, such as therapeutic agents or matrix components, can be studied in this way, for example, effects of fibrin and platelet-rich plasma on the reepithelialization rates. Another method to monitor keratinocyte migration in punch biopsies is the measurement of epiboly, the mass migration of keratinocytes over a dermal surface [24]. The relative straightforwardness in cultivating the biopsies after punching in culture medium makes this assay an attractive alternative. The positive effect of fibroblasts on reepithelialization was demonstrated in a comparative study on FT wounds created by freezing with a liquid nitrogen-cooled metal bar and on superficial incisional wounds created with a scalpel [25]. Differential effects were observed for exogenously administered growth factors: keratinocyte growth factor and epidermal growth factor were able to stimulate wound closure in the former but not in the latter model, indicating differences in the underlying biology and possibly wounding regimen. In contrast to cold injuries, short periods of heat treatment, for example, by using a soldering iron, can also be used to inflict wounds in human skin models. A comparative study revealed differences and commonalities between both methods, such as morphological and chemokine profiles [26]. For quantitative wound healing studies, it is crucial to generate reproducible wounds of defined depths with high accuracy in an automated manner. Computer-controlled, high-precision drills of defined diameters have been recently described, allowing the generation of standardized wounds in FT human skin equivalents under sterile conditions [27]. Upscaling of wound model generation would make this method a platform for efficacy testing of compound libraries.

Targeted, local tissue ablation from corneas can be achieved by lasers. Thus, lasers constitute attractive, noninvasive tools for the generation of wounds in human 3D FT skin equivalents. A method was described using a low-powered excimer laser, which allowed the generation of reproducible wounds of defined surface and depth with a visibly intact cut surface, replicating features of the in vivo wound healing demands on keratinocytes [28]. Assessment of wound closure upon topical application of nanoparticle formulated morphine in FT skin wound models generated by a CO_2 laser proved to be a viable option [29]. This laser wound model allows for optimization of drug candidate formulation in vitro, with respect to wound closure, toxicity, and irritation.

Skin Infections

Tissue-engineered human skin models also hold great promise for studying interactions between microbes and skin, and for assessing antimicrobial treatment options. This is of high medical importance, since bacterial infections of burn or chronic wounds can severely delay healing. Models were developed that allowed quantitative measurement of colonialization of skin equivalents by resident and transient microbes under conditions comparable to "in vivo" skin [30]. The authors then used gene expression analysis to further investigate whether commensals and transient pathogens elicit different innate immune responses [31]. While the resident strain only evoked minor gene expression changes, the transient pathogen caused multiple changes in a large number of transcripts, including many skin defense factors, possibly reflecting the mode of interaction of the respective microbes with skin. A wounded skin model in conjunction with bacterial infection can serve as a valuable drug development system. In this respect, a burn wound, which was introduced into an FT human skin equivalent using a heated metal rod, was subsequently challenged by bacteria [32]. This model was used to demonstrate that surface application and subsequent removal of polymers functionalized with antibiotics were able to reduce the bacterial load. Another study reported the use of a cold injury human FT equivalent for the testing of therapeutic substances against infection by multidrug-resistant bacteria. Reproducible wounds were inflicted with a liquid nitrogen-cooled metal device before methicillin-resistant *Staphylococcus aureus* (MRSA) strains were added to the wounds [33]. The authors were able to demonstrate that synthetic peptides designed on the human antimicrobial cathelicidin-derived peptide LL-37 were able to eradicate MRSA strains from the wounded skin model. Therefore, this model seems suitable for identifying locally applied drug candidates for the treatment of burn wound infections in patients. In addition, the biology of fungal infections can also be studied in human reconstituted skin models. For example, one study on a human 3D epidermal model revealed that *Candida albicans* elicited morphological features and a cytokine similar to irritant/allergic contact dermatitis [34].

Stimulus-Induced Disease Phenotypes

For diseases that are manifested by an imbalanced interaction of mediators produced by certain cell types and epithelial cells, human reconstituted skin models provide a suitable drug discovery platform. If the mediators are known, they can simply be added to in vitro skin models, and the functional consequences, such as manifestation of disease features, can be measured. It is important that the nontreated models do not or only marginally express these features so that the response to mediator treatment can be unambiguously quantified. If applicable, pharmacological inhibition of the induced disease mechanism should alleviate disease induction. In this regard, a reconstituted psoriatic skin model was produced by treating FT equivalents with psoriasis-associated cytokines, whereas co-treatment with an antipsoriatic agent directly acting on keratinocytes inhibited the psoriatic phenotype [35]. Furthermore, healthy donor-derived cells can be primed in vitro toward a phenotype associated with a particular pathology and then used as a stimulant to generate the corresponding human disease model, such as in vitro activated CD4+ T cells [36]. As a consequence, the model developed a psoriasiform inflammation phenotype, displaying many hallmarks of in vivo epidermal psoriasiform inflammation, which could be counteracted by anti-inflammatory drugs. This model therefore not only constitutes a system to study the pathogenesis of inflammatory skin diseases, but also a platform for the identification of antiinflammatory drug candidates.

Models Comprising Patient-Derived Material

An intriguing alternative to establishing disease context in tissue culture upon addition of disease-causing stimuli consists of the application of patient-derived diseased cells (Figure 1(C)). It is important that these cells are capable of maintaining their disease phenotype in culture for the duration of the whole experiment. Practical caveats of working with patient-derived cells are the prior identification of medical collaboration partners at a hospital, as well as the necessity to obtain ethical and legal approval before cells can be acquired. Furthermore, the establishment of cell cultures from patients is a delicate undertaking that requires expert skills. Special laboratory precautions also need to be taken in order to enable safe conditions for working with specimens potentially containing pathogens.

Patient-derived cells or materials can be selected for cellular models based on several criteria. Cells with abnormalities can either directly cause disease (e.g., due to genetic predisposition) or the disease environment imposes a phenotype on cells that indirectly induces them to cause disease (such as the hyperglycemic state on fibroblasts). A mild form of a skin blistering disease called "junctional epidermolysis bullosa" is caused by the absence of hemidesmosomes and defective epithelial adhesion due to a homozygous mutation leading to a lack of laminin-5 production. Organotypic skin cultures made up of patient-derived mutant keratinocytes resulted in improper attachment to an underlying dead de-epidermized dermis, the presence of rudimentary hemidesmosomes only, and gross disorganization along the basement membrane zone [37]. Interestingly, expression of the wild-type form of the mutant gene led to a reversion of these phenotypes, highlighting potential therapeutic intervention strategies for these extracellular matrix diseases. Similarly, keratinocytes derived from two patients suffering from a moderate and severe form of epidermolytic ichthyosis, respectively, both due to a mutation in the KRT10 gene, were immortalized and hence were capable of reproducing the disease phenotype in organotypic epidermal models, revealing suprabasal blistering [38]. A further example is given by the generation of DNA repair-deficient models xeroderma pigmentosum (XP) patient-derived fibroblasts and keratinocytes. In contrast to normal cells, the XP models displayed several morphological and biochemical features of human conditions [39]. Importantly, the DNA repair-deficiency in response to UVB exposure was restored upon retrovirus-mediated gene transfer of XP-C keratinocytes using wild-type XP-C complementary DNA (cDNA). Several FT skin models with high relevance to the human disease were generated with psoriasis patient-derived fibroblasts and/or keratinocytes. One study reported that the macroscopic, histological, and immunohistochemical characteristics of the pathology of psoriasis were partially maintained in the disease models [40].

Similar models proved utility in assessing drug candidates: delphinidin, a dietary anthocyanidin, displayed antiproliferative, antiinflammatory, and prodifferentiation effects in an FT human skin equivalent harboring psoriatric keratinocytes [41].

For several types of cancer, such as metastatic melanoma, there is no adequate animal model available that would allow the biology of the human disease as well as cell death resistance mechanisms to be investigated. A human 3D skin model provides the appropriate environment for the study of melanoma due to its architecture and cell type context. A recent publication has described the establishment of a 3D FT skin-melanoma model, where melanoma cell line-derived spheroids of defined size and cell number were injected into the dermal collagen/fibroblast compartment [42]. The embedded melanoma spheroids displayed several key characteristics of human skin melanoma metastasis, and the 3D environment had a major impact on the therapeutic response to sensitized drug treatment. Models such as these, potentially equipped with patient-derived melanoma cells, hold great promise for drug discovery screening approaches. Supplementing such models with other relevant cell types, for example, endothelial cells, will further increase the pathophysiologic context. Toward this aim, a microvascularized skin model was setup by inclusion of primary human endothelial cells, highlighting the functional interaction between them and aggressive metastatic melanoma cells [43]. 3D tumor skin models are not limited to melanoma, but can be extended to other cancer types, such as squamous cell carcinoma (SSC) and, in principle, even to precancerous lesions, such as actinic keratosis. Concerning the former, epidermal and dermo-epidermal models were generated in which an SSC cell line was introduced [44]. Both models displayed different molecular patterns, highlighting the crucial contribution of fibroblasts on SSC progression.

Diseased cells derived from patients suffering from multifactorial diseases, without any primary disease-driver alterations, are also suitable for experimental interrogation in human 3D skin models. A recent study has compared patient-derived fibroblasts from diabetic foot ulcers, a debilitating complication of diabetes mellitus, to nonulcerated, site-matched skin of diabetic and nondiabetic patients using a suite of 3D cellular models [45]. Fibroblasts derived from diabetic foot ulcers replicated multiple aspects of chronic ulcers, such as decreased wound closure, reduced angiogenesis, lowered ECM production, and elevated keratinocyte proliferation, in contrast to fibroblasts derived from nondiabetic patients. Concordantly, transplantation of diabetic foot ulcer fibroblasts in a mouse model of cutaneous wound healing led to slower wound closure compared to control fibroblasts. Fibroblasts isolated from diabetic foot ulcers seem to memorize the hyperglycemic conditions present in the patient's body even if the cells are cultivated under normal glycemic conditions, as is the case with the above-mentioned 3D skin models. This metabolic memory is due at least in part to epigenetic changes, such as altered DNA methylation patterns [46].

Human Skin Explants

Human skin biopsies have been used for drug discovery research for several decades. Healthy skin, derived from plastic surgery, limb amputation, or a cadaver, has been widely used for the assessment of formulation properties for drug penetration and permeation, and for phase II biotransformation of drug substances [47,48].

Thanks to the close-to-(patho-)physiological nature, human skin explants also serve as highly contextual models to interrogate biology, disease mechanisms, and drug action (Figure 1(D)). Explants can be cultivated partially or fully submerged in medium, treated with substances applied topically or into the medium, and analyzed for molecular signaling, cellular and biochemical events by particular techniques. Concerning biology, several co-culture systems have been established to study the interaction between distinct cell types and present in skin explants. Toward that goal, one study described the establishment of a neuro-cutaneous interaction model, whereby primary rat sensory neurons extracted from dorsal root ganglia were co-cultured with human skin explants [49]. Re-innervation led to increased epidermal thickness, cell density, and quality as well as lower apoptosis, resulting in higher epidermal integrity. As for reconstituted skin models, a disease phenotype can be either induced in culture in healthy skin explants or perpetuated in culture in disease-associated explants. With respect to the former, Jonkman and colleagues [50] (see also references therein) established models for pemphigus, a chronic mucocutaneous autoimmune bullous disease, by cultivating healthy human skin biopsies under ALI conditions on transwell inserts in contact with a solution containing pemphigus IgG stimulants to induce subcorneal acantholysis. For ethical reasons, it is often difficult to harvest biopsies from diseased skin for experimental purposes. One approach made use of biopsies taken from nonlesional skin from atopic dermatitis patients who were either wild types for FLG or homozygous mutants [51]. These different explants revealed that the FLG genotype–phenotype correlation was maintained in vitro, and that protein expression in vitro and in vivo remained mostly similar. Therefore, such models can be regarded as valid alternatives to the above-described reconstituted, genetically engineered FLG skin equivalents.

General Considerations for Patient-Derived Disease Models

In order to use patient-derived cells for the establishment of predictive and translatable skin models, a subset of the following criteria must be met as a minimum. First, cells derived from a patient suffering from a particular disease must perpetuate certain aspects of the disease phenotype in a relevant skin model, whereas cells derived from a site-matched control population need not. The architecture of the diseased skin model should represent the one of the diseased skin in vivo as far as experimentally possible. The more pronounced the disease phenotype is manifested, the easier it is to score effects of experimental intervention. Secondly, because an in vitro skin model must be predictive for a particular disease, the patient population from which the cells are derived must be as homogenous as possible. While this is generally the case for monogenic diseases, it is more difficult to accomplish for multifactorial diseases, such as chronic wounds in the elderly population. Thus, cells from a selected patient subpopulation (e.g., diabetic foot ulcers) should be applied for 3D skin models. Thirdly, to ensure that the phenotype observed in culture is due to the disease state of the patient-derived cells, multiple independent disease patient samples should be compared with multiple independent matched control samples. Once the disease and cell culture phenotype correlation has been firmly established, drug discovery experiments can be performed on a small number of disease and control cultures. To assess the validity of the hit findings, the hits need then to be validated on the larger cohort of disease and control samples. Fourthly, if applicable, transplantation of patient-derived diseased cells onto a suitable animal model should replicate traits of the human disease, in contrast to matched control cells. This scenario would allow for the assessment of experimental substances on either a human target cell population or on animal cells affected by the presence of human disease cells in an in vivo physiological context, ensuring in vivo and potentially cross-species translatability. Both reconstructed skin models and skin explants can be grafted onto animals [50,52]. Finally, the molecular, cellular, and biochemical insights gained in ex vivo models derived from patient biopsies must be validated in patients by performing comparative studies, substantiating model and data quality.

EXPERIMENTAL PARAMETERS INFLUENCING READOUT

In order to obtain realistic, relevant data and to enable correct model validation, parameters influencing the final end-point analysis have to be critically reviewed. The choice of vehicle, media composition, dose volume (finite/infinite), and saturation level (saturated/unsaturated) are critical parameters that can significantly affect penetration behavior, and hence accumulation in a distinct skin layer. Moreover, artificial application scenarios of in vitro experimental setups hardly replicate the in vivo situation of dermal drug application. Adsorption of a chemical compound across the skin is primarily accomplished by diffusion through the lipid envelope of the SC (lipophilic pathway). However, we need to bear in mind that other permeation routes exist through skin appendices such as hair follicles and sweat glands, which are generally missing in in vitro models. Nevertheless, hydrophobicity, hydrogen bonding, and molecular weight (MW) are the factors that are most likely to determine the process of penetration. Hence, it is not surprising that the first attempts at quantifying this process were based on the permeability coefficient ($\log k_p$) using lipophilicity (measured by the octanol–water partition coefficient, $\log K\text{o/w}$) and the MW [53]. The permeability coefficient essentially characterizes the steady-state permeation rate of a chemical from a specific vehicle through a particular membrane. Assuming infinite dose conditions, the concentration does not significantly change over time after a lag time following application of chemicals, and steady-state flux (*J*ss) is attained [54]. This mathematical derivation makes adjusting screws and the risk of false conclusions apparent.

First of all, it is questionable whether infinite dose setups actually correspond to real-life application procedures. It is more likely that repeated finite dose applications simulate drug application onto the skin as leave-on or rinse-off scenarios. Correct data acquisition and interpretation of finite dose experiments are challenging. Several experimental factors have to be considered: (a) homogeneous and accurate application of the substance, (b) mechanical effect during application, (c) hydration of the SC, and (d) drug recovery/detection problems due to detection limits of HPLC/UV–VIS analysis of the compound in the acceptor medium [55]. Besides permeation through the skin barrier and experimental evaluation of this process, typically in Franz cells, penetration and adsorption of the compound may be of interest especially in cases where the drug is intended to act locally in the area of topical application. Raman spectroscopy is a valuable method in cases where the penetration depth of a compound should be monitored. As this method is nondestructive, penetration can be followed over time, presupposed the compound is Raman active. However, until now only relative, but no absolute, measurements have been possible. For the determination of drug accumulation in distinct tissue zones, matrix-assisted laser desorption/ionization time-of-flight (MALDI-TOF) mass spectrometry imaging in combination with liquid chromatography with tandem mass spectroscopy (LC-MS/MS) analysis can be used to investigate the spatial distribution of a drug after oral application in the hair shaft. These techniques reveal two main drug incorporation sites in the hair [56]. In 2011, the usefulness of MALDI-MS imaging to study the absorption of a xenobiotic compound into a reconstructed human skin model was shown for imipramine detected in Straticell-RHE-EPI-001 [57]. Although the feasibility of this technique was presented, no conclusive evidence of the biotransformation of imipramine in the tissue model was observed.

Another critical parameter is represented by the choice of the vehicle, which is very likely to influence the penetration, accumulation, and permeation of a compound. The solubility, ionization, and permeation characteristics of ibuprofen, for example, were shown to be influenced by the amount of ethanol, and the flux of ibuprofen into the skin was optimal for 50:50 and 75:25 ethanol–water vehicles [58]. In most cases, the solubility of the applied drug dictates the solvent for penetration and permeation analysis, but several possible modes of interference have been identified: (a) fatty acids [59,60], pyrrolidones, and azone induce perturbation of the intercellular domains; (b) ethanols [61] and essential oils change the permeant's solubility in the SC and thus influence the concentration gradient; and (c) urea or DMSO [62] open the intracellular route by swelling or modification of the keratin conformation. Release and absorption of a drug from a delivery system is also adjustable by the choice of a defined polymer system. Permeation and penetration of salicylic acid (SA), for example, were evaluated under finite dose conditions in Franz cells and Raman spectroscopy [63]. Two noteworthy conclusions were drawn from the results: (1) A positive correlation between the amount of SA inside human skin in vivo at 2 μm depths, and the amount permeated through pigskin in vitro, and (2) the influence of the nature of the gelling agent on model drug skin permeation. Although the number of replications was very low, the results underline the importance of performing permeation as well as penetration analyses when using specific delivery systems.

Finally, the identity and nature of the skin tissue itself influences the readout, and especially, the fitness of the model is a critical issue with respect to PK data acquisition. For all biological test systems, the impact of the experimental conditions needs to be taken into account, regardless of whether in vitro created systems or ex vivo biopsies are used. For in vitro created skin models, the choice of the medium, supplementation, and cultivation conditions have a significant impact on barrier function, gene expression, and model integrity, thereby influencing PK and PD evaluations.

GATING STUDIES

Cross-Species Translatability

Ideally, in vitro skin models should enable assessment of whether drug candidates are suitable to progress into preclinical animal studies or even clinical testing (Figure 1(E) and (F)). The following questions are thereby addressed: Do drug candidates engage with their targets in order to achieve a proximal PD response? Are molecular and cellular pathways being modulated accordingly? Can the appropriate functional changes be measured? Do the applied drug candidates reach the destined target cell type? Can the doses of the applied drug candidates predict dosing regimen in animal models and even patients? Can skin models predict whether drug candidates will be effective in the respective animal models, thus allowing selection of and entry into preclinical models? Answers to such questions would affirm the value of in vitro skin models for drug discovery.

Nonhuman in vitro and ex vivo complex skin models intended for research are mostly derived from rodent or porcine species. If initial pilot studies are conducted using biochemical assays only or in cells from different species than those intended for in in vitro and ex vivo skin models, they often serve to address cross-species translatability. This information is of crucial importance, since it enables prediction of whether pathways of interest will be conserved and adequately respond to pathway modulation by genetic or small molecule interrogation in rodent or porcine models. Rodent in vivo models are widely used, since there are numerous skin disease models available which reflect particular traits of human disorders. The ability to manipulate the genome allows the generation of genetically altered mice predisposed to certain skin conditions. However, the anatomy, histology, and physiology of rodent skin greatly differ from human skin, resulting in differences in skin function, such as permeation or wound healing. The skin of domestic pigs is generally accepted to be the most similar to human skin [47]. Therefore, pigs are commonly used as large preclinical animal models for skin diseases with high predictive power for clinical trials. Porcine skin explants, in particular from pig ears, serve as valuable test systems on which to perform functional studies and the evaluation of skin formulations. Assessing the effects of drug candidates in in vitro and ex vivo models may provide a rough projection of which compound concentration should be used in the respective animal models, thereby allowing the application of more refined experimental conditions and eventually reducing animal experimentation.

PD Modulation

For drug discovery, it is essential to know whether drug candidates engage with their intended targets in order to achieve a molecular response in living systems. One could hereby distinguish between "proximal" and "distal" PD effects, such as target engagement or target-proximal effects and further downstream consequences caused by alterations of molecular pathway functions. Proximal PD measurements in cellular 3D models might allow predictions of dosing regimen of particular drug candidates for in vivo studies or even clinical trials. While the complexity of human skin certainly affects drug action, 3D models represent a good starting point for dosing. Numerous methods have been described to study target engagement, primarily for small molecules in cell cultures and in vivo models. While biomarkers for substrates or products of enzymatic reactions can readily be assessed in cell cultures provided that both substrates and products are present in the same cell type, this would not be feasible if they are present in different cell types [64]. For the latter, human reconstituted skin models provide attractive options.

Physiological molecular pathway signaling must be achieved and maintained in 3D models, so that the effects of drug candidates or exogenous stimuli can be measured. Several studies have been published that describe pathway activation at both gene expression and protein activation levels. For example, one report described a 2-day distal PD response to IL-17, a cytokine implicated in inflammation and immunity, in a reconstituted FT skin model [65]. The authors identified a large number of IL-17 pathway genes as well as cytokine and cell cycle-related genes, demonstrating the feasibility of 3D skin models to appropriately replicate biological signaling. The high relevance of the IL-17 gene signature in the 3D models for the IL-17 biology in psoriasis patients is reflected by the overlap with genes that are suppressed in vivo by a neutralizing antibody to IL-17A. Pathway activity can best be measured by cellular protein activity, and protein phosphorylation states provide a good proxy in this regard. A study by Yang et al. (2012) [66] using untreated and irradiated 3D FT reconstituted skin models identified >1000 phosphopeptides, of which 176 were quantitatively regulated in response to radiation dose, reflecting changes in cellular signal transduction pathway function. The fact that radiation-responsive proteins were identified in all layers of the skin model suggests that the entire skin is radiation sensitive and highlights the feasibility of performing quantitative phospho-proteomics approaches in reconstructed skin models. Therefore, phosphorylation states of proteins could serve as PD response markers for drug candidates.

Cellular Efficacy Studies

One advantage of human 3D reconstructed skin models is that the effect of a drug candidate on cell behavior can be assessed in cells of the same species, as subsequently intended in clinical trials. The higher the predictive value of these models, the higher the chances of the drug candidate being efficacious in humans. Clearly, other parameters, such as drug formulation, delivery, and exposure also strongly affect responsiveness to drugs in the body. We define cellular efficacy as changes in phenotypic cell behavior, such as increased cell migration leading to faster wound closure and healing.

Cell behavior is always a function of time. Events following damage, or the addition of a cellular stimulus, are governed by the intricate interplay of cellular pathways in a timely manner. It is often critical to follow cell behavior over time in a noninvasive manner. To do so, an elegant organotypic ALI murine skin model has been established, whereby primary mouse epidermal keratinocytes are cultivated on transwell polyester culture inserts in the presence or absence of dermal fibroblasts in the lower compartment [67]. The optical transparency of transwell cultures allows real-time interrogation using in situ imaging, such as reepithelialization processes following wounding and drug candidate treatment. The authors were able to discern two successive steps of epidermal regeneration: first, the formation of an intact monolayer of cells covering the wound and secondly, keratinocyte stratification in the center of the wound bed and subsequent differentiation, eventually leading to a fully stratified and terminally differentiated epidermis. A recent proof-of-principle study carried out in a human reconstructed FT skin model has demonstrated that Cx43 is down-regulated at the tip of artificially introduced wounds, in contrast to an unwounded model [68]. Importantly, addition of the Cx43 mimetic inhibitory peptide Gap27 maintained reduced levels of Cx43 at the wound margins, and this was accompanied by increased wound closure. These efficacy data and the fact that non-healing diabetic wounds display enhanced Cx expression at wound edges suggest that modulating Cx function might provide therapeutic opportunities [67].

FUTURE CHALLENGES

Cell Source

The availability of the recently developed induced pluripotent stem cell (iPSC) technology will allow additional layers of complexity and physiology to be included in the generation of human skin models. The iPSC technology enables, in principle, the generation of unlimited quantities of an expanding repertoire of differentiated cells from a limited number of patient-derived somatic cells. Therefore, models could be generated that harbor the appropriate genetic composition of either healthy people or patients suffering from particular diseases (Figure 1(B)). Toward this goal, a human 3D skin equivalent was generated consisting of iPSC-derived fibroblasts and keratinocytes [69]. As the technology matures, further iPSC-derived cell types present in normal skin will become available for the construction of skin models. Furthermore, human embryonic stem cells (hESC) have been shown to be capable of producing human 3D skin models without the need for reprogramming to induce pluripotency [70]. Combined with newly available, highly efficient gene-editing technologies such as clustered, regularly interspaced, short palindromic repeat (CRISPR) [71], which can be employed in somatic cells and iPSC/hESC, the generation of (patho-)physiological human skin models will be substantially facilitated in the years to come.

Fabrication Technology

Creation of reconstructed skin models is, in most cases, realized by manual positioning of cells and, in the case of FT models, extracellular matrix components (collagen I, xenobiotic origin) on a permeable membrane (polycarbonate or polyethylene). Differentiation of keratinocytes is induced by reaching confluence of the cell layer (contact inhibition), a shift of Ca^{2+} concentration, contact to air, and application of defined factors to promote epidermal differentiation. This procedure is well established and enables standard creation of RHE and also FT models. For FT models, contraction of the underlying collagen by fibroblasts can hinder standardized model creation. Procedures such as collagen hydrogel compression, collagen-cross-linking, or adaptation of serum concentration and others can decrease the contraction rate.

However, when it comes to more complex organotypic models, manual procedures soon become less feasible for standard medium-throughput screenings. Aligned with new approaches to replicate different stampings of the human skin, new technological opportunities need to be implemented. Especially for biopharmaceutics and the development of systemically acting drugs, implementation of subcutis, blood vessels, glandular structures, and other tissues will certainly become an important issue in this context. It is worthwhile taking a look at current developments in regenerative medicine, where junctional lymphatic capillaries have been generated using 3D hydrogels, and the formation of lymphatic capillaries was adjustable using both lymphangiogenic and antilymphangiogenic stimuli [72]. Recent developments in the technology of bioprinting, where the structuring of compartments and tissues is feasible, have shown that human skin models can also be created by this manufacturing method [73,74]. Although proof-of-concept is shown, such models are not yet deployed in standard procedures. Whether this method will revolutionize the field of organotypic skin model creation will be seen within the next 5 years.

Fluidic Cultivation

In order to reflect the systemic effects of a drug applied topically to the skin and to take dynamic cultivation into account, the complexity of the system has to be increased by enabling the connection and communication by a microfluidic system. This system might even connect organotypic tissues or explants to in vitro skin models, especially, when testing drug uptake, barrier tissues such as the skin can significantly reduce the bioavailability of drugs that are absorbed from the outside [75]. In order to connect different tissue types and to investigate the interactions, a dynamic two-tissue microcirculation chip was developed, which integrated a human skin explant into a chip system next to a human artificial liver model [76]. Inside the chip, the tissues were either exposed to a fluidic flow or shielded from it by a transwell system. Cultivation over a period of 28 days was feasible, which would allow a repeated finite dose application, as discussed earlier. Human FT skin models as well as biological vasculature have also been integrated into the multiorgan chip, thereby coming even closer to an in vivo microenvironment. By combining other tissue types of interest such as liver [77], multiorgan chips are likely to emerge as essential devices for PD and possibly efficacy studies.

CONCLUSION

Over the last decade a huge step in human skin model development has been realized. Models that replicate functional characteristics of healthy as well as diseased in vivo skin were developed. Nevertheless, validated skin models are currently only available for assessment of the irritation and corrosion potential of compounds. Incorporation of such 3D systems in biopharmaceutical processes is on the way, but systematic studies and correlation to animal models are still rare. The organotypic test systems have to be established in order to fulfill demands of PK, PD, as well as efficacy studies and to allow translatability to human skin. Thanks to the conscientious characterization of the plurality of human skin models, it is only a matter of time until some of these systems are part of drug developmental processes. Models of noninfected and infected wounds, as well as 3D human in vitro models replicating psoriasis or atopic dermatitis, are already well developed and have shown their feasibility for the assessment of efficacious compounds. For some targets and diseases, however, in vitro reconstituted skin models may not turn out to be relevant or feasible systems, and in such cases, other models, such as explants, will be preferred. This process of evaluation and cross-comparison is still in the early phase. With the establishment of new fabrication and cell manipulation techniques, we expect the development of a new generation of human skin models within the next few years.

REFERENCES

[1] Tfaili S, Gobinet C, Josse G, Angiboust JF, Manfait M, Piot O. Confocal Raman microspectroscopy for skin characterization: a comparative study between human skin and pig skin. Analyst 2012;137(16):3673–82.

[2] Yang R, Zheng Y, Li L, Liu S, Burrows M, Wei Z, et al. Direct conversion of mouse and human fibroblasts to functional melanocytes by defined factors. Nat Commun 2014;5:5807.

[3] Serrati S, Chilla A, Laurenzana A, Margheri F, Giannoni E, Magnelli L, et al. Systemic sclerosis endothelial cells recruit and activate dermal fibroblasts by induction of a connective tissue growth factor (CCN2)/transforming growth factor beta-dependent mesenchymal-to-mesenchymal transition. Arthritis Rheum 2013;65(1):258–69.

[4] Mathes SH, Ruffner H, Graf-Hausner U. The use of skin models in drug development. Adv Drug Deliv Rev 2014;69–70:81–102.

[5] Mildner M, Ballaun C, Stichenwirth M, Bauer R, Gmeiner R, Buchberger M, et al. Gene silencing in a human organotypic skin model. Biochem Biophys Res Commun 2006;348(1):76–82.

[6] Mildner M, Jin J, Eckhart L, Kezic S, Gruber F, Barresi C, et al. Knockdown of filaggrin impairs diffusion barrier function and increases UV sensitivity in a human skin model. J Invest Dermatol 2010;130(9):2286–94.

[7] Pendaries V, Le Lamer M, Cau L, Hansmann B, Malaisse J, Kezic S, et al. In a three-dimensional reconstructed human epidermis filaggrin-2 is essential for proper cornification. Cell Death Dis 2015;6:e1656.

[8] Van Gele M, Geusens B, Brochez L, Speeckaert R, Lambert J. Three-dimensional skin models as tools for transdermal drug delivery: challenges and limitations. Expert Opin Drug Deliv 2011;8(6):705–20.

[9] Wever BD, Kurdykowski S, Descargues P. Human skin models for research applications in pharmacology and toxicology: introducing NativeSkin, the ''missing link'' bridging cell culture and/or reconstructed skin models and human clinical testing. Appl In Vitro Toxicol 2015;1(1):26–32.

[10] Aardema MJ, Barnett BB, Mun GC, Dahl EL, Curren RD, Hewitt NJ, et al. Evaluation of chemicals requiring metabolic activation in the EpiDerm 3D human reconstructed skin micronucleus (RSMN) assay. Mutat Res 2013;750(1–2):40–9.

[11] Harris IR, Siefken W, Beck-Oldach K, Brandt M, Wittern KP, Pollet D. Comparison of activities dependent on glutathione S-transferase and cytochrome P-450 IA1 in cultured keratinocytes and reconstructed epidermal models. Skin Pharmacol Appl Skin Physiol 2002;15(Suppl. 1):59–67.

[12] Gotz C, Pfeiffer R, Tigges J, Blatz V, Jackh C, Freytag EM, et al. Xenobiotic metabolism capacities of human skin in comparison with a 3D epidermis model and keratinocyte-based cell culture as in vitro alternatives for chemical testing: activating enzymes (Phase I). Exp Dermatol 2012;21(5):358–63.

[13] Oesch-Bartlomowicz B, Oesch F. Mechanisms of toxification and detoxification which challenge drug candidates and drugs. Compr Med Chem 2007:193–214.

[14] Harris IR, Siefken W, Beck-Oldach K, Wittern KP, Pollet D. NAD(P)H:quinone reductase activity in human epidermal keratinocytes and reconstructed epidermal models. Skin Pharmacol Appl Skin Physiol 2002;15(Suppl. 1):68–73.

[15] Batz FM, Klipper W, Korting HC, Henkler F, Landsiedel R, Luch A, et al. Esterase activity in excised and reconstructed human skin–biotransformation of prednicarbate and the model dye fluorescein diacetate. Eur J Pharm Biopharm 2013;84(2):374–85.

[16] Oesch F, Fabian E, Guth K, Landsiedel R. Xenobiotic-metabolizing enzymes in the skin of rat, mouse, pig, guinea pig, man, and in human skin models. Arch Toxicol 2014;88(12):2135–90.

[17] Thomas AC, Tattersall D, Norgett EE, O'Toole EA, Kelsell DP. Premature terminal differentiation and a reduction in specific proteases associated with loss of ABCA12 in Harlequin ichthyosis. Am J Pathol 2009;174(3):970–8.

[18] Eckl KM, Alef T, Torres S, Hennies HC. Full-thickness human skin models for congenital ichthyosis and related keratinization disorders. J Invest Dermatol 2011;131(9):1938–42.

[19] van Drongelen V, Alloul-Ramdhani M, Danso MO, Mieremet A, Mulder A, van Smeden J, et al. Knock-down of filaggrin does not affect lipid organization and composition in stratum corneum of reconstructed human skin equivalents. Exp Dermatol 2013;22(12):807–12.

[20] Eckl KM, Weindl G, Ackermann K, Kuchler S, Casper R, Radowski MR, et al. Increased cutaneous absorption reflects impaired barrier function of reconstructed skin models mimicking keratinisation disorders. Exp Dermatol 2014;23(4):286–8.

[21] O'Shaughnessy RF, Choudhary I, Harper JI. Interleukin-1 alpha blockade prevents hyperkeratosis in an in vitro model of lamellar ichthyosis. Hum Mol Genet 2010;19(13):2594–605.

[22] Falanga V, Isaacs C, Paquette D, Downing G, Kouttab N, Butmarc J, et al. Wounding of bioengineered skin: cellular and molecular aspects after injury. J Invest Dermatol 2002;119(3):653–60.

[23] Laplante AF, Germain L, Auger FA, Moulin V. Mechanisms of wound reepithelialization: hints from a tissue-engineered reconstructed skin to long-standing questions. FASEB J 2001;15(13):2377–89.

[24] Luo S, Yufit T, Carson P, Fiore D, Falanga J, Lin X, et al. Differential keratin expression during epiboly in a wound model of bioengineered skin and in human chronic wounds. Int J Low Extrem Wounds 2011;10(3):122–9.

[25] El Ghalbzouri A, Hensbergen P, Gibbs S, Kempenaar J, van der Schors R, Ponec M. Fibroblasts facilitate re-epithelialization in wounded human skin equivalents. Lab Invest 2004;84(1):102–12.

[26] Breetveld M, Richters CD, Rustemeyer T, Scheper RJ, Gibbs S. Comparison of wound closure after burn and cold injury in human skin equivalents. J Invest Dermatol 2006;126(8):1918–21.

[27] Rossi A, Appelt-Menzel A, Kurdyn S, Walles H, Groeber F. Generation of a three-dimensional full thickness skin equivalent and automated wounding. J Vis Exp 2015;96.

[28] Vaughan MB, Ramirez RD, Brown SA, Yang JC, Wright WE, Shay JW. A reproducible laser-wounded skin equivalent model to study the effects of aging in vitro. Rejuvenation Res 2004;7(2):99–110.

[29] Kuchler S, Wolf NB, Heilmann S, Weindl G, Helfmann J, Yahya MM, et al. 3D-wound healing model: influence of morphine and solid lipid nanoparticles. J Biotechnol 2010;148(1):24–30.

[30] Holland DB, Bojar RA, Jeremy AH, Ingham E, Holland KT. Microbial colonization of an in vitro model of a tissue engineered human skin equivalent–a novel approach. FEMS Microbiol Lett 2008;279(1):110–5.

[31] Holland DB, Bojar RA, Farrar MD, Holland KT. Differential innate immune responses of a living skin equivalent model colonized by *Staphylococcus epidermidis* or *Staphylococcus aureus*. FEMS Microbiol Lett 2009;290(2):149–55.

[32] Shepherd J, Sarker P, Rimmer S, Swanson L, MacNeil S, Douglas I. Hyperbranched poly(NIPAM) polymers modified with antibiotics for the reduction of bacterial burden in infected human tissue engineered skin. Biomaterials 2011;32(1):258–67.

[33] Haisma EM, de Breij A, Chan H, van Dissel JT, Drijfhout JW, Hiemstra PS, et al. LL-37-derived peptides eradicate multidrug-resistant *Staphylococcus aureus* from thermally wounded human skin equivalents. Antimicrob Agents Chemother 2014;58(8):4411–9.

[34] Schaller M, Mailhammer R, Korting HC. Cytokine expression induced by *Candida albicans* in a model of cutaneous candidosis based on reconstituted human epidermis. J Med Microbiol 2002;51(8):672–6.

[35] Tjabringa G, Bergers M, van Rens D, de Boer R, Lamme E, Schalkwijk J. Development and validation of human psoriatic skin equivalents. Am J Pathol 2008;173(3):815–23.

[36] van den Bogaard EH, Tjabringa GS, Joosten I, Vonk-Bergers M, van Rijssen E, Tijssen HJ, et al. Crosstalk between keratinocytes and T cells in a 3D microenvironment: a model to study inflammatory skin diseases. J Invest Dermatol 2014;134(3):719–27.
[37] Vailly J, Gagnoux-Palacios L, Dell'Ambra E, Romero C, Pinola M, Zambruno G, et al. Corrective gene transfer of keratinocytes from patients with junctional epidermolysis bullosa restores assembly of hemidesmosomes in reconstructed epithelia. Gene Ther 1998;5(10):1322–32.
[38] Chamcheu JC, Pihl-Lundin I, Mouyobo CE, Gester T, Virtanen M, Moustakas A, et al. Immortalized keratinocytes derived from patients with epidermolytic ichthyosis reproduce the disease phenotype: a useful in vitro model for testing new treatments. Br J Dermatol 2011;164(2):263–72.
[39] Bernerd F, Asselineau D, Frechet M, Sarasin A, Magnaldo T. Reconstruction of DNA repair-deficient xeroderma pigmentosum skin in vitro: a model to study hypersensitivity to UV light. Photochem Photobiol 2005;81(1):19–24.
[40] Jean J, Lapointe M, Soucy J, Pouliot R. Development of an in vitro psoriatic skin model by tissue engineering. J Dermatol Sci 2009;53(1):19–25.
[41] Chamcheu JC, Pal HC, Siddiqui IA, Adhami VM, Ayehunie S, Boylan BT, et al. Prodifferentiation, anti-inflammatory and antiproliferative effects of delphinidin, a dietary anthocyanidin, in a full-thickness three-dimensional reconstituted human skin model of psoriasis. Skin Pharmacol Physiol 2015;28(4):177–88.
[42] Vorsmann H, Groeber F, Walles H, Busch S, Beissert S, Walczak H, et al. Development of a human three-dimensional organotypic skin-melanoma spheroid model for in vitro drug testing. Cell Death Dis 2013;4:e719.
[43] Gibot L, Galbraith T, Huot J, Auger FA. Development of a tridimensional microvascularized human skin substitute to study melanoma biology. Clin Exp Metastasis 2013;30(1):83–90.
[44] Brauchle E, Johannsen H, Nolan S, Thude S, Schenke-Layland K. Design and analysis of a squamous cell carcinoma in vitro model system. Biomaterials 2013;34(30):7401–7.
[45] Maione AG, Brudno Y, Stojadinovic O, Park LK, Smith A, Tellechea A, et al. Three-dimensional human tissue models that incorporate diabetic foot ulcer-derived fibroblasts mimic in vivo features of chronic wounds. Tissue Eng Part C Methods 2015;21(5):499–508.
[46] Park LK, Maione AG, Smith A, Gerami-Naini B, Iyer LK, Mooney DJ, et al. Genome-wide DNA methylation analysis identifies a metabolic memory profile in patient-derived diabetic foot ulcer fibroblasts. Epigenetics 2014;9(10):1339–49.
[47] Flaten GE, Palac Z, Engesland A, Filipovic-Grcic J, Vanic Z, Skalko-Basnet N. In vitro skin models as a tool in optimization of drug formulation. Eur J Pharm Sci 2015;75:10–24.
[48] Manevski N, Swart P, Balavenkatraman KK, Bertschi B, Camenisch G, Kretz O, et al. Phase II metabolism in human skin: skin explants show full coverage for glucuronidation, sulfation, N-acetylation, catechol methylation, and glutathione conjugation. Drug Metab Dispos 2015;43(1):126–39.
[49] Lebonvallet N, Boulais N, Le Gall C, Pereira U, Gauche D, Gobin E, et al. Effects of the re-innervation of organotypic skin explants on the epidermis. Exp Dermatol 2012;21(2):156–8.
[50] van der Wier G, Pas HH, Jonkman MF. Experimental human cell and tissue models of pemphigus. Dermatol Res Pract 2010;2010:143871.
[51] van Drongelen V, Haisma EM, Out-Luiting JJ, Nibbering PH, El Ghalbzouri A. Reduced filaggrin expression is accompanied by increased *Staphylococcus aureus* colonization of epidermal skin models. Clin Exp Allergy 2014;44(12):1515–24.
[52] Gache Y, Baldeschi C, Del Rio M, Gagnoux-Palacios L, Larcher F, Lacour JP, et al. Construction of skin equivalents for gene therapy of recessive dystrophic epidermolysis bullosa. Hum Gene Ther 2004;15(10):921–33.
[53] Potts RO, Guy RH. Predicting skin permeability. Pharm Res 1992;9(5):663–9.

[54] Buist HE, van Burgsteden JA, Freidig AP, Maas WJ, van de Sandt JJ. New in vitro dermal absorption database and the prediction of dermal absorption under finite conditions for risk assessment purposes. Regul Toxicol Pharmacol 2010;57(2–3):200–9.

[55] Selzer D, Abdel-Mottaleb MM, Hahn T, Schaefer UF, Neumann D. Finite and infinite dosing: difficulties in measurements, evaluations and predictions. Adv Drug Deliv Rev 2013;65(2):278–94.

[56] Kamata T, Shima N, Sasaki K, Matsuta S, Takei S, Katagi M, et al. Time-course mass spectrometry imaging for depicting drug incorporation into hair. Anal Chem 2015;87(11):5476–81.

[57] Avery JL, McEwen A, Flinders B, Francese S, Clench MR. Matrix-assisted laser desorption mass spectrometry imaging for the examination of imipramine absorption by Straticell-RHE-EPI/001 an artificial model of the human epidermis. Xenobiotica 2011;41(8):735–42.

[58] Watkinson RM, Herkenne C, Guy RH, Hadgraft J, Oliveira G, Lane ME. Influence of ethanol on the solubility, ionization and permeation characteristics of ibuprofen in silicone and human skin. Skin Pharmacol Physiol 2009;22(1):15–21.

[59] Aungst BJ. Structure/effect studies of fatty acid isomers as skin penetration enhancers and skin irritants. Pharm Res 1989;6(3):244–7.

[60] Tanojo H, Bouwstra JA, Junginger HE, Bodde HE. In vitro human skin barrier modulation by fatty acids: skin permeation and thermal analysis studies. Pharm Res 1997;14(1):42–9.

[61] Pershing LK, Lambert LD, Knutson K. Mechanism of ethanol-enhanced estradiol permeation across human skin in vivo. Pharm Res 1990;7(2):170–5.

[62] Williams AC, Barry BW. Penetration enhancers. Adv Drug Deliv Rev 2004;56(5):603–18.

[63] Mateus R, Moore DJ, Hadgraft J, Lane ME. Percutaneous absorption of salicylic acid–in vitro and in vivo studies. Int J Pharm 2014;475(1–2):471–4.

[64] Simon GM, Niphakis MJ, Cravatt BF. Determining target engagement in living systems. Nat Chem Biol 2013;9(4):200–5.

[65] Chiricozzi A, Nograles KE, Johnson-Huang LM, Fuentes-Duculan J, Cardinale I, Bonifacio KM, et al. IL-17 induces an expanded range of downstream genes in reconstituted human epidermis model. PLoS One 2014;9(2):e90284.

[66] Yang F, Waters KM, Webb-Robertson BJ, Sowa MB, von Neubeck C, Aldrich JT, et al. Quantitative phosphoproteomics identifies filaggrin and other targets of ionizing radiation in a human skin model. Exp Dermatol 2012;21(5):352–7.

[67] Kandyba EE, Hodgins MB, Martin PE. A murine living skin equivalent amenable to live-cell imaging: analysis of the roles of connexins in the epidermis. J Invest Dermatol 2008;128(4):1039–49.

[68] Pollok S, Pfeiffer AC, Lobmann R, Wright CS, Moll I, Martin PE, et al. Connexin 43 mimetic peptide Gap27 reveals potential differences in the role of Cx43 in wound repair between diabetic and non-diabetic cells. J Cell Mol Med 2011;15(4):861–73.

[69] Itoh M, Umegaki-Arao N, Guo Z, Liu L, Higgins CA, Christiano AM. Generation of 3D skin equivalents fully reconstituted from human induced pluripotent stem cells (iPSCs). PLoS One 2013;8(10):e77673.

[70] Petrova A, Celli A, Jacquet L, Dafou D, Crumrine D, Hupe M, et al. 3D in vitro model of a functional epidermal permeability barrier from human embryonic stem cells and induced pluripotent stem cells. Stem Cell Reports 2014;2(5):675–89.

[71] Sander JD, Joung JK. CRISPR-Cas systems for editing, regulating and targeting genomes. Nat Biotechnol 2014;32(4):347–55.

[72] Marino D, Luginbuhl J, Scola S, Meuli M, Reichmann E. Bioengineering dermo-epidermal skin grafts with blood and lymphatic capillaries. Sci Transl Med 2014;6(221):221ra14.

[73] Rimann M, Bono E, Annaheim H, Bleisch M, Graf-Hausner U. Standardized 3D bioprinting of soft tissue models with human primary cells. J Lab Autom 2015.

[74] Lee V, Singh G, Trasatti JP, Bjornsson C, Xu X, Tran TN, et al. Design and fabrication of human skin by three-dimensional bioprinting. Tissue Eng Part C Methods 2014;20(6):473–84.

[75] Esch MB, King TL, Shuler ML. The role of body-on-a-chip devices in drug and toxicity studies. Annu Rev Biomed Eng 2011;13:55–72.

[76] Wagner I, Materne EM, Brincker S, Sussbier U, Fradrich C, Busek M, et al. A dynamic multi-organ-chip for long-term cultivation and substance testing proven by 3D human liver and skin tissue co-culture. Lab Chip 2013;13(18):3538–47.

[77] Materne EM, Maschmeyer I, Lorenz AK, Horland R, Schimek KM, Busek M, et al. The multi-organ chip - a microfluidic platform for long-term multi-tissue coculture. J Vis Exp 2015;98.

[78] Schafer-Korting M, Bock U, Diembeck W, Dusing HJ, Gamer A, Haltner-Ukomadu E, et al. The use of reconstructed human epidermis for skin absorption testing: results of the validation study. Altern Lab Anim 2008;36(2):161–87.

[79] Schafer-Korting M, Bock U, Gamer A, Haberland A, Haltner-Ukomadu E, Kaca M, et al. Reconstructed human epidermis for skin absorption testing: results of the German prevalidation study. Altern Lab Anim 2006;34(3):283–94.

[80] Schafer-Korting M, Mahmoud A, Lombardi Borgia S, Bruggener B, Kleuser B, Schreiber S, et al. Reconstructed epidermis and full-thickness skin for absorption testing: influence of the vehicles used on steroid permeation. Altern Lab Anim 2008;36(4):441–52.

[81] Ackermann K, Borgia SL, Korting HC, Mewes KR, Schafer-Korting M. The Phenion full-thickness skin model for percutaneous absorption testing. Skin Pharmacol Physiol 2010;23(2):105–12.

[82] Haisma EM, Rietveld MH, de Breij A, van Dissel JT, El Ghalbzouri A, Nibbering PH. Inflammatory and antimicrobial responses to methicillin-resistant *Staphylococcus aureus* in an in vitro wound infection model. PLoS One 2013;8(12):e82800.

[83] Mohapatra S, Coppola D, Riker AI, Pledger WJ. Roscovitine inhibits differentiation and invasion in a three-dimensional skin reconstruction model of metastatic melanoma. Mol Cancer Res 2007;5(2):145–51.

[84] Commandeur S, van Drongelen V, de Gruijl FR, El Ghalbzouri A. Epidermal growth factor receptor activation and inhibition in 3D in vitro models of normal skin and human cutaneous squamous cell carcinoma. Cancer Sci 2012;103(12):2120–6.

[85] Commandeur S, de Gruijl FR, Willemze R, Tensen CP, El Ghalbzouri A. An in vitro three-dimensional model of primary human cutaneous squamous cell carcinoma. Exp Dermatol 2009;18(10):849–56.

[86] van Kilsdonk JW, van den Bogaard EH, Jansen PA, Bos C, Bergers M, Schalkwijk J. An in vitro wound healing model for evaluation of dermal substitutes. Wound Repair Regen 2013;21(6):890–6.

[87] Marquardt Y, Amann PM, Heise R, Czaja K, Steiner T, Merk HF, et al. Characterization of a novel standardized human three-dimensional skin wound healing model using non-sequential fractional ultrapulsed CO2 laser treatments. Lasers Surg Med 2015;47(3):257–65.

[88] Bernard FX, Morel F, Camus M, Pedretti N, Barrault C, Garnier J, et al. Keratinocytes under fire of proinflammatory cytokines: bona fide innate immune cells involved in the Physiopathology of chronic atopic dermatitis and psoriasis. J Allergy (Cairo) 2012;2012:718725.

[89] Raychaudhuri S, Mitra A, Datta-Mitra A. Immunomodulatory mechanisms of action of calcitriol in psoriasis. Indian J Dermatol 2014;59(2):116–22.

[90] Rueffner C. Psoriatic in vitro epidermis. Household Pers Care Today 2011;2.

[91] Cornelissen C, Marquardt Y, Czaja K, Wenzel J, Frank J, Luscher-Firzlaff J, et al. IL-31 regulates differentiation and filaggrin expression in human organotypic skin models. J Allergy Clin Immunol 2012;129(2):426–33. 433 e1–8.

[92] Wang S, Olt S, Schoefmann N, Stuetz A, Winiski A, Wolff-Winiski B. SPINK5 knockdown in organotypic human skin culture as a model system for Netherton syndrome: effect of genetic inhibition of serine proteases kallikrein 5 and kallikrein 7. Exp Dermatol 2014;23(7):524–6.

Chapter 19

Animal Models for Wound Healing

Glicerio Ignacio[1], Idris El-Amin[2], Vince Mendenhall[3]
[1]*David H. Murdock Research Institute, SOS Division, Working Buildings, LLC, Kannapolis, NC, USA;* [2]*Max Planck Florida Institute for Neuroscience, Jupiter, FL, USA;* [3]*Wake Forest University School of Medicine, Winston-Salem, NC, USA*

Chapter Outline

INTRODUCTION

The development of skin substitutes and other novel products to improve wound healing and scarring is usually conducted in a stepwise fashion, starting with in vitro testing, preclinical studies, and finally, clinical trials.

Animal models of wound healing are considered when computer models or in vitro techniques are not possible or when the complex physiological responses to wounding need to be analyzed. These preclinical studies facilitate investigation of the test article in the presence of exudates, blood, inflammatory cells, etc., which can have an effect on the activity of the skin

Skin Tissue Engineering and Regenerative Medicine. http://dx.doi.org/10.1016/B978-0-12-801654-1.00019-X

substitute or topical application of the agent. In addition to studying the effect of the test article within the conditions of the wound, animal studies are also integral to the investigation of the safety of the product. It is important to determine any deleterious effect on tissues that the test article may induce, thus potentially prolonging the wound healing period or resulting in an inferior end result. Animal trials to determine the efficacy of the test article in improving various aspects of wound healing are generally conducted after both in vitro and in vivo safety studies are completed. Successful completion of these preclinical studies will eventually allow the product to be evaluated in human clinical trials.

The research community should, however, continue to promote the discovery of alternatives to currently accepted in vivo methods. The author assumes that the reader has weighed all alternatives to animal usage prior to making the decision to integrate the application of animal models to wound healing studies and the development of skin substitutes and other novel products to be used to improve wound healing and scarring.

The physiological response to wounding is a very sophisticated process, briefly characterized by four well-defined, bridging phases: hemostasis, inflammation, proliferation, and remodeling. The selection of small and large animal models for the preclinical evaluation of skin substitutes hinge on the comparative anatomy of the different species with respect to their response to injury versus that of humans. The animal model chosen should also demonstrate and closely mimic the wound pathogenesis seen in the clinic. The data generated from preclinical studies and applications may vary greatly depending upon the animal model chosen.

This chapter discusses the selection of small and large animal models for evaluation of skin substitutes or the topical application of biological and chemical formulations to be used to hopefully improve natural wound healing and scarring. This chapter also discusses the criteria of animal selection and model development for both chronic and acute wound conditions.

COMPARATIVE ANIMAL MODELS OF WOUND HEALING

Most studies involved in wound healing involve both in vitro and in vivo experimentation. However, in vitro studies do not take into consideration the local and systemic contributions to wound healing, such as vascularization of the wound site, inflammation, immune responses, and infection. This portion of the chapter will focus on selection of the appropriate vertebrate and invertebrate animal species currently used to study various skin conditions. Many species have been used as wound healing models, but are beyond the scope of this chapter. When looking at translational models of wound healing, the vast majority of studies are performed either in rodents or swine. However, there are several key similarities and differences that exist, which influence the type of species considered for animal research.

Mice and Rats

Rodents are often used as the first choice due to the amount of scientific knowledge previously acquired across various body systems, their small size, and the ability to manipulate certain genetic, dietary, and environmental variables. Rats and mice have gestational periods between 19 and 22 days, which allows for the generation of large colonies as well as the evaluation of multigenerational effects. Comparing responses in juvenile, adult, and geriatric mice is possible given their life spans are typically 1–2 years for mice and 2–3 years for rats.

Care must be taken when manipulating the skin of rodents because of their large body surface area to weight ratio, making them prone to changes in environmental conditions. Perioperatively, these animals may require thermal support through the placement of heating blankets or pads around them, in order to prevent hypothermia. Postoperatively, these animals should be monitored for inappetence, dehydration, and pain. If needed, water and food supplementation can be given orally via syringe or oral gavage. Isotonic crystalloid fluids can also be administered subcutaneously if needed, but may alter wound healing if given near the surgical site. Postoperative pain should be adequately controlled with the use of appropriate analgesic agents, especially since rodents can self-mutilate when they perceive excessive pain. Opioids, such as buprenorphine, are the analgesic agents of choice, since nonsteroidal antiinflammatory drugs cannot be administered due to their inhibitory effects on inflammation.

Overall, the process of wound healing is quite different between rodents and man, primarily as a function of their skin type. Rodents have loose skin, whereas humans have tight skin. Although wound contraction occurs in both species following cutaneous injury, this process is slower in tight-skinned species, such as humans and pigs, where healing of wounds occurs primarily through reepithelialization.

Genetically, the mouse and human are very similar when comparing the protein-coding regions, which are approximately 85% identical [1]. However, significant differences exist between rodents and humans, as well as between rodent strains, that affect wound healing.

Anatomically, the skin of the mouse and rat is loosely attached to the subcutis and is densely covered in hair (with the exception of nude mice and other hairless strains) [2]. This loose attachment allows for the placement of subcutaneous devices such as bioengineered scaffolds. When considering use of haired mice or rats, it should be noted that the hair cycle in rodents is approximately 3 weeks compared with several years in humans [3]. One alternative is the outbred hairless (*Hr* gene mutation) mouse which is immunocompetent, unlike the nude (*Foxp1* gene mutation) mouse [4]. In relation to the skin in toto, both epidermal and dermal layers are considerably thinner than in humans, with both rodent species lacking apocrine and eccrine glands. Although not necessarily involved in the homeostasis of the epidermal layer, stem cells from sweat glands and hair follicle progenitors are thought to contribute to the reepithelialization of damaged skin in humans [5,6]. Rats and mice also possess a subcutaneous panniculus carnosus muscle, which is absent in humans. This muscle is believed to facilitate wound contraction and collagen formation [2].

When developing studies using both sexes of mice and rats, it should be noted that anatomical and functional differences exist between sexes that may impact results. The skin of male mice and rats is considerably stronger due to a thicker dermis, presumably, as a result of increased collagen. On the other hand, the female mice and rats have a thicker epidermis and hypodermis, due to greater proliferation [7]. Functionally, full-thickness wounds heal at different rates as well. Cutaneous wound healing was impaired in female BALB/c mice treated with aspirin versus male BALB/c mice in one study. This difference was attributed to elevated myeloperoxidase activity and altered levels of growth factors and macrophages [8].

It has also been speculated that microbiota affect wound healing. When comparing germ-free mice with wild-type Swiss mice, it was observed that germ-free mice had faster closure of excisional wounds and lower numbers of inflammatory cells present at wound sites. Also altered were cytokine levels in the wound tissue. Germ-free mice had higher levels of interleukin-10 and lower levels of TGFβ1, which promote angiogenesis and reduce fibrosis, respectively. These findings suggest a more regenerative repair than fibrotic repair when the microbiota is low. These findings also suggest that a reduction in microbial flora through the administration of antibiotics prior to or during the wound healing process, may positively affect the end result [9].

Swine

Pigs have been extensively utilized in wound healing studies over the past five decades. The pig's advantages are its large size, similar epidermal and dermal thickness, sparse hair coat, skin architecture, and healing process. Unlike rodents and other animals, the skin is firmly attached to the subcutis, as is humans [10]. Functionally, pig skin is also similar to humans in terms of lipid content, keratinous protein fractions, and epidermal turnover time [11]. Pigs are different from humans with regard to the proportion and distribution of glandular tissue. Apocrine glands, present in human sweat glands and restricted to the axillary and rectogenital region, are distributed throughout the entire body in swine. Body eccrine sweat glands are present in human and Catarrhini primates like macaques, but absent in swine [12].

The overall size of most adult pigs allows for various treatment groups and controls within the same animal. The relative skin thickness is advantageous to researchers when making partial- or full-thickness wounds. Typically, wounds are created on the dorsum of the animal, which allows for easy visualization. Wounds created on the ventrum may become contaminated, even when covered with adequate dressings, due to typical research housing conditions and the natural temperament of the pig. Also, the skin on the ventrum is much thinner and thus not as similar to humans as it is on the dorsum.

Like humans, the healing process for partial-thickness wounds is by reepithelialization and occurs in a similar fashion, but varies significantly with age. In a study involving Yucatan pigs of various ages, reepithelialization of full-thickness wounds was significantly more extensive in younger pigs (2-month-old) compared to adult pigs (24-month-old). Growth factors such as vascular endothelial growth factor (VEGF) and platelet-derived growth factor-BB were higher during the healing process in the younger pigs compared to the adult pigs as well [10,13]. It is believed that lower levels of growth factors, such as VEGF, are associated with delayed healing. These growth factors not only promote angiogenesis but also potentially activate macrophages and keratinocytes [14].

Certain breeds of pigs can be used to study various types of wound healing. For instance, deep dermal injuries to red Duroc or Yucatan pigs can elicit hypertrophic scarring, or keloid formation, a phenomenon that occurs in humans, especially prevalent in African Americans. Nevertheless, these lesions are visually different, in that human scars appear redder and have an abrupt edge. Nonetheless, hair growth, pigmentation, and collagen patterns are similar [15].

Other things to consider when using swine as the test system for preclinical studies involving wound healing are the costs associated with obtaining and housing pigs. In general, purchasing pigs from a USDA Class B dealer is less expensive than obtaining pigs from a specific pathogen-free facility. Also, housing costs are generally more for the larger pigs, such as Yorkshire pigs versus miniature pigs, due to the space requirements set forth by the Animal Welfare Act and associated regulations. Of course, miniature pigs cost more than Yorkshires to begin with. Thus, a researcher should attempt to perform initial studies in rodents, if scientifically justifiable, and then proceed to definitive studies in the appropriate swine breed.

In our facilities, we have also noted that swine, compared to rodent models of cutaneous wound healing, have a greater incidence of skin infection (unpublished observations). To prevent infection and injury to the study sites in swine to be used for the evaluation of wound healing differences according to the use of test article, the wounds should be protected by covering them with various types of dressings (unless a dressing is the test article), bandages, and jackets that are changed frequently. Like rodents, opioids should be given to control pain after injury, and nonsteroidal antiinflammatory drugs should be avoided. Postoperative topical or systemic antibiotics should not be administered prophylactically, but are indicated during the wounding procedure. Typically, if a wound becomes infected, any results seen in that wound are not valid.

Dogs and Cats

Dogs and cats are not usually utilized when studying wound healing. Like rodents, the skin is loosely adhered to the subcutis on the trunk, although tightly adhered on the extremities. As one would expect, full-thickness wound healing occurs by both reepithelialization and wound contraction. When comparing dogs and cats, dogs generally have a more rapid rate of full-thickness wound healing, and cats demonstrate a decreased amount of inflammation and subsequent granulation tissue present following a cutaneous full-thickness incisional injury. There is evidence that ultimate wound strength is double for dogs than for cats, thus suggesting a difference in the maturation and production of collagen [16,17].

As with pigs, research involving dogs and cats is regulated by the Animal Welfare Act and associated regulations. Specific housing, socialization, and enrichment guidelines are provided within these documents as well as the National Research Council's *Guide for the Care and Use of Research Animals*.

ANIMAL MODELS OF ACUTE WOUNDS

An acute skin wound is an injury to the skin that occurs suddenly rather than over time when uncomplicated, it heals at a predictable and expected rate according to the normal wound healing process and may result in a return to anatomic and functional integrity [18]. The return to anatomic and functional integrity is dependent on the type and location of the acute wound. Skin wounds whether subacute, acute, or chronic can be characterized by the mechanism of injury such as incised, contused, lacerated, puncture, avulsion, and burn wounds. Individual characteristics of animals will respond differently to each insult, so it is imperative that the correct animal model be used to study the effects of various test articles on each type of wound. Acute wound healing models generally fall into three categories that are used extensively. Those categories are partial- or full-thickness burn, incisional, and excisional wound models. The models are well characterized [19] and will be touched upon briefly below.

Incisional

Skin wounds caused by incisions are of particular interest to surgeons and clinicians who perform operations. The use of this model can be used to investigate the wound healing process in correlation with a novel product that may aid healing time and decrease scarring [20]. Other analyses of the outcomes of various treatments on such wounds would include tensile strength, histology, and aesthetics [21]. Incisional wounds in animal models are primarily created by scalpels and scissors. Tools such as electrocautery (Bovie knife) and LASERs are also frequently used in surgery and should also be investigated with animal models to further correlate the effect of the specific procedure. The use of electrocautery can cause thermal injury, which will add to tissue disruption and damage that is not present in wounds caused by sharp blades [21]. The final result must be weighed against the effect of better hemostasis with the use of electrocautery.

Excisional

Excisions that involve removing the epidermis and part of the dermis or epidermis and the entire dermal layers are generally termed partial- or full-thickness

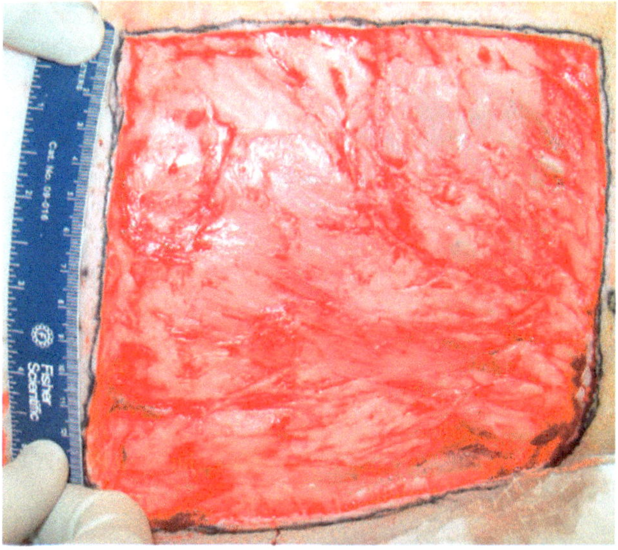

FIGURE 1 Full-thickness excisional wound in porcine model. Note tattoo along the edge of the wound to detect any wound contracture [23].

surgical excisions, respectively [22]. The general depths of an excision can range from 100 μm to 1500 μm thick, depending upon the animal species and the location of the wound on the body, i.e., the dorsal skin of the swine is much thicker than the ventrum, thus a full-thickness excision on the dorsal skin may be as much as 1500 μm, whereas on ventrum it may only be 250. The tools used to achieve an excision include a biopsy punch, surgical scissors, laser, scalpel, or dermatome. A partial- or full-thickness burn wound may be obtained with hot water, flame, or hot metal. The thickness of burns is time and temperature dependent. Differing from the incisional wound, the excision site can be harvested or biopsied to obtain cells, tissue, RNA, exudates, and histological specimens. Using the pig for such studies allows for a greater surface area, thus allowing for an increased amount of tissue area. Albanna, of the Wake Forest Institute of Regenerative Medicine, created an excisional wound in swine for in situ bioprinting of skin cells [23] (Figure 1).

Burns

Partial- and full-thickness burn models can be made by heating a weighted conductive object, by scalding the skin, or with flame. The key to studying treatments of burns is to create a depth that is as consistent as possible, thus flame methods are really not suitable. As with the excisional method, using a large animal is recommended, although many studies have utilized the rat. Burn injuries create a large zone of necrosis that extends over the point of contact injury, which progresses over a 2–3 day course (the "crust effect") [24]. Due to this progressive nature of the lesion, the burn researcher must adhere to very stringent procedures in terms of temperature and time of injury. If those conditions vary by even a small amount, the amount of injury can vary considerably, thus yielding nonreproducible results. In 1968, one of the first burn wound models was developed by Mason and Walker [25] using the scalding method. This hot water method was later validated histologically and refined by George Winter in 1975. Using 60 °C for 60 s producing a reproducible partial-thickness burn (about 300 μm deep), whereas 80 °C for 15–20 s produces a full-thickness burn wound that is immediately deep to the deepest hair follicles on the lateral surface, over the ribs of a 15-week-old Yorkshire pig, thus making it a full-thickness burn. Since that time, models of burn wounds include hot water (80 °C for 15–20 s), hot metal plates (250 °C), steam, hot oil, gas torches, radiant energy, and ignited alcohol gauze [24,26].

Mostly, all of these other methods suffer from lack of reproducibility. The gold standard remains hot water or a heated brass plate in a PEG:H_2O azeotrope of known temperature and weight.

ANIMAL MODELS OF SCARLESS WOUND HEALING

Axolotl and the African Clawed Frog

Although not a typical animal model of wound healing, axolotls have been extensively studied for their regenerative and healing capabilities. Axolotls and other salamanders have tight skin like humans, but unlike other vertebrates, have the ability to perfectly regenerate multiple internal and external tissues without the formation of scar tissue. Axolotls are described as a neotenic amphibian, indicating the species does not metamorphose from its aquatic form. When forced into metamorphosis, the animal can still regenerate limbs, although at a slower rate [27]. Elucidating the pathways that lead to regeneration, as opposed to scar formation and wound contraction, in salamanders will hopefully enable the development of better therapeutics in human medicine [28].

The African clawed frog, *Xenopus laevis*, like the axolotl, also has the ability to heal skin injuries without scar formation as a froglet. Like salamanders, blastema formation occurs during the healing process that contributes to the animal's ability to regenerate multiple tissues. The epidermis heals by regeneration, while the dermis appears to heal with repair. As the frog matures, the regenerative capacity diminishes, and wound repair occurs with tissue fibrosis. However, the healed skin of both axolotls and African clawed frogs following treatment with bleomycin results in the formation of fibrosis, suggesting the pathways that lead to scar formation are present but not active in both species [29].

Zebrafish

Zebrafish are an ideal model for wound healing, given their small size, cost, genetic screening potential, and environmental manipulation capabilities.

To study chemical inhibitors, one can simply add these agents directly to the water tank to evaluate their effects on wound healing.

Zebrafish are similar to salamanders in their response to cutaneous injury. In full-thickness wound models, reepithelialization occurs within hours and is completed by 24 h. As in mammals, granulation tissue forms at the injury site and is present between days 2 and 10. It is unclear at this time how the granulation tissue in zebrafish potentially contributes to reepithelialization without scar formation compared with humans. At ~1 month, the lesions are healed and are similar to unaffected skin, with normal epidermal and dermal layers with minimal scarring [30].

Spiny Mouse

The spiny mouse, *Acomys kempi*, has the natural ability to heal large wounds without scar formation, suggesting a possible mammal model for skin regeneration with minimal collagen deposition. In one study, small 4 mm wounds reepithelialized in 3 days. In *Mus* species, reepithelialization fails to occur in the same time frame. In addition, spiny mice also regenerated adnexal tissues, such as hair follicles and sebaceous glands [27].

Nude Mice

One commonly used mouse model is the FOXN1-deficient (nude) mouse, which lacks mature T lymphocytes and is deficient in hair. In partial-thickness wound healing projects, the nude mouse is often the recipient of grafts from other mouse strains or other species. In several studies by Barbul et al., it was shown that nude mice heal skin injuries differently than their wild-type counterparts. Tensile strength of the incisional wound repair site was stronger in nude mice than in wild-type mice [31]. Also, nude mice healed full-thickness incisional wounds faster, with earlier deposition of granulation tissue. At approximately 1 month post skin injury, the wounds of the nude mice were completely healed and deficient in collagen fibers compared to wild-type mice. Other factors thought to contribute to the nude mice's ability to promote scarless wound healing include higher levels of hyaluronic acid, lower levels of collagen, and altered levels of matrix metalloproteinases [32,33]. Finally, in another study by the same group, it was observed that in the punched ear model, nude mice formed blastema structures, commonly observed in limb regeneration models in salamanders and zebrafish [34]. It is unclear if rowett nude (RNU) rats with the same genetic mutation exhibit the same regeneration capabilities.

Select Mammalian Fetuses

Although most juvenile and adult forms of mammals are incapable of tissue regeneration, fetuses from certain mammalian species are capable of healing wounds without scar formation. After injury, a low level of inflammation is present, which is thought to promote regeneration as well. As in salamander blastema formation following injury, the mammalian fetus expresses high levels of collagen III during the repair process. Other factors thought to contribute in the fetus' regenerative potential include higher levels of tenascin C and TGFβ3 [35].

ANIMAL MODELS OF CHRONIC WOUNDS

A chronic wound is a wound that does not heal in the normal healing stages. Chronic wounds seem to be halted in one or more of the phases of wound healing. Chronic wounds are often characterized by hyperglycemia, persistent inflammation, and growth factor and cytokine receptor deficiencies, which lead to impaired progenitor cell recruitment, angiogenesis, and delayed epithelialization [36]. The resulting poor vascularization prevents oxygen and fibroblast infiltration that are required for reepithelialization. Chronic wounds are diverse in their etiology, but are usually assigned to one of three clinical categories: leg ulcers, diabetic foot ulcers, or pressure ulcers, commonly known as decubitus ulcers or bedsores [37]. There is no ideal animal model of this type of chronic wound [38]. The chronic wound is uncommon clinically in both domestic and laboratory animals; additional procedures are needed to attempt to reproduce the human comparable lesions in an animal model. Humans suffering from a chronic wound commonly suffer from other comorbidities that further complicate model development. Since local hypoxia is a cause of chronic wounds, most animal models attempt to recreate this condition by creating an acute wound which is then subjected to ischemia, pressure, reperfusion damage, or diabetic conditions.

The Pressure Ulcer Model

Pressure ulcers are chronic wounds, also known as decubitus ulcers or pressure sores. They generally occur over bony prominences, resulting in ischemia, cell death, and tissue necrosis [39]. Apart from ischemia as a result of applied external pressure, lymph flow in the local area is also inhibited. When lymph flow is inhibited, there is an accumulation of metabolic waste products and enzymes that result in tissue necrosis [40]. Animal models that allow wounded tissue to be reperfused with blood following hypoxia might better compare to human pressure ulcers or chronic wounds in which perfusion has been restored. The reperfusion of ischemic tissue is required for its survival; but secondary to the tissue damage, increased inflammatory mediators and free oxygen radicals are released into the area, thus causing additional injury. The most common animals that have been used for this model are rats, mice, rabbits, and pigs. The skin is much thinner in elderly humans; and thus, rats can be used due to their thin loose skin [40]. One of the more common methods used in the rat is the magnet ischemia–reperfusion model. A steel plate is inserted beneath the dorsal skin, and a magnet is applied over a number of cycles to generate the "ulcer" [37]. In the pig, a model was developed by applying pressure via a cast placed on the scapula [41]. Cast placement enabled the animal to move freely and cycle through ischemia and reperfusion which mimics bedsores.

Chronic Ischemic Wounds

Complications of the vascular system are responsible for ischemic conditions leading to the chronicity of the acute wound. A model that produces severe ischemia can be induced in the mouse by the transverse rectus abdominis myocutaneous flap method, which transects either the superior or inferior epigastric vessels, generating severe necrotic skin wounds within 1 week [42]. Less severe ischemic models have utilized the bipedal flap model which induces ischemia via surgical incisions that sever the blood flow to an excisional wound. The use of silicone sheets placed strategically in the wound in order to inhibit vascular ingrowth from underlying vessels can also be used to further create necrosis [43,44]. Validation of the hypoxic state of the tissues is characterized by an increase in inflammatory markers [44]. This procedure has been used in the rat to test the effects of topical medications [45]. In large animals, the pig is again the most practical test system for this procedure in preclinical studies. As mentioned above, porcine skin offers a closer anatomical comparison to human skin than do loose-skinned rodents and heals primarily through reepithelialization with minimal contraction [38].

The rabbit ear ulcer model has also been developed to mimic an ischemic wound. Ischemia is created by ligating two of the three arteries that supply the ear prior to making a 6-mm-punch biopsy surgical wound down to the cartilage to create a full-thickness wound that has very little lateral vascular supply [46].

Diabetic Animal Wound Models

There are many animal models of diabetes [47,48]. Three of the most commonly used strains of mice to model chronic wound healing in individuals with diabetes are the Akita, NONcNZ010, and *db/db* mice. Alternatively, diabetes can be chemically induced in rodents, nonhuman primates, rabbits, and pigs by streptozotocin or alloxan leading to death of pancreatic β cells [47]. Diabetes is naturally occurring in the cynomolgus macaque [48]. In a direct comparison of these models, a recent study showed that the NONcNZ010 mouse model had impaired healing subsequent to wounding [49]. By combining independent diabetes risk-conferring quantitative trait loci from two unrelated strains of New Zealand obese mice with nonobese nondiabetic mice, this new novel strain was created [50]. In only 1 year's time diabetic nephropathy was observed in some of the male mice [51]. After the animals are confirmed diabetic, the animals can be subjected to acute wound injuries for testing of various treatment modalities to increase healing time.

CONCLUSION

A summary of the key similarities and differences between animal models and humans is presented in Table 1. When comparing animal models, it is important to know that porcine models were comparable to humans 78% of the time. This is compared with 53% and 57% for small mammal and in vitro studies, respectively.

TABLE 1 Similarities and Differences in Wound Healing between Animals and Humans

Species	Common Name	Mammalian Skin	Type Second-Degree Closure
Mus musculus	Mouse	Loose	Contraction
Rattus norvegicus	Rat	Loose	Contraction
Sus scrofa	Pig	Tight	Reepithelialization
Canis familiaris	Dog	Trunk loose, extremities tight	Trunk—contraction; extremities—reepithelialization
Felis domesticus	Cat	Trunk loose, extremities tight	Trunk—contraction; extremities—reepithelialization
Ambystoma mexicanum	Axolotl	Not applicable	Regeneration
Danio rerio	Zebrafish	Not applicable	Regeneration
Xenopus laevis	African clawed frog	Not applicable	Regeneration in younger animals, tissue scarring in adults
Homo sapiens	Human	Tight	Reepithelialization

A number of factors influence these results including method of wound creation, treatment formulations, schedules and strategies, wound exposure, infection rates, and individual animal variability [10]. The correct selection of species for animal modeling will ultimately affect the translational outcome for each project.

REFERENCES

[1] Makalowski W, Zhang J, Boguski MS. Comparative analysis of 1196 orthologous mouse and human full-length mRNA and protein sequences. Genome Res 1996;6(9):846–57.

[2] Dorsett-Martin WA. Rat models of skin wound healing: a review. Wound Repair Regen 2004;12(6):591–9.

[3] Wong VW, Sorkin M, Glotzbach JP, Longaker MT, Gurtner GC. Surgical approaches to create murine models of human wound healing. J Biomed Biotechnol 2011;2011:969618.

[4] Benavides F, Oberyszyn TM, VanBuskirk AM, Reeve VE, Kusewitt DF. The hairless mouse in skin research. J Dermatol Sci 2009;53(1):10–8.

[5] Lu C, Fuchs E. Sweat gland progenitors in development, homeostasis, and wound repair. Cold Spring Harb Perspect Med 2014;4(2).

[6] Ito M, Liu Y, Yang Z, et al. Stem cells in the hair follicle bulge contribute to wound repair but not to homeostasis of the epidermis. Nat Med 2005;11(12):1351–4.

[7] Azzi L, El-Alfy M, Martel C, Labrie F. Gender differences in mouse skin morphology and specific effects of sex steroids and dehydroepiandrosterone. J Invest Dermatol 2005; 124(1):22–7.

[8] dos Santos JS, Monte-Alto-Costa A. Female, but not male, mice show delayed cutaneous wound healing following aspirin administration. Clin Exp Pharmacol Physiol 2013;40(2):90–6.

[9] Canesso MC, Vieira AT, Castro TB, et al. Skin wound healing is accelerated and scarless in the absence of commensal microbiota. J Immunol 2014;193(10):5171–80.

[10] Sullivan TP, Eaglstein WH, Davis SC, Mertz P. The pig as a model for human wound healing. Wound Repair Regen 2001;9(2):66–76.

[11] Weinstein G. Comparison turnover time of keratinous protein fractions in swine and human epidermis. In: Bustad L, editor. Swine in biomedical research. Seattle: Frayn; 1966. p. 287–97.

[12] Montagna W. Some particularities of human skin and the skin of nonhuman primates. G Ital Dermatol Venereol 1984;119(1):1–4.

[13] Yao F, Visovatti S, Johnson CS, et al. Age and growth factors in porcine full-thickness wound healing. Wound Repair Regen 2001;9(5):371–7.

[14] Johnson KE, Wilgus TA. Vascular endothelial growth factor and angiogenesis in the regulation of cutaneous wound repair. Adv Wound Care (New Rochelle) 2014;3(10):647–61.

[15] Zhu KQ, Carrougher GJ, Gibran NS, Isik FF, Engrav LH. Review of the female Duroc/Yorkshire pig model of human fibroproliferative scarring. Wound Repair Regen 2007;15(Suppl. 1):S32–9.

[16] Bohling MW, Henderson RA, Swaim SF, Kincaid SA, Wright JC. Cutaneous wound healing in the cat: a macroscopic description and comparison with cutaneous wound healing in the dog. Vet Surg 2004;33(6):579–87.

[17] Volk SW, Bohling MW. Comparative wound healing–are the small animal veterinarian's clinical patients an improved translational model for human wound healing research? Wound Repair Regen 2013;21(3):372–81.

[18] Spear M. Acute or chronic? What's the difference? Plast Surg Nurs 2013;33(2):98–100.

[19] Davidson JM. Animal models for wound repair. Arch Dermatol Res 1998;290(11):S1–11.

[20] Kieran I, Knock A, Bush J, et al. Interleukin-10 reduces scar formation in both animal and human cutaneous wounds: results of two preclinical and phase II randomized control studies. Wound Repair Regen 2013;21(3):428–36.

[21] Charoenkwan K, Chotirosniramit N, Rerkasem K. Scalpel versus electrosurgery for abdominal incisions. Cochrane Database Syst Rev 2012;6:CD005987.

[22] Kiwanuka E, Hackl F, Philip J, Caterson EJ, Junker JP, Eriksson E. Comparison of healing parameters in porcine full-thickness wounds transplanted with skin micrografts, split-thickness skin grafts, and cultured keratinocytes. J Am Coll Surg 2011;213(6):728–35.

[23] Albanna MZ, Murphy SV, Zhao W, et al. In situ bioprinting of autologous skin cells accelerates skin regeneration. J. Tissue Eng. Regener Med. 2012;(6):94–95.

[24] Davis SC, Mertz PM, Eaglstein WH. Second-degree burn healing: the effect of occlusive dressings and a cream. J Surg Res 1990;48(3):245–8.

[25] Walker HL, Mason Jr AD. A standard animal burn. J Trauma 1968;8(6):1049–51.

[26] Boykin JV, Eriksson E, Pittman RN. In vivo microcirculation of a scald burn and the progression of postburn dermal ischemia. Plast Reconstr Surg 1980;66(2):191–8.

[27] Seifert AW, Monaghan JR, Voss SR, Maden M. Skin regeneration in adult axolotls: a blueprint for scar-free healing in vertebrates. PLoS One 2012;7(4):e32875.

[28] Denis JF, Levesque M, Tran SD, Camarda AJ, Roy S. Axolotl as a model to study scarless wound healing in vertebrates: role of the transforming growth factor beta signaling pathway. Adv Wound Care (New Rochelle) 2013;2(5):250–60.

[29] Bertolotti E, Malagoli D, Franchini A. Skin wound healing in different aged *Xenopus laevis*. J Morphol 2013;274(8):956–64.

[30] Richardson R, Slanchev K, Kraus C, Knyphausen P, Eming S, Hammerschmidt M. Adult zebrafish as a model system for cutaneous wound-healing research. J Invest Dermatol 2013;133(6):1655–65.
[31] Barbul A, Shawe T, Rotter SM, Efron JE, Wasserkrug HL, Badawy SB. Wound healing in nude mice: a study on the regulatory role of lymphocytes in fibroplasia. Surgery 1989;105(6):764–9.
[32] Gawronska-Kozak B. Scarless skin wound healing in FOXN1 deficient (nude) mice is associated with distinctive matrix metalloproteinase expression. Matrix Biol 2011;30(4):290–300.
[33] Gawronska-Kozak B, Bogacki M, Rim JS, Monroe WT, Manuel JA. Scarless skin repair in immunodeficient mice. Wound Repair Regen 2006;14(3):265–76.
[34] Gawronska-Kozak B. Regeneration in the ears of immunodeficient mice: identification and lineage analysis of mesenchymal stem cells. Tissue Eng 2004;10(7–8):1251–65.
[35] Redd MJ, Cooper L, Wood W, Stramer B, Martin P. Wound healing and inflammation: embryos reveal the way to perfect repair. Philos Trans R Soc Lond B Biol Sci 2004;359(1445):777–84.
[36] Ansell DM, Holden KA, Hardman MJ. Animal models of wound repair: are they cutting it? Exp Dermatol 2012;21(8):581–5.
[37] Nunan R, Harding KG, Martin P. Clinical challenges of chronic wounds: searching for an optimal animal model to recapitulate their complexity. Dis Model Mech 2014;7(11):1205–13.
[38] Sisco M, Mustoe TA. Animal models of ischemic wound healing. Toward an approximation of human chronic cutaneous ulcers in rabbit and rat. Methods Mol Med 2003;78:55–65.
[39] Kosiak M. Etiology and pathology of ischemic ulcers. Arch Phys Med Rehabil 1959;40(2):62–9.
[40] Nguyen PK, Smith AL, Reynolds KJ. A literature review of different pressure ulcer models from 1942–2005 and the development of an ideal animal model. Australas Phys Eng Sci Med 2008;31(3):223–5.
[41] Sundin BM, Hussein MA, Glasofer S, et al. The role of allopurinol and deferoxamine in preventing pressure ulcers in pigs. Plast Reconstr Surg 2000;105(4):1408–21.
[42] Fang RC, Mustoe TA. Animal models of wound healing: utility in transgenic mice. J Biomater Sci Polym Ed 2008;19(8):989–1005.
[43] Roy S, Biswas S, Khanna S, et al. Characterization of a preclinical model of chronic ischemic wound. Physiol Genomics 2009;37(3):211–24.
[44] Chen C, Schultz GS, Bloch M, Edwards PD, Tebes S, Mast BA. Molecular and mechanistic validation of delayed healing rat wounds as a model for human chronic wounds. Wound Repair Regen 1999;7(6):486–94.
[45] Canapp Jr SO, Farese JP, Schultz GS, et al. The effect of topical tripeptide-copper complex on healing of ischemic open wounds. Vet Surg 2003;32(6):515–23.
[46] Ahn ST, Mustoe TA. Effects of ischemia on ulcer wound healing: a new model in the rabbit ear. Ann Plast Surg 1990;24(1):17–23.
[47] King AJ. The use of animal models in diabetes research. Br J Pharmacol 2012;166(3):877–94.
[48] Wagner JD, Cline JM, Shadoan MK, Bullock BC, Rankin SE, Cefalu WT. Naturally occurring and experimental diabetes in cynomolgus monkeys: a comparison of carbohydrate and lipid metabolism and islet pathology. Toxicol Pathol 2001;29(1):142–8.
[49] Fang RC, Kryger ZB, Buck 2nd DW, De la Garza M, Galiano RD, Mustoe TA. Limitations of the db/db mouse in translational wound healing research: Is the NONcNZO10 polygenic mouse model superior? Wound Repair Regen 2010;18(6):605–13.
[50] Cho YR, Kim HJ, Park SY, et al. Hyperglycemia, maturity-onset obesity, and insulin resistance in NONcNZO10/LtJ males, a new mouse model of type 2 diabetes. Am J Physiol Endocrinol Metab 2007;293(1):E327–36.
[51] Brosius 3rd FC, Alpers CE, Bottinger EP, et al. Mouse models of diabetic nephropathy. JASN 2009;20(12):2503–12.

Chapter 20

Human Skin Bioprinting: Trajectory and Advances

Kyle Binder[1], Aleksander Skardal[2,3]

[1]*Department of Neurology, Wake Forest University, Winston-Salem, NC, USA;* [2]*Wake Forest Institute for Regenerative Medicine, Wake Forest School of Medicine, Winston-Salem, NC, USA;* [3]*Virginia Tech-Wake Forest School of Biomedical Engineering and Sciences, Wake Forest University Health Sciences, Winston-Salem, NC, USA*

Chapter Outline

OVERVIEW OF BIOPRINTING

Tissue engineering seeks to construct biological substitutes to restore and maintain normal function in diseased or injured tissues [1,2]. Normal human tissues are composed of multiple different cell types and extracellular matrix (ECM) components that exist in precise locations to facilitate the biological functions of that tissue. Therefore, creating biological substitutes to restore diseased or injured tissues requires creating complex arrangements of cell types and ECM in ways that allow tissue and organ morphogenesis [3,4]. Traditional tissue engineering techniques, such as manual cell seeding or cell spraying, are incapable of achieving the proper level of spatial resolution to place specific cells and ECM in specific locations. As a result, the field of biofabrication has arisen to meet these challenges.

Biofabrication is a burgeoning field that seeks to combine living and nonliving components of biological structures in a controlled manner. Techniques for controlled microdeposition of materials have been available for decades, as the first dot matrix printers were released for commercial use in the 1970s with the advent of the LA30 (Digital Equipment Corporation, Maynard, MA). In the

Skin Tissue Engineering and Regenerative Medicine. http://dx.doi.org/10.1016/B978-0-12-801654-1.00020-6

1980s, robotic spotting systems were developed for precise microscale deposition of biological substrates for medical diagnostics (MicroFab Technologies Inc., Plano, TX). The field of biofabrication developed in large part due to the development of cytoscribing, which laid the groundwork for precise controlled deposition of proteins and cells [5,6]. In the late 1990s, desktop inkjet printers attained orifice diameters of 50 μm and drop resolutions of 80 μm. With this technological advent, researchers earnestly began to examine the use of ink-jet printing as a low-cost alternative to robotic spotting systems for controlled delivery of both living and nonliving components of biological structures. This gave rise to the field of bioprinting, which seeks to leverage drop-on-demand delivery systems to create biological structures [7–9]. Most bioprinting systems create tissue substitutes by layering two-dimensional (2-D) constructs along a *Z*-axis to create a three-dimensional (3-D) tissue structure. Several examples of 3-D printers and bioprinters are shown in Figure 1. Skin can be modeled as a 3-D structure consisting of multiple 2-D constructs: subcutaneous tissue, dermis, and epidermis, each consisting of multiple cell types arranged in precise spatial configurations. As a result, skin bioprinting is a natural evolution of bioprinting technology.

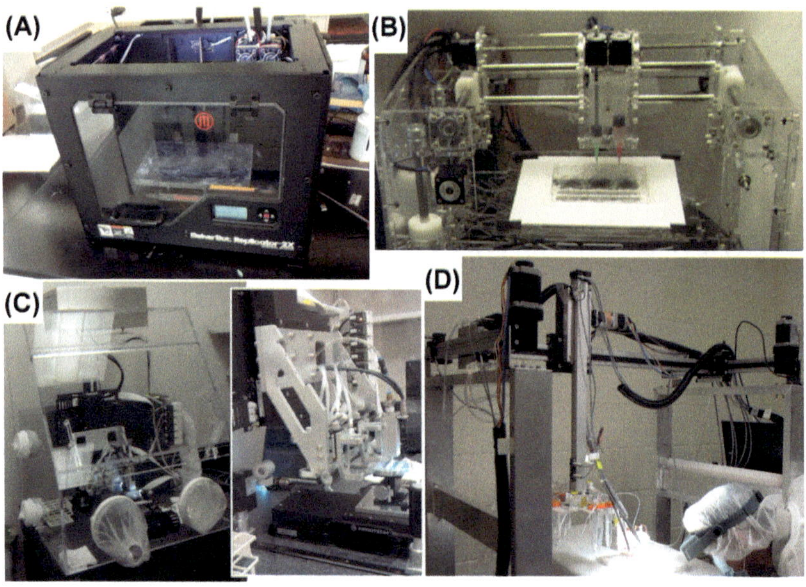

FIGURE 1 Examples of 3-D printers and bioprinters. (A) The MakerBot Replicator 2X, a widely available, low-cost 3-D polymer/resin printer for creating nonbiological 3-D structures. (B) The open source Fab@Home Model 2 3-D printer, which has demonstrated use in a variety of applications, including printing of nonliving organ models for teaching and training purposes [57,58] and living 3-D tissue constructs [33–35]. (C and D) A custom designed and built (C) 3-D bioprinter and (D) skin bioprinter from the Wake Forest Institute of Regenerative Medicine [10,37].

SKIN BIOPRINTER DEVELOPMENT

Development of a skin bioprinter begins by clearly defining the primary purpose of the bioprinter. Examples include in vitro generation of skin substitutes, in situ small cosmetic repair, and in situ large wound repair. The primary purpose informs the design goals and system requirements for the two major components of the skin bioprinter: the movement system and delivery system. These components are inextricably linked. The authors have developed skin bioprinters for in vivo wound repair [10]. The primary purpose of our first bioprinter was proof-of-concept delivery of fibroblasts and keratinocytes in small wounds. Our second bioprinter was designed to deliver fibroblasts and keratinocytes for rapid repair of large wounds. Despite their similar purposes, these bioprinters had unique requirements for their movement and delivery systems, such as overall working ranges in each axis. The relationship of a device's primary purpose to its movement and delivery systems is explored in this section. The principle design parameters for a skin bioprinter can be divided in terms of the delivery and movement systems.

Delivery Systems

Parameters for the delivery system include spatial resolution, throughput, integration with materials to be printed, and the delivery matrix. *Spatial resolution* is the smallest 2-D area that can be bioprinted; in some applications this is also the *droplet size*. *Throughput* is the amount of material that can be delivered through the delivery system in a specified amount of time. *Bio-ink* is the main substance that is delivered into a specific area. Bio-ink is composed of materials that can range from cells to growth factors to biomaterials. If the bio-ink is composed of living cells, *viability* is an important parameter, especially with regard to skin bioprinting. *Single cell precision* refers to the ability of a delivery system to reliably place single cells in a specific location. The *delivery matrix* is the carrier material in which the bioprinted materials are delivered, which in some cases can be the bio-ink itself. A delivery matrix often requires a *gelation agent* or chemical reaction capability to form cross-links so it can become semisolid and maintain its 3-D shape.

There are three main types of delivery systems that can be used for skin bioprinting. The first is inkjet bioprinting, of which there are two major categories: thermal and piezoelectric. Both of these systems have a similar overall structure in which a cartridge is filled with bio-ink that is then forced through a chamber to an output orifice. Thermal inkjet bioprinters are typically derived from commercially available desktop inkjet printers. In these applications, commercially available ink cartridges act as the delivery system. The ink from these cartridges can be removed and replaced with a bio-ink [8,9]. Thermal inkjets use a heating element that creates a bubble in the bio-ink, generating pressure that forces the bio-ink through the output orifice. The parameters of the delivery system are highly dependent on the type of ink cartridge used. Spatial resolution, in particular, is highly dependent on the type of inkjet printer and the size of the output orifice. Commercially available desktop inkjet printers are not designed to repeatedly deliver ink in specific locations on a micrometer scale. Instead, they are designed to deliver multiple ink drops in a defined spatial distribution within a specified area. The distribution of ink drops appears solid to the human eye. With regard to bioprinting, this property of inkjet printers means that single cell precision cannot be guaranteed using a modified commercially available inkjet printer. Spatial precision can be manipulated to a certain degree by altering the properties of the bio-ink in relation to the average drop volume of the cartridge. For an HP26 cartridge, a concentration of 7.7×10^6 cells/mL corresponds on average to one cell per droplet [11].

Throughput in modified commercial inkjet printers is highly variable. One of the major difficulties with modifications to inkjet cartridges is their tendency to rapidly clog. This is due in large part to salt deposition within the output orifice as water evaporates due to the thermal heating. It is extremely difficult to clean an ink cartridge after it has become clogged. Some groups have experimented with adding chemical chelators to the bio-ink in an effort to alleviate this issue [11]. Throughput is the major concern that prevents thermal inkjet printing from being used for skin bioprinting. Cell viability is also a concern due to the mechanics of thermal inkjet printing. The thermal element within the cartridge heats the ink to 300 °C for about 2 μs which raises the temperature of the bio-ink by about 4–10 °C [12]. Cell survival has been found to be 70–90%, although some groups have determined that cells require a recovery period after bioprinting to restore membrane integrity [12–15]. Compromise of cell membrane integrity during the printing process can actually be leveraged for gene transfection [16], but it is not desirable for skin bioprinting. Thermal inkjet printers are useful for their low cost and ease of modification; however, the issues with spatial resolution and throughput prevent their use as skin bioprinters.

In contrast to thermal inkjet printing, piezoelectric inkjet printing does not alter the bio-ink to create pressure inside the cartridge. Piezoelectric printers often use either acoustic waves or a material that changes shape in order to generate pressure inside the cartridge that forces bio-ink through the output orifice [17,18]. These types of systems are generally more controllable than thermal inkjet printers. Acoustic waves in particular are controllable through the manipulation of the wave parameters including duration, amplitude, and frequency. Thus, spatial resolution in these systems is variable to a certain extent but is far more controllable than modified thermal inkjet printers. Single cell precision is excellent due to the ability to control cell droplets. Throughput is also improved: these systems can be modified to remove the output nozzle, thus avoiding the issue of clogging that is prevalent with thermal inkjet printers [18]. In addition, there is no deposition of salts within the output orifice as the system does not evaporate the water in the bio-ink. Throughput is also improved compared to thermal inkjet printers through the use of parallel cartridges, which can deliver multiple cell types simultaneously. Cell viability has been 80–90% in studies of encapsulated cells delivered through acoustic bioprinting [19].

Even though piezoelectric inkjet bioprinting has several advantages over thermal inkjets for skin bioprinting, both types of inkjet printing suffer from limitations regarding delivery matrices [20,21]. The greatest limitation is the viscosity of the delivery matrix. In order to form a droplet, the delivery matrix must be liquid inside the cartridge and become solid once it has exited the printhead. This transformation is achieved using a gelation agent. Practically, this limitation means that delivery matrices for inkjet bioprinting are constrained to a maximum viscosity of about 10 cP [20]. Delivery matrices are also constrained by the gelation agent required to form a solid or semisolid gel. Gelation agents typically form cross-links between molecules of the matrix and can range from chemical reactions to ultraviolet radiation to pH changes. However, some gelation agents are toxic to cells, which further limits the available delivery matrices [22]. Despite these limitations, inkjet bioprinters are popular due to their wide availability, as well as their relatively low cost. A method for modifying desktop inkjet printers is described in this chapter.

Laser-assisted bioprinting (LAB) has also been used for skin bioprinting. Originally designed to pattern metals for electronic circuits, these systems comprise a variety of techniques that transmit bio-ink from a reservoir to the receiving substrate. Typically these devices include a laser with pulse and focus control, a laser-transparent print ribbon, and a receiving substrate. The laser-absorbing layer usually consists of glass with a covering of gold or titanium on which the bio-ink is attached, termed the ribbon. Laser pulses are focused into this absorbing layer, which evaporates, causing a buildup of pressure that transfers the bio-ink from the ribbon to the receiving substrate [23–25]. This receiving substrate is the equivalent of the delivery matrix used for inkjet bioprinters. One of the main advantages of LAB is the lack of a nozzle that most other delivery systems require. Without a nozzle, the clogging issues that plague inkjet delivery systems are completely obviated. This also allows for a wider range of bio-inks over inkjet bioprinters, as viscosities up to 300 cP have been bioprinted. Spatial resolution can be controlled with adjustments to the laser pulse characteristics including wavelength, pulse duration, repetition rate, as well as energy and beam focus diameter. In addition, bio-ink properties including viscosity, thickness, and surface tension can be adjusted [26]. Throughput is low in these systems for two reasons. First, each ribbon needs to be independently prepared for LAB, which lowers the throughput. Second, the printing speed is highly dependent on the laser pulse rate. This can be improved with multiple LAB systems in parallel, but this adds significantly to the expense of the system. Unfortunately, single cell precision suffers from limitations inherent to LAB. The spatial distribution of cells is random, similar to the way in which inkjet printers deliver ink in a spatial distribution. However, single cell precision can be approximated by using a high concentration of cells on the ribbon so that on average, the statistical distribution of bioprinted cells is one cell per droplet.

This is easier to achieve than inkjet bioprinting because LAB is capable of positioning femtoliters of cell suspensions with high resolution [27]. Viability is excellent with LAB, as cells are not harmed by the printing procedure [27–30]. However, metallic residue is a byproduct of the printing mechanism, and metallic particles have been found in the bioprinted constructs, which may be toxic to cells [31]. LAB has several advantages over inkjet bioprinting; however, its main disadvantage is its high cost compared to desktop inkjet printers, which are readily available.

The final type of delivery system is the broad category of extrusion delivery systems. There are too many different types of extrusion delivery systems to discuss individually [32]; thus, mechanical and pneumatic delivery systems are examined in this section as they relate to skin bioprinting. In general, extrusion delivery systems operate by dispensing a controlled amount of bio-ink either continuously or in discrete droplets. Both mechanical and pneumatic delivery systems can delivery bio-ink continuously or discretely through the use of force applied to the bio-ink to push it through the printhead. Mechanical systems typically use a screw or piston to apply force to the bio-ink. Pneumatic systems use air or hydrostatic pressure to apply a constant force to the bio-ink where the extrusion is controlled by valves that open to allow bio-ink to pass through the printhead. Spatial resolution is limited by the droplet size in these systems and is highly dependent on the type of system used. Bio-ink is virtually limitless as extrusion delivery systems can be tailored to deliver virtually any cell type or biomaterial. In addition, the cell density of extrusion delivery systems is higher than what can be performed with inkjet or LAB. Compared to other types of delivery systems, throughput in extrusion-based systems is excellent. Viability is dependent on the type of system but is generally high, as cells typically are exposed to lower levels of shear stress during the printing process. Single cell precision is dependent on the type of system but is typically not as good as inkjet or LAB. This is the main disadvantage of extrusion bioprinting compared to other types of delivery systems. Delivery matrices and gelation agents are also virtually limitless with extrusion-based delivery systems due to the customization available with these types of systems, and the wide variety of biomaterials that can be customized to integrate with bioprinting systems, be it for tissue construct fabrication [33–35] or healing of skin [36,37]. The delivery system can be tailored to virtually any desired delivery matrix or gelation agent. Furthermore, extrusion delivery systems can deposit large quantities of cells at once to self-assemble into a bioprinted structure [38–42].

As described above, the primary purpose of our skin bioprinter was to deliver fibroblasts and keratinocytes for rapid repair of large wounds. We chose a form of extrusion bioprinting, because throughput was paramount to all other design parameters for our primary purpose [10]. Specifically, from a clinical standpoint, it was important to have a system with the capability to deliver enough bio-ink to cover wounds in a relatively short amount of time. The delivery system used pneumatic pressure and a series of valves connected to different cartridges.

Our initial experiments used only cartridges of fibroblasts, keratinocytes, and gelation agent; however, the cartridge-based design allows bioprinting of virtually any cell type or biomaterial that can be packaged into a cartridge. The same system was extended to bioprint amniotic fluid stem cells for skin regeneration without any modifications to the delivery system [37] and has been further tested for compatibility with a variety of hydrogel vehicles [36].

In summary, each type of delivery system has advantages and disadvantages. Skin bioprinting can be performed in many different ways, and the choice of delivery system influences the capabilities of the bioprinting system. The delivery system, in turn, is influenced heavily by the delivery matrix chosen for the skin bioprinting application.

Delivery Matrices as Applied to Skin Bioprinting

Delivery matrices vary greatly by the delivery system. These matrices fall within the general moniker of *biomaterials*, as they must coexist with biological components such as cells, growth factors, or cytokines, and in vivo environments such as wounds. The term *biomaterials* comprises a wide variety of materials that is constantly evolving, ranging from cell-supportive pliable hydrogels, to robust metal or ceramic implants, from extremely small nanoparticles for drug delivery and imaging, to advanced medical devices such as pacemakers, hearing aids, and artificial hearts. As research in materials chemistry and biological sciences continues to expand, so will the number classifications of biomaterial types [43,44].

Clearly, only a subset of these biomaterials are suitable as delivery matrices and bio-inks for bioprinting. In the context of bioprinting and biofabrication, biomaterials generally are limited to two primary categories. The first category is that of curable polymers that result in mechanically robust materials that act as scaffolding to printed constructs. These are typically employed in 3-D construct fabrication, rather than bioprinting of skin. Many such materials typically require high temperatures or toxic solvents to facilitate printing, and therefore are not appropriate for printing together with cells or in situ into cellularized environments. Cells are typically seeded onto and into the scaffolds after fabrication. The second category of biomaterials is that of soft biomaterials such as hydrogels, generally with a high water content, inside of which cells are capable of residing. These can be comprised of synthetic or natural polymers, and do not possess the same levels of mechanical properties as curable support polymers. The inherent characteristics of these different printing materials, including mechanical properties, melting points, and available chemistries for cross-linking and functionalization make up the factors responsible for successful bioprinting. Additionally, these parameters can be harnessed to yield a virtually limitless toolbox of delivery matrix and bio-ink varieties, most of which are hydrogel materials, and many of which are suitable for bioprinting.

However, a common hurdle with respect to the use of biomaterials in bioprinting is that most biomaterials were not originally designed specifically for use in bioprinting. As such, a great amount of work has focused on adapting more traditional materials by chemical and mechanical manipulations to better integrate with bioprinting processes and hardware, while maintaining non-toxic and cell-supportive characteristics. To be considered cell supportive, these hydrogels must not induce toxicity in cells and should provide cell-binding motifs to allow for cell adherence, be they innate to the base materials or through chemical modification steps. Techniques to generate cross-linking reactions can be designed to be noncytotoxic, allowing 3-D encapsulation of cells within the hydrogel polymer networks at the time of gelation. This is becoming exceedingly important as there is an increasing movement from 2-D to 3-D systems in regenerative medicine research and application [45]. All in vivo tissues are 3-D in nature, and bioprinting should echo this characteristic, implementing a 3-D approach [46]. Hydrogel biomaterials fall into one of two major categories: synthetic hydrogels, which employ polymers that are synthesized in the laboratory, or naturally derived hydrogels, which employ polymers, often polysaccharides, but can also be comprised of peptides or proteins, purified from natural sources and are often further manipulated in the laboratory by chemical means. Common examples of synthetic hydrogels include polyethylene glycol (PEG)-based materials, such as PEG diacrylate (PEGDA), and polyacrylamide (PAAm)-based gels. Examples of naturally derived hydrogels that are commonly used in the laboratory, and have been explored for bioprinting, include collagen, hyaluronic acid, gelatin, alginate, and fibrin. In general, synthetic materials allow for increased control over molecular weights and distributions, as well as cross-linking densities, supporting precise control of mechanical properties such as elastic modulus E'. Conversely, naturally derived materials may be more difficult to manipulate into specific ranges of physical properties, but often have an innate bioactivity through biological peptide sequences or conformational motifs that cells interact with, improving tissue integration and biocompatibility. An overview of some of the materials used as delivery matrices and bio-inks in bioprinting are summarized in Table 1. However, it should be noted that this list is not exhaustive; many other materials suitable for bioprinting exist, and others are currently being developed.

Matrices specific for skin bioprinting have been evaluated. As described above, the skin bioprinter developed by our group employs pneumatic pressure and a series of valves connected to different cartridges to facilitate delivery matrix deposition. Initial experimental procedures employed a two-part fibrin–collagen delivery matrix. Specifically, the fibrin-precursor fibrinogen was mixed with soluble collagen Type I yielding the first part, and a solution of thrombin comprised the second. These solutions were printed separately, gelling upon deposition after coming into contact with one another [37,47]. While useful, this fibrin–collagen gel was likely not the optimal material for this application. It required two components to be delivered, relying on diffusion for mixing, and as such, suffered from not polymerizing immediately in situ. In a wound environment containing irregularities in structure or convex surfaces, this resulted

TABLE 1 Examples of Commonly Used Biomaterials Employed in Tissue Engineering and Bioprinting Applications. Also Described Are Specific Modification of Some of the Base Materials, Crosslinking Methods, Crosslinking Times, Pros, Cons, Common Applications, and Ease of Adaption to Bioprinting

Hydrogel	Modification	Cross-linking Method	Cross-linking Speed	Pros	Cons	Common Applications	Ease of Adaption to Bioprinting
Collagen	–	Hydrophobic bonding	0.5–1 h	• Naturally cell adherent • Major component of native ECM	• Slow gelation • Can be associated with tissue fibrosis	• Implants • Cell encapsulation • Substrate coating • Dermal substitutes	Difficult
Fibrin	–	Thrombin-catalyzed fibrin polymerization	Seconds	• Fast gelation • Cell adherent	• Difficult to control geometry due to fast gelation	• Cell delivery • Cell encapsulation • Surgical glue	Easy (two-part delivery)
Hyaluronic acid	Thiolated	Thiol group cross-link	15–30 min	• Commercially available in kit form with gelatin for cell adherence • Mechanical properties can be modulated by cross-linker geometry and MW	• Generally low mechanical properties	• Cell encapsulation • Cell delivery • Wound healing	Difficult (but alternative cross-linking methods are available; see next)

Continued

TABLE 1 Examples of Commonly Used Biomaterials Employed in Tissue Engineering and Bioprinting Applications. Also Described Are Specific Modification of Some of the Base Materials, Crosslinking Methods, Crosslinking Times, Pros, Cons, Common Applications, and Ease of Adaption to Bioprinting—cont'd

Hydrogel	Modification	Cross-linking Method	Cross-linking Speed	Pros	Cons	Common Applications	Ease of Adaption to Bioprinting
Hyaluronic acid	Thiolated	UV Photopolymeration (thiol-ene)	Seconds	• Easily controllable fast gelation • Mech. Properties can be modulated by cross-linker geometry and MW	• Generally low mechanical properties	• Cell delivery • Bioprinting • Wound healing	Easy (UV control)
Hyaluronic acid	Methacrylated	UV Photopolymeration	Minutes	• Gelation speed modulated by UV intensity	• Low mechanical properties	• Cell encapsulation	Easy (UV control)
Gelatin	–	Temperature- based hydrophobic bonding	Seconds - minutes	• Naturally cell adherent	• Unstable at physiological temperature	• Cell encapsulation	Easy (cooled stage)
Gelatin	–	Glutaraldehyde	Hours	• Naturally cell adherent • Stable after cross-linking	• Cross-linking must be accomplished prior to addition of cells	• Scaffolds and films for cell seeding	Difficult

Sodium alginate	–	$CaCl_2$ ion exchange	Seconds	• Easy to create gel microspheres • Inexpensive	• Difficult to control geometry due to fast gelation • Not cell adherent without modification • $CaCl_2$ can induce toxicity with prolonged exposure to cells	• Cell encapsulation in microspheres	Easy (two-part delivery)
Silk (fibroin)	–	Hydrophobic formation of semi- crystalline structures	Minutes to hours	• Stable • Strong mechanical properties	• Slow degradation limits use to more permanent implantations	• In vivo implantation	Medium (can rely on addition of other materials for gelation)
Polyethylene glycol (PEG)	Acrylated (PEGDA)	UV Photopolymerization	Minutes	• Easy to control mechanical properties by changing MW	• Not cell adherent without modification	• Cell encapsulation • Cell delivery	Easy
Silicone	–	Temperature-accelerated curing	Hours	• High stability	• Lacks natural components	• Nonadherent dressings • Microfluidic devices	Difficult

in some sloughing off of the delivery matrix. Furthermore, collagen typically contracts during gelation, and the Type I isoform is also the primary component in scarring, which could potentially act as a direct adversary for healthy, non-scarred skin regeneration. Therefore, we assessed a panel of hydrogel types commonly used in various applications in regenerative medicine for characteristics such as gelation times, ease of use, biocompatibility, immunogenicity, and bioprinter compatibility. As expected, the different materials exhibited differing pros and cons, again suggesting that a particular material should be chosen based on the specific target application. However, we did identify additional hydrogel formulations that were suitable for integration in our skin bioprinter and likely good choices for use as delivery matrices in wound healing treatments [36]. One such hydrogel, a UV–photopolymerizable hyaluronic acid and gelatin material is currently being employed in a variety of applications in our laboratory, including skin printing.

Movement Systems

Parameters for the movement system in a bioprinter include movement range, spatial precision, spatial accuracy, and movement speed. *Movement range* is the total distance the movement system can travel in any direction. *Spatial accuracy* is the smallest distance that the movement system can travel. *Spatial precision* refers to the error in spatial accuracy that is influenced by a variety of factors that depend on the type of movement system used. Typical sources of spatial error when creating micrometer-scale constructs include vibrations, drop distance error, and weight applied to the system. Vibrations originating from the movement system can create micrometer errors in a bioprinted construct. These vibrations can be ignored in most applications, but constructs requiring micrometer precision can be significantly altered by such vibrations. Drop distance error refers to a printhead being held a certain distance above its intended target. Any errors in the direction of the droplet will be magnified as the distance from the printhead to the target increases. Finally, the weight applied to the system can affect the distance the movement system travels with a given input. This error depends on the type of movement system used. Certain types of movement systems have the capability to overcome this error. *Movement speed* is the velocity at which a movement system can traverse a given distance in a specified period of time.

The movement system can be any device that positions a delivery system at a specific location. Virtually any movement system can be used; however, the primary purpose of the system will determine the spatial precision required for the movement system. Examples of movement systems include inkjet printers [7,15], stationary delivery systems with a moving stage, and mobile delivery systems.

Skin bioprinting can be accomplished with modified desktop inkjet printers. There are several drawbacks to using a modified desktop inkjet printer as a movement system for a skin bioprinter. First, the movement range of a modified inkjet printer is limited because the printhead is fixed on two axes. The length of the skin construct is limited to the length of the printhead because the printhead is fixed and cannot be altered. Second, the width of the construct axis is limited to the width of the inkjet printer. Again, this axis cannot be altered. Theoretically, there is no limit to the height of the bioprinted construct; however, in practice, the delivery matrix and gelation agent used to create the construct limits the height of the construct. Spatial accuracy depends on the type of inkjet printer used because it is dependent on the inkjet cartridge used. Spatial precision also depends on the type of inkjet cartridge used. Inkjet cartridges are designed to place drops of material in a specific area, such that multiple drops create an intended spatial distribution. Such cartridges are not designed to repeatedly deliver material to an exact point. Researchers must account for this error if using modified inkjet printers to bioprint skin constructs. Movement speed is excellent along one axis because this type of system leverages the preexisting hardware of the inkjet printer. Given these limitations, we have found that modified inkjet printers are most useful for proof-of-concept testing of new methods for skin bioprinting but are generally not useful for creating in vitro or in vivo skin constructs.

We have previously described a method for inexpensive modifications of commercially available desktop inkjet printers, which is briefly discussed here [7]. Desktop inkjet printers can create 2-D skin constructs with virtually no modifications. However, creating a 3-D construct requires the addition of a *Z*-axis in the paper tray [8,15,48,49]. The addition of a *Z*-axis allows the printer to create multiple 2-D layers in succession to create a 3-D construct. This is typically performed by adding an electronic elevator to the paper tray that lowers upon receiving a signal to print another layer. Printing successive 2-D layers requires overriding the normal function of the printer, which is not designed to print without paper loaded in the machine. Different printers have different mechanisms for detecting paper; however, generally printers will use a sensor located in the rear of the printer that triggers the next page. It is possible to override the sensor, which precludes the need for reverse engineering the printer hardware. However, this approach is highly dependent on the model of printer used. The major advantage of modifying a desktop inkjet printer is that the hardware is inexpensive and is already quality tested by the manufacturer. However, it is impossible to separate the movement and delivery systems for a modified desktop inkjet printer. Thus, this approach suffers from all of the limitations described above for desktop inkjet delivery systems. To improve the capabilities of the movement and delivery systems, we recommend separating the systems so they can be independently tailored to the primary purpose of the skin bioprinter.

It is possible to create a skin bioprinting system in which the delivery system is stationary and the skin construct moves beneath the delivery system. In this type of system the *X*- and *Y*-axes are attached to a stage that contains the bioprinting area. The delivery system is mounted with a movable *Z*-axis above this stage. The movement system is calibrated to the location of the delivery system and moves the bioprinting area along the *X*- and *Y*-axes until it reaches the desired location. The delivery system activates and bioprints a material at this location. The movement system then transports the printing area to the next desired location beneath the delivery system. Movement range for this type of bioprinting system is improved compared to the modified desktop inkjet printers because the movement system is theoretically not constrained along any axis. Spatial precision is determined by the type of motor used to drive the movement stage. Our initial prototype used stepper motors to move the print area in 1.57 μm increments which could provide single cell precision if necessary. Spatial accuracy is excellent with this type of system because most sources of error can be easily controlled. Vibrations can be minimized because the delivery system is fixed. Drop distance error can be adjusted using the *Z*-axis and can be minimized as necessary. Weight applied to the system did not affect our constructs because we attached an encoder to the stepper motors to ensure that the motors reached their intended destination. Movement speed for this system depends on the type of motor system used. Our system was designed to minimize micrometer errors and thus movement speed was decreased [7,10]. In addition, the entire construct needs to move to a new location for each cell drop, which increased build times compared to other prototypes. When applied to skin bioprinting, we found this system most useful for testing delivery system prototypes because we could switch the delivery system without needing to design an entirely new movement system for each delivery system prototype. However, we chose to focus on delivery systems mounted to a movement system, due to our need for throughput for repair of large skin wounds.

In vivo skin bioprinting currently requires a delivery system mounted to a movement system that is capable of moving in 3-D. The type of movement system required depends on the primary purpose of the skin bioprinter. The primary purpose of our first bioprinter was proof-of-concept testing of in vivo bioprinted fibroblasts and keratinocytes. This device was designed for wounds created in a murine wound model. We did not require micrometer precision for deposition of fibroblasts and keratinocytes in a murine model, as these cell types can self-arrange into appropriate 3-D skin constructs. We used a pressure-based microextrusion delivery system mounted to a movement system comprised of three stepper motors connected to threaded rods. The movement range of this type of system is limited only to the length of each axis, which can be changed according to the primary purpose. Spatial precision is dependent on the types of motors used. Our stepper motors had a spatial precision of 1.57 μm per step. Spatial accuracy in this type of system is dependent on the primary purpose. Again, we did not require micrometer precision for delivery of fibroblasts and keratinocytes. Therefore, we sacrificed spatial accuracy for increased movement speed. We did not need to include a spatial encoder or account for micrometer vibrations. The resolution of 1.57 μm per step was three orders of magnitude smaller than the size of the cell drops; therefore, the positioning errors introduced by the lack of an encoder did not influence the final construct. Movement speed was excellent because we designed our bioprinting system to move to specific locations in small wounds quickly rather than achieving micrometer accuracy. This example illustrates how the primary purpose of the bioprinter influences the design parameters of the movement system.

The first skin bioprinter prototype performed admirably for repairing small skin wounds. We endeavored to create a second bioprinter with the primary purpose of rapid repair of large skin wounds in clinical settings. However, the threaded stepper motors were unable to scale to larger wounds because they were unable to support the weight of a larger printhead while maintaining adequate movement speed. Thus, we developed a larger skin bioprinter using belt-driven stepper motors instead of threaded rods. This type of system allowed us to create large in vivo constructs quickly [10]. Movement range and movement speed were the paramount parameters in this system. We created a mobile lightweight frame where dimensions of the system are large enough to cover the torso of an average patient but small enough to easily pass through most door frames. The movement system consisted of belt-driven stepper motors capable of 100 μm movements. The *X*- and *Y*-axes were attached to the frame and the *Z*-axis was attached to the *X*- and *Y*-axes. This independent *Z*-axis allows the delivery system to track with the curvature of a patient's body. Movement range and movement speed were excellent with this configuration. However, spatial precision and accuracy were sacrificed in this system. We were able to bioprint 100 cm^2 skin constructs consisting of one layer of fibroblasts and one layer of keratinocytes in 45 min, including gelation time [50].

SKIN BIOPRINTING

Laser-Assisted Bioprinting

LAB has been used in several studies to develop skin substitutes. In one study, fibroblasts and keratinocytes were arranged using LAB in two layers in vitro to develop a bilayered skin structure. The delivery matrix in this case was a sheet of Matriderm™ (Dr Suwelack Skin & Health Care, Billerbeck, Germany). The researchers labeled the fibroblasts and keratinocytes using fluorescent cell membrane markers and were able to show that their bilayer model generates a dermis and epidermis [29]. In a separate study, similar bilayer constructs were created in vitro and implanted in vivo in a dorsal skin fold chamber in mice. These skin constructs formed dermis, as well as epidermis, and the implanted cells could be detected in the mouse skin several weeks after implantation [30].

Extrusion Bioprinting

Other researchers have chosen, as we have, to pursue extrusion delivery systems for skin bioprinting. A recent study by Lee et al. has developed an eight-valve delivery system with a three-axis movement system. The delivery system used pneumatic valves with an adjustable dispensing volume. In this system the droplet size can be adjusted by varying the pneumatic pressure and the valve-opening time. This study created in vitro skin substitutes by bioprinting fibroblasts and keratinocytes on collagen layers as the delivery matrix. Nebulized sodium bicarbonate was used as the gelation agent to create collagen gel. These constructs were able to generate dermis and epidermis structures [51].

In addition to bioprinting experience, our laboratory has a track record in pioneering isolation, characterization, and applications of therapeutically potent amniotic fluid-derived stem (AFS) cells [52–56]. However, until recently, no one had explored the use of these human AFS cells in skin bioprinting for wound healing. We have recently demonstrated the use of AFS cells for this application using a full-thickness wound model in mice (Figure 2) [37]. We employed the skin printer described above [10] to deliver AFS cells to wounds within a fibrin–collagen gel vehicle, and compared this treatment to MSCs or no cells delivered in the same gel delivery vehicle. AFS cells and MSCs showed similar potency in topographical wound closure, significantly accelerating wound closure compared to the cell-free control. Upon histological inspection, we observed that the delivery of the stem cells induced significantly greater vascularization in the new tissue,

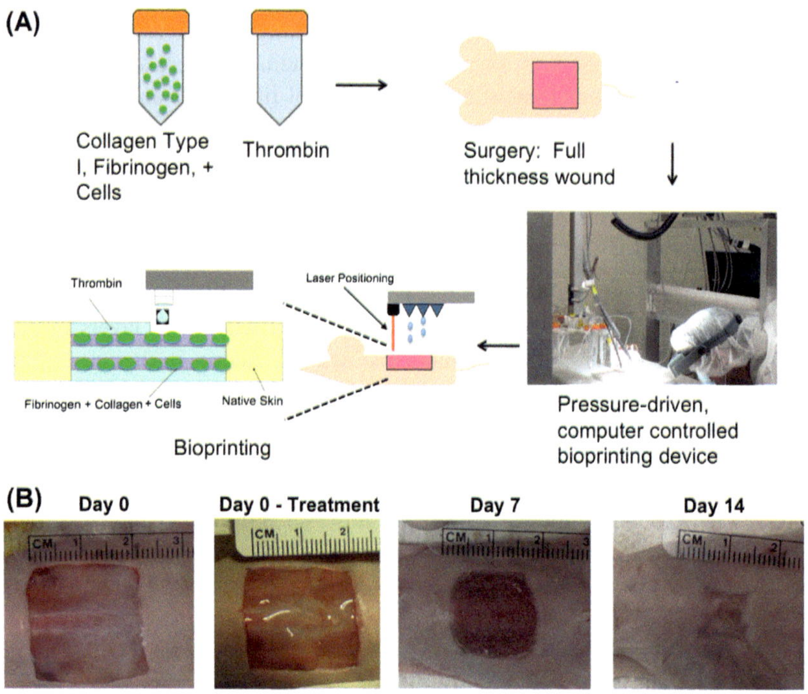

FIGURE 2 Bioprinting of cells and hydrogel materials for skin wound regeneration. (A) A schematic describing the approach by which stem cells are bioprinted in situ order to accelerate healing of a full-thickness skin wound. (B) Skin wound treatment, wound closure, and wound healing over time.

leading to formation of mature vessels, displaying robust smooth muscle actin and a lack of extravasated red blood cells in the surrounding tissue. These results suggested that AFS cells have the potential to be an effective therapy for treating skin wounds, and skin bioprinting is an efficient and effective manner by which to deliver the cells to the wounds. Our team is currently exploring additional delivery matrices to improve the deposition procedure during bioprinting and to provide an environment that can maximize the therapeutic potential of the delivered cells.

CONCLUSION

Extensive skin wounds, including severe burns, can be devastating to patients, even when treated, and place a huge financial burden on society. These patients require a treatment that results in protection of the wound during the healing process and closure of the wound in as short a time as possible. Skin bioprinting is a rapidly growing technology that stands to make a significant impact on wound healing treatments. To succeed, device design and development, delivery matrix and biomaterial optimization, and consideration of biological components such as cells and other bioactive factors need to be combined in a streamlined and efficient manner, merging the expertise of researchers and clinicians from a wide range of backgrounds. As viability, functionality, and safety concerns of skin bioprinting are satisfied from a regulatory standpoint, demand for bioprinted skin as a viable option for patients will likely expand, leading to an important paradigm shift in therapeutic treatments.

REFERENCES

[1] Atala A. Tissue engineering and regenerative medicine: concepts for clinical application. Rejuvenation Res 2004;7(1):15–31.

[2] Atala A. Engineering organs. Curr Opin Biotechnol 2009;20(5):575–92.

[3] Mikos AG, Herring SW, Ochareon P, Elisseeff J, Lu HH, Kandel R, et al. Engineering complex tissues. Tissue Eng 2006;12(12):3307–39.

[4] Yeatts A, Fisher JP. Tubular perfusion system for the long term dynamic culture of human mesenchymal stem cells. Tissue Eng Part C 2010;17(3):337–48.

[5] Klebe RJ. Cytoscribing: a method for micropositioning cells and the construction of two- and three-dimensional synthetic tissues. Exp Cell Res 1988;179(2):362–73.

[6] Klebe R, Thomas C, Grant G, Grant A, Gosh P. Cytoscription: computer controlled micropositioning of cell adhesion proteins and cells. J Tissue Cult Methods 1994;16(3):189–92.

[7] Binder K, Arthur A, Yoo J, Atala A. Drop-on-demand inkjet bioprinting: a primer. Gene Ther Regul 2011;6(1):33–49.

[8] Boland T, Xu T, Damon B, Cui X. Application of inkjet printing to tissue engineering. Biotechnol J 2006;1(9):910–7.

[9] Roth EA, Xu T, Das M, Gregory C, Hickman JJ, Boland T. Inkjet printing for high-throughput cell patterning. Biomaterials 2004;25(17):3707–15.

[10] Yoo JJ, (Winston-Salem, NC, US), Atala A, (Winston-Salem, NC, US), Binder KW, (Winston-Salem, NC, US), Zhao W, (Winston-Salem, NC, US), Dice D, (Yadkinville, NC, US), Xu T (El Paso, TX, US), inventors. Delivery system. United States patent 20110172611. 2011.

[11] Parzel CA, Pepper ME, Burg T, Groff RE, Burg KJ. EDTA enhances high-throughput two-dimensional bioprinting by inhibiting salt scaling and cell aggregation at the nozzle surface. J Tissue Eng Regen Med 2009;3(4):260–8.
[12] Cui X, Dean D, Ruggeri ZM, Boland T. Cell damage evaluation of thermal inkjet printed Chinese hamster ovary cells. Biotechnol Bioeng 2010;106(6):963–9.
[13] Chang R, Nam J, Sun W. Effects of dispensing pressure and nozzle diameter on cell survival from solid freeform fabrication-based direct cell writing. Tissue Eng Part A 2008;14(1):41–8.
[14] Nair K, Gandhi M, Khalil S, Yan KC, Marcolongo M, Barbee K, et al. Characterization of cell viability during bioprinting processes. Biotechnol J 2009;4(8):1168–77.
[15] Xu T, Gregory CA, Molnar P, Cui X, Jalota S, Bhaduri SB, et al. Viability and electrophysiology of neural cell structures generated by the inkjet printing method. Biomaterials 2006;27(19):3580–8.
[16] Xu T, Rohozinski J, Zhao W, Moorefield EC, Atala A, Yoo JJ. Inkjet-mediated gene transfection into living cells combined with targeted delivery. Tissue Eng Part A 2009;15(1):95–101.
[17] Saunders RE, Gough JE, Derby B. Delivery of human fibroblast cells by piezoelectric drop-on-demand inkjet printing. Biomaterials 2008;29:193–203.
[18] Demirci U, Montesano G. Single cell epitaxy by acoustic picolitre droplets. Lab Chip 2007;7(9):1139–45.
[19] Tasoglu S, Demirci U. Bioprinting for stem cell research. Trends Biotechnol 2013;31(1):10–9.
[20] Kim JD, Choi JS, Kim BS, Choi YC, Cho YW. Piezoelectric inkjet printing of polymers: stem cell patterning on polymer substrates. Polymer 2010;51(10):2147–54.
[21] Murphy SV, Atala A. 3D bioprinting of tissues and organs. Nat Biotechnol 2014;32(8):773–85.
[22] Hennink WE, van Nostrum CF. Novel crosslinking methods to design hydrogels. Adv Drug Deliv Rev 2002;54(1):13–36.
[23] Chrisey DB. Materials processing: the power of direct writing. Science 2000;289(5481):879–81.
[24] Colina M, Serra P, Fernandez-Pradas JM, Sevilla L, Morenza JL. DNA deposition through laser induced forward transfer. Biosens Bioelectron 2005;20(8):1638–42.
[25] Dinca V, Kasotakis E, Catherine J, Mourka A, Ranella A, Ovsianikov A, et al. Directed three-dimensional patterning of self-assembled peptide fibrils. Nano Lett 2008;8(2):538–43.
[26] Guillotin B, Guillemot F. Cell patterning technologies for organotypic tissue fabrication. Trends Biotechnol 2011;29(4):183–90.
[27] Barron JA, Krizman DB, Ringeisen BR. Laser printing of single cells: statistical analysis, cell viability, and stress. Ann Biomed Eng 2005;33(2):121–30.
[28] Koch L, Kuhn S, Sorg H, Gruene M, Schlie S, Gaebel R, et al. Laser printing of skin cells and human stem cells. Tissue Eng Part C 2009;16(5):847–54.
[29] Koch L, Deiwick A, Schlie S, Michael S, Gruene M, Coger V, et al. Skin tissue generation by laser cell printing. Biotechnol Bioeng 2012;109(7):1855–63.
[30] Michael S, Sorg H, Peck CT, Koch L, Deiwick A, Chichkov B, et al. Tissue engineered skin substitutes created by laser-assisted bioprinting form skin-like structures in the dorsal skin fold chamber in mice. PLoS One 2013;8(3):e57741.
[31] Schiele NR, Corr DT, Huang Y, Raof NA, Xie Y, Chrisey DB. Laser-based direct-write techniques for cell printing. Biofabrication 2010;2(3):032001.
[32] Smith CM, Stone AL, Parkhill RL, Stewart RL, Simpkins MW, Kachurin AM, et al. Three-dimensional bioassembly tool for generating viable tissue-engineered constructs. Tissue Eng 2004;10(9–10):1566–76.
[33] Skardal A, Zhang J, McCoard L, Oottamasathien S, Prestwich GD. Dynamically crosslinked gold nanoparticle – hyaluronan hydrogels. Adv Mater 2010;22(42):4736–40.

[34] Skardal A, Zhang J, McCoard L, Xu X, Oottamasathien S, Prestwich GD. Photocrosslinkable hyaluronan-gelatin hydrogels for two-step bioprinting. Tissue Eng Part A 2010;16(8):2675–85.
[35] Skardal A, Zhang J, Prestwich GD. Bioprinting vessel-like constructs using hyaluronan hydrogels crosslinked with tetrahedral polyethylene glycol tetracrylates. Biomaterials 2010;31(24):6173–81.
[36] Murphy SV, Skardal A, Atala A. Evaluation of hydrogels for bio-printing applications. J Biomed Mater Res A 2013;101(1):272–84.
[37] Skardal A, Mack D, Kapetanovic E, Atala A, Jackson JD, Yoo J, et al. Bioprinted amniotic fluid-derived stem cells accelerate healing of large skin wounds. Stem Cells Transl Med 2012;1(11):792–802.
[38] Mironov V, Kasyanov V, Drake C, Markwald RR. Organ printing: promises and challenges. Regen Med 2008;3(1):93–103.
[39] Mironov V, Visconti RP, Kasyanov V, Forgacs G, Drake CJ, Markwald RR. Organ printing: tissue spheroids as building blocks. Biomaterials 2009;30(12):2164–74.
[40] Mironov V, Kasyanov V, Markwald RR. Organ printing: from bioprinter to organ biofabrication line. Curr Opin Biotechnol 2011;22(5):667–73.
[41] Jakab K, Norotte C, Damon B, Marga F, Neagu A, Besch-Williford CL, et al. Tissue engineering by self-assembly of cells printed into topologically defined structures. Tissue Eng Part A 2008;14(3):413–21.
[42] Norotte C, Marga FS, Niklason LE, Forgacs G. Scaffold-free vascular tissue engineering using bioprinting. Biomaterials 2009;30(30):5910–7.
[43] Williams D. The continuing evolution of biomaterials. Biomaterials 2011;32(1):1–2.
[44] Williams DF. On the nature of biomaterials. Biomaterials 2009;30(30):5897–909.
[45] Prestwich GD. Evaluating drug efficacy and toxicology in three dimensions: using synthetic extracellular matrices in drug discovery. Acc Chem Res 2008;41(1):139–48.
[46] Mironov V, Reis N, Derby B. Review: bioprinting: a beginning. Tissue Eng 2006;12(4): 631–4.
[47] Xu T, Binder KW, Albanna MZ, Dice D, Zhao W, Yoo JJ, et al. Hybrid printing of mechanically and biologically improved constructs for cartilage tissue engineering applications. Biofabrication 2013;5(1):015001.
[48] Nakamura M, Kobayashi A, Takagi F, Watanabe A, Hiruma Y, Ohuchi K, et al. Biocompatible inkjet printing technique for designed seeding of individual living cells. Tissue Eng 2005;11(11–12):1658–66.
[49] Xu C, Inai R, Kotaki M, Ramakrishna S. Electrospun nanofiber fabrication as synthetic extracellular matrix and its potential for vascular tissue engineering. Tissue Eng 2004;10(7–8): 1160–8.
[50] Binder K. In situ bioprinting of the skin [Dissertation]. Wake Forest University; 2011.
[51] Lee V, Singh G, Trasatti JP, Bjornsson C, Xu X, Tran TN, et al. Design and fabrication of human skin by three-dimensional bioprinting. Tissue Eng Part C 2014;20(6):473–84.
[52] De Coppi P, Bartsch Jr G, Siddiqui MM, Xu T, Santos CC, Perin L, et al. Isolation of amniotic stem cell lines with potential for therapy. Nat Biotechnol 2007;25(1):100–6.
[53] Chun SY, Mack DL, Moorefield E, Oh SH, Kwon TG, Pettenati MJ, et al. Pdx1 and controlled culture conditions induced differentiation of human amniotic fluid-derived stem cells to insulin-producing clusters. J Tissue Eng Regen Med 2012;9(5):540–9.
[54] Guan X, Delo DM, Atala A, Soker S. In vitro cardiomyogenic potential of human amniotic fluid stem cells. J Tissue Eng Regen Med 2010;5(3):220–8.
[55] Rodrigues MT, Lee SJ, Gomes ME, Reis RL, Atala A, Yoo JJ. Amniotic fluid-derived stem cells as a cell source for bone tissue engineering. Tissue Eng Part A 2012;18(23–24):2518–27.

[56] Skardal A, Mack D, Atala A, Soker S. Substrate elasticity controls cell proliferation, surface marker expression and motile phenotype in amniotic fluid-derived stem cells. J Mech Behav Biomed Mater 2013;17:307–16.

[57] Abdel-Sayed P, Kalejs M, von Segesser LK. A new training set-up for trans-apical aortic valve replacement. Interact Cardiovasc Thorac Surg 2009;8(6):599–601.

[58] Kalejs M, von Segesser LK. Rapid prototyping of compliant human aortic roots for assessment of valved stents. Interact Cardiovasc Thorac Surg 2009;8(2):182–6.

Chapter 21

Translational Research of Skin Substitutes and Wound Healing Products

Mark E. Furth

Wake Forest Innovations, Wake Forest Baptist Medical Center, Winston-Salem, NC, USA

Chapter Outline

TRANSLATIONAL RESEARCH AND THE COMMERCIALIZATION PATHWAY

"Translational research" can be operationally defined as work that "transforms scientific discoveries arising from laboratory, clinical or population studies into new clinical tools and applications that improve human health by reducing disease incidence, morbidity, and mortality" [1]. Efforts to create skin-like constructs to treat acute and chronic wounds and burns trace back at least four decades [2]. An ideal skin substitute remains as a yet unattained "Holy Grail" [3] or Grails—because distinct indications require diverse solutions [4]. This chapter does not attempt to survey the extensive history of translational research in skin substitutes, a field in which the author has limited direct experience and which is covered in depth elsewhere in this book. The purpose, rather, is to highlight guiding principles common to the advancement of all regenerative medicines from the laboratory to the clinic and the health-care marketplace [5]. Comparable issues arise for skin substitutes as in the broad context of organ bioengineering and regeneration [6].

All translational projects hinge on the precise definition of an unmet medical need. Most often the successful introduction of a new regenerative medicine

Skin Tissue Engineering and Regenerative Medicine. http://dx.doi.org/10.1016/B978-0-12-801654-1.00021-8

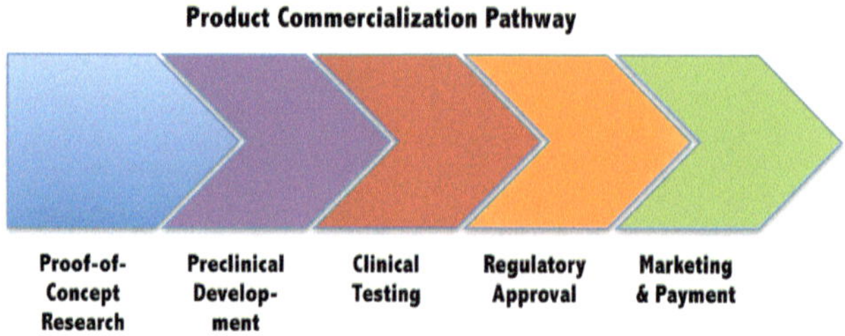

FIGURE 1 Stages in translational research and development leading to the marketing of an approved product.

technology to meet that need and become an accepted clinical practice will rest on the creation of a commercially viable service or product. One can, therefore, envision a "Commercialization Pathway" comprising five stages (Figure 1). The ultimate goal is to deliver the new technology to all patients likely to benefit. This in turn often depends on securing acceptance by payers, matched to appropriate Current Procedural Terminology (CPT®) codes and Healthcare Common Procedure Coding System codes. The Commercialization Pathway may prove a "bumpy road," as noted recently by Nancy Parenteau, one of the pioneers of tissue-engineered skin substitutes [7]. Nonetheless, the journey can be fruitful, especially when those in the driver's seat maintain a clear focus on the end point of approval of a product that successfully addresses the targeted medical problem. The solution should not only be safe and effective and mark an improvement over the existing standard of care, but should also make economic sense. That is, the investment in the development of the product, together with its pricing and reimbursement, must align with the magnitude and severity of the demand and the willingness of the society and payers to bear the cost of the solution to the medical problem.

TARGET PRODUCT PROFILE AND REGULATORY PATH

Achievement of regulatory approval by the United States Food and Drug Administration (FDA), and/or equivalent bodies in other nations, represents a crucial stage in the Commercialization Pathway. An important step early in the development process is the formulation of a Target Product Profile (TPP), which, in the FDA's words, "provides a statement of the *overall intent* of a drug [or regenerative medicine] development program, and gives information about the drug [or regenerative medicine product] *at a particular time* in development" [8]. Effectively, the TPP enables both the developer and the FDA to conceptualize and communicate about product specifications and indications for use, and to agree on the data that will be required by the Agency to support approval.

The TPP is a living document that evolves as more information is obtained through the stages of the Commercialization Pathway. Proof-of-concept research includes both basic mechanistic studies and testing in available animal models, ideally those best suited to support the desired TPP. The specific regulatory requirements that must be met to obtain approval, likewise, should be clearer once the TPP is defined. This includes in the United States the need to comply with current good manufacturing practice, the rules that govern the environment, conditions, and documentation associated with making product that will be utilized in humans [9]. Similarly, adherence to rules of current good laboratory practice, formal quality control standards distinct from standard academic peer review, is essential when obtaining much of the later preclinical data demanded by the FDA, notably formal toxicology studies, as well as relevant clinical lab data [10]. Finally, FDA approval rests on the submission of a compelling data package that demonstrates control over every facet of a reproducible manufacturing process, the product's safety (first and foremost in the Agency's mandate), and its efficacy. The exact regulatory requirements that must be met and the center(s) within FDA responsible for product approval depend on specific features of the product [11]. Part of the art in TPP formulation and product development resides in understanding the FDA's rules sufficiently well to create and test innovative technologies without the encumbrance of extra, unnecessary regulatory hurdles. A recent review elucidates in depth the Agency's perspective on regenerative medicine products [12].

ECONOMIC CHALLENGES AND STRATEGIC DEVELOPMENT FRAMEWORK

The TPP constitutes a record of the product that is most likely to launch, pending FDA approval. Its important elements focus on the indications and usage of a therapeutic drug, device, or combination product and include such practical parameters as dosing and administration; contraindications, warnings, and adverse reactions; clinical and pharmacological studies; and storage and handling. A more comprehensive view of product development builds on a Strategic Development Framework that encompasses, in addition to what is required merely for regulatory approval, the needs of the marketplace and the *value proposition* that underpins and justifies product development [13]. To guide translation leading toward commercialization, Tebbey and Rink propose adding two components beyond the TPP to generate this framework. The first is a Target Market Profile that employs key information about a disease or therapeutic area to understand unmet medical needs, patient populations, drivers of use, competition, and economic cost of the disease. The second is a Strategic Target Profile (STP) that captures "a vision for a product that will meet the needs of the market" [13]. Content of the STP includes value drivers, patient share, and economic factors such as pricing, revenue and profitability, the cost of investment in development, cost of goods, and obligations to pay license fees and royalties. The STP also incorporates a pharmacoeconomic evaluation of the financial costs and consequences of therapy to the health-care system and society.

Some of the first cell-based skin substitute products to achieve FDA approval illustrate the significance of the broad Strategic Development Framework to an emerging field such as regenerative medicine. At the beginning, in the late 1990s, the FDA approved several tissue-engineered human skin equivalents (HSEs), including Transcyte® (1997), Apligraf® (1998), Dermagraft® (2001), and Orcel® (2001/2008) [14–18]. The HSEs represented a disruptive technological change. They utilize human cells (fibroblasts, with or without keratinocytes) combined with a matrix scaffold in a three-dimensional construct to treat serious wounds such as nonhealing diabetic foot ulcers.

The development and regulatory approval of these revolutionary products fueled considerable optimism and commercial investment in tissue engineering and regenerative medicine technologies, along with other Advanced Therapeutic Medicine Products such as gene therapies [19,20]. However, enthusiasm quickly waned in the face of a difficult economic climate and the failure of the "breakthrough" HSE products to attain profitability due to manufacturing and distribution challenges, the high cost of goods, and slow adoption in the medical marketplace [21]. Indeed, Michael Lysaght, a periodic chronicler of both the technology and economics of regenerative medicine, opined that in 2000–2002 "tissue engineering was in a shambles because of disappointing product launches, failed regulatory trials, and the general investment pullback following the dot-com crash" [22]. Among many disappointing events, Advanced Tissue Sciences, the developer of Dermagraft, was liquidated in early 2003 [23]. Yet by 2007, Lysaght could point to a remarkable fivefold growth over 5 years in economic activity in tissue engineering, regenerative medicine, and stem cell therapeutics, and he and others returned to predicting a rosy future for regenerative medicine [22,24,25].

Where do we stand today, another 8 years down the road? Academic publication in tissue engineering and regenerative medicine is at an all-time high, as is confidence that this research will indeed translate eventually into many commercially viable therapies approved for human use. However, the business of translation into commercially successful products remains difficult and cyclical. New product approvals remain scant and consequently some high-profile startup companies have hit hard times or failed (e.g., Tengion, Winston-Salem, North Carolina).

Nonetheless, sales of regenerative products continue to grow. For example, Apligraf and Dermagraft, now both marketed by Organogenesis (Canton, Massachusetts), have risen phoenix-like from the economic ashes and become commercially successful products. "Next-generation" tissue-engineered dermal products in clinical development, for example, StrataGraft® (Stratatech Corporation, Madison, Wisconsin) and the autologous skin substitute PermaDerm™ (Regenicin, Inc., Little Falls, New Jersey), appear poised for regulatory decision and possible market entry in the relatively near future [14,26,27]. Some crystal ball gazers foresee that by 2019, the target population in the United States for bioengineered skin substitutes and replacements will exceed six million individuals, corresponding to a potential market of about $24 billion [14]. Modeling of the economics of tissue-engineered therapies supports realistic product development and rational pricing [28]. Moreover, medical practitioners currently training in wound healing and related fields increasingly seem to be responsive to evidence supporting both the medical efficacy and the cost-effectiveness of skin substitutes (e.g., see a "Resident's Corner" article in *Cutis*) [29].

FUTURE TRANSLATIONAL NEEDS

A 2014 Industry Symposium organized by the European Union chapter of the Tissue Engineering and Regenerative Medicine International Society (TERMIS-EU) focused on two issues perceived to be crucially rate limiting for translation from basic research findings to successful Advanced Therapeutic Medicine Products—namely, "(1) entrepreneurial exploitation of breakthrough ideas and innovations and (2) regulatory market approval" [30]. Drawing on the history of the biotechnology industry, since its inception in the early 1970s, the authors of a report on the Symposium point to the importance of companies achieving financial critical mass, in some instances through consolidation and merger. Another route over the key hurdle of accessing capital [31] may be partnering with early adapters among major companies in the pharmaceutical industry. However, the increasingly risk-averse behavior of venture capital and other investors, and of many potential large corporate partners, currently raises barriers to funding and to alliances with preclinical-stage companies. While some have dangled toes in the regenerative medicine waters, to a great extent the "large pharmaceutical firms are still in the monitoring mode regarding the new arena of cytotherapeutics rather than in a deployment mode" [30].

One possible point of entry for pharma is the use of regenerative medicine and tissue engineering technologies as platform research tools for the discovery and development of products that fall within the industry's current "sweet spot," that is, small molecule or monoclonal antibody therapeutics. The use of living skin equivalents for such research could thus have the added benefit of easing the entrance of pharmaceutical companies into commercialization of Advanced Therapeutic Medicine Products [14,32,33].

In the regulatory domain an earlier TERMIS working group emphasized the critical need for "clarity and predictability" to support translational research, product development, and investment [34]. The group encouraged communication between the United States FDA and the European Medicines Agency (EMA). It identified three major themes to clarify the rules for development and approval of tissue engineering/regenerative products and to eliminate international inconsistencies: establishment of a consistent regulatory framework for these products; globalization to include the agencies responsible for regulation outside the United States and Europe; and harmonization, that is, development of a unified strategy to align national practices and achieve consensus on regulatory requirements.

FUTURE DIRECTIONS

The development of skin substitutes, like regenerative medicine as a whole, remains a dynamic field. Translation of new technological breakthroughs could potentially revolutionize the materials and methods used to assemble tissue and organ constructs and also the sourcing and capacities of their living cellular components [35].

Current scaffolds may give way to a variety of "smart" biomaterials capable of self-assembly, guiding or programming cellular differentiation and function, or being modulated by subtle changes in the environment such as temperature or salt concentration [36–41]. Similarly, laser-assisted bioprinting offers new levels of precision for the generation of authentic three-dimensional organization in tissue-engineered skin and other tissues and organs [42–45].

Cell sourcing for skin products today rests almost entirely on methods developed decades ago to isolate and expand a few cell types, in particular, dermal fibroblasts, keratinocytes, and mesenchymal stromal cells (often referred to, somewhat dubiously, as stem cells). Advances in stem cell biology have greatly expanded possibilities to isolate and expand certain committed lineage-specific stem and progenitor cells for therapeutic products [46–48]. Even more profoundly, the isolation of pluripotent human embryonic stem (ES) cells offered, at least in theory, the unrestricted ability to produce large number of cells of any lineage or stage of development [49,50]. While human ES cells encountered some practical difficulties and became the subject of ethical debate, the development of cellular reprogramming technology enabled the production of induced pluripotent stem cells from any human donor [51]. Pharmacological tools to facilitate reprogramming and the directed differentiation of pluripotent stem cells enhance the potential to produce sufficient quantities of clinical-grade cells for translational programs in skin substitutes, among numerous potential applications [52–55].

Taken together with increasingly powerful tools for genetic manipulation and gene editing [56–61] of cultured human cells, the pluripotent stem cell technology opens remarkable new vistas for translational regenerative medicine. For example, individualized constructs could be prepared to treat genetic defects affecting the skin, as exemplified by the correction of a specific collagen defect in epidermolysis bullosa [62,63]. Similarly, cells might be armed with novel functions such as the ability to secrete antiinfective or immunomodulatory peptides as therapy for life-threatening nonhealing wounds [64,65]. A more distant but still potentially realizable dream would be the use of genetic tools to dampen inflammatory responses and/or eliminate histocompatibility antigens that mediate immune rejection. This may enable the production of "universal" off-the-shelf skin products that could potentially last far longer than current genetically unmatched grafts, which buy time for regeneration of the recipient's own skin but cannot persist because they activate the host immune response.

REFERENCES

[1] Wang X. A new vision of definition, commentary, and understanding in clinical and translational medicine. Clin Transl Med 2012;1(1):5.

[2] Tavis MJ, Thornton J, Danet R, Bartlett RH. Current status of skin substitutes. Surg Clin North Am 1978;58(6):1233–48.

[3] Nathoo R, Howe N, Cohen G. Skin substitutes: an overview of the key players in wound management. J Clin Aesthet Dermatol 2014;7(10):44–8.

[4] Nyame TT, Chiang HA, Orgill DP. Clinical applications of skin substitutes. Surg Clin North Am 2014;94(4):839–50.

[5] Translational regenerative medicine. London: Elsevier Inc.; 2015.

[6] Orlando G, Soker S, Stratta RJ. Organ bioengineering and regeneration as the new Holy Grail for organ transplantation. Ann Surg 2013;258(2):221–32.

[7] Parenteau NL. Driving the bumpy road to commercialization. Tissue Eng Part A 2014;20(11–12): 1563–4.

[8] Guidance for Industry and Review Staff. Target product profile – a strategic development process tool, draft. In: U.S. Department of health and human services food and drug administration CfDEaRC. 2007. Rockville (MD).

[9] Current good manufacturing practice and investigational new drugs intended for use in clinical trials. Final rule. Fed Regist 2008;73(136):40453–63.

[10] McCarty LS, Borgert CJ, Mihaich EM. Information quality in regulatory decision making: peer review versus good laboratory practice. Environ Health Perspect 2012;120(7):927–34.

[11] Levi B, Lisiecki J, Rubin P, et al. Demystifying the U.S. Food and Drug Administration: I. Understanding agency structure and function. Plast Reconstr Surg 2014;133(6):1495–501.

[12] Lee MH, Au P, Hyde J, et al. Translation of regenerative medicine products into the clinic in the United States: FDA perspective. In: Atala A, Allickson JG, editors. Translational regenerative medicine. London: Elsevier Inc.; 2015. p. 70–95.

[13] Tebbey PW, Rink C. Target product profile: a renaissance for its definition and use. J Med Mark 2009;9(4):301–7.

[14] Zhang Z, Michniak-Kohn BB. Tissue engineered human skin equivalents. Pharmaceutics 2012;4(1):26–41.

[15] Bello YM, Falabella AF, Eaglstein WH. Tissue-engineered skin. Current status in wound healing. Am J Clin Dermatol 2001;2(5):305–13.

[16] Streit M, Braathen LR. Apligraf–a living human skin equivalent for the treatment of chronic wounds. Int J Artif Organs 2000;23(12):831–3.

[17] Edmonds ME, Foster AV, McColgan M. 'Dermagraft': a new treatment for diabetic foot ulcers. Diabet Med 1997;14(12):1010–1.

[18] Pham C, Greenwood J, Cleland H, Woodruff P, Maddern G. Bioengineered skin substitutes for the management of burns: a systematic review. Burns 2007;33(8):946–57.

[19] Lysaght MJ, Reyes J. The growth of tissue engineering. Tissue Eng 2001;7(5):485–93.

[20] Lysaght MJ, Hazlehurst AL. Private sector development of stem cell technology and therapeutic cloning. Tissue Eng 2003;9(3):555–61.

[21] Lysaght MJ, Hazlehurst AL. Tissue engineering: the end of the beginning. Tissue Eng 2004;10(1–2):309–20.

[22] Lysaght MJ, Jaklenec A, Deweerd E. Great expectations: private sector activity in tissue engineering, regenerative medicine, and stem cell therapeutics. Tissue Eng Part A 2008;14(2):305–15.

[23] Pangarkar N, Pharoah M, Nigam A, Hutmacher DW, Champ S. Advanced tissue sciences inc.: learning from the past, a case study for regenerative medicine. Regen Med 2010;5(5):823–35.

[24] Mason C, Dunnill P. The strong financial case for regenerative medicine and the regen industry. Regen Med 2008;3(3):351–63.
[25] Hussain A, Rivers PA. The economic value of investing in regenerative medicine. J Health Care Finance 2009;36(2):45–54.
[26] Schurr MJ, Foster KN, Centanni JM, et al. Phase I/II clinical evaluation of StrataGraft: a consistent, pathogen-free human skin substitute. J Trauma 2009;66(3). 866–873; discussion 73–4.
[27] Centanni JM, Straseski JA, Wicks A, et al. StrataGraft skin substitute is well-tolerated and is not acutely immunogenic in patients with traumatic wounds: results from a prospective, randomized, controlled dose escalation trial. Ann Surg 2011;253(4):672–83.
[28] Tan TE, Peh GS, Finkelstein EA, Mehta JS. A practical model for economic evaluation of tissue-engineered therapies. Wiley Interdiscip Rev Syst Biol Med 2015;7(2):91–100.
[29] Han G. State-of-the-art wound healing: skin substitutes for chronic wounds. Cutis 2014;93(1): E13–6.
[30] Bayon Y, Vertes A, Ronfard V, et al. Turning regenerative medicine breakthrough ideas and innovations into commercial products. Tissue Eng Part B Rev 2015;21(6):560–71.
[31] Bertram TA, Tentoff E, Johnson PC, Tawil B, Van Dyke M, Hellman KB. Hurdles in tissue engineering/regenerative medicine product commercialization: a pilot survey of governmental funding agencies and the financial industry. Tissue Eng Part A 2012;18(21–22):2187–94.
[32] Paolini C, Orecchia A, Failla CM, Gallinari P, Zambruno G, Steinkuhler C. Skin equivalents: a tool for the discovery and validation of pharmacodynamic biomarkers. Cancer Invest 2013;31(1):60–6.
[33] Mathes SH, Ruffner H, Graf-Hausner U. The use of skin models in drug development. Adv Drug Deliv Rev 2014;69–70:81–102.
[34] Bertram T, Hellman KB, Bayon Y, Ellison S, Wilburn S. The regulatory imperative: international perspective. Tissue Eng Part B Rev 2013;19(3):191–3.
[35] Furth ME, Atala A. Tissue engineering: future perspectives. In: Lanza R, Langer R, Vacanti J, editors. Principles of tissue engineering. 4th ed. London: Elsevier, Inc.; 2014. p. 83–126.
[36] Furth ME, Atala A, Van Dyke ME. Smart biomaterials design for tissue engineering and regenerative medicine. Biomaterials 2007;28(34):5068–73.
[37] Mieszawska AJ, Kaplan DL. Smart biomaterials – regulating cell behavior through signaling molecules. BMC Biol 2010;8:59.
[38] Aguirre A, Gonzalez A, Navarro M, Castano O, Planell JA, Engel E. Control of microenvironmental cues with a smart biomaterial composite promotes endothelial progenitor cell angiogenesis. Eur Cell Mater 2012;24:90–106. [discussion].
[39] Williams DF. On the nature of biomaterials. Biomaterials 2009;30(30):5897–909.
[40] Tekin H, Sanchez JG, Tsinman T, Langer R, Khademhosseini A. Thermoresponsive platforms for tissue engineering and regenerative medicine. AIChE J 2011;57(12):3249–58.
[41] Appel EA, Tibbitt MW, Webber MJ, Mattix BA, Veiseh O, Langer R. Self-assembled hydrogels utilizing polymer-nanoparticle interactions. Nat Commun 2015;6:6295.
[42] Koch L, Kuhn S, Sorg H, et al. Laser printing of skin cells and human stem cells. Tissue Eng Part C 2010;16(5):847–54.
[43] Schiele NR, Chrisey DB, Corr DT. Gelatin-based laser direct-write technique for the precise spatial patterning of cells. Tissue Eng Part C 2011;17(3):289–98.
[44] Koch L, Deiwick A, Schlie S, et al. Skin tissue generation by laser cell printing. Biotechnol Bioeng 2012;109(7):1855–63.
[45] Michael S, Sorg H, Peck CT, et al. Tissue engineered skin substitutes created by laser-assisted bioprinting form skin-like structures in the dorsal skin fold chamber in mice. PLoS One 2013;8(3):e57741.

[46] Schmelzer E, Zhang L, Bruce A, et al. Human hepatic stem cells from fetal and postnatal donors. J Exp Med 2007;204(8):1973–87.
[47] Lanzoni G, Oikawa T, Wang Y, et al. Concise review: clinical programs of stem cell therapies for liver and pancreas. Stem Cells 2013;31(10):2047–60.
[48] Wang Y, Lanzoni G, Carpino G, et al. Biliary tree stem cells, precursors to pancreatic committed progenitors: evidence for possible life-long pancreatic organogenesis. Stem Cells 2013;31(9):1966–79.
[49] Thomson JA, Itskovitz-Eldor J, Shapiro SS, et al. Embryonic stem cell lines derived from human blastocysts. Science 1998;282(5391):1145–7.
[50] Amit M, Carpenter MK, Inokuma MS, et al. Clonally derived human embryonic stem cell lines maintain pluripotency and proliferative potential for prolonged periods of culture. Dev Biol 2000;227(2):271–8.
[51] Takahashi K, Tanabe K, Ohnuki M, et al. Induction of pluripotent stem cells from adult human fibroblasts by defined factors. Cell 2007;131(5):861–72.
[52] Christ GJ, Saul JM, Furth ME, Andersson KE. The pharmacology of regenerative medicine. Pharmacol Rev 2013;65(3):1091–133.
[53] Aberdam D. Derivation of keratinocyte progenitor cells and skin formation from embryonic stem cells. Int J Dev Biol 2004;48(2–3):203–6.
[54] Guenou H, Nissan X, Larcher F, et al. Human embryonic stem-cell derivatives for full reconstruction of the pluristratified epidermis: a preclinical study. Lancet 2009;374(9703):1745–53.
[55] Shamis Y, Hewitt KJ, Carlson MW, et al. Fibroblasts derived from human embryonic stem cells direct development and repair of 3D human skin equivalents. Stem Cell Res Ther 2011;2(1):10.
[56] Gasiunas G, Siksnys V. RNA-dependent DNA endonuclease Cas9 of the CRISPR system: Holy Grail of genome editing? Trends Microbiol 2013;21(11):562–7.
[57] Hou Z, Zhang Y, Propson NE, et al. Efficient genome engineering in human pluripotent stem cells using Cas9 from *Neisseria meningitidis*. Proc Natl Acad Sci USA 2013;110(39):15644–9.
[58] Mali P, Aach J, Stranges PB, et al. CAS9 transcriptional activators for target specificity screening and paired nickases for cooperative genome engineering. Nat Biotechnol 2013;31(9):833–8.
[59] Mali P, Yang L, Esvelt KM, et al. RNA-guided human genome engineering via Cas9. Science 2013;339(6121):823–6.
[60] Walsh RM, Hochedlinger K. A variant CRISPR-Cas9 system adds versatility to genome engineering. Proc Natl Acad Sci USA 2013;110(39):15514–5.
[61] Wei C, Liu J, Yu Z, Zhang B, Gao G, Jiao R. TALEN or Cas9-rapid, efficient and specific choices for genome modifications. J Genet Genomics 2013;40(6):281–9.
[62] Melo SP, Lisowski L, Bashkirova E, et al. Somatic correction of junctional epidermolysis bullosa by a highly recombinogenic AAV variant. Mol Ther 2014;22(4):725–33.
[63] Sebastiano V, Zhen HH, Derafshi BH, et al. Human COL7A1-corrected induced pluripotent stem cells for the treatment of recessive dystrophic epidermolysis bullosa. Sci Transl Med 2014;6(264):264ra163.
[64] Schurr MJ, Foster KN, Lokuta MA, et al. Clinical evaluation of NIKS-based bioengineered skin substitute tissue in complex skin defects: Phase I/IIa clinical trial results. Adv Wound Care 2012;1(2):95–103.
[65] Thomas-Virnig CL, Allen-Hoffmann BL. A bioengineered human skin tissue for the treatment of infected wounds. Adv Wound Care 2012;1(2):88–94.

Index

'*Note*: Page numbers followed by "f" indicate figures, "t" indicate tables, and "b" indicate boxes.'

B

C

D

R

S

T

X

Y

Z

Edwards Brothers Inc.
Ann Arbor MI. USA
December 1, 2017